Modern Trends in Biothermokinetics

Modern Trends in Biothermokinetics

Edited by

Stefan Schuster
University of Bordeaux II
Bordeaux, France

Michel Rigoulet
Institute of Cellular Biochemistry, CNRS
Bordeaux, France

and

Rachid Ouhabi and Jean-Pierre Mazat
University of Bordeaux II
Bordeaux, France

Springer Science+Business Media, LLC

Library of Congress Cataloging-in-Publication Data

Modern trends in biothermokinetics / edited by Stefan Schuster ... [et
al.].
       p.   cm.
   "Proceedings of the fifth International Meeting on
Biothermokinetics, held September 23-26, 1992, in Bordeaux
-Bombannes, France"--Copr. p.
   Includes bibliographical references and index.
   ISBN 978-1-4613-6288-3   ISBN 978-1-4615-2962-0 (eBook)
   DOI 10.1007/978-1-4615-2962-0
   1. Thermodynamics--Congresses. 2. Bioenergetics--Congresses.
3. Metabolism--Congresses. 4. Biological transport--Congresses.
I. Schuster, Stefan. II. International Meeting on Biothermokinetics
(5th : 1992 : Bordeaux, France)
QP517.T48M63 1993
574.19'121--dc20   93-34950
                                                           CIP

Proceedings of the Fifth International Meeting on Biothermokinetics, held September 23–26, 1992, in
Bordeaux–Bombannes, France

ISBN 978-1-4613-6288-3

© 1993 Springer Science+Business Media New York
Originally published by Plenum Press New York in 1993
Softcover reprint of the hardcover 1st edition 1993

PREFACE

This book includes articles relating to presentations given in a variety of forms (lectures, posters, contributions to round tables, software presentations) at the 5th International Biothermokinetics Meeting held in Bordeaux-Bombannes, September 23-26, 1992. The fact that not just lectures were considered for these proceedings reflects the aims of BTK meetings to instigate discussion, promote scientific cooperation and confront as many different ideas as possible with each other (at best heretical ones). BTK conferences have expanded more and more; 130 participants came to the 1992 meeting from 20 countries. It was therefore necessary to hold the round tables in parallel sessions.

It is difficult to have an unbiased feeling of what should be selected as the salient features of the meeting. As the name suggests, Biothermokinetics embraces thermodynamic and kinetic approaches to experimental and theoretical investigations of biological processes, in particular at the cellular level. This "classical" point of view is mainly represented in the chapter "Thermodynamics and Kinetics of Transport Processes and Biological Energy Transduction".

In recent years, Metabolic Control Analysis has become established as a major part of the field. Accordingly, a separate chapter is devoted to this analysis. The problem of how to distinguish between regulation and control is of ever increasing interest, as evidenced by developments since the conference. A new aspect is the extension to larger parameter changes, developed by the ever young fathers of Metabolic Control Analysis. Their paths are again slightly different, and again, the decision as to which approach is more appropriate seems to depend on the situation under study and is perhaps also a matter of taste.

The chapter about Modeling of Cell Processes shows that kinetic approaches are at present more widely used than thermodynamic ones. The topic "Investigation of Cell Processes" is rather a miscellaneous one; it concerns NMR studies, HPLC measurements, modulated gene expression, electrophysiological studies, neural network approaches, etc.

Mitochondria are still a favorite subject of biothermokinetic research; chapter IV is devoted to them.

To meet with the ever increasing demand for specific software for simulating biological processes by computer, software demonstrations have become part of BTK conferences since the Noorwijkerhout meeting in 1990. We are happy to be able to incorporate six articles on this topic into the present book.

Following the discussion at the Round Table on software development, John Woods and Dr David Fell (Oxford Brookes University) arranged an electronic mail distribution group linking the conference delegates who possessed email addresses. This is about to go to a vote for establishment on a permanent basis as the BioThermoKinetics/bionet.metabolic-reg. section of the BIOSCI division of the USENET news system on the Internet, which also allows subscription by electronic mail. In addition, they have established an FTP server on the Internet (address 161.73.104.10) that holds documents and computer programs in this subject area for free public access (with user name 'anonymous').

In the First Announcement, the organizing committee proposed, as one of various topics, "Theoretical Oenology (with practical exercises)", refering to local traditions of the Bordeaux region. This was mainly meant as a joke, but serious contributions were nevertheless welcome. And there was one (apart from various discussions in night sessions in the bar, that were not found entirely suitable for inclusion in the proceedings). Dr Henny Daams presented a poster about champagne making in The Netherlands, and we are pleased to include the corresponding article as an oenological epilogue in this book, similar to the epilogue of the meeting, which was an excursion to the vineyards of the Médoc region.

The editors

ACKNOWLEDGEMENTS

We gratefully acknowledge that the 5th International Meeting on Biothermokinetics was subsidized by:

Université de Bordeaux II,
Ministère de la Recherche et de la Technologie (Programme ACCES),
Ministère de l'Education Nationale,
Conseil Régional d'Aquitaine,
Fondation Fourmentin-Guilbert (France)
Hansatech Instruments Ltd. (Norfolk, England),
Labsystems SA (Paris, France),
Anton Paar KG (Graz, Austria)/Cyclobios (University of Innsbruck, Austria).

The Organizing Committee is indebted to the Château Aney (Cussac-Fort-Médoc, France), the Société Jean Guillot (Sainte-Eulalie, France), the Syndicat Viticole des Premières Côtes de Bordeaux (Quinsac, France) and the Société Maxime Bernier (Bourg-sur-Gironde, France) for making the wine tasting possible. The companies A33 (Bordeaux) and Mécabureau (Bordeaux) supported the conference by lending computer and copying facilities.

The organization of the conference was aided by a great many people. Among these were a great number of colleagues at the University Bordeaux II, the CNRS and INRA institutes (Bordeaux). The members of the International Study Group on Biothermokinetics (we wish to mention Drs H.V. Westerhoff, R. Heinrich, D.B. Kell and G.R. Welch) as well as Dr D.A. Fell (Oxford) interacted with the organizers in a fruitful way. The technical assistance both in organizing the conference and in editing the present proceedings of Mr R. Gaüzère, Mrs M. Malgat, Mr J.-C. Portais, Mrs F. Rallion and Ms M. Sanchez (Bordeaux), and Drs R. Arndt and R. Schuster (Berlin) was very helpful. We owe particular thanks to Dr A. de Waal (Amsterdam) for his invaluable assistance in managing the processing of manuscripts on computer.

One of the editors (S.S.) gratefully acknowledges grants from the Science Council of NATO (under the auspices of the DAAD, Germany) and the NWO (The Netherlands).

Last but not least, it is a true pleasure to express our gratitude to the staff of the Plenum Publishing Company, in particular to Ms N. Clark, Ms A. Hackett and Mr G. Safford for friendly and efficient cooperation.

The editors and organizing committee

CONTENTS

II. Modeling of Cell Processes with Applications to Biotechnology and Medicine

III. Metabolic Control Theory: New Developments and Applications

IV. Control and Regulation of Oxidative Phosphorylation

V. Investigation of Cell Processes

VI. Computer Programs for Modeling Metabolic Systems

VII. Oenological Epilogue

I. THERMODYNAMICS AND KINETICS
   OF TRANSPORT PROCESSES
   AND BIOLOGICAL ENERGY TRANSDUCTION

FRONTIERS IN
THE MATHEMATICAL DESCRIPTION
OF "BIOTHERMOKINETIC" PHENOMENA

G. Rickey Welch

Department of Biological Sciences
University of New Orleans
New Orleans, LA 70148, U.S.A.

Despite an abundant and diverse literature on the application of physico-mathematical methods to particular biological problems, it can hardly be said that there exists nowadays a "theoretical biology" analogous to theoretical physics. Many Natural Philosophers would argue that biology is predominantly an empirical science, too young — or, perhaps, just not epistemologically appropriate — for the kind of theoretical lawfulness (*viz.*, abstract mathematical formulae, fundamental equations) which characterizes pure physics. The evolution of "mathematical biology" has followed rather much of a utilitarian bent. One tends to view a given biological phenomenon in isolation, assign to it a particular physical mechanism, explicate a mathematical model, and submit to empirical verification. If there is *quantitative* agreement with experimental data, the model is considered "good." Missing from this montage of quantitative models, which defines the realm of today's "theoretical biology," are the *qualitative* (or *relational*) aspects of biological processes, as might be expected of a "physical theory." Notwithstanding, there is mounting concern in philosophical circles over the unification of biology with physics. The call for such unity of science is coming from theoretical advances on the physical side, as it has come to interface with the biotic character of the Universe[1]. If biology is to fulfill its potential in this unification, it must re-examine the role of mathematics as a language for exploring the rich veneer of the living world.

In bringing mathematics to bear in our dialogue with the "living state," one must ask: What do we actually *want* from our understanding of life (short of trying to define "what is life"!)? If, for example, the desire is just to have an empirical grasp of the distribution of "valves" operative on metabolic fluxes in the living cell (say, for some biotechnological purpose), then a kinetic paradigm like "metabolic control analysis" (based on such simple mathematical relations as the Michaelis-Menten equation) suffices. Indeed, enzyme kinetics has become the dominant theme in the attempt to "quantify" the behavior of life at the cellular level. Fostered by the "molecular biology" era in recent decades, cell metabolism has come to be regarded merely as a linear superposition of enzyme-kinetic states. Save for identifying (in principle) all of the individual enzymes in the metabolic pathways of the cell and marking their ligand-saturation properties — albeit generating utilitarian value in areas such as pharmacology and genetic engineering, one can rightly wonder: Has the kinetic (or "kinematic") representation provided a physically-consistent picture of the cell in the *dynamic* context? Without invoking any kind of Aristotelian teleology, one might legitimately question: Where are the holistic notions of "force" and "causality" in the existing biological view? For answers to these deeper questions, one must discern the edifice of cell metabolism beyond the superficial overlay of kinetic analysis.

In applying mathematics to scientific analysis in areas such as biology, there are a number of philosophical roads one might take. Following the lineage of Western (Greek) thought, the modern form of *physicalism* (emanating from the Vienna Circle) would seem to be apropos. The idea here is not necessarily to reduce the biological system to a collection of physical objects; rather, one distills ("reduces," if you like) the formal description of the system-behavior down to the language of pure physics. A unique physical feature of the living world is the system-isomorphism amongst the various levels of biological complexity, from the microcosm of the cell to the macrocosm of socio-ecosystems[2]. This *status rerum* demands that the general mathematical characterization of the "living state" manifest a hierarchical symmetry, or "relational invariance"[3].

We suggest to follow the train of thought, enunciated a century ago by the patriarchal physiologist Claude Bernard[4], called "physiological determinism." While recognizing the uniqueness of the "living state," he sought to provide biology with a more "physicalist" foundation. Bernard espoused the "protoplasmic theory of life" (which, by the *fin-de-siècle*, rivaled the "cell theory" in importance[5]) and viewed the "relational" properties of cellular protoplasm (which he called "life in the naked state") as the basis for understanding the system-behavior at all biological levels. Such a perspective rationalizes the development of invariant mathematical relationships spanning the gamut of life's hierarchy. The cell seems a fitting place to begin the construction of a generalized mathematical description — not simply because it is the smallest functional "living unit," but due to the fact that the individual "events" in cellular metabolism are amenable to exact depiction by well-known *physical* concepts. Such a course of study has a long and colorful history[5].

The development of a mathematical description of natural phenomena demands that we — the sentient, cogitative beings — adopt a metaphorical/analogical construct derived ultimately from the macroscopic world of our sensory perception[6]. Blending Einsteinian relativity with quantum mechanics, the evolution of physical theory has come to regard the *field* as the primal entity in our understanding of material events, energy flow, *etc.*, with the atomistic character of the material world ascribable (according to the very latest thought!) to the metaphor of infinitesimally-small, resonating "strings"[1]. The "field," in turn, has been reduced to pure *geometry*, with symmetry-relations taking on a *dynamical* role.

The "field" idea is not new to biology. It appeared (under the guise of diffusion-gradients of morphogen substances) in the vernacular of developmental biology (*e.g.*, embryology) early in the 20th century and continues today to be used by workers in this subdiscipline. Moreover, the "field" notion is involved, implicitly, whenever one applies *nonequilibrium thermodynamics* to the characterization of specific biological phenomena[7]. Biology may benefit (and, indeed, be drawn closer to today's unity movement in Natural Philosophy), though, by incorporating the "field" principle more fundamentally into its ontological fabric. Such assertion is certainly not intended to conjure any aura of "vitalism" in modern-day biology; nor is any new "force of Nature" being proposed. In the vein of philosophical "physicalism," what we proffer is that biology should cross the analogical/metaphorical bridge to theoretical physics — thereby using the contemporary concepts, language, and mathematical formalism of physical "fields" in viewing the system-relations in the "living state."

Looking specifically at the subcellular (protoplasmic) level, the potential worthiness of a field-geometric approach is realized from an empirical appreciation of the organization of cell metabolism. Living cells (especially in eukaryotes) contain a complex and diverse particulate infrastructure, encompassing extensive membranous reticulation and a cytoplasmic phase laced with a dense array of various proteinaceous fibrillar elements[8]. The protein density in association with cytomembranes and organelles is almost crystal-like. There is a remarkable phylogenetic homology in the surface area-to-volume ratio for all cytological substructures in the biological world[9]. Accumulating evidence indicates that the majority of the enzymes of intermediary metabolism operate in organization with these structures[10,11]. Within the localized microenvironments engendered by this subcellular heterogeneity, the thermodynamic-kinetic character of metabolic processes (particularly those which are functionally juxtaposed to membranous electric fields and "protonmotive" forces) may differ significantly from a homogeneous-solution system, as ascertained from *in vitro* analysis of isolated enzymes[12]. There is a clear need for a more holistic framework — a "field" substratum — which treats metabolism in a mathematically "local" and "affinely-connected" manner.

Before pursuing theoretically any formal "field" approach to cell metabolism, one must consider the basic question: Are there any localized entities which manifest field-type phenomena? In the affirmative, such can be found in the dynamic action of *enzymes*. We have come to appreciate, that the enzyme-protein is a fluid, anisotropic *energy-transducer*, which can couple the bound chemical-subsystem, not only to local physical fields in the macromolecular matrix, but also to external fields in cellular microenvironments[13]. The anisotropy of protein function is translated into higher-order metabolic organization in the living cell.

Ultimately, the construction of a "field" formalism demands mathematical elaboration of the nature of "space-time," "observables," "point-events," "process-trajectories," *etc.*, for the phenomenon of interest. Cell metabolism involves, basically, a bimodal "flow," *viz.*, (electro)chemical reaction and mass diffusion, under heterogeneous and anisotropic circumstances. This situation is subsumable to a generalized *phase-space diffusion* approach, allowing a rather simplified "physicalist" description of metabolic flow[5].

One factor accounting for the unity of physical forces (and fields) in Nature is a transcending (perhaps metaphysical) precept known as the *Principle of Least Action*. The mathematical formulation of this "superlative" usually entails the condition, that the integral of some function (*e.g.*, the "Lagrangian") of the system-behavior has a smaller (or larger) value for the *actual* operation of the system than it would have for any other "imagined" operation under the given situation. Accordingly, a generalized analytical procedure is provided for selecting the path of a physical process and, by derivation, the equation of motion.

The potential applicability of the *Principle of Least Action* — and the question of an appropriate "Lagrangian" — for the "living state" has been a subject of contention in theoretical biology since the early part of this century[5]. Nonequilibrium thermodynamics offers the most suitable picture of a "force" operative on macroscopic (ensemble) flow-processes, such as those relevant to biological systems. Holistically speaking, the "living state" appears as an epiphenomenon of the so-called "thermodynamic arrow" governing the global behavior of the Cosmos. Stemming therefrom, the Rayleigh-Onsager *dissipation function* [7] emerges as a candidate for the biotic "Lagrangian."

In order to incorporate the *relational* features of the "metabolic field," regarding the "affinely-connected" processes in heterogeneous, anisotropic media *in vivo*, one must take the logical step to a "geometrodynamical" field description. Here, we look to contemporary theoretical physics for pure analogies. The "classical" geometrical field-construction is seen in Einstein's *theory of general relativity*, which endows the "curvature" of space-time with dynamical effects on matter and energy; whereby, motion on a space-time manifold follows the "shortest distance between two points" defined by *geodesic paths*. One indication of the biophysical tract of pursuit is realized in the striking mathematical resemblance between the "geodesic" Lagrangian of relativity theory and the "dissipation-function" Lagrangian of nonequilibrium thermodynamics. Extending such reasoning to the cellular level then leads to the postulation that the putative "metabolic field" has a *tensorial form of rank=2*, with a central role for the *diffusion-coefficient tensor* [5].

Pushing the "geometrodynamical" description into the micro-domain of cellular metabolism, we find potential analogies/metaphors also from *gauge field theory*. This theory (which is at the heart of particle physics today) strikes at the "local" character of the very measurement-basis of physical phenomena, employing the notion of "gauge invariance" as a dynamical principle for the operation of fields of force. Applying this idea to the thermodynamics/kinetics of microenvironmental transduction processes in the living cell yields novel insights into the "physicalist" character of bioenergetics[5].

Finally, we note the growing importance of *fractal geometry* in the mathematical depiction of biological phenomena — spanning the entire hierarchy of life-forms. Applying "fractal-geometric" principles, in particular, to the understanding of the metabolic workings in structured microenvironments at the cellular level, one finds that the nonequilibrium flow-processes assume a local, non-Euclidean form. Moreover, the *thermodynamic potentials* themselves are seen to be coupled to the "fractal dimensionality" of the system[14].

In conclusion, we draw attention to the fact that the science of biology, for centuries, has wrestled with the philosophical import of the concept of *organization* in the understanding of the "living state." The situation remains vague and lacking in objectivity and quantification. Perhaps the "geometrodynamical" approach to *fields*, so prevalent in physics today, will open new vistas in biology — as portended early in this century by such prescient mathematical biologists as D'Arcy Thompson, Alfred Lotka, Nicholas Rashevsky, and Ludwig von Bertalanffy.

References

1. P. Davies (ed.). "The New Physics," Cambridge University Press, Cambridge (1989).
2. N.Rashevsky. "Mathematical Biophysics," Dover, New York (1938).
3. G.R. Welch, The living cell as an ecosystem: hierarchical analogy and symmetry, *Trends Ecol. Evol.* 2:305-309 (1987).
4. C. Bernard. "Leçons sur les phénomènes de la vie communs aux animaux et aux végétaux," (1878). English translation by H.E. Hoff, R. Guillemin and L. Guillemin, Charles C. Thomas Publisher, Springfield (Ill.) (1974).
5. G.R. Welch, An analogical "field" construct in cellular biophysics: history and present status, *Prog. Biophys. Mol. Biol.* 57:71-128 (1992).
6. E.T. Bell. "Mathematics: Queen and Servant of Science," American Mathematical Association, Washington (D.C.) (1951).
7. I. Gyarmati. "Non-equilibrium Thermodynamics: Field Theory and Variational Principles," Springer, Heidelberg (1970).
8. K.R. Porter, ed., The cytomatrix, *J. Cell Biology* 99:1s-248s (1984).
9. P. Sitte, General principles of cellular compartmentation, *in:* "Cell Compartmentation and Metabolic Channelling," L. Nover, F. Lynen and K. Mothes, eds, Elsevier, Amsterdam (1980).
10. P. Srere, Complexes of sequential metabolic enzymes, *Ann. Rev. Biochem.* 56:89-124 (1987).
11. G.R. Welch and J.S. Clegg (eds). "The Organization of Cell Metabolism," Plenum, New York (1987).
12. G.R. Welch and T. Keleti, On the "cytosociology" of enzyme action *in vivo*: a novel thermodynamic correlate of biological evolution, *J. Theor. Biol.* 93:701-735 (1981).
13. G.R. Welch, ed. "The Fluctuating Enzyme," Wiley, New York (1986).
14. G.R. Welch, Bioenergetics and the cellular microenvironment, *Pure and Applied Chemistry*, in press.

MONITORING A HETEROGENEOUS COUPLED SYSTEM USING SOLUBLE AND BIOLOGICALLY LOCALIZED FIREFLY LUCIFERASE AS A PROBE FOR BULK AND LOCAL [ATP]

Claude Aflalo

Dept. of Life Sciences
Ben Gurion University of the Negev
P.O.Box 653, Beer Sheva 84105, Israel

1. Introduction

Classical biochemistry has provided a great deal of information about isolated catalytic units (from enzymes to organelles), by studying their properties in dilute solution or suspension. The results are often directly projected on their functionality *in vivo*, implicitly assuming a homogeneous behavior *in situ*. However, a critical look at the cellular environments[1,2] strongly refutes the latter assumption and indicates that special kinetic treatments would rather be required to describe these situations, taking into account physical steps (diffusion, partition, etc.) inherent to heterogeneous systems, in addition to the intrinsic properties of the catalysts as observed in homogeneous phase[3]. Such treatment has been successfully applied for immobilized enzyme systems[4], stressing that the activity of the enzymes is exclusively determined by the *local* concentration of reactants in their own microenvironment. However, in these systems only the *bulk* concentrations of reactants have been assessed. It is thus important to have means to monitor also local events which reflect more directly the activity of enzymes *in situ*.

Light emission by firefly luciferase in the presence of luciferin and molecular oxygen, is commonly used to monitor the concentration of ATP in solution with high sensitivity and selectivity[5]. The concept of local enzymic probes[6] has thus been applied in artificial systems, in which firefly luciferase was coimmobilized in agarose beads with kinases producing or consuming ATP. The basic principles for monitoring local [ATP] with luciferase were delineated, and the heterogeneous system was mathematically modeled, accounting for diffusion of reactants between the porous support and the bulk solution[7].

Modern Trends in Biothermokinetics, Edited by
S. Schuster *et al.*, Plenum Press, New York, 1993

2. Outline of the Mitochondria-Luciferase-Hexokinase Model System

Firefly luciferase has been expressed in yeast and specifically localized at the cytoplasmic face of the outer membrane of mitochondria by genetic engineering[8]. An experimental model system has been set up in which oxidative phosphorylation in isolated yeast mitochondria was coupled to the hexokinase reaction in solution, enabling dual measurements of ATP concentration near the mitochondrial surface and in solution. The former can be measured experimentally using the localized probe, while the latter is estimated with soluble luciferase added to the mitochondrial suspension[11]. This heterogeneous biological system has been modeled as a surface source for ATP (mitochondria) coupled to a bulk sink (consumption by hexokinase) through diffusion. For that, a spherical geometry was assumed, and each mitochondrion surrounded by an average volume of bulk solution was treated as a closed system (sustaining a zero-flux condition at the outer boundary). This yields a second order differential equation describing the steady state. The solution, obtained using appropriate boundary conditions[10], provides an expression for the concentration of ATP at any point in space. Of particular interest are the concentration at the surface ($[ATP]_s$) and the average $[ATP]$ in the medium ($\langle ATP \rangle$), as compared to that expected for a homogeneous system ($[\widetilde{ATP}]$), in relation to the total concentration of nucleotides ($[AXP]$). The final expressions are:

$$\frac{[ATP]_s}{[AXP]} = \frac{k_{ox}A\,[\text{mito}]}{k_{ox}A\,[\text{mito}] + \eta\,k\,''[HK]} \quad , \quad \frac{\langle ATP \rangle}{[AXP]} = \frac{k_{ox}A\,[\text{mito}]}{\dfrac{k_{ox}}{\eta}A\,[\text{mito}] + k\,''[HK]} \quad ,$$

$$\frac{[\widetilde{ATP}]}{[AXP]} = \frac{k_{ox}A\,[\text{mito}]}{k_{ox}A\,[\text{mito}] + k\,''[HK]} \quad , \quad \text{with} \quad \eta = \frac{\langle ATP \rangle}{[ATP]_s} = \frac{k_{HK}^{app}}{k\,''[HK]} \quad ,$$

where k_{ox} and k'' stand for the microscopic second order rate constants for ATP production at the surface and its consumption in solution, respectively; A is the specific area for mitochondria (cm^2/mg prot.); and k_{HK}^{app} is the apparent rate constant for consumption as derived from experimental data assuming a homogeneous ATP production in solution. The effectiveness factor η ($0 \leq \eta \leq 1$) depends on the relation between the concentrations of mitochondria and hexokinase, and that between the rates of diffusion and catalysis in solution. Lower η (excess hexokinase and/or slow diffusion) indicates a larger deviation from a homogeneous behavior.

3. Results and Discussion

Low concentrations of ATP emerging from mitochondria (produced by oxidative phosphorylation and adenylate kinase) can be determined by directly monitoring the light output with exogeneous soluble luciferase or the outer membrane-bound enzyme. The steady state concentration of ATP generated by the mitochondria and consumed by exogeneous soluble hexokinase has been assessed in both the systems (Fig. 1). While in the control system (Fig. 1A) the steady state [ATP] can be progressively reduced to undetectable concentrations at sufficiently high hexokinase, the light output from mitochondria-bound luciferase (Fig. 1B) is reduced to a finite value under identical conditions.

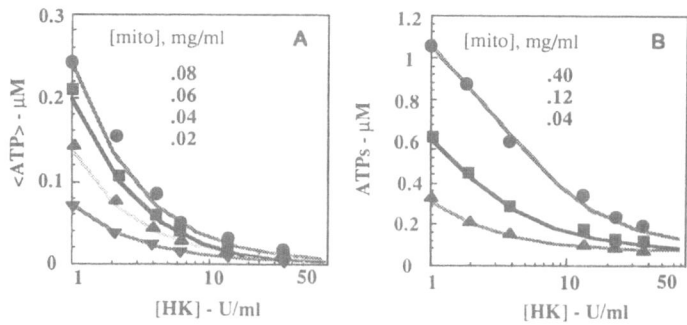

Figure 1. Fitting of experimental measurements to the solutions for the unstirred suspension [Copyright Acad. Press]. The data were acquired as the steady state light output from luciferase in solution (A) or anchored to the mitochondria membrane (B), in the presence of substrates (succinate, ADP, P_i, glucose and luciferin), mitochondria and hexokinase at the indicated concentrations. The second order rate constant for hexokinase ($k = 3.5$ ml/U·min) and the apparent rate constant for ATP formation in mitochondria ($k_{ox}A = 8\text{-}25$ ml/mg·min, variable with different preparations of isolated mitochondria) were assessed from the time course of ATP consumption by hexokinase alone, and ATP production by energized mitochondria alone, respectively. The experimental setup permits the determination of endogenous nucleotides, initial rate of phosphorylation in the absence of hexokinase, and titration of ATP with hexokinase at each concentration of mitochondria[9]. The symbols represent data points and the lines were drawn using the equations derived from the model and the parameters estimated by fitting the data to the equations. Reasonable values for the diffusion constant, the rate of phosphorylation were obtained from the fitting procedure (for details, see Ref. 10).

These results indicate that when ATP is efficiently depleted in the medium by hexokinase (as detected by soluble luciferase), the ATP emerging from the mitochondria is still available to the localized probe at the surface of the membrane. The steady state value of local [ATP] in the presence of excess hexokinase represents about 10% of the total nucleotides concentration added to mitochondria. This ratio depends strongly on the specific phosphorylation ability of the mitochondria but is independent of their concentration, indicating that a true local event is being monitored by the probe. The light output from this isolated system with localized luciferase exposed to the medium can easily be calibrated against limiting concentrations of ATP added to the mitochondrial suspensions, in the absence of ATP-producing (or consuming) reactions.

The model predicts the kinetic behavior of the system observed experimentally at steady state when the concentrations of mitochondria, hexokinase (Fig.1) and nucleotide (not shown) are varied. The results indicate that diffusion of reactants (ATP and ADP) between the membrane and the bulk solution represents a rate limiting step in the coupled system. Consequently, gradients of nucleotides are established at steady state, in which ATP (or ADP) accumulates (or is depleted) near the membrane relatively to the bulk solution. A similar situation may occur in the cell[12], where ATP - significantly depleted in the cytosol - may still be available to enzymes in close contact with the mitochondria.

The results emphasize the importance of diffusion (and other physical steps) in cellular catalysis which is heterogeneous in essence. The model provides therefore a more realistic view of biological processes *in situ* than the conventional approach, in which a homogeneous behavior (i.e. no gradients) is assumed. Additional insight is provided by the development of new equations which take into account the effect of mixing, non-specific adsorption of hexokinase to the membrane, and its specific binding to porin[13], resulting in channeling of nucleotides between their respective sites of production and utilization. All these situations, likely to occur in cells[14] are shown to further modulate the behavior of the coupled system beyond that expected for the simple heterogeneous behavior initially assumed.

References

1. J.S. Clegg, Properties and metabolism of the aqueous cytoplasm and its boundaries, *Am. J. Physiol.* **246**:R133-R151 (1984).
2. K.R. Porter, The cytomatrix: a short history of its study, *J. Cell Biol.* **99**:3s-12s (1984).
3. J.-M. Engasser and C. Horvath, External and internal diffusion in heterogeneous enzymes systems, *Appl. Biochem. Bioeng.* **1**:127-220 (1976).
4. R. Goldman and E. Katchalski, Kinetic behavior of two-enzymes membrane carrying out a consecutive set of reactions, *J. Theor. Biol.* **32**:243-257 (1971).
5. M. DeLuca and W.D. McElroy, Purifications and properties of firefly luciferase, *Methods Enzymol.* **57**: 3-15 (1978).
6. H. Seis and B. Chance, The steady-state level of catalase compound A in isolated hemoglobin-free perfused rat liver, *FEBS Lett.* **11**:172-176 (1970).
7. C. Aflalo and M. DeLuca, Continuous monitoring of adenosine 5'-triphosphate in the microenvironment of immobilized enzymes by firefly luciferase, *Biochemistry* **26**:3913-3919 (1987).
8. C. Aflalo, Targeting of cloned firefly luciferase to yeast mitochondria, *Biochemistry*, **29**: 4758-4766, (1990).
9. C. Aflalo, Biologically localized firefly luciferase: a tool to study cellular processes, *Int. Rev. Cytol.* **130**:269-323 (1991).
10. C. Aflalo and L.A. Segel, Local probes and heterogeneous catalysis: a case study of a mitochondria-luciferase-hexokinase coupled system. *J. Theor. Biol.* **158**:67-108 (1992).
11. J.J. Lemasters and C.R. Hackenbrock, Continuous measurement of Adenosine triphosphate with firefly luciferase luminescence, *Methods Enzymol.* **56**:530-544 (1979).
12. P.D. Jones, Intracellular diffusion gradients of oxygen and ATP, *Am. J. Physiol.* **250**:C663-C675 (1986).
13. F.N. Gellerich, M. Wagner, M. Kapischke and D. Brdiczka, Effect of Macromolecules on ADP-Channeling into Mitochondria, this volume.
14. P.A. Srere, Complexes of sequential metabolic enzymes, *Annu. Rev. Biochem.* **56**:89-124 (1987).

CONTROL OF CYTOCHROME *C* OXIDASE:
KINETIC, THERMODYNAMIC OR ALLOSTERIC?

Peter Nicholls

Dept. of Biological Sciences
Brock University
St. Catharines, Ontario, Canada, L2S 3A1

1. Introduction

Cytochrome *c* oxidase generates a membrane potential and a pH gradient by translocating charges across the membrane in which it is embedded. These gradients inhibit the enzyme in the process of respiratory control. The charge translocation can involve either electrons or protons, the former a consequence of the topochemical arrangement of redox centres, the latter a result of some mechanism of alternating access to the two sides of the membrane by proton-donating and accepting groups, either involving the catalytic machinery of oxygen reduction or an 'indirect' conformational process within the protein complex. Respiratory control can be exerted thermodynamically upon the overall process, kinetically upon one or more charge-translocating steps, or allosterically at a non-charge-translocating step, affected by ΔpH and/or $\Delta\Psi$.

Where is such control found experimentally? Fig. 1 summarizes the sequence of electron transfer from cytochrome *c* to oxygen. As shown in the figure, the first and last steps of the sequence seem unlikely candidates to be control sites; they are very rapid and show no sign of response to membrane energization. More likely are the steps involving reduction and oxidation of cytochrome *a*. What is the experimental evidence for ΔpH and/or $\Delta\Psi$ effects at either of these two sites?

Copeland *et al.*[1-3] suggest different conformational states of cytochrome *a* that function during steady state turnover of cytochrome *c* oxidase. Despite the importance of the reaction involved[4], no consensus yet exists as to the pathway of electron transfer. However,

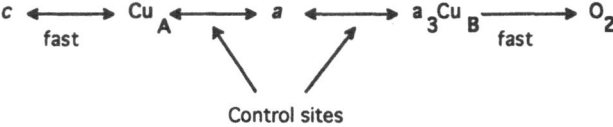

Figure. 1. Electron transfer pathway: cytochrome *c* oxidase.
A linear pathway[5,6] is assumed and the sites of ΔpH and/or $\Delta\Psi$ control[7,8] are indicated.

Hill[5] as well as Chan et al.[6] have developed a scheme for the initial reaction steps. In this scheme (see Fig. 1), cytochrome a lies between the initial electron acceptor, Cu_A, and cytochrome a_3Cu_B. The central position of cytochrome a suggests a direct involvement in energy conservation[1,9,10]. Reduction of the binuclear center to an O_2-reactive form requires two electrons. Malmström[11] believes that initial reduction of the oxidized enzyme alters its conformation and facilitates reduction of a_3Cu_B. Alternatively Moody and coworkers[12] find that single electron reduction is sufficient to bring the enzyme into what they term its 'fast' or reactive state. Copeland[1] suggests that the reduced cytochrome a present during steady state is different from that at full reduction. Understanding the steady state behavior of the enzyme is thus essential to unravel both control and mechanism.

2. Materials and Methods

Cytochrome oxidase was isolated from beef heart according to the method of Kuboyama et al. [13], with Tween-80 instead of Emasol. DOPC (dioleylphosphatidylcholine) and DOPE (dioleylphosphatidyl-ethanolamine) were products of Avanti Lipids; sodium ascorbate, cytochrome c (type VI, horse heart), TMPD (N,N,N',N'-tetramethyl-p-phenylene-diamine) and valinomycin were from Sigma Chemical Co. Nigericin was from Calbiochem. Cytochrome c oxidase-containing vesicles (COV) were prepared as described by Wrigglesworth et al.[14]. In brief, 100 mg DOPE plus 100 mg DOPC were dissolved in chloroform. The organic solvent was evaporated under nitrogen. The resulting lipid film was suspended in 5 ml 100 mM HEPES buffer, pH 7.4, with 1.5% sodium cholate, by vigorous vortexing, and the suspension sonicated for 7 min in the pulse mode at 30% duty cycle (Heat Systems - Ultrasonics W-375 sonicator) on ice under nitrogen, and centrifuged for 10 min at 20,000 g to remove undispersed lipid and titanium particles. Cytochrome oxidase was added to a final concentration of 5 μM cyt. aa_3 (50 μl 500 μM or ≈100 mg/ml stock solution). Dialysis was then carried out at 4 °C against 100 mM HEPES for 4 hr., followed by 10mM HEPES, 40 mM KCl, 50 mM sucrose pH 7.4 for 2 days (and at least 3 buffer changes). The final stock suspension of COV usually contained about 65 mM lipid and 5 μM cytochrome oxidase. Respiratory control ratios (RCR = respiration rate in presence of valinomycin plus nigericin divided by the rate in absence of ionophores) varied from 7 to 10. Absorption spectra were recorded on a Beckman DU-7 HS™ spectrophotometer interfaced with an Apple II-GS™ microcomputer. Steady state observations were made either in this system or with an Aminco DW-2™ instrument linked to a Compaq™ 286 computer with Olis™ data acquisition hardware and software. Cytochrome aa_3 concentrations in COV were determined from difference spectra, fully reduced minus oxidized at 605 - 630 nm, using a ΔE of 27 mM^{-1} cm^{-1}.

The ratio of outward-facing to total enzyme was determined by the ratio of the reduction at 605-630 nm with ascorbate and cyt. c alone to that with TMPD present. Steady state reductions of cytochromes c, a and a_3 during aerobic oxidation of ascorbate by COV were monitored in both controlled (initial) and uncontrolled (after addition of valinomycin and nigericin) states. The contributions of cytochromes a and a_3 are calculated using the relative extinctions given in eqs (1) and (2),

$$\text{\% redn at 605-630 nm = \% redn cyt. } a* 0.84 + \text{\% redn cyt. } a_3*0.16 \tag{1}$$

$$\text{\% redn at 445-470 nm = \% redn cyt. } a* 0.55 + \text{\% redn cyt. } a_3*0.45 \tag{2}$$

assuming that cytochrome a contributes 84% of the absorbance in the visible (a) region and 55 % of that in the Soret region, with cytochrome a_3 contributing the remainder[15,16]. The higher than usual[10] contribution assumed for the Soret region reflects the shift in the Soret spectrum of oxidized cytochrome a_3 .

3. Results

If steady state data are analyzed according to eqs (1) and (2), cytochrome a_3 is found to remain almost fully oxidized until anaerobiosis, as in classical studies[15]. Addition of uncoupler produces a substantial change in the cytochrome c steady state (e.g. from $\approx 12 \%$ reduced to $\approx 3 \%$ reduced), while the steady state of cytochrome a is almost unaffected by the big flux change as controlling gradients are abolished. If [ascorbate] is increased, the flux and cytochrome c reduction level increase in parallel. Although turnover is thus proportional to % cytochrome c reduction over a wide range, it is not proportional to % reduction of cytochrome a ; the latter reaches a maximum value long before the enzyme achieves V_{max}, whatever the energy state. But membrane energization has a double effect - it modulates the equilibrium between cyt. c and cyt. a, reflecting changes in the redox potential of cyt. a, and it alters the maximal reduction of cyt. a, reflecting controlled steps lying between cyt. a and the binuclear center (see data in Table 1).

Table 1. Reduction of cyt. a by cyt. c in the COV steady state.

COV state	% redn cyt. c @ 50% max redn. cyt. a	% redn cyt. a (maximal)	E_m (cyt. a) apparent*
controlled	35	28	+275 mV
+ valinomycin	30	52	+282 mV
+ nigericin	$\approx$25	25	$\approx$+288 mV
+ val and nig**	6.5	30	+328 mV

*assuming E_m for cytochrome c = +260 mV under these conditions.

**valinomycin and nigericin abolish $\Delta\Psi$ and ΔpH, respectively.

(pH 7.4, 50 mM phosphate buffer, 30 °C.)

The spectrum of cytochrome aa_3 undergoes changes as the steady state continues, especially if a relatively poor electron donor such as TMPD, rather than cytochrome c, is employed as substrate. This involves a state change in the bimetallic centre, as it takes place not only under normal conditions but also in the presence of low spin terminal inhibitors such as cyanide[17]. Only in presence of a ligand such as formate does the enzyme stay 'high-spin' throughout the aerobic steady state. With COV (and ascorbate plus TMPD as substrate), cytochrome a becomes progressively less reduced (shown by a decrease in absorbance at 445nm) and the ferric haem (cyt. a_3) shifts its Soret maximum to the red. Upon anaerobiosis both centers go fully reduced, but appropriate deconvolution shows that cytochrome a need not change its spectral form after cytochrome a_3 has become reduced[16].

4. Discussion

In the COV, cytochrome a responds both to electron input from cytochrome c and to membrane energization, either ΔpH or $\Delta\Psi$. There are no obvious spectral changes in cytochrome a, reduced or oxidized, in response to the redox state of the binuclear center. The spectral state of cytochrome a seems to remain constant as that of cytochrome a_3 varies. Nevertheless the redox potential of cytochrome a changes with changing environment. It is modulated both by reduction of the binuclear center, and by membrane energization. Each of these processes decreases this redox potential, from a value of about +330 mV to values $\approx$60-100 mV more negative. In accordance with the 'neo-classical' model[18,19] the accompanying spectral effects are minimal.

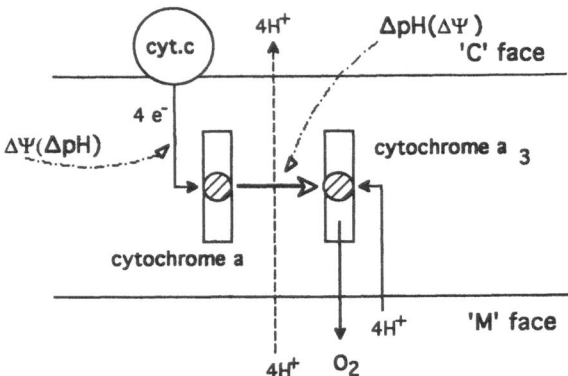

Figure 2. Membrane arrangement and control: cytochrome c oxidase.
A summary of current ideas (cf. Ref. 21) concerning the topochemical arrangement of the metal centres within the membrane. All charge separation occurs either in reduction of cytochrome a or as a result of proton flux to and from $a_3 Cu_B$. Proton translocation is coupled to electron transfer between a and $a_3 Cu_B$. These electrons pass between the two haem groups at an almost constant membrane depth.

Ferric cytochrome a_3 exists in various presumed 'spin states' which respond to the redox states of cytochrome a and Cu_B. Reduction of cytochrome a is associated with a transition in which the ferric cytochrome a_3 Soret spectrum is red-shifted. The change occurs both in free enzyme[17] and in COV (this paper), and accounts for all the spectral changes that occur during the steady state. Neither substantial reduction of the binuclear center nor spectral changes in reduced cytochrome a are involved.

Steady state cytochrome a reduction reaches a maximum when flux is still increasing. The simplest explanation is that the enzyme component whose reduction controls that flux is not cytochrome a. A probable candidate is the visible copper, Cu_A. If it is the reduction of this center but not that of cytochrome a which controls turnover, some electron transfer from cytochrome c to O_2 in the steady state must proceed directly from Cu_A to the binuclear center without involving cytochrome a.

I therefore propose the model of electron transfer shown in Fig. 2. Initial reduction of cytochrome aa_3 by cytochrome c occurs at Cu_A. Cytochrome a is reduced via Cu_A in a step that is both ΔpH and $\Delta \Psi$ dependent (Table 1) and transfers both its electron and that of Cu_A to rereduce the binuclear centre when oxygen is added to the fully reduced enzyme[7,8]. This electron transfer is not transmembranous but almost parallel to the membrane plane. It is highly ΔpH sensitive[7,8] (see also Table 1). During the steady state the spectrum of the oxidized haem (cytochrome a_3) changes so that the Soret band is more red-shifted. This is linked to spectral changes in reduced cytochrome a in the presence of ligands such as azide[14, 20]. But both cytochrome a and Cu_A themselves are simple electron transfer centers.

The formulation of the charge transfer patterns as in Fig. 2 does not necessarily determine the location or nature of proton pumping by the enzyme. Consider the processes of Eqs. (3)-(6):

$$a^{2+}a_3 Cu_B^{n+} + H^+_{in} \longleftrightarrow a^{3+}a_3 Cu_B^{(n-1)+} + H^+_{out} \tag{3}$$

$$H^+_{out}a^{2+}a_3 Cu_B^{n+} + H^+_{in} \longleftrightarrow H^+_{in}a^{3+}a_3 Cu_B^{(n-1)+} + H^+_{out} \tag{4}$$

$$H^+_{in}a^{2+}a_3 Cu_B^{n+} + H^+_{out} \longleftrightarrow H^+_{out}a^{3+}a_3 Cu_B^{(n-1)+} + H^+_{in} \tag{5}$$

$$H^+{}_{in}a^{2+}a_3Cu_B{}^{n+} \longleftrightarrow H^+{}_{out}a^{3+}a_3Cu_B{}^{(n-1)+} \qquad (6)$$

Note that eq. (6) is equivalent to the sum of eqs (3) and (5).

The electron transfer in Eq. (4) is linked to an 'M' side protonation and a 'C' side deprotonation. The thermodynamic effect is identical to that felt by a pump (eq. 3). But no charge is moved. Conversely, if a 'pump' (eq. 3) is linked to a reversed pattern of protonation/deprotonation (eq. 5), the net process (eq. 6) will show no ΔpH dependence even though protons are being moved. Proton pumping can therefore only reliably be measured directly, as by the classical oxygen or cytochrome c pulse experiments.

Nevertheless it now seems probable that transmembranous charge movement in cytochrome c oxidase must be of protons rather than electrons; there is therefore no clear analogy with processes involved in the operation of either bacterial and plant photosynthetic centres or the cytochrome bc_1 complex. Secondly, either or both direct and indirect mechanisms of proton pumping may be involved; current evidence is inadequate to rule out one or the other. Thirdly it seems that the control exerted by electrochemical gradients is not thermodynamic in origin but either kinetic or allosteric, because the effects of ΔpH and $\Delta\Psi$ are not equivalent.

Acknowledgments

I acknowledge the skilled technical assistance of Ms. Brenda Tattrie in preparing cytochrome oxidase and proteoliposomes. I thank Drs. John Wrigglesworth, Peter Butko and Chris Cooper for ongoing discussions of proton pumping and the mechanism of cytochrome c oxidase, and Dr. Bruce Hill for preprint versions of some of his work. Experimental studies described here were supported by Canadian NSERC operating grant #A-0412.

References

1. R. A. Copeland, Conformational switching at cytochrome a during steady-state turnover of cytochrome c oxidase, *Proc. Natl. Acad. Sci. USA* **88**:7281-7283 (1991).
2. D. Sherman, S. Kotake, N. Ishibe and R. A. Copeland, Resolution of the electronic transitions of cytochrome c oxidase: evidence for two conformational states of ferrous cytochrome a, *Proc. Natl. Acad. Sci. USA* **88**:4265-4269 (1991).
3. S. R. Lynch, D. Sherman and R. A. Copeland, Cytochrome c binding affects the conformation of cytochrome a in cytochrome c oxidase, *J. Biol. Chem.* **267**:298-302 (1992).
4. D. Keilin and E. F. Hartree, Cytochrome and cytochrome oxidase, *Proc. Roy. Soc.* **B127**:167-191 (1939).
5. B. C. Hill, The reaction of the electrostatic cytochrome c-cytochrome oxidase complex with oxygen, *J. Biol. Chem.* **266**:2219-2226 (1991).
6. S. Han, Y. Ching and D. L. Rousseau, Cytochrome c oxidase: decay of the primary oxygen intermediate involves direct electron transfer from cytochrome a, *Proc. Nat. Acad. Sci.* **87**:8408-8412 (1990).
7. L. Gregory and S. Ferguson-Miller, Independent control of respiration in cytochrome c oxidase vesicles by pH and electrical gradients, *Biochemistry* **28**:2655-2662 (1989).
8. S. Papa, N. Capitanio, G. Capitanio, E. De Nitto and M. Minuto, The cytochrome chain of mitochondria exhibits variable H^+/e^- stoichiometry, *FEBS Lett.* **288**:183-186 (1991).
9. M. Wikström, Energy-dependent reversal of the cytochrome oxidase reaction, *Proc. Natl. Acad. Sci. USA* **78**:4051-4054 (1981).
10. M. Wikström, K. Krab and M. Saraste. "Cytochrome Oxidase - A Synthesis," Academic Press, New York and London (1981).
11. B. G. Malmström, Cytochrome c oxidase as a redox-linked proton pump, *Chem. Rev.* **90**:1247-1260 (1990).
12. A. J. Moody, U. Brandt and P. R. Rich, Single electron reduction of 'slow' and 'fast' cytochrome c oxidase, *FEBS Lett.* **293**:101-105 (1991).

13. M. Kuboyama, F.C. Yong and T.E. King, Studies on cytochrome oxidase VIII. Preparation and some properties of cardiac cytochrome oxidase, *J. Biol. Chem.* **247**:6375-6383 (1972).

14. J. M. Wrigglesworth, C.E. Cooper, M. Sharpe and P. Nicholls, The proteoliposomal steady state: effects of size, capacitance and membrane permeability on cytochrome oxidase-induced ion gradients, *Biochem. J.* **270**:109-118 (1990).

15. B. Chance and G.R. Williams, The respiratory chain and oxidative phosphorylation, *Adv. Enzymol.* **17**:65-134 (1956).

16. P. Nicholls and J. M. Wrigglesworth, Routes of cytochrome a_3 reduction: the neoclassical model revisited, *Ann. N.Y. Acad.Sci.* **550**:59-67 (1988).

17 .P. Nicholls and V. A. Hildebrandt, Binding of ligands and spectral shifts in cytochrome c oxidase, *Biochem. J.* **173**:65-72 (1978).

18. P. Nicholls and L.C. Petersen, Haem-haem interactions in cytochrome aa_3 during the anaerobic-aerobic transition, *Biochim. Biophys. Acta* **357**:462-467 (1974).

19. M. K. F. Wikström, H. J. Harmon, W. J. Ingledew and B. Chance, A reevaluation of the spectral, potentiometric and energy-linked properties of cytochrome c oxidase in mitochondria, *FEBS Lett.* **65**:259-277 (1976).

20. G. Goodman and J. S. Leigh, The distance between cytochromes a and a_3 in the azide compound of bovine heart cytochrome oxidase, *Biochim. Biophys. Acta* **890**:360-367 (1987).

21. G. T. Babcock and M. Wikström, Oxygen activation and the conservation of energy in cell respiration, *Nature* **356**:301-309 (1992).

MECHANISM OF BUPIVACAINE UNCOUPLING
IN RAT HEART MITOCHONDRIA

F. Sztark[1,2], P. Schönfeld[3], P. Dabadie[1,2] and J.-P. Mazat[2]

[1]Département des Urgences, Hôpital Pellegrin
  33076 Bordeaux Cedex, France
[2]Université Bordeaux II
  146, rue Léo Saignat, 33076 Bordeaux Cedex, France
[3]Institut für Biochemie
  Medizinische Akademie Magdeburg
  Leipziger Strasse 44, 3090 Magdeburg, Germany

1. Introduction

Local anesthetics have diverse actions on isolated mitochondria and the cardiotoxicity of the tertiary-amine bupivacaine could be explained by interactions with cellular energy metabolism[1]. Bupivacaine has been reported to affect the mitochondrial metabolism by uncoupling, but the mechanism is still under discussion[2-6]. Different uncoupling modes have been proposed. Dabadie and coworkers[3] showed that bupivacaine can act alone as an uncoupler by increasing the proton permeability of membranes through a shuttling mechanism. More recently, Terada and coworkers[4] have concluded that bupivacaine acts as a decoupler without significant reduction of the protonmotive force. Furthermore, the reinforcement of bupivacaine uncoupling by lipophilic anions like tetraphenylboron (TPB⁻)[§], has been taken by these authors as an indication for a third uncoupling mechanism by formation of ion pairs within the membrane and increase of proton conductance (proton-leak inducer).

Tertiary amine-type local anesthetics exist under physiological conditions in the protonated and deprotonated form. On the other hand, quaternary-amine analogs are always in the positively charged form, and cannot therefore act as protonophore. We used the quaternary-amine QX 572 as a reference compound for elucidation of the bupivacaine uncoupling mechanism in rat heart mitochondria.

[§]Abbreviations: CCCP, carbonylcyanide m-chlorophenylhydrazone; TPB⁻, tetraphenylboron; VAL, valinomycin; BUPI, bupivacaine, 1-butyl-N-(2,6-dimethylphenyl)-2-piperidinecarboxamide; QX 572, dimethyldiphenyl-acetamido-ammonium; BCECF/AM, acetoxymethyl ester of 2,7-biscarboxy-ethyl-5(6)-carboxyfluorescein; TPP⁺, tetraphenyl-phosphonium; RHM, rat-heart mitochondria.

It is concluded that uncoupling by bupivacaine or [QX 572 + TPB⁻] is due to a protonophore-like mechanism or to the dissipation of the electrochemical proton gradient (by electrophoretic uptake) respectively.

2. Materials and Methods

2.1. Chemicals

Bupivacaine was a gift from Laboratoire Roger Bellon (France). Bupivacaine HCl was dissolved in distilled water at 50 mM concentration. QX 572 was from Astra (Sweden) and used under the same conditions. Acetoxymethyl ester of BCECF (BCECF/AM) was purchased from Molecular Probes Inc. (U.S.A.), [^{3}H]TPP$^+$ was from Amersham (F.R.G.).

2.2. Preparation of mitochondria

Mitochondria were isolated from rat heart as described by Makinen and coworkers[7], in a medium containing 75 mM sucrose, 225 mM mannitol, 0.1 mM EDTA and 10 mM Tris-HCl (pH 7.2). Proteins were measured by the biuret method.

2.3. Rate of respiration

Mitochondrial respiration was measured polarographically at 30 °C using a Clark-type electrode connected to a micro-computer giving an on-line display of rate meter values. The incubation medium contained 25 mM sucrose, 75 mM mannitol, 100 mM KCl, 10 mM Tris-phosphate, 50 mM EDTA, 10 mM Tris-HCl (pH 7.4) and 5 mM glutamate plus 5 mM malate as substrates.

2.4. Measurements of transmembrane potential

The transmembrane potential was estimated from the equilibration of tritiated tetraphenyl-phosphonium ([^{3}H]TPP$^+$) between mitochondria and medium. Mitochondria were separated from the medium by rapid centrifugation through a silicone oil layer. Calculation of the transmembrane potential was based on a matrix volume of 1 µl/mg protein and on a correction for unspecific binding of [^{3}H]TPP$^+$ (experiments with deenergized mitochondria).

2.5. ATP synthesis

The rate of ATP synthesis was determined during mitochondrial respiration at State 3 (after addition of ADP) by HPLC measurement of the ATP content in the medium at various time intervals.

2.6. Swelling

Passive proton-permeability of the inner membrane in the presence and absence of local anesthetics was estimated by means of the swelling of non-respiring mitochondria in 100 mM potassium acetate and 10 mM Tris-maleate buffer, pH 7.2, in the presence of 50 ng/mg valinomycin, 0.2 µg/ml antimycin and 25 µg/mg oligomycin[8]. Swelling was measured as a decrease in optical absorption at 550 nm.

2.7. Fluorescence measurements of matrix pH

The effect of local anesthetics on the matrix pH was estimated using 2,7-biscarboxy-ethyl-5(6)-carboxyfluorescein (BCECF) as a pH-dependent fluorescent probe[9]. After loading (30 min, with the membrane permeable acetoxymethyl ester of BCECF), mitochondria were suspended in an incubation medium (pH 7.0) containing 5 mM glutamate plus 5 mM malate, but without phosphate. Fluorescence of BCECF-loaded mitochondria was recorded at 530 nm, using 509 nm as excitation wavelength. For calibration, fluorescence was recorded at different pH, after addition of Triton X100 (0.1 % v/v).

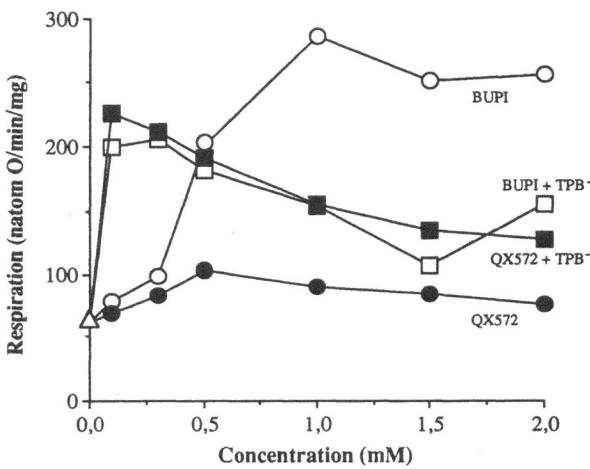

Figure 1. Dose-response effect of bupivacaine and QX 572 on state 4 mitochondrial respiration. Bupivacaine and QX 572 were added with and without 2 μM TPB⁻ in the incubation medium containing rat heart mitochondria (1 mg/ml) and 5 mM glutamate plus 5 mM malate as respiratory substrates.

3. Results

3.1. Effects of local anesthetics on State 4 respiration

Fig. 1 shows the dose-dependence effects of bupivacaine and QX 572 on State 4 respiration of rat heart mitochondria in the absence and presence of TPB⁻ (2 μM). Bupivacaine stimulated respiration with and without TPB⁻. On the other hand, QX 572 did not stimulate respiration without TPB⁻. In the presence of TPB⁻, the respiration responded with a similar sensitivity to addition of both local anesthetics. This synergistic effect of TPB⁻ was attributed to the facilitated permeation of cationic local anesthetics across the inner membrane, resulting from formation of an electroneutral ion pair[2].

3.2. Transmembrane potential measurements

Table 1 summarizes the effects of the two local anesthetics on transmembrane potential. Stimulation of respiration by bupivacaine (without TPB⁻) was paralleled by a decrease in transmembrane potential. A similar decrease was observed for both local anesthetics in the presence of TPB⁻, at the point of maximal stimulation of respiration.

Table 1. Effects of local anesthetics with and without 2 μM TPB⁻ on transmembrane potential.

Concentration (mM)	Membrane Potential (mV)			
	Bupivacaine		QX 572	
	- TPB⁻	+ TPB⁻	- TPB⁻	+ TPB⁻
0.0	167	165	167	165
0.1	160	135	162	131
1.0	124	67	145	56

3.3. Effects on ATP synthesis

Results are similar to those as to transmembrane potential (Table 2). Bupivacaine inhibited ATP synthesis and TPB⁻ reinforced this action. The effect of [QX 572 + TPB⁻] was the same as [bupivacaine + TPB⁻].

Table 2. Effects of local anesthetics with and without 2 μM TPB⁻ on mitochondrial ATP synthesis.

Concentration (mM)	ATP synthesis (nmol/min/mg)			
	Bupivacaine		QX 572	
	- TPB⁻	+ TPB⁻	- TPB⁻	+ TPB⁻
0.0	828	828	828	828
0.1	-	431	-	398
1.0	477	0	805	0

3.4. Swelling in the potassium acetate/valinomycin system

When rat heart mitochondria are incubated in 100 mM potassium acetate, swelling necessitates simultaneous permeabilization to K^+ (by valinomycin) and to H^+ (to maintain electroneutrality). H^+ permeabilization is obtained with a protonophore such as CCCP. Fig. 2 shows that bupivacaine induced swelling of rat heart mitochondria just like CCCP, whereas [QX 572 + TPB⁻] was unable to mediate such an effect.

3.5. Effects on matrix pH

The ability of bupivacaine to increase the proton permeability of the inner membrane can also be demonstrated with energized mitochondria. The expected effect was acidification of the matrix space. Variations of the matrix pH were followed by the fluorescence change of BCECF-loaded rat-heart mitochondria. It can be seen in Fig. 3 that bupivacaine induced a clear acidification of the matrix pH such as CCCP. No effect was observed with QX 572, even in the presence of TPB⁻.

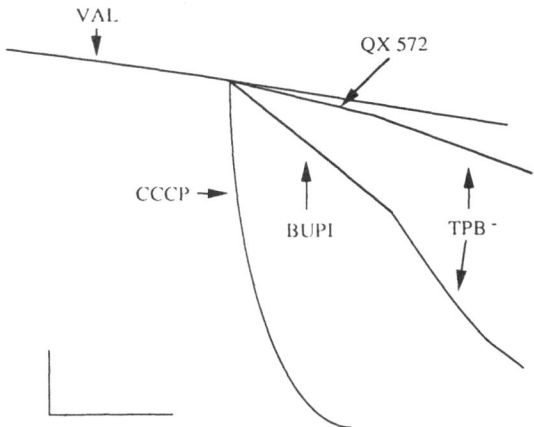

Figure 2. Effect of CCCP, bupivacaine and QX 572 on iso-osmotic swelling in potassium acetate. Mitochondria were incubated in a solution of 100 mM potassium acetate and 10 mM Tris-maleate buffer, pH 7.2, in the presence of 0.2 µg/ml antimycin and 25 µg/mg oligomycin. Addition of valinomycin (50 ng/mg) and CCCP (1 µM) led to a swelling by K^+ influx and H^+ efflux. Effects of bupivacaine (1 mM) and QX 572 (1 mM) with and without TPB^- (2 µM) were shown. Bars indicate (vertically) 0.05 A and (horizontally) 1 min.

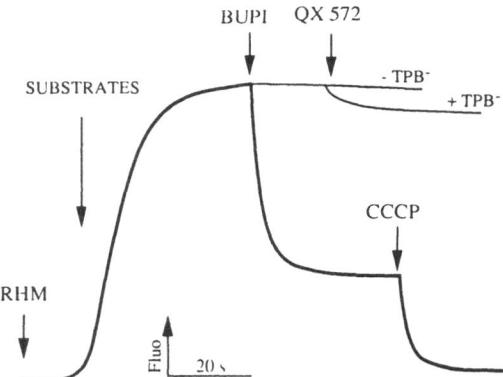

Figure 3. Change of matrix pH in respiring mitochondria by bupivacaine. Shown are fluorescence (Fluo.) record of BCECF-loaded mitochondria in medium described under "Materials and Methods". 1 mM local anesthetic (bupivacaine or QX 572) and 1 µM CCCP were added to the cuvette as indicated by the arrows.

4. Discussion

Stimulation of State 4 respiration, together with a paralleled decrease of transmembrane potential and inhibition of ATP synthesis, corresponds to a classical uncoupling mechanism and clearly excludes a decoupling mechanism[10].

Uncoupling of oxidative phosphorylation can be explained by three modes of action:

(1) an electrophoretic uptake of local anesthetics into the matrix space of the mitochondria with a dissipation of transmembrane potential,

(2) an increase in proton permeability of the inner membrane by a protonophoric mechanism, and

(3) a proton-leak by incorporation of ion-pairs, e.g. [bupivacaine-TPB⁻] or [QX 572-TPB⁻] into the phospholipid membrane[4].

Comparing the effects of QX 572 (which cannot be protonated and cannot act as a protonophore) and bupivacaine can lead to a better understanding of the uncoupling effect of the latter. QX 572 uncoupling (in the presence of TPB⁻) could be explained only by the first mechanism (electrophoretic accumulation). The missing effect on proton permeability excludes the third possibility (proton-leak inducer). Bupivacaine uncoupling could be rationalized by its electrophoretic uptake, facilitated by TPB⁻ and similar to the case of QX 572. Nevertheless, in some experimental conditions we have proved that bupivacaine acts as a protonophore-like compound (swelling, acidification of matrix pH) and there is no reason to assume a different bupivacaine uncoupling mechanism with or without TPB⁻.

In summary, bupivacaine is an uncoupler of oxidative phosphorylation and this uncoupling effect could be explained by a protonophore-like mechanism.

Acknowledgements

This work was supported by the Fondation pour la Recherche Médicale. The authors thank Dr Rune Sandberg of Astra Pharmaceuticals, Sweden for providing the local anesthetics.

References

1. J.E. de La Coussaye, B. Bassoul, B. Albat, P.A. Peray, J.P. Gagnol, J.J. Eledjam and A. Sassine, Experimental evidence in favor of role of intracellular actions of bupivacaine in myocardial depression, *Anesth. Analg.* **74**:698-702 (1992).
2. K.D. Garlid and R.A. Nakashima, Studies on the mechanism of uncoupling by amine local anesthetics, *J. Biol. Chem.* **258**:7974-7980 (1983).
3. P. Dabadie, P. Bendriss, P. Erny and J.P. Mazat, Uncoupling effects of local anesthetics on rat liver mitochondria, *FEBS Lett.* **226**:77-82 (1987).
4. H. Terada, O. Shima, K. Yoshida and Y. Shinohara, Effects of the local anesthetic bupivacaine on oxidative phosphorylation in mitochondria, *J. Biol. Chem.* **265**:7837-7842 (1990).
5. X. Sun and K.D. Garlid, On the mechanism of bupivacaine-induced uncoupling of mitochondria, *Biophys. J.* **59**:136a (1991).
6. K. van Dam, Y. Shinohara, A. Unami, K. Yoshida and H. Terada, Slipping pumps or proton leaks in oxidative phosphorylation - The local anesthetic bupivacaine causes slip in cytochrome c oxidase of mitochondria, *FEBS Lett.* **277**:131-133 (1990).
7. M.W. Makinen and C.P. Lee, Biochemical studies of skeletal muscle mitochondria, *Arch. Biochem. Biophys.* **126**:75-82 (1968).
8. P.J.F. Henderson, J.D. McGivan and J.B. Chappell, The action of certain antibiotics on mitochondrial, erythrocyte and artificial phospholipid membranes - The role of induced proton permeability, *Biochem. J.* **111**:521-535 (1969).
9. D.W. Jung, M.H. Davis and G.P. Brierley, Estimation of matrix pH in isolated heart mitochondria using a fluorescent probe, *Analyt. Biochem.* **178**:348-354 (1989).
10. H. Rottenberg, Decoupling of oxidative phosphorylation and photophosphorylation, *Biochim. Biophys. Acta* **1018**:1-17 (1990).

MOSAIC NON-EQUILIBRIUM THERMODYNAMICS AS A TOOL IN THE STUDY OF OXIDATIVE PHOSPHORYLATION

B.H. Groen[1,2], J.A. Berden[1] and K. van Dam[1]

[1]E.C. Slater Institute for Biochemical Research
University of Amsterdam
Plantage Muidergracht 12, Amsterdam, The Netherlands
[2]Faculty of Physics and Astronomy
The Free University
De Boelelaan, 1081 HV Amsterdam, The Netherlands

1. Introduction

One approach to study energy metabolism is to focus on its thermodynamic aspects. Because life is incompatible with a situation of equilibrium, classical equilibrium thermodynamics does not cover the most important aspects of energy metabolism, since it neglects the flows as well as the time dependence. In fact, it can only be used for systems with vanishing net flows. Non-Equilibrium Thermodynamics (NET), introduced by Kedem and Caplan in 1965[1] is a first step towards a realistic quantitative description of (biological) energy transduction. However, this theory is phenomenological, it regards energy transducing systems as a black box. For a better understanding of oxidative phosphorylation we need to open the black-box. This is the aim of the Mosaic Non-Equilibrium Thermodynamics (MNET) developed by Westerhoff and Van Dam[2], which is an extension of NET. MNET contains parameters directly linked to the characteristics of the energy-transducing system, thus forming a bridge between purely phenomenological thermodynamic models and kinetic models.

2. MNET Description of Oxidative Phosphorylation

The MNET-model can describe oxidative phosphorylation in relation to the underlying mechanism, as it proceeds according to the chemi-osmotic model, with three flows and three forces. The flows are the rate of respiration (J_o), the rate of ATP synthesis ($-J_p$), and the net proton flow (J_H). The forces are the oxidation potential (ΔG_o), the phosphate potential (ΔG_p) and the proton motive force ($\Delta \tilde{\mu}_{H^+}$). The relationship between each flow-force pair is expressed in the phenomenological coefficients (L_0, L_p and L_H), which are a measure for the specific activity of the enzymes. However, each flow does not necessarily depend on each force, as in the NET model. The model includes factors for three possible processes that may regulate the coupling between respiration and ATP synthesis: increased permeability of the inner mitochondrial membrane to protons (leak, L_H^l), slip in the redox-pumps (L_0^s), and slip in the ATP synthase (L_p^s). Of these processes only proton leak is generally considered. The stoichiometries of the redox pumps (n_0) and the ATP synthase (n_p) are also included. Furthermore, it is possible to account for saturation of the enzymes or pathways, in which case the relation between a flow and its conjugated force is assumed to be linear until the flow reaches its maximum (or minimum) value. Above (or under) this "threshold" value for the force, a further increase in the force no longer results in an increase of the flow. In practice, this is more relevant for the oxidation process than for the phosphorylation process. To account for this saturation we may replace the measured value for the oxidation potential by the "threshold" value ($\Delta G_0^£$, which is equal to $\Delta G_0 - \Delta G_0^{\neq}$) (for definitions of these terms see Ref. 3).

$$J_0 = \left(L_0 + L_0^s\right)\Delta G_0^£ + n_0 L_0 \Delta\tilde{\mu}_{H^+} \, ,$$

$$-J_p = \left(L_p + L_p^s\right)\Delta G_p + n_p L_p \Delta\tilde{\mu}_{H^+} \, ,$$

$$J_H = n_0 L_0 \Delta G_0^£ + n_p L_p \Delta G_p + \left(n_0^2 L_0 + n_p^2 L_p + L_H^l\right)\Delta\tilde{\mu}_{H^+} \, .$$

In steady state, without net proton flow, we may simplify these equations by eliminating the proton-motive force and rewriting:

$$J_0 = L_{oo}\Delta G_0^£ - L_{op}\Delta G_p \, ,$$

$$-J_p = L_{op}\Delta G_0^£ - L_{pp}\Delta G_p \, .$$

The thermodynamic efficiency is defined as the quotient of the output of the system (the rate of phosphorylation multiplied by the phosphate potential) and the input of the system (the rate of respiration multiplied by the oxidation potential). The mutual dependence of respiration and ATP synthesis, expressed in the form of the coupling coefficient q, can be calculated from the cross-coefficient L_{op}, normalized by dividing it by $\sqrt{L_{oo}L_{pp}}$. The phenomenological stoichiometry, Z, equal to $\sqrt{L_{pp}/L_{oo}}$, is comparable - but not identical - to the classical P/O.

We have focused on two problems:

(1) Is there a difference in the efficiency and/or mechanism of oxidative phosphorylation between organ(ism)s, and
(2) what is the precise effect of substances that affect oxidative phosphorylation, such as uncouplers.

All mitochondria have basically the same structure, but it is not *a priori* clear that no differences exist between mitochondria isolated from different sources. On the contrary, why should all (isolated) mitochondria have the same value for e.g. the coupling coefficient? Furthermore, some characteristics of oxidative phosphorylation may very well depend on the specific circumstances in the cell, such as the amount of substrate available or the energy demand of the cell or tissue. To compare the efficiency of oxidative phosphorylation in different tissues, we have isolated mitochondria from rat liver and potato tubers and measured the efficiency of oxidative phosphorylation with succinate as substrate. Despite the differences in organism and function of the tissue, we found similar values for important parameters: the maximal efficiency and the value for the coupling coefficient were comparable in both tissues. This indicates that both kinds of mitochondria are probably optimized for the same output function[6]. However, in some respects these mitochondria were not alike. The rates of the flows and the values for the phenomenological coefficients were higher for rat liver mitochondria, indicating a higher specific activity of the enzyme complexes. This is not surprising, considering that the tubers are a storage tissue.

Although both kinds of mitochondria seemed optimized for the same output function, the contribution of slip and leak appeared to be different. Whereas we deduced that rat liver mitochondria have at least some ATP synthase slip, we found that in potato tuber mitochondria redox-slip is an important factor. Furthermore, we have determined the efficiency of oxidative phosphorylation in potato tuber mitochondria, using two substrates with the same P/O value but a large difference in oxidation potential, namely succinate and NADH. This is possible in potato tuber mitochondria, because plant mitochondria contain a NADH dehydrogenase located at the outside of the inner mitochondrial membrane, that directly oxidizes NADH in the inter-membrane space. We found that potato tuber mitochondria had approximately the same value for the coupling coefficient with both substrates, but a lower value for the maximal efficiency, due to the much higher oxidation potential[4].

Furthermore, the results confirmed our hypothesis concerning the "threshold" value for the oxidation potential. Although the oxidation potential for NADH oxidation is much higher than for succinate oxidation, the value for the "threshold" is approximately the same. This means that with both substrates a large percentage of the input oxidation potential is not used for ATP synthesis.

What does the MNET analysis teach us about oxidative phosphorylation? Firstly, mitochondria from tissues as diverse as rat liver and potato tuber are probably optimized for the same output function, under the chosen conditions of unlimited supply of substrate. Secondly, mitochondria may be well coupled even when the specific activity of the enzyme complexes is low. It has been reported that the coupling coefficient of rat-liver mitochondria depends on the availability and nature of the substrate for oxidative phosphorylation[5]. Therefore, we do expect that mitochondria are capable of regulating their efficiency by varying the amount of slip and leak, to adapt to the energy demand of the cell. An extreme example of this adaptation would be the uncoupling protein present in brown adipose mitochondria, which enables them to switch from ATP synthesis to heat production[6]. Another possibility to change the energy output of oxidative phosphorylation could be a switch to another substrate. On a longer time scale, a cell may enhance its respiratory capacity by increasing the amount of respiratory complexes. This seems to be the effect of endurance training, as measured in rat hind limbs[7].

As the MNET model includes parameters for slip and leak, it is possible to discriminate between these processes. Two kinds of experiments clearly show whether a substance causes redox slip. Firstly, this can be done by measurement of the effect of a compound on the phosphate potential during State 4 respiration. In State 4 an increased proton leak or ATP synthase slip should decrease the phosphate potential, whereas an increased redox slip has no effect on this phosphate potential. Secondly, by measuring the effect of a compound on the rate of ATP synthesis as a function of the rate of respiration. All protonophoric uncouplers increase State 4 respiration. But whereas leak or ATP synthase slip lead to an increase in dJ_p/dJ_o, redox-slip does not cause a change in this parameter. Furthermore, it is possible to predict the effect of a change in stoichiometry (n_o or n_p): a decreased stoichiometry of the redox pumps, or an increased stoichiometry of the ATP synthase both appear to give rise to a decreased value of dJ_p/dJ_o. By such criteria the MNET model can discriminate between slip and a change in stoichiometry.

We have used the MNET model to determine the effect of several uncouplers, such as 2,4-dinitrophenol, gramicidin and palmitic acid. The mechanism of uncoupling by fatty acids is still disputed. Proposed mechanisms are classical protonophoric uncoupling, decoupling, and involvement of the nucleotide translocator. Uncoupling by fatty acids (in rat liver mitochondria) was not accompanied by a change in the phosphate potential in State 4 (see Ref. 8), nor a change in dJ_p/dJ_o, from which we concluded that according to the MNET model palmitic acid causes redox slip.

3. Conclusions

We conclude that the MNET model is a very useful tool to discriminate between leak, slip or a change in stoichiometry. Furthermore, the model suggests experiments that indicate whether slip and/or leak occur in isolated mitochondria in the absence of uncouplers.

References

1. O. Kedem and S.R. Caplan, Degree of coupling and its relation to efficiency of energy conversion, *Trans. Faraday Soc.* **21**:1897-1911 (1965).
2. H.V. Westerhoff and K. van Dam. "Thermodynamics and Control of Biological Free-Energy Transduction," Elsevier, Amsterdam (1987).
3. B.H. Groen, J.A. Berden and K. van Dam, Differentiation between leaks and slips in oxidative phosphorylation, *Biochim. Biophys. Acta* **1019**:121-127 (1990).
4. B.H. Groen, H.G.J. van Mil, J.A. Berden and K. van Dam, The efficiency of oxidative phosphorylation in potato-tuber mitochondria is different for succinate and external NADH, *Biochim. Biophys. Acta*, in press.
5. S. Soboll and J.W. Stucki, Regulation of the degree of coupling of oxidative phosphorylation in intact rat liver, *Eur. J. Biochem.* **109**:269 (1980).
6. B. Cannon and J. Nedergaard, The biochemistry of an inefficient tissue: brown adipose tissue, *in*: "Essays in Biochemistry," Vol. 20, P.N. Campbell and R.D. Marshall, eds, Academic Press, London, pp. 110-164 (1985).
7. K.J.A. Davies, L. Packer and G.A. Brooks, Biochemical adaptation of mitochondria, muscle, and whole-animal respiration to endurance training, *Arch. Biochem. Biophys.* **209**:539-544 (1981).
8. B.H. Groen, J.A. Berden and K. van Dam, Do all uncouplers of oxidative phosphorylation cause leak?, Abstract 4th BTK Meeting Noordwijkerhout (The Netherlands), (1990).

THE REDOX STATE OF CYTOCHROMES AS A NEW TOOL TO ASSESS THE THERMODYNAMIC STATE OF MITOCHONDRIA

Siro Luvisetto, Ibolya Schmehl,
Marcella Canton, and Giovanni Felice Azzone

C.N.R. Unit for Studies in Mitochondrial Physiology
and Department of Experimental Biomedical Sciences
University of Padova
Via Trieste 75, 35121-Padova, Italy

1. Introduction

In the present work we have followed the changes of the absorption spectra of mitochondrial suspensions during transitions from anaerobiosis to steady state and after addition of uncouplers. From these spectra we have selected proper wavelengths to follow the kinetics of the spectroscopic changes of the various cytochrome components of the redox chain. We focused our attention to a group of cytochromes, those of the cc_1, providing a response similar in significance to that of the electrodes or of the dyes assessing the membrane potential.

Furthermore, by titrations of the redox state of cytochromes with malonate or by titrations of the membrane potential with FCCP or with atractyloside in the presence of ADP, we have measured the flow-force and force-force relationships between the respiratory rate or the membrane potential and the cytochromes response converted in a thermodynamic force[1].

A comparative analysis with the flow-force and force-force relationships as obtained during titrations of the overall electron transfer reaction from succinate to oxygen was also conducted. From the comparison between the different flow-force and force-force relationships we conclude that the cytochrome response may be taken as an endogeneous probe providing direct information on the energy level of the system.

2. Spectral and Kinetic Analysis

Fig. 1 shows that the transition of mitochondria from anaerobiosis to steady state was

Modern Trends in Biothermokinetics, Edited by
S. Schuster *et al.*, Plenum Press, New York, 1993

accompanied by a decrease of absorbance in the Soret region of the cytochromes at 423 and 440 nm, and in the visible region at 520, 550 and 605 nm, corresponding to an oxidation of the cytochromes of the cc_1 and aa_3 groups[2]. The oxidation of cytochromes absorbing at 423 and at 550 nm (cytochromes cc_1) showed the following features:

i) it increased proportionally to the size of the oxygen pulse and to the rise of the membrane potential up to steady state level;

ii) in steady state the cytochromes were in general 50-60 % oxidized;

iii) further oxidation was obtained after addition of uncoupler (FCCP) or respiratory inhibitor (malonate or antimycin) and was maximal in the presence of excess of antimycin;

iv) additions of small amounts of ADP, inducing State4-State3-State4 transitions, were accompanied by proportional cycles of cytochrome oxidation.

In contrast, the absorbance decrease induced by oxygen at 440 and 605 nm (cytochromes aa_3) indicated an almost maximal oxidation (90-95 % in steady state) and addition of uncouplers or ADP did not cause further detectable changes. In mitochondria partially inhibited with KCN, uncoupling was followed by increased reduction of cytochromes aa_3, and additions of ADP were accompanied by proportional cycles of cytochromes aa_3 reduction. These results suggest that in intact mitochondria, at variance from the reconstituted system, the responses of cytochromes cc_1 (423-403 nm) reflect largely a thermodynamic control while those of cytochromes aa_3 (605-630 nm) reflect mainly a kinetic control.

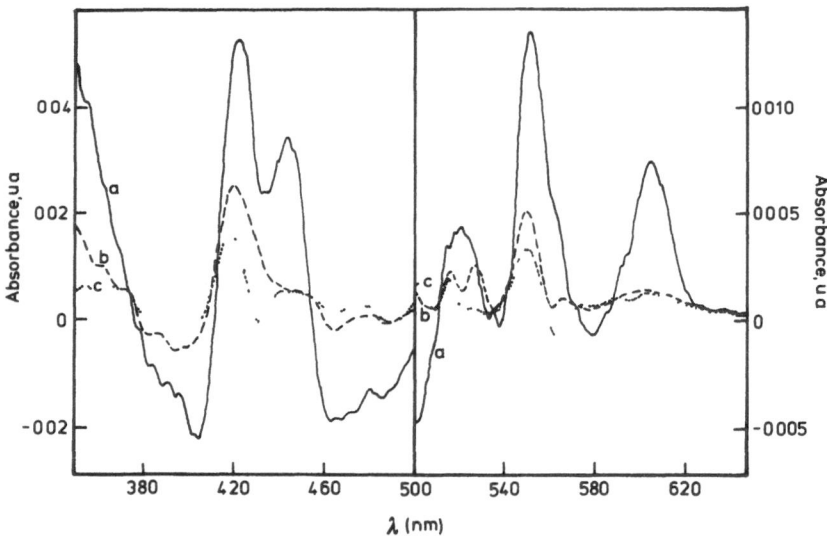

Figure 1. Difference spectra of anaerobic (trace a), aerobic (trace b) or uncoupled (FCCP, trace c) mitochondria with respect to antimycin inhibited mitochondria. RLM (3 mg/ml) were incubated in the presence of excess oxygen. After 2 min antimycin-A (50 ng/mg) was added and the spectrum between 360 and 640 nm was recorded. This spectrum was used as reference. Medium: 0.175 M sucrose, 5 mM succinate, 30 mM MOPS/Tris, 0.2 mM EGTA, 5 mM $MgCl_2$, 5 mM P_1, 5 µM rotenone, catalase, T 25 °C, pH 7.4.

To distinguish between the different cytochromes contributing to the absorbance change at 423-403 nm, sulfite was used as electron donor substrate[3]. Sulfite is oxidized by the sulphite oxidase located in the intermembrane space. The reduced form of sulfite oxidase is then able to reduce cytochrome c, which is then oxidized by cytochrome oxidase.

Mitochondria were first let respiring aerobically in the presence of excess oxygen and a limiting amount of succinate. Under these conditions, exhaustion of succinate caused the electron carriers of the redox chain to go in a fully oxidized state. The successive addition of sulfite in the presence of antimycin caused an increase in the reduction state of cytochrome. The same extent of oxidation, as in the case of mitochondria respiring with succinate as electron donor substrate, was obtained. In control experiments, consumption of oxygen due to autoxidation of sulfite to dithionite, or contribution of sulfite to the absorbance at 423-403 nm were found to be negligible. From these results and from the comparison between cytochrome responses obtained at 550-530 nm we conclude that the changes of absorbance at 423-403 nm in mitochondria during respiration, are predominantly due to changes in the redox state of cytochrome c.

3. Thermodynamic Analysis

The nature of the thermodynamic control over the cytochrome c redox state has been studied by applying the analysis of the flow-force and force-force relationships. To this purpose, the absorbance changes at 423-403 nm have been first converted into a parameter denoted as cytochrome redox affinity, A^*, and then the relationships between A^* and respiration or membrane potentials have been analyzed. The above flow-force and force-force relationships have also been compared with the relationships based on the redox affinity A_e of the overall electron transfer from succinate to oxygen.

The affinity of the overall redox reaction, A_e, has been calculated by using the equation

$$A_e = RT \ln K \sqrt{[Succ]/[Fum]}$$

with $K=1.31 \cdot 10^{13}$, at pH=7.4, pO_2=0.2 atm, RT=0.592 Kcal/mol at 25 °C.[4]

The cytochrome redox affinity, A^*, has been calculated from the 423-403 nm cytochrome c response by using the following equation:

$$A^* = RT \ln K^* [Red]/[Ox] ,$$

where K^* has been estimated from the difference in the redox potential between cytochrome c and oxygen. A^* thus represents the redox affinity for the span of the redox chain from cytochrome c to oxygen. [Red] and [Ox] represent the percentage of reduction and oxidation of cytochrome c. 100 % of cytochrome oxidation and 100 % of cytochrome reduction levels were determined as the absorbance levels after addition of antimycin or after exhaustion of oxygen, respectively.

The left-hand panels in Fig. 2 show the flow-force relationships between J_0 and A_e (panel A) or A^* (panel B) as obtained by titrations of A_e with decreasing succinate concentration (20 mM - 200 μM) or by titrations of A^* with increasing amounts of malonate in the presence of succinate (20 mM), respectively. The right-hand panels in Fig. 2 show the force-force relationships between membrane potentials and A_e (panel C) or A^* (panel D) as obtained under the same conditions.

Both the flow-force and the force-force relationships, with respect to A_e and to A^*, were clearly biphasic, i.e., the respiratory rate and the membrane potential remained roughly constant in a wide range of values of A_e and of A^* and then began to decline markedly below certain values of the two redox affinities. The results of Fig. 2 show that the pattern of the flow-force and of the force-force relationships, found during the malonate titrations using the experimental parameter A^* as the input force, corresponds to the flow-force and the force-force relationships obtained using the parameter A_e as the input force. This parallelism gives a thermodynamic significance to the parameter A^* and suggests a possible role for the cytochrome redox state during energy conversion.

Fig. 3 shows another relevant feature of the relationship between the membrane potential and cytochrome redox affinity, A^*. When the system was titrated with "inhibitors" acting at the level of membrane potential, a strict proportionality was maintained between the two thermodynamic forces. In the experiment of Fig. 3 the membrane potential was varied either by increasing the concentration of uncouplers (FCCP) or by varying the rate of phosphorylation. This was obtained by adding excess ADP and then titrating back with atractyloside. The two procedures gave similar results.

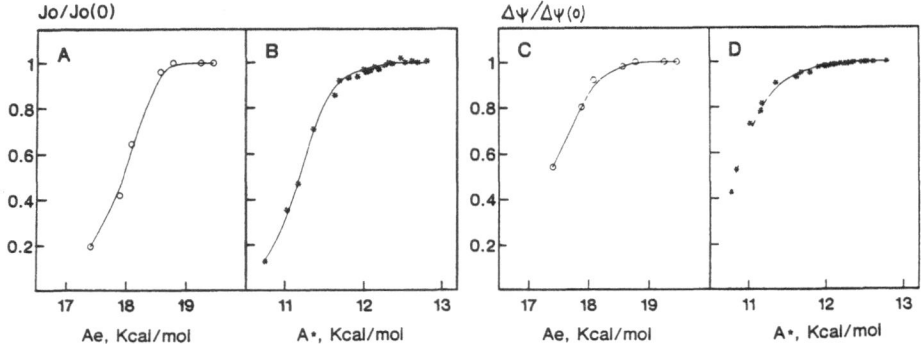

Figure 2. Flow-force (left panels) and force-force (right panels) relationships between normalized respiratory rate or membrane potentials and redox affinity (panels A, C) or cytochrome redox affinity (panels B, D). $J_o(0)$ and $\Delta\Psi(0)$ represent the rate of oxygen consumption (15 nmol/mg min) and the membrane potential (190 mV) as obtained with 20 mM succinate in the absence of malonate. Fumarate (50 μM) and malate (215 μM) were also added. The concentration of malate was the equilibrium concentration with fumarate, assuming an equilibrium constant of 4.3 for the fumarate hydratase reaction.

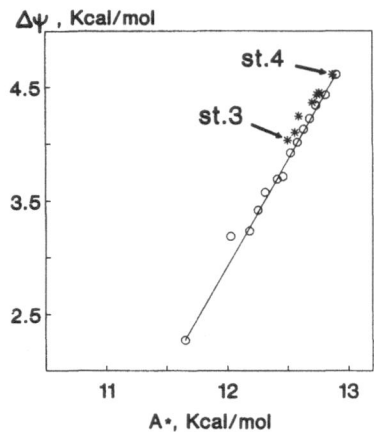

Figure 3. Force-force relationships between membrane potentials and cytochrome redox affinity during titrations of mitochondria in State 4 and State 3. Titrations with FCCP (0-200 pmol/mg, circles). Titrations with atractyloside (0-20 μM, stars) in the presence of ADP (1 mM). Arrows indicate the State 4 and the State 3 as the starting points of the two titrations.

A relevant question is that of the role of the cytochrome response in the reaction leading to the generation of membrane potential. Consider first the correspondence between the decline of membrane potential and of A^* during titrations performed by adding increasing uncoupler concentrations (Fig. 3). Given that the membrane potential is a reliable parameter of the mitochondrial energy level, the correspondence between cytochrome response and membrane potential favor the view that the 423-403 nm cytochrome response is a probe of the thermodynamic control on mitochondrial respiration. A feasible explanation for the observed correspondence is that the decline of the membrane potential affects the redox equilibria of the cytochromes. The 423-403 nm cytochrome response would then follow the changes of the membrane potential being a function of it.

However, the analysis of the behavior of the phenomenological redox affinity A^* and of the redox affinity A_e during the flow-force and the force-force titrations performed with malonate indicate that the changes of the 423-403 nm cytochrome response may not follow in a proportional manner the changes of the membrane potential. Clearly, if that would be the case one would expect a constant parallelism between the cytochrome response and the trend of the membrane potential. Instead the titrations with malonate indicate a decline of the phenomenological redox affinity A^* under conditions where there is a constant level of membrane potential (see Fig. 2).

The results of these titrations suggest that the cytochrome response is a quantitative internal probe of the membrane potential either when the system is titrated against variations of the membrane potential (as with uncouplers or atractyloside, or under conditions when there are no restrictions of the electron supply to the cytochromes. On the other hand, restriction of the electron supply causes a variation of the phenomelogical cytochrome affinity A^* at constant membrane potential. This may indicate a behavior of the cytochrome as a buffer of electrons at equivalent membrane potential.

In a thermodynamic view, if we consider the cytochrome redox affinity, A^*, as the input force and the membrane potential as the output force in energy conversion, the results of Figs 2 and 3 indicate that during titrations with agents acting on the output force a close proportionality exists between changes in the input and output forces. On the contrary, during titrations with agents acting on the input force the relationship between input and output forces is highly non-proportional. This difference in behavior of the input and output forces are significant in order to clarify the nature of both the energy-converting and the energy-dissipating processes.

4. Conclusions

The present results indicate that, under static-head conditions, titration of the input force, say A_e or A^*, is not accompanied by significant changes in the level of the output force, say the membrane potential. On the other hand, depression of the output force is accompanied by a parallel decline of the input force. The question arises as to the reasons for the different patterns of the force ratio under the two sets of titrations.

If the mitochondria were close to equilibrium under State-4 conditions one would expect a strict proportionality between output and input forces under both types of titrations. However, these results are in accord with the general concept that, at static head, mitochondria are in a condition not of thermodynamic equilibrium but rather of energy dissipation. This energy dissipation increases with the increase of the input force. The titration with malonate diminishes the input force in the region where there is energy dissipation and therefore is not accompanied by decline of the output force. The energy dissipation during State-4 conditions has been attributed either to proton leaks in the membrane and/or slips in the proton pumps. In both cases, however, either leaks or slips lead to energy dissipation by means of futile cycles, and the rate of energy dissipation increases with the membrane potential. The present discussion on the different effects caused by titrations of the output or input forces is independent of the nature of the reaction leading to energy dissipation during static head.

References

1. D. Walz, Thermodynamics of oxidation-reduction reactions and its application to bioenergetics, *Biochim. Biophys. Acta* **505**:279 (1979).
2. R. Lemberg and J. Barrett. "Cytochromes," Academic Press, New York (1973).
3. H.J. Cohen, S. Betcher-Lange, D.L. Kessler and K.V. Rajagopalan, Hepatic sulfite oxidase. Congruency in mitochondria pf prosthetic groups and activity, *J. Biol. Chem.* **247**:7759 (1972).
4. D. Pietrobon, M. Zoratti, G.F. Azzone and S.R. Caplan, Intrinsic uncoupling of mitochondrial proton pumps. 2. Modeling studies, *Biochemistry* **25**:767 (1986).

THERMODYNAMIC RESPONSE PARADIGM
AND ITS APPLICATION
TO OXIDATIVE PHOSPHORYLATION

Bernard Korzeniewski and Wojciech Froncisz

Institute of Molecular Biology
Jagiellonian University
al. Mickiewicza 3
31-120 Kraków, Poland

1. The Thermodynamic Response Coefficient

If we change the concentration of the substrate or product of a metabolic system, the global thermodynamic span of the whole system is changed as well. It is followed by changes in the local thermodynamic spans of particular reactions. After some time, the system stabilizes at a new steady state, with new global and local Gibbs free energy changes. If we perform an infinitesimally small change in the substrate or product concentration, we can determine a parameter which we call *thermodynamic response coefficient*:

$$T_{ijk} \overset{\text{def}}{=} \delta\Delta G_i \,/\, \delta\Delta G_j * r_{ij} \,, \tag{1}$$

where k, j, and i denote the indices of the external metabolite (substrate or product), M_k, whose concentration is changed, of the considered system U_j, and of an enzyme e_i (being a component of this system), respectively. $\delta\Delta G_i$ and $\delta\Delta G_j$ stand for the infinitesimal changes in the Gibbs free energy difference of the reaction catalyzed by enzyme e_i, and in the Gibbs free energy difference of the whole considered system U_j, where

$$\Delta G_j = \sum_{i=1}^{n} \Delta G_i * r_{ij} \,. \tag{2}$$

r_{ij} denotes the relative flux through enzyme e_i in system U_j; this parameter is equal to unity in linear pathways and can have a value between zero and unity in branched pathways. In other words, the thermodynamic response coefficient is an infinitesimal change in the thermodynamic span of a given reaction (process) divided by an infinitesimal change in such a span of the entire considered sytem. In the present paper, i indicates a particular component of the oxidative phosphorylation system, j indicates this system itself (or a subsystem thereof) and k refers to oxygen.

The coefficient defined above is a global parameter: it depends not only on properties of a given enzyme, but also on features of the whole system. Moreover, its value can alter if the concentrations of different external metabolites (substrates or products) are changed. The thermodynamic response coefficient indicates which reactions compensate for external perturbations in the thermodynamic span and which of them keep their Gibbs free energy change relatively constant. The latter type of behavior would be advantageous for reactions utilizing "top" metabolites of a pathway (for example ATP) in order to maintain their concentrations essentially unchanged.

Of course, thermodynamic response coefficients exhibit a summation property, because the total change in thermodynamic span is the sum of such changes for the particular reactions. So the *thermodynamic summation property* can be expressed as follows:

$$\sum_{i=1}^{n} T_{ijk} = 1 . \tag{3}$$

This property allows us to easily assess to what extent a given enzyme shares the compensation of the change in thermodynamic span of the whole system caused by a change in the concentration of an external metabolite M_k.

2. Application to the Oxidative Phosphorylation System

On the basis of the semi-quantitatively verified dynamic model of the oxidative phosphorylation system developed previously[1-3], we intend to calculate thermodynamic response coefficients with respect to changes in oxygen concentration for the main reactions (processes) taking part in the process of oxidative ATP synthesis. These coefficients are presented in Table 1. The values are calculated for a saturating oxygen concentration of 240 μM.

It can be seen that the ATP/ADP carrier and, to less an extent, the cytochrome oxidase, proton leak, substrate dehydrogenation, and external ATP utilization bear the thermodynamic cost of a decrease in oxygen pressure. These components of the system also have, at least under some conditions, high flux control coefficients (the flux control

Table 1. Thermodynamic response coefficients for the components of the oxidative phosphorylation system with respect to the oxygen concentration for the saturating oxygen concentration of 240 μM compared with flux control coefficients for the same conditions.

enzyme (process)	thermodynamic response coefficient	flux control coefficient
substrate dehydrogenation	0.10	0.14
cytochrome oxidase	0.24	0.04
proton leak	0.14	0.18
ATP synthase	0.00	0.00
internal ATP consumption	0.02	0.02
ATP/ADP carrier	0.37	0.13
phosphate carrier	0.03	0.01
external ATP utilization	0.10	0.48

coefficients calculated in Ref. 2 for Mode 1 of cell activity are shown for comparison). External ATP utlization can be regarded as an exception to this rule, because it has the greatest flux control coefficient under most conditions[2], but exhibits a relatively small thermodynamic response coefficient. It is in very good agreement with what can be expected. The ATP utilization process is a "purpose" of the whole system. If it is to ensure an optimal response for the cell energy demand it should exhibit two features. First, it should control the metabolic flux through the system. Second, it should keep the concentration of its substrates and products as unchanged as possible. The high flux control coefficient and the low thermodynamic response coefficient reflects the above features, respectively.

Some other reactions with much weaker a control of the flux through the system compensate the changes of the thermodynamic span of the system more strongly. Reactions which do not control the flux (ATP synthase, phosphate carrier, internal ATP consumption) have a weak response to changes in the total Gibbs free energy change. The case of cytochrome oxidase, which is apparently an exception, will be discussed below. Thus, the cellular bioenergetic system seems to be assembled very purposefully and exhibits efficient mechanisms of regulation. The thermodynamic response coefficients presented in Table 1 are calculated for Mode 1 of cell activity (β-oxidation of fatty acids are being used as a source of respiratory substrates, and ATP utilization is "resting").

The thermodynamic response coefficients of some chosen enzymes (processes) at different oxygen concentrations are shown in Table 2.

Here the picture is not as clear as for saturating oxygen concentrations. First of all, we can observe a fall of the thermodynamic response coefficient for cytochrome oxidase below zero at lower oxygen concentrations. The values of this parameter for other components of the system rise (since they fulfill the thermodynamic summation property). The reason for such a thermodynamic response pattern is most probably the kinetics of cytochrome oxidase. It has been shown both in experiments[4-7] and simulations[1] that this enzyme is able to maintain an essentially constant rate of oxygen consumption in almost the entire physiological range of the oxygen concentration (by a very low Michaelis-Menten constant with respect to oxygen). For doing so, the concentration of the other substrate of cytochrome oxidase, namely the reduced form of cytochrome c, must be strongly enlarged in order to compensate for the decreased oxygen concentration. Therefore, the resultant effect is a rise in the thermodynamic span of the cytochrome oxidase-catalyzed reaction, while the span through the whole system falls down. Thus, the negative thermodynamic

Table 2. Thermodynamic response coefficients for the chosen components of the oxidative phosphorylation system with respect to oxygen concentration for various oxygen concentrations. (The coefficient R denotes the thermodynamic response coefficient for the external ATP utilization calculated for the phosphorylation subsystem.)

enzyme (process)	thermodynamic response coefficient					
$[O_2] =$	240μM	100μM	30μM	10μM	3μM	1μM
substrate dehydrogenation	0.10	0.18	0.29	0.45	0.53	0.65
cytochrome oxidase	0.24	-0.42	-1.64	-2.57	-3.02	-3.09
proton leak	0.14	0.27	0.51	0.68	0.76	0.75
ATP/ADP carrier	0.37	0.70	1.23	1.42	1.21	0.85
external ATP utilization	0.10	0.19	0.43	0.75	1.16	1.44
coefficient R	0.19	0.19	0.23	0.31	0.42	0.54

response coefficient for cytochrome oxidase is the thermodynamic "price" of the kinetic regulation (maintaining the respiratory rate essentially constant).

Taking this into account, the control of the thermodynamic response of external ATP utilization can better be discussed for lower oxygen concentrations if we take into account the phosphorylation subsystem (ATP synthase, ATP/ADP carrier, phosphate carrier and external ATP utilization) alone. The second reason is that an accurate comparison between different reactions can be performed only for the reactions with the same value of the coefficient r_{ij}.

The coefficient R in Table 2 denotes the thermodynamic response coefficients for external ATP utilization calculated for the phosphorylation subsystem. We can see that the value of this coefficient is relatively small and changes rather slightly with oxygen concentration at higher, physiological values. Above 10 μM of oxygen concentration the ATP/ADP carrier is a main factor bearing the thermodynamic cost of the decrease in oxygen concentration pressure. At lower oxygen concentrations this cost decreases and the kinetic regulation of cytochrome oxidase is borne to a comparable extent also by the substrate dehydrogenation, external ATP utilization and proton leak.

The high thermodynamic response coefficient for the ATP/ADP carrier connected with the low one for external ATP utilization implies that the internal phosphorylation potential is much less stable during the decrease in oxygen concentration than the external one. Such a case is presented in Fig. 1. This figure shows a simulated dependence of three thermodynamic forces on time during oxygen consumption until anaerobiosis by suspension of cells in a closed chamber (see Ref. 1).

The oxygen concentration reaches zero after 240 - 250 seconds. It can be clearly seen that the external phosphorylation potential is much more independent of the oxygen concentration than the internal phosphorylation potential and proton-motive force. The two latter forces behave very similarly, as they are near equilibrium with each other and there is no "thermodynamic barrier" between them. This theoretical supposition should be verified experimentally. If it is true, the ATP/ADP carrier seems to be a very important factor

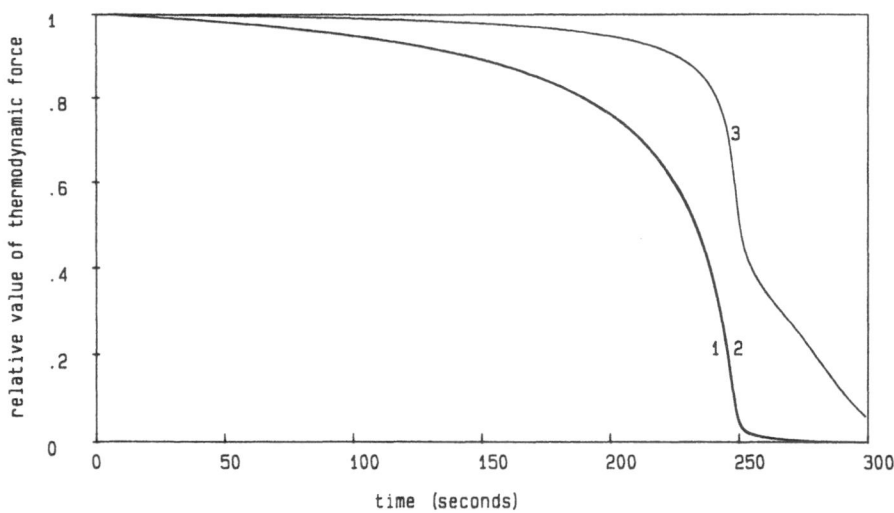

Figure 1. Relative changes of three thermodynamic forces during oxygen consumption by a cell suspension in a closed chamber. Oxygen is completely removed after approximately 245 seconds.

(1) $10^{n_A \Delta p / Z}$ (Δp - proton-motive force, n_A - number of protons used for the synthesis of one ATP molecule); (2) internal [ATP]/[ADP][P_i] ratio; (3) external [ATP]/[ADP][P_i] ratio.

buffering the changes in the thermodynamic span of the system caused by a decrease in oxygen concentration (and other perturbations). Thus, this carrier represents a "thermodynamic barrier" between the internal and external phosphorylation potential, compensating for changes in the former by changing its own displacement from equilibrium, so the latter is kept essentially constant. Such a behavior is possible because, as is seen from our kinetic model of the ATP/ADP carrier[1,2], this carrier is very sensitive to the external [ATP]/[ADP] ratio, but almost independent of the internal one. So, it is able to respond adequately to changes in the ATP demand of the cell, being relatively insensitive to changes in the internal phosphorylation potential.

Our model[1] exhibits a good agreement with experiments performed by Wilson's group[4-7] for the aerobiosis-anaerobiosis transition. The changes in parameter values (metabolite concentrations, thermodynamic forces) at higher oxygen concentrations are insignificant. However, it is not in contradiction to large changes in the thermodynamic response coefficient values, as they are quotients of two small relative changes.

The values of thermodynamic response coefficients cannot be obtained from the values of flux control coefficients because one additional term is needed: the elasticity coefficient for cytochrome oxidase with respect to oxygen concentration.

Acknowledgement

This work has been supported by the grant KBN 4 4374 9102.

Note

After preparation of this manuscript we found that a coefficient similar to the presented thermodynamic response coefficient has been proposed earlier ('free-energy control coefficient'[8]). The mathematical formalism is similar; however, we explicitly take into account three factors not expressed clearly in the definition of the free-energy control coefficient:

1. The value of the thermodynamic response coefficient for a given enzyme is not a unique function of the change in the thermodynamic span through the whole system - this value depends on *which* external metabolite concentration is changed (index k in T_{yjk}). For example, if we assume that k corresponds to the respiratory substrate and the substrate dehydrogenation process is almost independent of the concentration of respiratory fuel and we change this concentration, we will obtain a thermodynamic response coefficient near unity for the substrate dehydrogenation and near zero for other processes.

2. If the considered metabolic pathway is branched, the value of the thermodynamic response coefficient for a given enzyme will depend on the relative flux through this enzyme (coefficient r_{ij} in our definition).

3. The thermodynamic summation property can be formulated as a consequence of the two above-mentioned features.

Unfortunately, in the cited publication there is practically no metabolic interpretation of the free-energy control coefficient, so we are not able to make any further comparison.

References

1. B. Korzeniewski and W. Froncisz, An extended dynamic model of oxidative phosphorylation, *Biochim. Biophys. Acta* **1060**:210-223 (1991).

2. B. Korzeniewski and W. Froncisz, Theoretical studies on the control of the oxidative phosphorylation system, *Biochim. Biophys. Acta* **1102**:67-75 (1992).

3. B. Korzeniewski and W. Froncisz, A dynamic model of mitochondrial respiration, *stud. biophys.* **132**:173-187 (1989).

4. D.F. Wilson, M. Erecinska, C. Drown and J.A. Silver, The oxygen dependence of cellular energy metabolism, *Arch. Biochem. Biophys.* **195**:485(1979).

5. D.F. Wilson, C.S. Owen and M. Erecinska, Quantitative dependence of mitochondrial oxidative phosphorylation on oxygen concentration: A mathematical model, *Arch. Biochem. Biophys.* **195**:495 (1979).

6. T. Kashiwagara, D.F. Wilson and M. Erecinska, Oxygen dependence of cellular metabolism: The effect of O_2 tension on gluconeogenesis and urea synthesis in isolated rat hepatocytes, *J. Cell Physiol.* **120**:13 (1984).

7. D.F. Wilson, W.L. Rumsey, T.J. Green and J.M. Vanderkooi, The oxygen dependence of mitochondrial oxidative phosphorylation measured by a new optical method for measuring oxygen concentration, *J. Biol. Chem.* **263**:2712 (1988).

8. H.V. Westerhoff, P.J.A.M. Plomp, A.K. Groen and R.J.A. Wanders, Thermodynamics of the control of metabolism, *Cell Biophys.* **10**:239-267 (1987).

A MODEL STUDY ON THE INTERRELATION BETWEEN THE TRANSMEMBRANE POTENTIAL AND pH DIFFERENCE ACROSS THE MITOCHONDRIAL INNER MEMBRANE

Stefan Schuster and Jean-Pierre Mazat

Université Bordeaux II
Dépt. de Biochimie Médicale
146, rue Léo Saignat, 33076 Bordeaux, France

1. Introduction

It has often been assumed that the transmembrane potential ($\Delta\Psi$) across the mitochondrial inner membrane be proportional to the pH difference (ΔpH)[1-4]. It is intuitively clear that there is a monotonic interrelation between these quantities, since both of them are established by proton extrusion by the redox pumps. The exact relationship is not, however, obvious.

Whereas $\Delta\Psi$ and ΔpH are thermodynamically equivalent (cf. Ref. 5, ch. 2.3.), they are kinetically inequivalent[6] because the rates of the processes involved in biological energy transduction depend on $\Delta\Psi$ in a different way than on ΔpH. Therefore, construction of kinetic models of oxidative phosphorylation necessitates studies on the interrelation between the two quantities in question.

Traditional models based on linear irreversible thermodynamics[5,7,8] cannot be used to cope with this problem, because $\Delta\Psi$ and ΔpH only enter the equations via the proton-motive force, which is a linear combination of these quantities.

Reich and Rohde[9] studied the problem in question by numerically calculating the steady-state solutions of a dynamic model. They concluded that over a wide range of parameters the ratio between $\Delta\Psi$ and ΔpH is approximately constant. The kinetic equations and reaction scheme of this model are, however, rather outdated.

2. The Basic Model

We consider five groups of ions: protons, H^+, other cations, C^+, hydroxyl ions, OH^-, other inorganic anions, A^-, and negatively charged sites of impermeant proteins, phospholipids and other polyelectrolytes, P^-. Since the matricial proteins cover a considerable part of the inner volume, one has to decide whether the concentrations are to refer to the whole volume or to the aqueous phase only. We prefer the latter option, because it allows to have $H_i^+ = 10^{-7}$ M at pH 7. In a basic model, we assume that the distribution of all permeant inorganic anions and cations except protons and hydroxyl ions with respect to the two sides of the membrane follows the Nernst equation:

$$C_i^+ / C_o^+ = \exp(-u) \ , \quad A_i^- / A_o^- = \exp(u) \ , \tag{1a,b}$$

with $\quad u = F\Delta\Psi/RT.$ $\hfill (2)$

We take into account that protons can bind to negative sites, P$^-$, of proteins and phospholipids. As mitochondria contain a multitude of polyelectrolytes, we decompose the set of ions P$^-$ according to their dissociation constant, $K_P^{(k)}$, into n classes $P^{(k)-}$. Since every protonation process

$$P^{(k)}H_i \ \rightleftharpoons \ P^{(k)-} + H_i^+$$

is in equilibrium whenever the whole system is at steady state, we have

$$P^{(k)-}H_i^+ / P^{(k)}H_i = K_P^{(k)} \ . \tag{3}$$

With the conservation relation $P^{(k)-} + P^{(k)}H_i = P_T^{(k)}$ we obtain

$$P^{(k)-} = \sum_k \frac{K_P^{(k)} P_T^{(k)}}{H_i^+ + K_P^{(k)}} \ , \ P^{(k)}H_i = \sum_k \frac{H_i^+ P_T^{(k)}}{H_i^+ + K_P^{(k)}} \ . \tag{4a,b}$$

In spite of the electrogenic extrusion of protons, the net electric charge on either side of the membrane is negligibly small (quasi-electroneutrality condition),

$$C_i^+ - A_i^- + H_i^+ - OH_i^- - \sum_{k=1}^{n} P^{(k)-} = 0 \ . \tag{5}$$

Quasi-electroneutrality is very frequently invoked in the modeling of diffusion in the presence of electric fields[10,11].

In many situations, the external volume is much greater than the matrix volume of mitochondria, in particular in the case of resuspension experiments. Thus, the external concentrations can be considered approximately constant. Under the condition of constant volume, water plays, formally, a similar role as the polyelectrolytes. Therefore, the hydroxyl ions can be included as $(n+1)$st class of ions $P^{(k)-}$. Thus, eqs (4a) and (5) give

$$C_o^+ \exp(-u) - A_o^- \exp(u) = \sum_{k=1}^{n+1} \frac{K_P^{(k)} P_T^{(k)}}{H_i^+ + K_P^{(k)}} - H_i^+ \ . \tag{6}$$

The term on the r.h.s. of this equation is bounded above. Since the left-hand side of eq. (6) is a monotonic increasing function of $\Delta\Psi$, the latter is bounded above as well, which is in accordance with experimental findings[12,13]. If the external concentrations of anions and cations are equal to each other, $A_o^- = C_o^+$, eq. (6) can then be simplified to

$$2C_o^+ \sinh(-F\Delta\Psi/RT) = \sum_{k=1}^{n+1} \frac{K_P^{(k)} P_T^{(k)}}{H_i^+ + K_P^{(k)}} - H_i^+ \; . \tag{7}$$

This equation can be solved for $\Delta\Psi$ in closed form:

$$-\Delta\Psi = \frac{RT}{F} A\sinh\left[\frac{\displaystyle\sum_{k=1}^{n+1} \frac{K_P^{(k)} P_T^{(k)}}{H_i^+ + K_P^{(k)}} - H_i^+}{2C_o^+}\right] \tag{8}$$

with $\quad H_i^+ = 10^{-(\Delta pH + pH_o)} \; . \tag{9}$

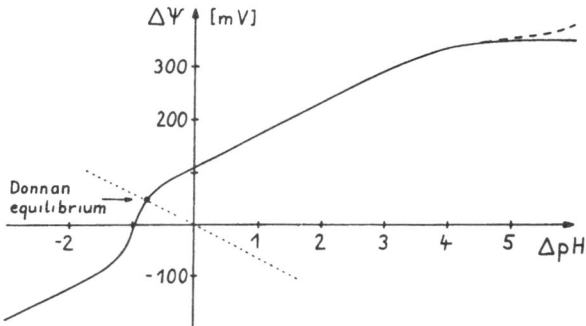

Figure 1. Graph of the relationship $\Delta\Psi$ vs. ΔpH as calculated by numerical simulation. Parameter values: $n=1$, $K_P(1)=1E\text{-}11$, $P_T(1)=1$ M, $pH_o =7.0$, $C_o^+ =1E\text{-}4$ mM. The straight line represents the Nernst equation as for protons (which is not normally fulfilled).

Function (8) is monotonic increasing throughout, as can be proved by implicit differentiation of eq. (6). Furthermore, it can be shown that $\Delta\Psi$ is, for small internal pH, a linear function of ΔpH.

An interesting reference state is the situation where the proton concentrations are in equilibrium (Donnan state). Note that in this state, $\Delta\Psi$ is not, in general, zero, due to the non-permeant polyelectrolytes which cause a non-equal distribution of protons.

3. Inclusion of the K+/H+ Exchanger

We now study the effect of the K+/H+ exchanger[14]. This exchanger implies the K+ gradient to be slightly shifted from the equilibrium state. For simplicity's sake, we make no distinction between cations that are exchanged against protons and cations that are not. Using a simple linear ansatz, we may write

$$v_1 = k_1 H_o^+ C_i^+ - k_1 H_i^+ C_o^+ , \tag{10}$$

$$v_2 = k_2 C_o^+ \exp\left(-\frac{u}{2}\right) - k_2 C_i^+ \exp\left(\frac{u}{2}\right), \tag{11}$$

41

with v_1, v_2 denoting the net rates of proton-cation exchange and of passive cation uniport, respectively. (For a more complex rate equation see Ref. 15). The forward and backward rate constants can be equalized for symmetry reasons. In steady state, eqs (10,11) imply

$$C_i^+ = \frac{k_2\, exp(-u/2) + k_1 H_i^+}{k_2\, exp(u/2) + k_1 H_o^+}\ C_o^+ .$$
(12)

The electroneutrality condition leads to

$$C_o^+\ \frac{k_2\, exp(-u/2) + k_1 H_i^+}{k_2\, exp(u/2) + k_1 H_o^+}\ -\ A_o^-\, exp(u) = \sum_{k=1}^{n+1} \frac{K_P^{(k)} P_T^{(k)}}{H_i^+ + K_P^{(k)}} - H_i^+ .$$
(13)

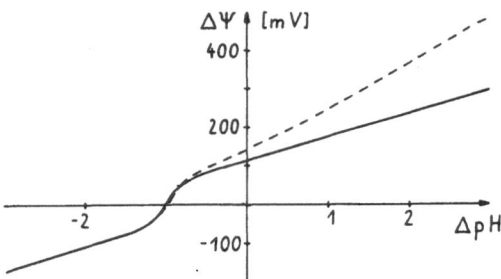

Figure 2. Schematic representation of the $\Delta\Psi$ vs. ΔpH curves with and without activity of a K^+/H^+-exchanger.

The qualitative effect of the K^+/H^+ exchanger is shown in Fig. 2.

In contrast to eq. (7), eq. (13) cannot be solved for $\Delta\Psi$ in closed form. Nevertheless, one can derive interesting qualitative assertions on the interrelation between ΔpH and $\Delta\Psi$. Although activation of the K^+/H^+ exchanger is likely to change the transmemembrane potential, since a new steady state will be attained, we may discuss the location of the modified curve by a thought experiment in which $\Delta\Psi$ is kept constant. Inspection of the first term on the left-hand side of eq. (13) reveals that for negative $\Delta\Psi$ values, this term is smaller than the corresponding term $C_o^+ exp(-u)$ in eq. (6). Thus, the pH difference is, for fixed $\Delta\Psi$ values, diminished by this exchange. It can further be seen that the exchanger has almost no effect for small internal pH. This is because the cation concentration (first term on the l.h.s. of eq. (13)) is then negligibly small.

4. Volume Effects

In the case of external volumes comparable to the matrix volume, we have to take into account a conservation relation for the internal and external concentrations,

$$C_i^+ + \rho\, C_o^+ = (\rho+1)\, C^+, \quad A_i^- + \rho\, A_o^- = (\rho+1)\, A^- ,$$
(14a,b)

where C^+ and A^- are the concentrations of cations and anions, respectively, in the state with zero transmembrane potential. ρ denotes the ratio of external and matrix volumes. Together with eqs (1), (4a) and (5), this gives

$$\frac{(\rho+1)C^+\exp(-u)}{\rho+\exp(-u)} - \frac{(\rho+1)A^-\exp(u)}{\rho+\exp(u)} = \sum_{k=1}^{n+1} \frac{K_P^{(k)}P_T^{(k)}}{H_i^+ + K_P^{(k)}} - H_i^+ . \tag{15}$$

Since also the cytosolic space contains sites of polyelectrolytes susceptible to deprotonation, we introduce the symbol $Q^{(j)}$ and $K_Q^{(j)}$ to denote their concentrations respectively dissociation constants. Under consideration of the dissociation equilibrium and conservation relation for protons, we obtain

$$\rho\, H_0^+ + \sum_{j=1}^{m} \frac{\rho H_0^+ Q_T^{(j)}}{H_0^+ + K_Q^{(j)}} = P_T + \rho Q_T - H_i^+ - \sum_{k=1}^{n+1} \frac{H_i^+ P_T^{(k)}}{H_i^+ + K_P^{(k)}} . \tag{16}$$

This equation links the inner and outer concentrations of protons. Eqs (15) and (16) are, though not amenable to solution in closed form, sufficient to express ΔpH as a function of $\Delta\Psi$. It can be shown by implicit differentiation that the monotonicity of the relationship is not altered, and that the smaller the external volume is, the larger will be H_0^+, provided $\Delta\Psi$ is negative. This effect, which tends to increase ΔpH in the physiological region, is, however, negligibly small if $\Delta\Psi$ is near zero. As for the asymptotic behavior of the curve, we have to distinguish two cases:

(i) $(\rho+1)C^+ < P_T$; there is no upper limit for $-\Delta\Psi$, while ΔpH is bounded above;
(ii) $(\rho+1)C^+ > P_T$; $-\Delta\Psi$ is bounded above, while ΔpH is unbounded.

6. Discussion

We derived a nonlinear equation giving the potential difference across the mitochondrial inner membrane as a function of the ΔpH. It turns out that this function is quasi-linear in several ΔpH regions, in particular in the physiological region $0<\Delta$pH<2. This result shows that earlier hypotheses stating that $\Delta\Psi$ is approximately proportional to ΔpH[1-4] should be modified in that the function is only piecewise quasi-linear and not homogeneous (i.e., it does not cross the origin of co-ordinates).

The validity of the simplifying model assumptions might be dubious in some cases, but construction of any model has to be based on a compromise between exactitude of the assumptions and tractability of the model. Since we obtained a refinement to earlier models, which is in accordance with experimental findings and nevertheless analytically tractable, a good compromise appears to have been found.

The presented basic model contains, besides universal constants (F, R), only conservation sums and thermodynamic parameters such as equilibrium constants. In particular, the present model is invariant to the extent of inhibition or slipping of ATPase or the redox pumps. On the other hand, the present model does not allow to calculate the actual values of $\Delta\Psi$ or ΔpH, but only a curve on which these values are situated.

The case of complete thermodynamic equilibrium (complete inhibition of redox pumps and ATPase) does not necessarily imply zero $\Delta\Psi$. Such a Donnan equilibrium (with $\Delta\Psi$ values of up to 80 mV) can experimentally be achieved[16]. This phenomenon has not been taken into account in earlier models starting from a proportionality between ΔpH and $\Delta\Psi$.

We believe that the present model is also applicable to all membranes containing ATP-dependent proton pumps, such as yeast and plant vacuolar membranes, bacteriorhodopsin liposomes, and lysosome membranes[17].

It is worthwhile stressing the difference between those biological membranes where the $\Delta\Psi$ is essentially a diffusion potential (e.g. the cell membrane of erythocytes) and those membranes where the potential gradient is solely established by proton pumps. The former situation has frequently been modeled by the Goldman equation[11]. Since this equation is obtained by equating passive ion fluxes, it is not appropriate for mitochondria and related situations, where the passive proton flux cannot be equated to any other ion flux.

Acknowledgements

S. S. acknowledges financial support from the Science Council of NATO under the auspices of the DAAD (Germany). The authors wish to thank Drs E. Donath, R. Heinrich (Berlin), M. Rigoulet and R. Ouhabi (Bordeaux) for stimulating discussions.

References

1. R. Bohnensack, Control of energy transformation in mitochondria. Analysis by a quantitative model, *Biochim. Biophys. Acta* **634**:203-218. (1981)
2. R. Bohnensack, Mathematical modeling of mitochondrial energy transduction, *Biomed. Biochim. Acta* **44**:853-862 (1985).
3. H.G. Holzhütter, W. Henke, W. Dubiel and G. Gerber, A mathematical model to study short-term regulation of mitochondrial energy transduction, *Biochim. Biophys. Acta* **810**:252-268 (1985).
4. B. Korzeniewski and W. Froncisz, An extended dynamic model of oxidative phosphorylation, *Biochim. Biophys. Acta* **1060**:210-223 (1991).
5. H.V. Westerhoff and K. Van Dam. "Thermodynamics and Control of Biological Free-energy Transduction," Elsevier, Amsterdam (1987).
6. D. Pietrobon and S.R. Caplan, Flow-force relationships for a six-state proton pump model: Intrinsic uncoupling, kinetic equivalence of input and output forces, and domain of approximate linearity, *Biochemistry* **24**:5764-5776 (1985).
7. S.R. Caplan and A. Essig. "Bioenergetics and Linear Nonequilibrium Thermodynamics," Harvard University Press, Cambridge (Mass.) (1983).
8. J.W. Stucki, M. Compiani and S.R. Caplan, Efficiency of energy conversion in model biological pumps, *Biophys. Chem.* **18**:101-109 (1983).
9. J.G. Reich and K. Rohde, On the relationship between $Z\Delta pH$ and $\Delta\Psi$ as components of the proton-motive potential in Mitchell's chemiosmotic system, *Biomed. Biochim. Acta* **42**:37-46 (1983).
10. S.G. Schultz. "Basic Principles of Membrane Transport," Cambridge University Press, Cambridge (1980).
11. R.W. Glaser, A. Wagner and E. Donath, Volume and ionic composition changes in erythrocytes after electric breakdown. Simulation and experiment, *Bioelectrochem. Bioenerget.* **16**:455-467 (1986).
12. J. Duszynski, K. Bogucka and L. Wojtczak, Homeostasis of the protonmotive force in phosphorylating mitochondria, *Biochim. Biophys. Acta* **767**:540-547 (1984).
13. R. Ouhabi, M. Rigoulet, J.L. Lavie and B. Guérin, Respiration in non-phosphorylating yeast mitochondria - roles of non-ohmic proton conductance and intrinsic uncoupling, *Biochim. Biophys. Acta* **1060**:293-298 (1991).
14. K. Garlid, On the mechanism of regulation of the mitochondrial K^+/H^+ exchanger, *J. Biol. Chem.* **255**:11273-11279 (1980).
15. B.N. Kholodenko and L.I. Erlikh, Mathematical model of the regulation of the steady state values of volume and ΔpH in mitochondria, *Biophysics* (Moscow, English translation) **30**:911-919 (1985).
16. D. Nicholls. "Bioenergetics. An Introduction to the Chemisosmotic Theory," Academic Press, London (1982).
17. I. Mellman, R. Fuchs and A. Helenius, Acidification of the endocytic and exocytic pathways, *Ann. Rev. Biochem.* **55**:663-700 (1986).

COMPARISON OF RETINAL-BASED AND CHLOROPHYLL-BASED PHOTOSYNTHESIS: A BIOTHERMOKINETIC DESCRIPTION OF PHOTOCHEMICAL REACTION CENTERS

Klaas J. Hellingwerf[1], Wim Crielaard[1] and Hans V. Westerhoff[2]

[1]Department of Microbiology
[2]Department of Biochemistry
and E.C. Slater Institute for Biochemical and Microbiological Research
University of Amsterdam, Amsterdam, The Netherlands

1. Introduction

Life on earth depends on the transformation of sunlight (electromagnetic radiation) into biologically useful chemical energy. The first step in this process is the conversion of the free energy from sunlight into electrochemical ion gradients across biological membranes. Free energy derived from these, usually proton- but in some cases chloride-gradients, is subsequently stored in the form of the phosphorylation potential of ATP. In addition, free energy is stored in the form of redox potentials of coenzymes such as NAD(P)H.

In nature two mechanistically very different processes have evolved that catalyze the basic reaction of conservation of radiant free energy: retinal-based- and (bacterio)chlorophyll-based photosynthesis. Examples of pigment/protein complexes that carry out these processes are bacteriorhodopsin from the purple membranes of *Halobacterium halobium* and photochemical reaction centers from purple bacteria like *Rhodobacter sphaeroides*, respectively.

In both forms of photosynthesis light energy is converted into free energy of a proton gradient, i.e. into chemiosmotic free energy. However, the two mechanisms are completely opposite. In bacteriorhodopsin light absorption triggers a cis/trans isomerization of a protein-linked retinal group, which subsequently leads to a conformational change in the pigment/protein complex that moves the proton-carrying Schiff base towards the extracellular medium. In the photochemical reaction centers of purple bacteria light energy excites an electron to a much more negative redox potential. This electron is subsequently transfered along a chain of electron-accepting prosthetic pigment groups, until stable charge separation has been achieved, essentially without movement of atoms within the pigment/protein complex. In the intact bacterial membrane this series of reactions is followed by electron transfer in a cyclic electron transfer chain, with the ultimate result of proton translocation. Maximally two protons can be translocated for each electron that is excited in the reaction centers.

Also with respect to their spectral characteristics these two types of photosynthesis are completely divergent (Fig. 1). Retinal-based photosynthesis uses the green part of the visible spectrum, i.e. between 500 and 600 nm, whereas chlorophyll-based photosynthesis is typified by an absorbance spectrum that is nearly fully complementary to this. It has strong absorbance bands in both the blue and the red part of the spectrum (note, however, that carotenoids broaden this latter spectrum into the green). This complementarity has led to speculations about the evolutionary roots of these two types of photosynthesis. Goldworthy[1] has argued that chlorophyll-based photosynthesis is much more efficient than retinal-based photosynthesis and this assumption, together with the complementarity in their absorbance characteristics, has led him to postulate that in the course of evolution first retinal-based photosynthesis evolved and that it was subsequently outcompeted by chlorophyll-based photosynthesis, because of the much higher efficiency of the latter process.

In this contribution we want to address the issue of efficiency of free energy conversion in retinal-based- and chlorophyll-based photosynthesis. This question is not only of interest because of its evolutionary implications, but also because nowadays ecosystems can be found in which both types of photosynthesis appear to proceed in parallel.

Because of all other energy conversion processes that take place simultaneously, it is impossible to perform such a comparison using data collected from experiments with intact cells in which either type of photosynthesis is operative. Fortunately, both types of photosynthesis can be studied in a model system of liposomes, in which either

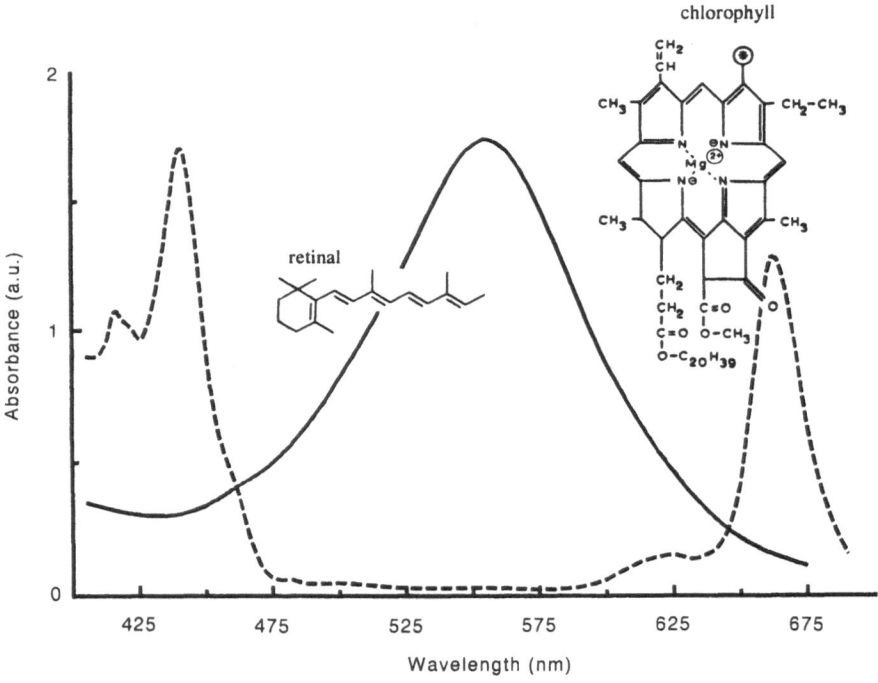

Figure 1. Typical visible absorption spectrum of a retinal-based- and a chlorophyll-based light energy converter and the chemical structure of their respective chromophoric groups. The star in the structure of chlorophyll is a methyl group for chlorophyll a and an aldehyde-group for chlorophyll b.

bacteriorhodopsin[2] or reaction centers from purple bacteria[3] have been incorporated. In the latter case, light-induced proton pumping can be accomplished by the addition of two soluble redox mediators: cytochrome c and ubiquinone[4].

During the last few years, the three-dimensional structure, at a few Ångströms resolution[5,6] of a representative of either type of proton pump has been resolved. In addition, a large amount of thermodynamic and kinetic information on partial reactions in photochemical reaction centers has been accumulated[7]. For retinal-based photosynthesis the information available is more restricted, predominantly because of remaining uncertainties about branching reactions in the terminal parts of the photocycle[8]. Also this information is, nevertheless, of much use for the development of a biothermokinetic description of the performance of these two types of pumps.

The criteria that can be used in comparing retinal-based- and chlorophyll-based photosynthesis are manifold. We will limit this discussion to three aspects:

(1) mechanistic efficiency (i.e. quantum yield),
(2) maximal thermodynamic conversion efficiency,
(3) kinetic efficiency.

2. Mechanistic Efficiency

The efficiency at which light quanta are used for photochemical processes is usually refered to as the quantum yield. For the retinal-based proton pump, bacteriorhodopsin, this value has been the subject of intense discussions in the literature. In older literature most values cited for this parameter are quite low (approx. 0.3), but recently a consensus has been reached that this value has been underestimated and in fact is between 0.6 and 0.7[9,10].

For chlorophyll-based photosynthesis the maximal mechanistic efficiency can be very high, i.e. close to unity. However, in the relevant ecosystems, the available light will be largely restricted to a wavelength range around 500 nm[11], so that light harvesting is dominated by carotenoids. In that case the quantum yield will be significantly lower and in most bacteria range from 0.3 to 0.7.[12]

It is clear from this discussion that it is not possible to make a firm statement as to which type of photosynthesis is most efficient in terms of quantum yield.

3. Thermodynamic Conversion Efficiency

A comparison of the two types of proton pumps, with respect to maximal thermodynamic conversion efficiency, can be made in the model system of a reconstituted proton pump by comparing the maximal proton gradient that can be generated by the pumps, when saturating amounts of light are available, and relating this value with the amount of free energy available per mole of photons. Multiplication of these two forces, for each pump, by the flux of translocated protons (J_H) and of absorbed photons (J_n), respectively, and determination of the ratio of the resulting two terms would yield the true thermodynamic efficiency η of the pumps. However, it is extremely difficult to measure J_n and the assumption of complete coupling between the two fluxes is most likely not valid for either of these pumps[7,13]. We will therefore limit the comparison to what is called here the thermodynamic conversion efficiency.

For this comparison between the two pumps we use the simplifying approach of describing the functioning of the pumps by mosaic non-equilibrium thermodynamics[4, 14-17]. The magnitude of the proton-motive force can be measured using ion-selective electrodes[2,4]. In this approach we have to take into account that during steady state illumination, the net rate of light-driven proton translocation is balanced by passive proton leak. Therefore, in order to determine the *maximal* proton gradient that can be generated by the pump (i.e. the hypothetical proton motive force at which the light-driven turnover of the pump is blocked by the counteracting proton gradient), we have to extrapolate to a situation without passive proton leak. This can be done by performing measurements of the proton gradient at various concentrations of a protonophoric uncoupler and various light intensities, followed by extrapolation to zero proton permeability. For the evaluation of such an experiment the following equation[2,15] can be used:

$$1/\Delta\mu_H = L_H^l/(L_v A_v) + 1/A_v , \tag{1}$$

in which $\Delta\mu_H$ is the proton motive force, L_H^l and L_v are proportionality constants for the passive proton leak and the light-intensity dependence of the turnover of the pump, respectively, and A_v is the intrinsic *maximal* capacity of the pump to generate a proton gradient. It follows directly from this equation that $\Delta\mu_H$ will equal A_v when the passive proton leak is zero.

The data available from this type of experiment are slightly variable for bacteriorhodopsin: values for A_v in the order of 0.2 to 0.25 Volt have been reported[2,18]. For reaction centers of *Rb. sphaeroides* a value of 0.21 V was determined[4]. As one mole of photons of a wavelength of 560 nm represents 2.3 eV of free energy, we calculate a maximal thermodynamic conversion efficiency for bacteriorhodopsin of 0.09 to 0.11. For the reaction centers this calculation is less direct. If we assume that a 500 nm photon enters the reaction centers via one of the carotenoids, an efficiency of 0.08 is obtained. However, when making use of the free energy of infrared photons, for instance of 880 nm, the efficiency is increased to 0.14. When the assumption is made that the stoichiometry of the cyclic electron transfer chain in the intact cell operates with a fixed stoichiometry of 2 protons per cycling electron and that the reaction center is able to generate under those conditions the same *maximal* proton gradient as in liposomes, a maximal value for the conversion efficiency of photosynthesis in *Rb. sphaeroides* of 0.3 can be calculated.

The above discussion illustrates that also on basis of thermodynamic efficiency, it is difficult to make a firm statement as to which type of photosynthesis is most efficient. However, the conclusion seems warranted that bacteriochlorophyll-based photosynthesis has the *potential* to display the highest maximal thermodynamic efficiency of the two, provided that infrared light is available plus an elaborate antenna system.

4. Kinetic Efficiency

In order to compare the kinetic performance of the two pumps it is important to have information about

   (i) the turnover rate under "level-flow" conditions, *i.e.* when no counteracting back-pressure is present and
   (ii) the modulating effect of a proton gradient on this turnover rate.

For bacteriorhodopsin, spectroscopic data suggest a turnover rate of 100 sec[-1] for the photocycle under light saturation[19]. Direct measurements of the rate of proton translocation by bacteriorhodopsin, however, have resulted[2,20,21] in numbers not higher than 10 sec[-1]. This latter rate is comparable to the turnover rate of halorhodopsin[22]. It is relevant to note here that particularly the terminal part of the photocycle of bacteriorhodopsin is not well-resolved[8]. Furthermore, the best definition of completion of a photocycle is the ability to re-enter such a cycle for a second time. When this approach was used, evidence for the involvement of a slow step in this photocycle was obtained[23]. Apart from these uncertainties, evidence is also available to conclude that a proton gradient *does* inhibit the turnover of bacteriorhodopsin[24,25]. A quantitation of these effects is difficult to give in view of the uncertainties, alluded to above.

The maximal turnover rate of the reconstituted reaction center proton pump, under level-flow conditions and saturating amounts of quinone and cytochrome c, may be infered from studies of reaction centers solubilized in detergent. This rate is limited by quinol release from the reaction centers. Measurements indicate a maximal turnover rate of 200 to 300 sec^{-1} for the bacteriochlorophyll-based proton pump[26]. In the intact cell, however, this rate may become severely restricted by

(i) additional rate-control by the cytochrome b/c$_1$ complex,
(ii) sub-saturating amounts of quinone and cytochrome c, and
(iii) back-pressure by the proton motive force.

In order to obtain quantitative information about the kinetic performance of the reconstituted reaction center proton pump, we have developed a thermokinetic scheme that describes light-driven proton pumping in this system. The model is based on the large amount of kinetic and thermodynamic data (microscopic parameters), available from the

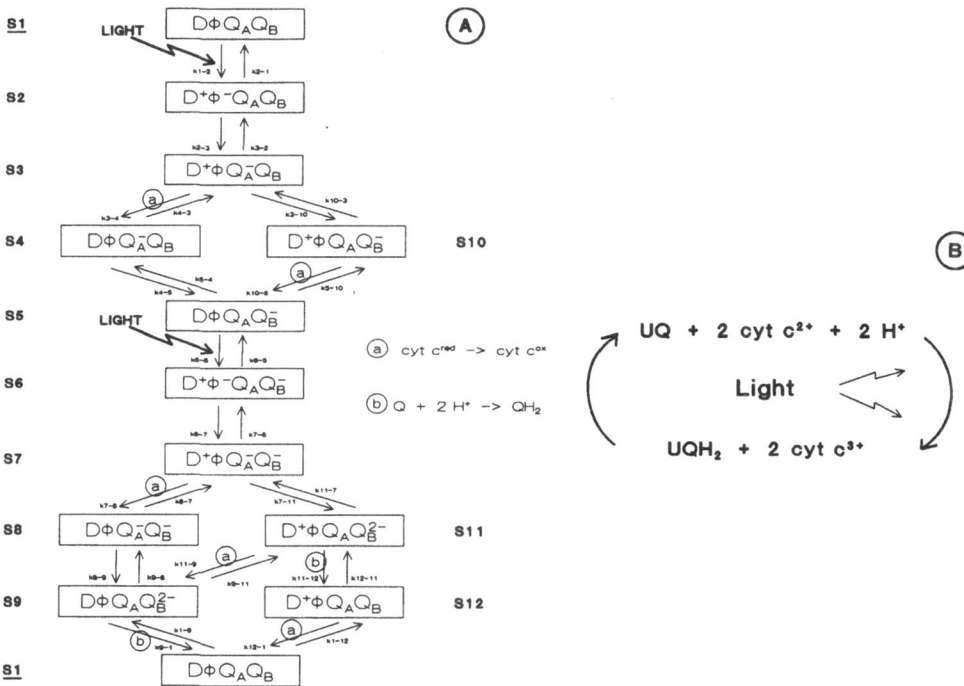

Figure 2. (A) Schematic representation of the 9 and 12 intermediate states of the reaction center and their mutual transitions. The ground-state (S1) has been indicated twice for reasons of clarity. The involvement of cytochrome and ubiquinone in some of the reactions is shown via a and b. (B) Summary of the two partial reactions that form the basis of the reaction center proton pump: Light-driven cytochrome c oxidation and spontaneous reduction of oxidized cytochrome c by quinol.

literature for reaction centers from *Rb. sphaeroides*. These data have been used to set up two parallel models, taking into account 9 and 12 different intermediate states of the reaction center, respectively (see Fig. 2A). A Jacobian matrix was defined, containing all microscopic rate equations of the partial reactions that are relevant for the two models.

The matrix-inversion method was subsequently used to predict the macroscopic steady-state parameters for these two models. Next, the total amount of quinone and cytochrome c plus the total number of electrons present in these two redox mediators were defined. The reduction level of the cytochrome was varied stepwise, from fully oxidized to completely reduced (600 steps) and for every step the matrix inversion was performed. The concentrations of the various intermediates thus obtained were used to calculate:

(i) the rate of light-driven cytochrome c oxidation and
(ii) the rate of the spontaneous reaction between quinol and oxidized cytochrome c (Fig.2B).

At the redox level of cytochrome c at which these two rates match, the steady-state turnover rate of the reaction center based proton pump is obtained.

Initial measurements of the macroscopic parameters of the reaction center based proton pump in detergents have led to the conclusion that at high quinol concentration the rate of spontaneous reduction of oxidized cytochrome c has to be described by a pseudo first-order reaction, rather than a second-order reaction. Furthermore, the results show a higher ratio of oxidized over reduced primary donor of the reaction centers than predicted by either one of the two models. From this it is clear that additional experiments and modeling will have to be performed to further refine the thermokinetic description. Future efforts will be specifically aimed at resolving the energetic coupling between proton and electron transfer. This approach will result in a mathematical description for this proton pump that can be used under conditions ranging from level-flow, all the way up to the static head.

It is clear that only a detailed thermokinetic description of both types of pumps will ultimately allow us to make a meaningful comparison with respect to their kinetic efficiency. At present, the available data is insufficient to give such a description.

5. Conclusions

We conclude that the present-day representatives of retinal-based- and chlorophyll-based photosynthesis perform rather similarly, with respect to efficiency. However, an aspect that may have been, and possibly still is, very important in the competition between these two processes of radiant energy conversion, is the possibility to use antenna pigments. Such pigments may allow considerable savings on the investment in protein synthesis, compared to an organism that contains a photo-converter without antenna pigments.

Perhaps the evolutionary relation between retinal-based- and chlorophyll-based photosynthesis is much more complex than a simple competition in energetic efficiency, if such a competition has been relevant at all. Arguments can be advanced to suggest that retinal-based photosynthesis evolved as a derivative of a procaryotic photosensory system[27].

References

1. A. Goldsworthy, Phycobilins, *in*: "Photoreceptor Evolution and Function," M.G. Holmes, ed., Academic Press, London (1991).
2. K.J. Hellingwerf, "Structural and Functional Studies on Lipid Vesicles Containing Bacteriorhodopsin," PhD Thesis, University of Amsterdam, Wm. Veenstra, Groningen (1979).

3. K.J. Hellingwerf, Reaction centers from *Rhodopseudomonas sphaeroides* in reconstituted phospholipid vesicles. II: Light-induced proton translocation, *J. Bioenerg. Biomembr.* **19**:225 (1987).

4. D. Molenaar, W. Crielaard and K.J. Hellingwerf, Characterization of protonmotive force generation in liposomes reconstituted from phosphatidylethanolamine, reaction centers with light-harvesting complexes isolated from *Rhodopseudomonas palustris*, *Biochemistry* **27**:2014 (1988).

5. R. Henderson, J.M. Baldwin, T.A. Ceska, F. Zemlin, E. Beckmann and K.H. Downing, Model for the structure of bacteriorhodopsin based on high-resolution electron cryo-microscopy, *J.Mol. Biol.* **213**: 899 (1990).

6. J. Deisenhofer and H. Michel, The photosynthetic reaction center from the purple bacterium *Rhodopseudomonas viridis*, *Science* **245**:1463 (1989).

7. G. Feher, J.P. Allen, M.Y. Okamura and D. Rees, Structure and function of bacterial photosynthetic reaction centres, *Nature* **339**:111 (1989).

8. G. Váró and J.K. Lanyi, Thermodynamics and energy coupling in the bacteriorhodopsin photocycle, *Biochemistry* **30**:5016 (1991).

9. G. Schneider, R. Diller and M. Stockburger, Photochemical quantum yield of bacteriorhodopsin from resonance Raman scattering as a probe for photolysis, *Chem. Phys.* **131**:17 (1989).

10. A. Xie, Quantum efficiencies of bacteriorhodopsin photochemical reactions, *Biophys. J.* **58**:1127 (1990).

11. C.E. Gibson and D.H. Jewson, The utilisation of light by microorganisms, *in*: "Aspects of Microbial Metabolism and Ecology," G.A. Codd, ed., Academic Press, London (1984).

12. T. Noguchi, H. Hayashi and M. Tasumi, Factors controlling the efficiency of energy transfer from carotenoids to bacteriochlorophyll in purple bacteria, *Biochim. Biophys. Acta* **1017**:280 (1990).

13. H.V. Westerhoff and Zs. Dancsházy, Keeping a light-driven proton pump under control, *Trends in Biochem. Sci.* **9**:112 (1984).

14. H.V. Westerhoff and K. van Dam, "Thermodynamics and Control of Free-Energy Transduction", Elsevier, Amsterdam (1987).

15. H.V. Westerhoff, B.J. Scholte and K.J. Hellingwerf, Bacteriorhodopsin in liposomes. I. A description using irreversible thermodynamics, *Biochim.Biophys.Acta* **547**:544 (1979).

16. H.V. Westerhoff, K.J. Hellingwerf, J.C. Arents, B.J. Scholte and K. van Dam, Mechanistic non-equilibrium thermodynamics describes biological energy transduction, *Proc.Natl.Acad.Sci. USA* **78**: 3554 (1981).

17. K.J. Hellingwerf, L.J. Grootjans, A.J.M. Driessen, W. de Vrij and W.N. Konings, A functional comparison of proton motive force generating systems, Abstract for the 13th International Congress of Biochemistry, Amsterdam, FR-456 (1985).

18. P.W.M. van Dijck, K. Nicolay, J. Leunissen-Bijvelt, K. van Dam and R. Kaptein, ^{31}P-Nuclear magnetic resonance and freeze fracture electron microscopic studies of reconstituted bacteriorhodopsin vesicles, *Eur. J. Biochem.* **117**:639 (1981).

19. W. Stoeckenius, R.H. Lozier and R.A. Bogomolni, Bacteriorhodopsin and the purple membrane of halobacteria, *Biochim. Biophys. Acta* **505**:215 (1979).

20. K.J. Hellingwerf, B.J. Scholte and K. van Dam, Bacteriorhodopsin vesicles. An outline of the requirements of light-dependent H$^+$ pumping, *Biochim.Biophys.Acta* **513**:66 (1978).

21. B. Höjeberg, C. Lind and H.G. Khorana, Reconstitution of bacteriorhodopsin vesicles with *Halobacterium halobium* lipids. Effects of variations in lipid composition, *J. Biol. Chem.* **257**:1690 (1982).

22. R. Melhorn, B. Schobert, L. Packer and J.K. Lanyi, ESR studies of light-dependent volume changes in cell envelope vesicles from *Halobacterium halobium*, *Biochim. Biophys. Acta* **809**:66 (1985).

23. L.A. Drachev, A.D. Kaulen, V.P. Skulachev and V.V. Zorina, The mechanism of H$^+$ transfer by bacteriorhodopsin, *FEBS Lett.* **226**:139 (1987).

24. K.J. Hellingwerf, J.J. Schuurmans and H.V. Westerhoff, Demonstration of coupling between the proton motive force across bacteriorhodopsin and the flow through its photochemical cycle, *FEBS Lett.* **92**: 181 (1978).

25. Zs. Dancsházy, S.L. Helgerson and W. Stoeckenius, Coupling between the bacteriorhodopsin photocycle kinetics and the proton motive force, *Photobiochem. and Photobiophys.* **5**:347 (1983).

26. E.J. Bylina, R. Jovine and D.C. Youvan, Quantitative analysis of genetically altered reaction centers using an *in -vitro* cytochrome oxidation assay, *in*: "The Photosynthetic Bacterial Reaction Center: Structure and Dynamics," J. Breton and A. Vermeglio, eds, NATO ASI Series Vol. 149, Plenum Press, New York (1988).

27. D.G. Stavenga, J. Schwemer and K.J. Hellingwerf, Visual pigments, bacterial rhodopsins and related retinoid-binding proteins, *in*: "Photoreceptor evolution and function", M.G. Holmes, ed., Academic Press, London (1991).

REGULATION OF ENERGY FLUX
OF YEAST DURING STEADY STATE
AND OSCILLATORY GROWTH

Regine Ölz[1], Katrin Larsson[1], Christer Larsson[1],
Erich Gnaiger[2] and Lena Gustafsson[1]

[1]Department of General and Marine Microbiology
 University of Göteborg
 Carl Skottsbergs Gata 22, S-413 19 Göteborg, Sweden
[2]Department of Transplant Surgery, Research Division
 University Hospital of Innsbruck
 Anichstrasse 35, A-6020 Innsbruck, Austria

1. Introduction

Baker's and brewer's yeast *Saccharomyces cerevisiae* is a microorganism of major industrial importance, both within traditional and new branches of industry. For effective control of biotechnological yeast processes, an improved understanding of metabolic regulation in yeast is necessary in combination with development of effective control variables.

S. cerevisiae is a fermentative organism, which has the metabolic capability of a mixed respiratory and fermentative catabolism during aerobic conditions. Non-optimized metabolic regulation of this rather complex catabolism may cause undesirable effects such as oscillatory growth in continuous cultures of *S. cerevisiae*. This growth behavior is widely documented and different hypotheses for the underlying mechanisms have been proposed[1-6]. Oscillations were observed in terms of biomass, carbon and energy substrate, ethanol, dissolved oxygen, pH, carbon dioxide evolution rate and oxygen uptake rate. In addition, intracellular rhythmic changes were demonstrated, including respiratory activity, enzyme activities, storage carbohydrate levels, together with partial synchronization of the cell population. Since metabolic changes, including the type of metabolism and rate changes are registered as a change in the rate of microbial heat production[7], we have used direct calorimetry as a method to monitor steady state *versus* oscillatory growth in continuous culture of yeast.

The potential of *S. cerevisiae* to regulate between several catabolic pathways (Fig. 1), is further accentuated at a reduced external osmotic potential. This may cause problems on an industrial scale. This feature, however, makes it an interesting organism for studies of metabolic regulation and of thermodynamic (enthalpy) and ergodynamic (Gibbs energy) efficiencies.

2. Materials and Methods

Organism and Continuous Culture Experiment

The yeast strain used was *Saccharomyces cerevisiae* Y41 (ATCC 38531)[8]. The culture medium consisted of yeast nitrogen base (YNB) without amino acids (Difco) and was

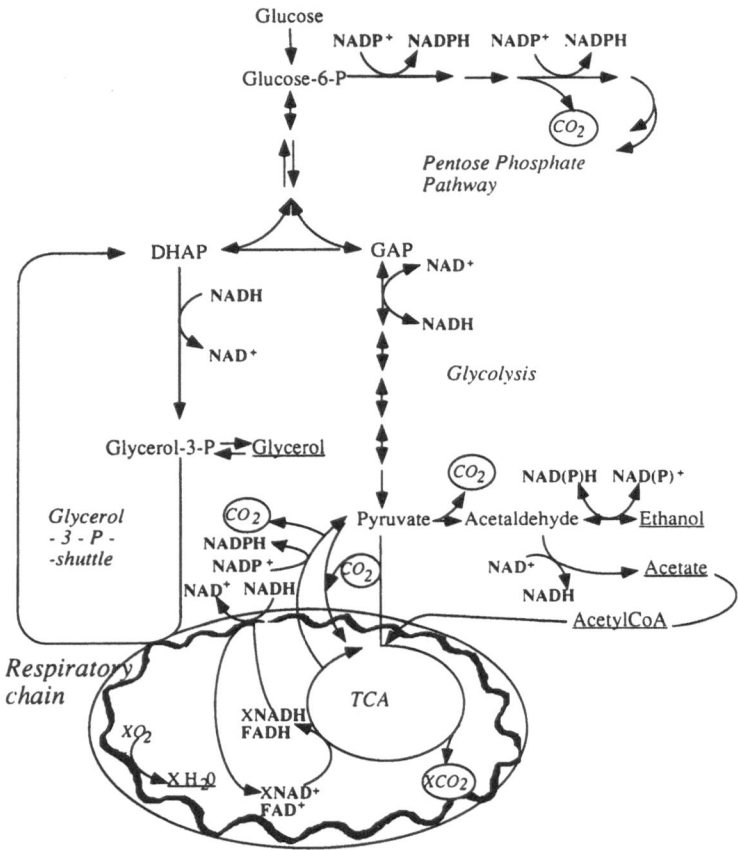

Figure 1. Simplified scheme of intermediary metabolism of *S. cerevisiae*. For details see Refs 19, 20.

supplemented with sterile glucose after autoclaving to a final concentration of 0.5% (w/v). For experiments at a reduced external osmotic potential, NaCl was added to a final concentration of 5%(w/v); referred to as high salt experiments, in contrast to experiments without NaCl, referred to as 0% NaCl. The experimental design and substrate and product analyses, as well as microcalorimetric measurement and energy balance calculations were described previously[9].

3. Results and Discussion

3.1. Energy flux and balances

S. cerevisiae catabolized glucose by pure respiration (Fig. 2 and Table 1) during steady state continuous cultivation on glucose at a low dilution rate (0.09 h[-1]), with and without NaCl. The only low-molecular product produced to a relatively high degree was glycerol during growth in salt[9]. Glycerol is the main osmoregulator used by *S. cerevisiae*[10].

Oscillatory growth was induced by decreasing the stirring and aeration rate (oxygen tension was always above 60% of air saturation). During oscillatory growth, there was a long-term periodic shift between respiratory and respiro-fermentative catabolism (Fig. 2 and Table 1). In experiments carried out at various dilution rates, the most pronounced oscillations were observed at intermediary dilution rates of salt-grown cells. There was not only oscillatory changes in the proportion of fermentation and respiration (Fig. 2), but oscillations were also observed at the level of pure respiration associated with periodic changes of the growth yield (data not shown). In addition, respiration was also accompanied by a varying production of catabolic products such as acetate and glycerol. It must also be kept in mind that glucose is not the only energy and carbon source during oscillatory growth. Periodically produced ethanol, acetate and surplus glycerol intermittently also serve as sources of energy and carbon. These periodic changes put additional constraints on the intracellular redox (NAD(P)/NAD(P)H) and energy (ATP/ADP) balances.

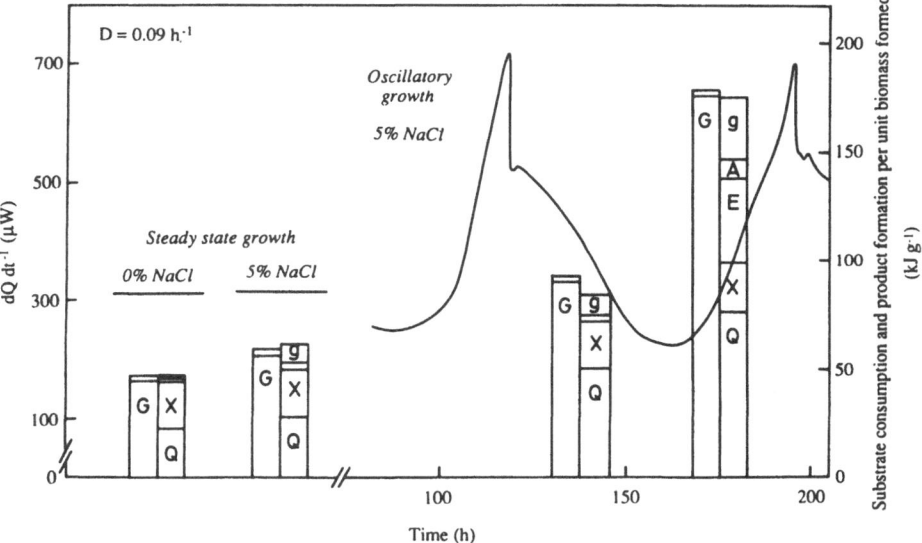

Figure 2. Energy balances during steady state and oscillatory growth of *S. cerevisiae* in 0 and 5% NaCl on glucose (monitored by the rate of heat production (—) (μW). The sum of the energy contents of the substrates consumed (left columns) and the products formed (including biomass) (right columns) per unit of ash-free biomass formed (kJ/g): glucose (G), biomass (X), integrated heat (Q), ethanol (E), acetate (A), glycerol (g). Unlabeled boxes refer to substrate (nitrogen source) and products (ethanol, acetate, and/or glycerol) of low amounts.

Table 1. Major parts of the carbon flux (a = anabolism; k = catabolism; g = glycerol; F_{EtOH} = ethanol fermentation; R = respiration) and thermodynamic properties of *Saccharomyces cerevisiae* at different physiological states during oscillatory growth in 5 % NaCl and during steady state growth in 0% and 5% NaCl at a dilution rate of 0.09 h[-1].[&]

	Oscillatory 5 % NaCl	Oscillatory 5 % NaCl	Steady state 5 % NaCl	Steady state 0 % NaCl
Fraction of total carbon flux[$]				
a	0.11	0.22	0.35	0.44
k	0.89	0.78	0.65	0.55
g	0.16	0.10	0.13	0.03
F_{EtOH}	0.23	0.03	0	0
R	0.45	0.66	0.47	0.50
$\Delta_k G_S$ [kJ/mol]	-1573.6	-2452.4	-2101.6	-2637.9
dQ_m/dt [J/(s·g)]	-1.78	-1.18	-0.68	-0.55
$d_r H_m/dt$ [J/(s·g)]	-1.88	-1.36	-0.58	-0.51
$d_r G_m/dt$ [J/(s·g)]	-2.05	-1.42	-0.63	-0.53
$Y_{x/s}$ [g/g]	0.10	0.19	0.31	0.42
ΔQ_X [kJ/g]	-75.7	-50.2	-27.8	-22.5
ε [%]	0.8	1.1	2.5	2.5
ε^* [%]	19	25	45	47
η [%]	20	26	45	47

[&]All expressions of biomass are ashfree, except for growth yield, $Y_{x/s}$. ε, ε^* and η are calculated according to eqs (3), (4), and (5), respectively.

[$]a and k are equal to $J_a/(J_a+J_k)$ and $J_k/(J_a+J_k)$, respectively. J_a and J_k are the fluxes of anabolism and catabolism, respectively, according to eqs (1) and (2).

For the energy flux analysis and efficiency calculations, growth equations were partitioned into the anabolic and catabolic half reactions. This is illustrated below by an anabolic (eq. 1) and a catabolic (eq. 2) growth equation during oscillatory growth of *S. cerevisiae* in 5% NaCl medium:

$$\alpha,\beta\text{-D } C_6H_{12}O_6(aq) + 0.946\ NH_4^+(aq) \longrightarrow$$

$$0.088\ CO_2(g) + 5.912\ CH_{1.64}O_{0.55}N_{0.16}(s) + 0.946\ H^+(aq) + 2.571\ H_2O(l) \qquad (1)$$

$$\alpha,\beta\text{-D } C_6H_{12}O_6(aq) + 3.023\ O_2(g) \longrightarrow$$

$$0.524\ C_2H_6O(aq) + 0.305\ C_3H_8O_3(aq) + 0.169\ C_2H_4O_2(aq)$$

$$+ 2.871\ H_2O(l) + 3.700\ CO_2(g) \qquad (2)$$

These equations were constructed using the concept of the degree of reduction[11]. The anabolic half reaction was balanced by CO_2 production. This is biochemically relevant, since anabolism requires NADPH which is produced via the pentose phosphate pathway. This NADPH production will result in anabolically produced CO_2 (Fig. 1). In addition, biosynthesis of precursors for anabolism via the catabolic pathways will result in production of CO_2. However, CO_2 production is at least in part compensated by anapleurotic CO_2

fixation. The net CO_2 production of these processes is substrate dependent, as is the net production or consumption of anabolically produced NADH. During growth on substrates more reduced than the biomass, such as ethanol, there is a net production of NADH in the anabolic pathways. The excess NADH is oxidized by the respiratory chain. This NADH oxidation seems to involve the glycerol-3-phosphate shuttle. The activity of the NAD-dependent cytoplasmatic enzyme (GPDH), which reduces dihydroxy-acetone-phosphate (DHAP) to glycerol-3-phosphate, was highly increased during growth on ethanol compared to growth on glucose[12].

The molar Gibbs energy change of the catabolic half reaction ($\Delta_k G_s$; Table 1) indicates the relative fermentative contribution. During oscillatory growth, fermentative metabolism gave rise to a molar Gibbs energy less negative than that of pure respiration, which would yield about -2800 kJ/mole of glucose oxidized. However, during growth at steady state without salt in the external medium, a virtually pure respiratory catabolism was indicated by the molar Gibbs energy change of -2637.9 kJ/mole. This is close to the value obtained in full oxidation of glucose. A comparison between the experimentally obtained mass specific heat flux (dQ_m/dt) and the theoretically calculated mass specific enthalpy flux ($d_r H_m/dt$) (Table 1), gave energy recoveries ($ER=dQ_m/dt/d_r H_m/dt$) between 0.87 and 1.17. This shows the accuracy of direct calorimetry in the continuous measurements of energy flux during growth processes. In pure respiratory catabolism the enthalpy and Gibbs energy changes are nearly identical[13]. This is also seen by comparing the mass specific fluxes of heat and enthalpy with mass specific power (equal to mass specific Gibbs energy flux, $d_r G_m/dt$) (Table 1). When the aerobic carbon flux was only 51 % of the total catabolic carbon flux (first column, Table 1), then the aerobic Gibbs energy flux amounted to about 90% of the total catabolic power. Therefore, the Gibbs energy and enthalpy fluxes were still nearly identical. However, at even higher proportions of fermentation relative to respiration, an increased deviation is expected.

3.2. Efficiencies

An unsolved problem in all calculations of growth efficiencies is the thermodynamically and biochemically appropriate definition of the input and output processes. To illustrate the inherent difficulties, three different modes of efficiency calculations were compared (Table 1):

(a) In the first approach microbial metabolism is partitioned into anabolism (a) and catabolism (k) according to eqs (1) and (2) (see also Fig. 3). The anabolic power output, $J_a \Delta_a G_s$, is divided by the catabolic power input, $J_k \Delta_k G_s$. This yields the Gibbs energy efficiency (ε) or ergodynamic efficiency[13], or thermodynamic efficiency[14,15] (the term thermodynamic efficiency is in this paper referred to as enthalpy efficiency):

$$\varepsilon = \frac{-J_a \Delta_a G_s}{J_k \Delta_k G_s} . \tag{3}$$

The Gibbs energy of formation and enthalpy of formation of biomass used in the calculations were obtained from the experimentally determined enthalpy of combustion of biomass[9]. The values used in this study are -45.2 and -40.4 kJ/C-mole biomass ($\Delta_f G_x$) and -88.3 and -79.5 kJ/C-mole biomass ($\Delta_f H_x$) for cells grown in 0% and 5% NaCl, respectively.

(b) An engineer's way of defining Gibbs energy efficiency (eq. 4), which has no biochemical meaning but rather a practical biotechnological value, was proposed by Roels[11] and recently by von Stockar et al.[16]. This definition gives the flow of Gibbs energy leaving the system as biomass divided by the difference in the flow of Gibbs energy between the substrates and products other than biomass. Written in another way, this definition reads:

$$\varepsilon^* = \frac{\Delta_c^* G_X^o}{\Delta_c^* G_X^o + \Delta_r G_X^o} \ . \tag{4}$$

This type of definition is based on the use of a reference state, in this case the products of combustion (c). However, the asterisk denotes that, instead of $N_2(g)$, $NH_3(g)$ is the reference state for nitrogen. This excludes the need for incorporating the N-substrate in the equation, when the N-source is ammonia or ammonium ions.

(c) By analogy to eq. (4), the enthalpy (thermodynamic) efficiency of growth is obtained (eq. 5):

$$\eta = \frac{\Delta_c^* H_X^o}{\Delta_c^* H_X^o + \Delta_r H_X^o} \ . \tag{5}$$

Gibbs energy efficiencies calculated according to eq. (3) are very low (Table 1). This originates from the fact that the Gibbs energy change of the anabolic reaction becomes small when cells grow on an energy and carbon source which shows a degree of reduction similar to that of the biomass[17]. Oscillatory growth seems to be less efficient than steady state growth, especially when there is a high degree of fermentation included (Table 1). Similar results were obtained with efficiency calculations according to eq. (4). Thermodynamic (enthalpy) efficiency (eq. 5) gave very similar results, showing that calculations of Gibbs energy efficiency may be replaced by calorimetric determinations. However, care has to be taken when the catabolism is not purely respiratory and to the fact that the Gibbs energy changes are concentration-dependent.

It is plausible to believe that growth during steady state conditions compared to oscillatory growth, permits cells to direct their metabolism towards an optimal efficiency at a specific rate of growth. Increased maintenance energy requirements, predicted by the growth yield ($Y_{x/s}$) and the ΔQ_x values (Table 1), for growth in salt compared to basic medium[9] do not seem to decrease the Gibbs energy efficiency. This can be explained by a decrease in the Gibbs energy change of the catabolic reaction (less negative value) in salt compared to that in basic medium because of glycerol production, which compensates for the lowered growth yield in salt and the thereby lowered flow of Gibbs energy in the anabolic half reaction.

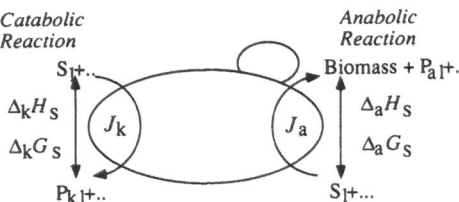

Figure 3. Subdivision of the growth reaction (r) into the catabolic (k) and anabolic (a) reactions during growth on one energy and carbon source.

The presented study indicated a reduced efficiency during oscillatory growth, but no change in efficiency because of external salt stress for the reasons explained above. Such efficiency analyses treat the cell as a black box or one-compartmental structure. This puts limits to metabolic variation in terms of maximal possible energy outflow from the cell in relation to the energy inflow to the cell for each specific type of metabolism. However, the one-compartmental approach provides limited information about the specific strategies of cells in response to changing constraints and loads. The two-compartmental approach distinguishes the catabolic and anabolic compartments energetically, with the ATP/ADP cycle as the general energy mediator. The two-compartmental analysis considers the overall efficiency as the product of the catabolic and anabolic efficiencies[18]. Uncoupling decreases the power output and efficiency. Maintenance energy requirements may increase owing to a work load imposed by external stress. This diminishes the growth efficiency but not necessarily the catabolic efficiency. A two-compartmental efficiency study allows conclusions to be obtained about the metabolic capabilities and optimization strategies at the various energy coupling sites from catabolism to anabolism. Future work will focus on catabolic and anabolic efficiencies considering various constraints and work loads due to changed maintenance energy requirements.

Acknowledgements

This work was supported by grants from the Swedish National Board for Natural Sciences, the Swedish Board for Industrial and Technical Development and FWF Austria P7162-B10. Grateful thanks to Ms. M. Jehler for patiently improving the manuscript layout.

References

1. M.T. Kuenzi and A.Fiechter, Change in carbohydrate composition and trehalase activity during the budding cycle of *Saccharomyces cerevisiae*, *Arch. Microbiol.* **64**:396-407 (1969).
2. H.K. von Meyenburg, Stable synchrony oscillations in continuous cultures of S. *cerevisiae* under glucose limitation, *in*: "Biological and Biochemical Oscillators," B. Chance, E.K. Pye, T.K. Ghosh and B. Hess, eds, Academic Press, New York (1973), p. 411-417.
3. C.-I. Chen, K.A. McDonald and L.Bisson, Oscillatory behaviour of *Saccharomyces cerevisiae* in continuous culture: I. Effects of pH and nitrogen levels, *Biotech. Bioeng.* **36**:19-27 (1990).
4. C.-I. Chen and K.A. McDonald, Oscillatory behaviour of *Saccharomyces cerevisiae* in continuous culture: II. Analysis of cell synchronization and metabolism, *Biotech. Bioeng.* **36**:28-38 (1990).
5. R. Grosz and G. Stephanopoulos, Physiological, biochemical, and mathematical studies of micro-aerobic continuous ethanol fermentation by *Saccharomyces cerevisiae*. I: Hysteresis, oscillations and maximum specific ethanol productivities in chemostat culture, *Biotech. Bioeng.* **36**:1006-1019 (1990).
6. E. Martegani, D. Porro, B.M. Ranzi and L.Alberghina, Involvement of a cell size control mechanism in the induction and maintenance of oscillations in continuous cultures of budding yeast, *Biotech. Bioeng.* **36**:453-459 (1990).
7. L. Gustafsson, Microbiological calorimetry, *Thermochim. Acta* **193**:145-171 (1991).
8. J.C. Anand and A.D. Brown , Growth rate patterns of the so-called osmophilic and non-osmophilic yeasts in solutions of polyethylene glycol, *J. Gen. Microbiol.* **52**:205-212 (1968).
9. R. Ölz, K. Larsson, L. Adler and L. Gustafsson, Energy flux and osmoregulation of *Saccharomyces cerevisiae* grown in chemostats under NaCl stress, *J. Bacteriol.*, in press.
10. A. Blomberg and L. Adler, Physiology of osmotolerance in fungi, *Adv. Microbial Physiol.* **33**:145-212 (1992).
11. J.A. Roels. "Energetics and Kinetics in Biotechnology," Elsevier, Amsterdam (1983).
12. A. Blomberg, C. Larsson and L.Gustafsson, Microcalorimetric monitoring of growth of *Saccharomyces cerevisiae*: Osmotolerance in relation to physiological state, *J. Bacteriol.* **170**:4562-4568 (1988).
13. E. Gnaiger, Concepts on efficiency in biological calorimetry and metabolic flux control, *Thermochim. Acta* **172**:31-52 (1990).
14. O. Kedem and S.R. Caplan, Degree of coupling and its relation to efficiency in energy conversion, *Trans. Faraday Soc.* **61**:1897-1911 (1965).
15. H. Westerhoff and K.van Dam. "Thermodynamics and Control of Biological Free-Energy Transduction," Elsevier, Amsterdam (1987).
16. U. von Stockar, Ch. Larsson and I.W. Marison, Calorimetry and energetic efficiencies in aerobic and anaerobic microbial growth, *Pure Appl. Chem.*, in press.
17. M. Rutgers, H.M.L. van der Gulden and K. van Dam, Thermodynamic efficiency of bac-terial growth

calculated from growth yield of *Pseudomonas oxalaticus* OX1 in the chemostat, *Biochim. Biophys. Acta* **973**:302-307 (1989).

18. E. Gnaiger, Optimum efficiencies of energy transformation in anoxic metabolism. The strategies of power and economy, *in*: "Evolutionary Physiological Ecology," P. Calow, ed., Cambridge Univ. Press, London (1987), pp. 7-36.

19. D.G. Fraenkel, Carbohydrate metabolism, *in*: "The Molecular Biology of the Yeast *Saccharomyces*: Metabolism and Gene Expression," N.J. Strathern, E.W. Jones and J.R. Broad, eds, Cold Spring Harbor Laboratory, Cold Spring Harbor, New York (1982), pp. 1-37.

20. C. Wills, T. Martin and T. Melham, Effect on gluconeogenesis of mutants blocking two mitochondrial transport systems in the yeast *Saccharomyces cerevisiae*, *Arch. Biochem. Biophys.* **246**:306-320 (1986).

Appendix: List of Symbols

J	Flux (mol s^{-1}g^{-1})
$Y_{x/s}$	Growth yield (g g^{-1})
ε	Gibbs energy efficiency
η	Enthalpy efficiency
$\Delta_r G_s$	Molar Gibbs energy change of reaction expressed per unit of substrate (kJ mol^{-1})
$\Delta_c{}^*G^o{}_x$	Modified standard molar Gibbs energy change of combustion yielding N in the form of NH$_3$, expressed per unit of biomass (kJ C-mol^{-1})
$\Delta_c{}^*H^o{}_x$	Modified standard molar enthalpy change of combustion yielding N in the form of NH$_3$, expressed per unit of biomass (kJ C-mol^{-1})
$\Delta_r G^o{}_x$	Standard molar Gibbs energy change of the growh reaction expressed per unit of biomass formed (kJ C-mol^{-1})
$\Delta_r H^o{}_x$	Standard molar enthalpy change of the growh reaction, expressed per unit of biomass formed (kJ C-mol^{-1})
ΔQ_x	Experimentally determined molar enthalpy change of the growth reaction, expressed per unit of ashfree biomass formed (kJ g^{-1})
$d_r G_m/dt$	Mass (m) specific Gibbs energy flux, expressed per unit of ashfree biomass in the system (J s^{-1}g^{-1})
$d_r H_m/dt$	Mass (m) specific enthalpy flux, expressed per unit of ashfree biomass in the system (J s^{-1}g^{-1})
dQ_m/dt	Mass (m) specific heat flux, expressed per unit of ashfree biomass in the system (J s^{-1}g^{-1})
Subscript a	Anabolism
Subscript k	Catabolism
Subscript c	Combustion
Subscript f	Formation

FREE-ENERGY COUPLING
MEDIATED BY DYNAMIC INTERACTIONS
BETWEEN MEMBRANE PROTEINS

R. Dean Astumian[1], Baldwin Robertson[2] and Martin Bier[3]

[1]Department of Surgery
 and Department of Biochemistry and Molecular Biology
 University of Chicago
 Chicago, Illinois 60637, U.S.A.
[2]Biotechnology Division
 National Institute of Standards and Technology
 Gaithersburg, Maryland 20899, U.S.A.
[3]Department of Biology
 Vrije Universiteit
 de Boelelaan 1087, 1081 HV Amsterdam, The Netherlands

1. Introduction

The influence of an external oscillating electric field on membrane enzymes has been experimentally observed on different systems. The rate of transport of Na^+ and K^+ by Na^+,K^+-ATPase in erythrocytes[1] and the synthesis of ATP in plant protoplasts[2] have been experimentally studied as a function of frequency and amplitude of the applied oscillating field.

In Section 2, we show with a simple model how it can be understood that an oscillating electric field can drive an enzyme-mediated reaction away from equilibrium, even if substrate and product do not interact with the electric field. We thus have a mechanism by which free energy can be absorbed from an a.c. field and transduced to a chemical potential.

In Section 3, we show how two nearby membrane enzymes, both whose catalytic cycle involves a movement of a charge, can interact in such a way that one conversion, going energetically downhill, can drive another conversion energetically uphill. We present a model of a coporter that works through this electroconformational coupling.

2. An Enzyme Mediated Reaction in an Applied Oscillating Field

In the presence of an electric field a rate constant is changed in the following way:

$$k_i = k_i^0 \exp\left(\frac{z_i \Psi}{kT}\right),$$

where k_i^0 is the transition rate in the absence of an electric field and z_i denotes the charge that has to be displaced across an electric potential Ψ for the activation of the i-th transition. In biological membranes even *in vivo* electric fields can be strong enough (of the order of megavolts per meter) that the effect on the transition rates is significant. If we let Ψ oscillate, i.e.

$$\Psi(t) = \Psi_0 \sin(\omega t),$$

then the transition rates will oscillate.

Fig. 1 shows the energy levels of a catalytic conversion of substrate S into product P via one intermediate complex ES, as shown in the following scheme.

$$E + S \underset{k_{-1}}{\overset{k_{+1}}{\rightleftharpoons}} ES \underset{k_{-2}}{\overset{k_{+2}}{\rightleftharpoons}} E + P.$$

If the relation $z_1 + z_2 + z_{-1} + z_{-2} = 0$ holds then there is a situation where the energy levels of E+S and E+P remain constant under the oscillating electric field, while the levels of the two

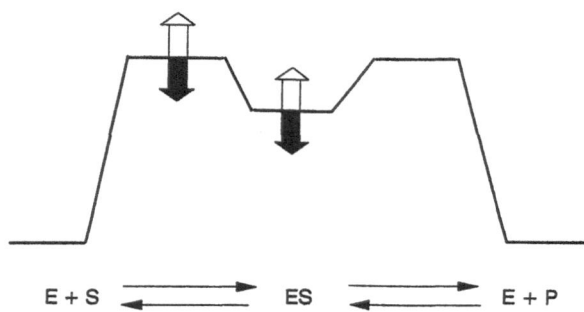

Figure 1. Free-energy diagram of an enzyme catalyzed conversion of substrate S into product P. The arrows illustrate how an oscillating electric field can affect the energy levels. Thick arrows indicate the effect of an increase in potential ($+\Psi$), open arrows indicate the effect of a decrease in potential ($-\Psi$).

activation barriers and the level of the intermediate complex oscillate. For Fig. 1 we moreover took $k_{-1}^0 = k_2^0$, $k_1^0 = k_{-2}^0$, $-z_{+1} = z_{+2}$ and $z_{-1} = z_{-2} = 0$; a choice that leads to an equilibrium where the concentration of P equals the concentration of S and that leaves the activation barrier between ES and E+P at a constant level while the remaining two levels oscillate.

During the phase of the field where Ψ is positive (filled black arrows) the energy level of the complex ES is lower. Binding of S and P by the enzyme will then be favorable. Because the activation barrier between E+S and ES is also down, S will initially bind faster than P. If the positive field were left on long enough, then the height of the activation barriers would not matter and eventually just as much S as P would bind. But the field switches before such an equilibrium is reached.

When the field is negative (open arrows) the energy of the intermediate complex ES is up and dissociation will become favorable. Dissociation in the direction of P will initially be faster because the activation barrier between ES and E+P is lower than the one between ES and S. So the net effect of one cycle is that S is pumped to P away from the equilibrium where the concentrations of S and P are equal.

Figs 2a and 2b show how the effect accumulates over many cycles of the field until the chemical gradient counterbalances the pumping force of the oscillating electrical field.

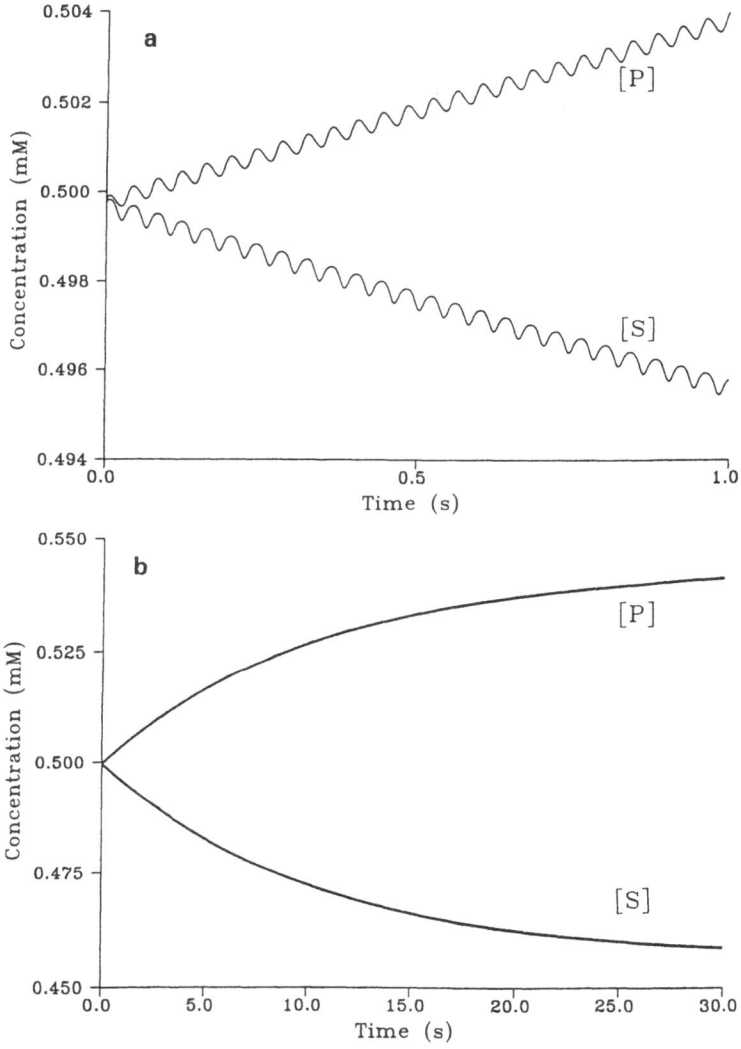

Figure 2. a) The change of S and P in a closed system exposed to an oscillating potential $\Psi = \Psi_0 \cos(ft/(2\pi))$. The parameters we used were $f = 25$ Hz, $\Psi_0 = 25$ mV, $k_1 = k_{-2} = 10^4$ 1/(mol sec), $k_{-1} = k_2 = 10^2$ sec^{-1}, $z_{+1} = -z_{+2} = 1$ elementary charge and $z_{-1} = z_{-2} = 0$. The starting concentrations were $S = P = 0.5$mM and the enzyme concentration (bound and unbound together) was 10 μM.
b) The same Figure on a larger timescale.

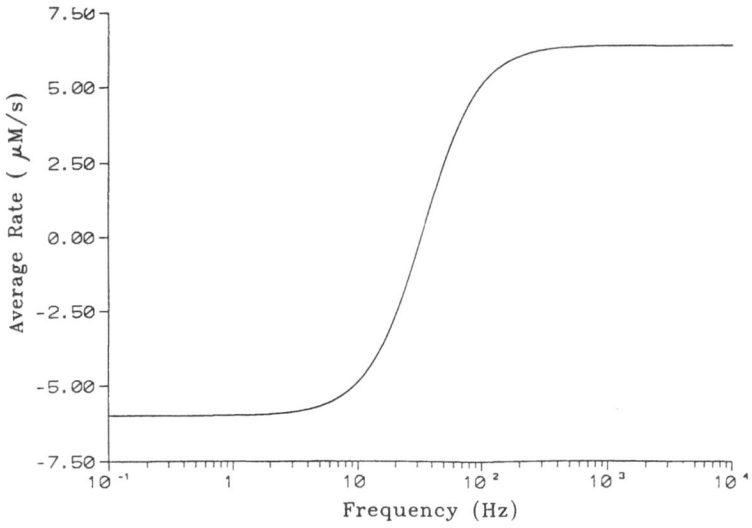

Figure 3. The average rate of transport vs. the frequency for the same parameter values as in Fig. 2, but for $S=0.44$mM and $P=0.56$mM.

Fig. 3 shows how the magnitude of the effect depends on the frequency of the applied field. At low frequency it is impossible to get away from the $S=P$ equilibrium. For higher frequencies a maximum efficiency is asymptotically reached. With this model the frequency and amplitude dependence of ATP synthesis[3] have been explained.

3. Electroconformational Coupling Between Enzymes

Two enzyme mediated processes can be coupled if the enzymes are close enough to interact electrically[4]. Fig. 4 schematically shows two enzymes that are embedded in a membrane. One enzyme is catalyzing the transport of a substrate in the direction of the potential, every time it transports a molecule its dipole moment is "flipping up and down," thus creating an oscillating electric field around itself. In this oscillating electric field the other enzyme can transport its substrate across the membrane against the potential. Each of the two enzymes in Fig. 4 can be in four states, so there are 16 states for the system as a whole. If we assume that the binding or release of substrate is slow compared to the conformational changes, then the mathematics that describes the system can be simplified. The situation would not be different if one of the enzymes or both of them would catalyze a reaction instead of transport of a substrate.

We simulated a system as depicted in Fig. 4 to describe a coporter. For each enzyme we assumed that binding or release of substrate does not involve any movement of charge but that the conformational change (the reorientation of the binding site to the other side of the membrane) involves the movement of two elementary charges across a 50 Å membrane. With the enzymes being 35 Å apart and a value for the dielectric constant of $\varepsilon=5$ this leads to $\Delta E=2kT$ between parallel and antiparallel dipoles. Fig. 5 was obtained through simulation and shows how the coporter is maximally efficient when the two enzymes are at a distance of 40 Å. For distances that are too small the two enzymes remain locked in the antiparallel conformation and no coupled transport is possible. For larger distances the coupling disappears.

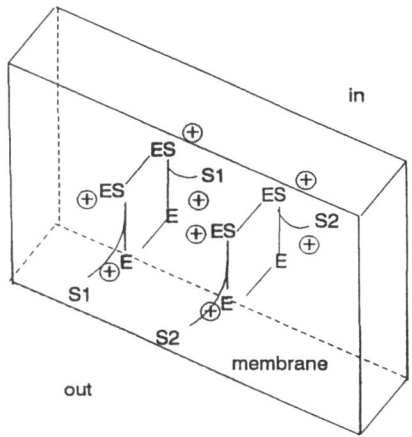

Figure 4. Two substrate transporting enzymes in a membrane. The vertical transitions stand for the binding and release of substrate. The horizontal transitions involve a 180 degree reorientation of the binding site.

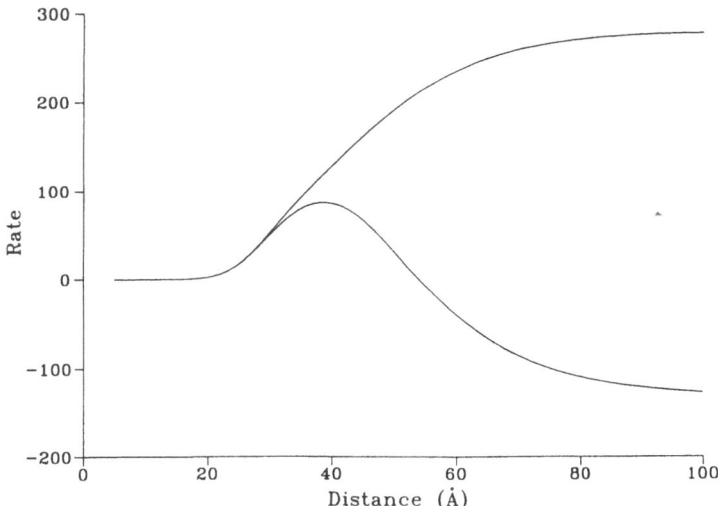

Figure 5. Transport rates for both substrates as a function of the distance between the two enzymes.

4. Conclusions

We have discussed the interaction between a transmembrane electric field and a membrane protein and also the possible electrostatic origin of the coupling of enzyme functions. If the catalytic cycle of an enzyme contains some displacement of charge, then the application of an external oscillating electric field can result in the enzyme transducing energy from the field oscillation to do chemical or transport work. This idea allows for an understanding of how two unrelated processes can be coupled when the catalyzing enzymes are close enough to interact electrically.

References

1. D. S. Liu, R. D. Astumian and T. Y. Tsong, Activation of Na^+ and K^+ pumping modes of (Na,K)-ATPase by an oscillatory electric field, *J. Biol. Chem.* **265**:7260-7267 (1990).
2. A. Graziana, R. Ranjeva and J. Teissie, External electric fields stimulate the electrogenic calcium/sodium ex-change in plant protoplasts, *Biochemistry* **29**:8313-8318 (1990).
3. B. Robertson and R. D. Astumian, Interpretation of the effect of an oscillatory electric field on membrane enzymes, *Biochemistry* **31**:138-141 (1992).
4. F. Kamp, R. D. Astumian and H. V. Westerhoff, Coupling of vectorial proton flow to a biochemical reaction by local electric interaction, *Proc. Natl. Acad. Sci. USA* **85**:3792-3796 (1988).

CHARACTERIZATION OF ENERGY-DEPENDENT PHOSPHATE UPTAKE BY BLUE-GREEN ALGA *ANACYSTIS NIDULANS*: ADAPTATION TO EXTERNAL PHOSPHATE LEVELS OBEYS A LINEAR FLOW-FORCE RELATION

Gernot Falkner, Ferdinand Wagner and Renate Falkner

Institute of Molecular Biology
Austrian Academy of Sciences
5020 Salzburg, Billrothstrasse 11, Austria

The incorporation of phosphate by the blue-green alga *Anacystis nidulans* proceeds against the potential difference between the external phosphate concentration $[P_e]$ and the internal polyphosphate pool, i.e. $2.3RT\log(K'[P_e])$ ($K'=10^{-5}$ M^{-1} represents the equilibrium constant of polyphosphate formation under the external conditions). The energy $2.3RT\Delta pH$ that drives this process originates, in the light, from photophosphorylation (ΔpH is the pH-gradient across the thylakoid membrane). Near equilibrium the uptake rate can be assumed to be proportional to the overall affinity of polyphosphate formation[1],

$$J_p = L_p\left[\log\left(K'[P_e]\right) + q^2 n_p \Delta pH\right] .$$ (1)

Accordingly, J_p is linearly dependent on $\log[P_e]$. In expression (1), the term $2.3RT$ is included in the proportionality factor L_p. This factor represents a conductivity coefficient that is proportional to the activity of the uptake system. n_p is a stoichiometric coefficient and q is a coefficient that expresses the degree of coupling between the input energy and the potential difference for polyphosphate formation. As can be seen from this relationship, a plot of J_p versus $\log[P_e]$ (Thellier plot[2]) should give a straight line. The slope of this line reflects the conductivity coefficient L_p, and the intercept on the $\log[P_e]$-axis the equilibrium phosphate concentration $[P_e]_A$ where uptake ceases (threshold for uptake). As can be seen from eq. (1), when $J_p = 0$,

$$\log[P_e]_A = -n_p q^2 \Delta pH - \log K' .$$ (2)

Hence eq. (1) can be rewritten in the simple form

Modern Trends in Biothermokinetics, Edited by
S. Schuster *et al.*, Plenum Press, New York, 1993

$$J_p = L_p\left(\log[P_e] - \log[P_e]_A\right) .$$ (3)

According to the principles of nonequilibrium thermodynamics this relationship should be valid only near equilibrium. We found, however, that the concentration range over which this linear relationship was obeyed, was a function of the previous growth history. Growth on batch culture, where the external concentration is much higher than the equilibrium concentration, led to the expression of wide linear regions (see Fig.1). The situation is different under phosphate-limited growth which causes an activation of the uptake system. Under these conditions the external concentration fluctuates near the equilibrium (threshold) value and linearity is confined to a small concentration range (see Fig.1).

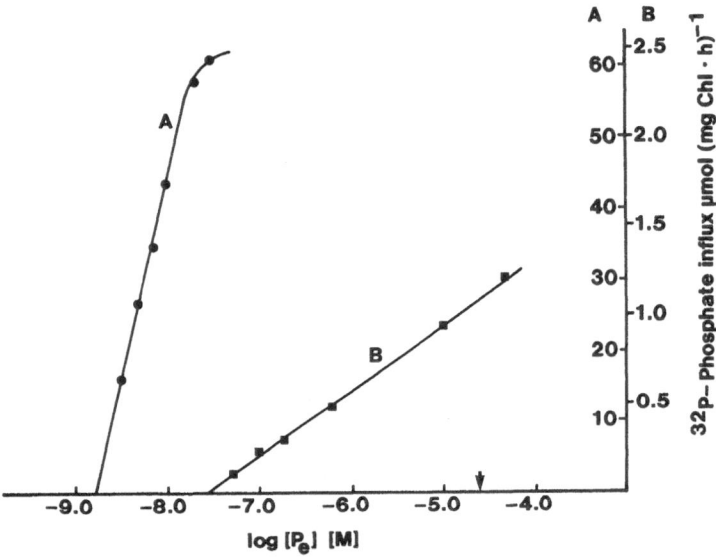

Figure 1. Change in the behavior of phosphate uptake before (A) and after (B) transition from phosphate-limited to unlimited growth, presented by Thellier plots of uptake rates versus the logarithm of the external phosphate concentration. For details see text. The methods employed have been described previously[1].

These findings are most readily explained in terms of a tendency of the uptake system to operate with optimal efficiency[3] at the prevailing growth concentration. This can be formulated as follows. The state of optimal efficiency obeys the function[1,4]

$$\log[P_e]_{opt} = \left(\sqrt{1-q^2} - 1\right)n_p \Delta pH - \log K' .$$ (4)

If uptake proceeds during growth with optimal efficiency at a steady state rate J_p^{ss}, a combination of eqs (2) and (4) with the linear function (3) leads to

$$J_p^{ss} = -L_p\left(\frac{\log\left(K'[P_e]_{opt}\right)}{n_p \Delta pH} + 1\right)\log\left(K'[P_e]_{opt}\right) .$$ (5)

Hence a transition of the same batch of algae from phosphate-limited growth on a low phosphate concentration to an unlimited growth at a high concentration should be accompanied by a change of the conductivity coefficient as described by eq. (5).

The experiment of Fig.1 shows that this is actually the case. The two lines represent Thellier plots of the concentration dependence of uptake rates. Graph A refers to phosphate-limited culture conditions and graph B to the same algae confronted with a high growth concentration of 25 μM phosphate (arrow on the $\log[P_e]$-axis). For A, growth rate in the chemostat was 0.041 h^{-1} and the total phosphate content of the algae, P_{tot}, was 25.9 μmol/mg chlorophyll; thus $J_p^{ss} = \mu P_{tot} = 1.06$ μmol/(mg chlorophyll h). Hence with eq. (3) an external phosphate concentration of 1.7 nM ($\log[P_e] = -8.77$) can be estimated from the slope (55 μmol/(mg chlorophyll h)) and intercept of Fig.1. This allows us to calculate the "input energy" $n_p\Delta pH$ in eq. (5), and from this, the conductivity coefficient at a growth concentration of 25 μM, yielding a value of 0.377 μmol/(mg chlorophyll h). This value corresponds closely to the experimentally observed conductivity coefficient of 0.365 μmol/(mg chlorophyll h) (cf. Fig.1B), which indicates that eq. (5) adequately describes the adaptation of the uptake system to the change of the external concentration employed in the experiment.

As can be seen from Fig.1, linearity extends over a concentration range of three orders of magnitude, contrary to the expectations of nonequilibrium thermodynamics. However, proportional relations between flows and forces have occasionally shown to be valid in biological systems far from thermodynamic equilibrium[5]. A special type of regulation of the uptake system must be postulated to account for this uptake behavior.

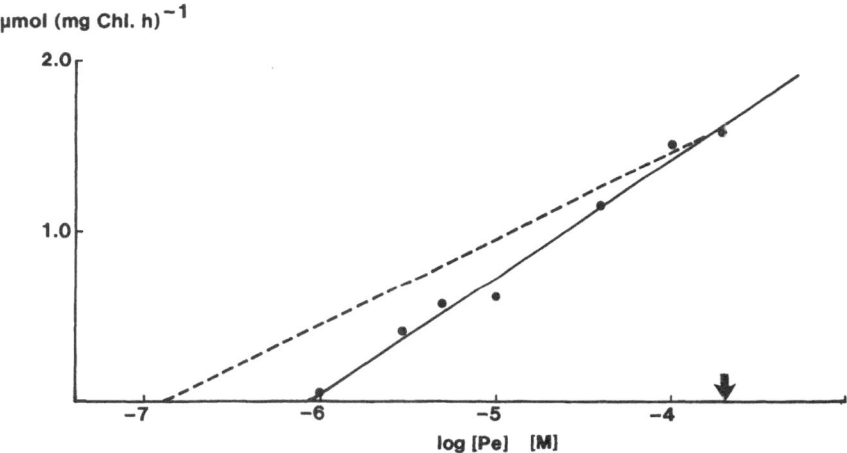

Figure 2. Uptake behavior of algae originating from a batch culture with high phosphate concentration (200 μM). Conditions are as in Fig.1.

When the growth concentration in the same culture was further increased from 25 μM to 200 μM, the uptake behavior developed completely new properties, such that the conductivity coefficient could no longer be calculated from the input energy in the phosphate-limited growth state. As can be seen from Fig.2, the actual equilibrium concentration was higher than the calculated one, indicating that less energy is utilized for incorporation at these high phosphate concentrations. This drop in energy conversion can be explained by an alteration of the H$^+$/ATP stoichiometry of the F-ATPase at the thylakoid membrane. Prolonged cultivation in batch culture changes this stoichiometry from about 4 to 3 and increases the apparent K_M for phosphate in photophosphorylation from 34 to 225 μM.[6] These modifications lead to an increase in the equilibrium concentration of cytoplasmic phosphate, seen as an upward shift of the external equilibrium concentration (threshold value) towards the growth concentration.

These results reveal new features of the adaptation properties of blue-green algae in an environment of fluctuating phosphate concentrations. Thus, a highly efficient phosphorylating machinery is required for incorporation under phosphate-limited growth conditions, resulting in a low equilibrium concentration. During transition to higher growth concentrations this machinery is maintained only as long as the uptake system can be regulated at optimal efficiency, such that linearity between J_p and $\log[P_e]$ extends from the equilibrium concentration to the growth concentration. If, however, the growth concentration rises above the linear range, the whole energy conversion system is reorganized, leading to a lower energy supply for phosphate uptake and in consequence to an increase of the equilibrium concentration.

Acknowledgements

The authors are grateful to Mrs. M. Schmittner for technical assistance and to the Austrian "Fonds zur Förderung der wissenschaftlichen Forschung" for financial support.

References

1. G. Falkner, R. Falkner and A.J. Schwab, Bioenergetic characterization of transient state phosphate uptake by the cyanobacterium *Anacystis nidulans*, *Arch. Microbiol.* **152**:353-361 (1989).
2. M. Thellier, An electrokinetic interpretation of the functioning of biological systems and its application to the study of mineral salts absorption, *Ann. Bot.* **34**:983-1009 (1970).
3. J.W. Stucki, M. Compiani and S.R. Caplan, Efficiency of energy conversion in model biological pumps: optimization by linear nonequilibrium thermodynamic relations, *Biophys. Chem.* **18**:101-109 (1983).
4. J.W. Stucki, The optimal efficiency and the economic degrees of coupling of oxidative phosphorylation, *Eur. J. Biochem.* **109**: 269-283 (1980).
5. H.V. Westerhoff and K. van Dam. "Thermodynamics and Control of Biological Free-energy Transduction," Elsevier, Amsterdam-New York-Oxford (1987).
6. F. Wagner and G. Falkner, Concomitant changes in phosphate uptake and photophosphorylation in the blue-green alga *Anacystis nidulans* during adaptation to phosphate deficiency, *J. Plant Physiol.* **140**:163-167 (1992).

BROWNIAN DYNAMICS SIMULATION OF CHARGED MEMBRANE COMPONENTS: IMPLICATIONS FOR THEIR LATERAL DISTRIBUTION AND VOLTAGE DEPENDENCE OF ION TRANSPORT THROUGH MEMBRANES

Dirk Walther[1], Peter Kuzmin[2] and Edwin Donath[1]

[1]Institut für Biophysik
  Humboldt-Universität zu Berlin
  Invalidenstrasse 42, D-O-1040 Berlin, Germany
[2]Frumkin-Institute of Electrochemistry, Moscow, Russia

1. Introduction

The modern understanding of biological membranes is based on the experimentally proved assumption that their components (proteins and lipids) behave like a two-dimensional fluid and thus are in general able to diffuse over the membrane. However, aggregation of protein molecules and inhomogeneous distribution of other membrane components are often observed (for instance, dense packings of bacteriorhodopsin in the purple membrane of halobacteria). Different physical principles have been held responsible to cause non-random distribution[1]. The objective of the present paper is to assess the contribution of conceivable electric interactions of possible discrete membrane charges carried by membrane proteins or lipids to lateral membrane order. Because electric charges are thought to play an important role in membrane processes it is worth considering the impact of the lateral electric potential on the diffusion of membrane particles.

Secondly, we applied lateral electric interactions of membrane charges to the ion exchange transport across the membrane. This was stimulated by experimental results of a special transport system in human red cells — the band-3 anion transport protein. Although ions are at least partially translocated as net charges, no detectable dependence on the transmembrane potential could be observed[2]. We postulated a simple transport model and analyzed whether lateral electric interaction of neighboring tranport units could explain this independence.

The paper is divided into two sections. We start with the presentation of the results of the Brownian Dynamics Simulations of the lateral distribution of charged membrane particles. Afterwards we refer to the transport characteristics of our model transport protein.

Modern Trends in Biothermokinetics, Edited by
S. Schuster *et al.*, Plenum Press, New York, 1993

2. Theory - Simulation Technique

For our simulations, the underlying electric interaction potential of discrete membrane charges was taken from a recently found analytical solution of the linearized Poisson-Boltzmann Equation. It allows an arbitrary position of a fixed charge in the system {electrolyte solution - membrane - electrolyte solution}. The applicability of this approach is ensured because of the high mobility of the ions in the adjacent solution compared to the mobility of discrete membrane charges. The membrane is considered to be homogeneous with a dielectric constant $\varepsilon_m=3$. In addition the screening effects of the adjacent ionic solution can be considered. The temperature was 310 K. The membrane thickness was set to 8 nm.

3. Brownian Dynamics Simulation of Membrane Charges

3.1. Simulation

Charges with charge numbers α were located on the rotation axis of cylindric membrane particles of radius R and density σ.

Our Brownian Dynamics Simulation takes into account direct deterministic electric interactions between the cylindric particles together with random forces. The parameters of our system allowed the application of the "Poisson-Langevin Equation" introduced by Lax[4] and Zwanzig[5]. It was integrated by using an algorithm proposed by Ermak[7]:

$$\vec{r}(t+\delta t) = \vec{r}(t) + \frac{D}{k_B T}\vec{F}_d(t)\delta t + \delta \vec{r}^G \;, \tag{1}$$

where $\vec{r}$, t and $\vec{F}_d$ are the position vector, time, and deterministic electric force, respectively. $\delta\vec{r}^G$ is a random additive term obtained from a Gaussian distribution with $\langle \delta r^G \rangle = 0$ and variance $\langle \left(\delta r^G_{x,y} \right)^2 \rangle = 2D^o\delta t$ with D^o being the diffusion coefficient of a single freely moving particle.

Starting from a random distribution of particles and after equilibration several quantities of the system were determined. To analyze the lateral order the Radial Distribution Function $g(r)$ was calculated,

$$g(r) = \frac{\left\langle \sum_i \sum_{j \neq i} \delta(r - r_{ij}) \right\rangle}{g_{corr}(r)} \;, \tag{2}$$

where $g_{corr}(r)$ is the correlation function for randomly distributed particles.

The diffusion coefficient and properties of the particle aggregation (the cluster size distribution function $H(N)$) and the probability of aggregation were also calculated.

3.2. Lateral distribution

We simulated particle systems consisting of 250 charged particles. Our simulations revealed that the lateral order was very sensible to the position of the charge with respect to the solution-membrane interface. Charge positions directly on the membrane surface or even slightly above led to no considerable deviations from the randomly distributed system.

However, locating the charges into the membrane caused remarkable lateral correlations. For oppositely charged particles the closest possible charge-charge distance, i.e. the radius of the cylindrical particles, became the crucial parameter. In Fig. 1, the Radial Correlation Function $g(r)$ is shown for *equally charged particles*. A significant lateral correlation was only observable when the charges were placed into the center of the membrane.

In the following we briefly present the results obtained for *oppositely charged particles*. Again charge locations above or on the membrane had only negligibly small effects on the lateral distribution even when a small particle radius ($R=0.5$ nm) was assumed (Fig. 2). This behavior changed when the charges entered the membrane. While even a charge number of $\alpha=1$ was sufficient to lead to a separation into different particle clusters when a radius of

Figure 1. Radial Distriibution Function $g(r')$ for equally charged particle systems with radius $R=1.5$ nm. A) surface charges; 1, $\alpha=1$, $I=100$ mM; 2, $\alpha=3$, $I=100$ mM; 3, $\alpha=1$, $I=1$ mM; 4, $\alpha=5$, $I=100$ mM; 5, $\alpha=3$, $I=100$ mM; 6, $\alpha=5$, $I=1$ mM. B) Charge location in the middle plane of the membrane and different charge numbers (marked at the curves), $D^o=10^{-9}$ cm^2/s, α denotes the lattice constant. of the obtained lattice.

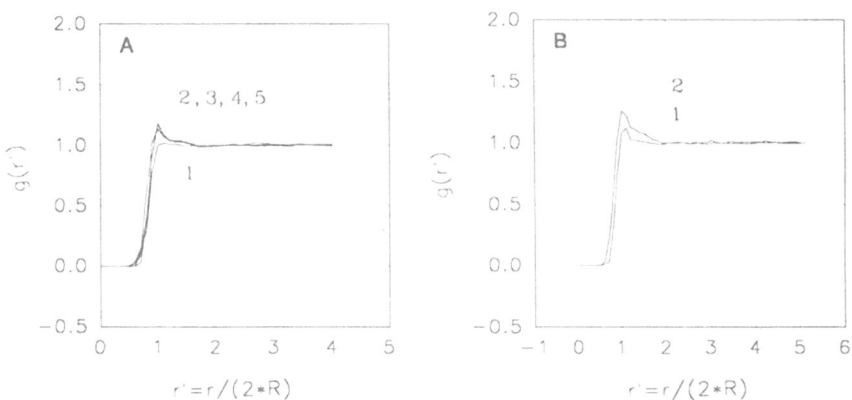

Figure 2. Radial Distribution Function for oppositely charged particles, charge position directly on the membrane-solution interface and different radius. A) parameters: $\sigma=0.02$ nm^{-2}, $D^o=10^{-9}$ cm^2/s, $R=1.5$ nm; 1, $\alpha=1$, $I=100$ mM; 2, $\alpha=1$, $I=1$ mM; 3, $\alpha=1$, $I=0$ mM; 4, $\alpha=2$, $I=100$ mM; 5, $\alpha=3$, $I=100$ mM. B) $R=0.5$ nm, $\alpha=1$, $D^o=10^{-8}$ cm^2/s, $\sigma=0.02$ nm^{-2}; 1, $I=100$ mM, 2, $I=1$mM.

R=0.5 nm was assumed, an increase of the radius to 1.5 nm reduced this effect strongly (snapshots in Fig. 3). Now at least a charge number of 3 was required to lead to stable aggregation. The difference of the lateral distribution for these two radii is shown in Fig. 4. The non-normalized version of $g(r)$ indicates that the lateral correlation for smaller particles is increased by a factor of 2 as compared with the 1.5 nm particles.

The dielectric constant of the membrane was critical for the electric interaction of membrane charges. At ε_m=2, the lateral correlation was expanding up to the second neighbor. Only a slight correlation to the first neighbor was obtained for ε_m=10. (R=1.5 nm). The simulated diffusion coefficient decreased for both repulsive and attractive forces. Only when the electric interaction energy was 10 times higher than the thermal energy, the diffusion coefficient decreased dramatically.

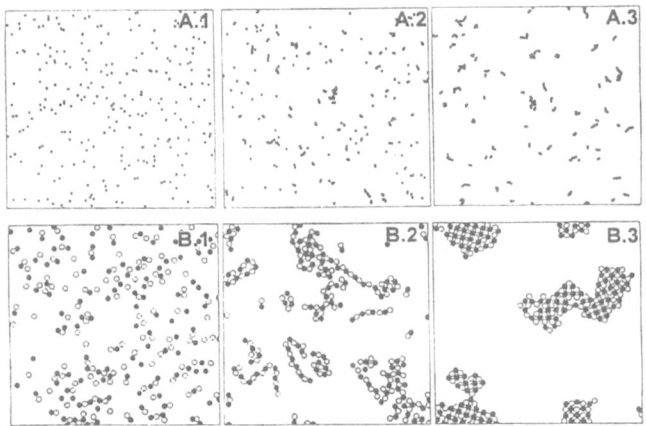

Figure 3. Snapshots of an oppositely charged particle system. Filled and open circles represent positive and negative charges. A) radius R =0.5 nm, I =100 mM, σ = 0.02nm^{-2}, D^0=10^{-8}cm^2/s. A.1) charge directly on the surface; A2) charge 0.5 nm beneath the surface; A.3) charge in the middle plane of the membrane. B) charge posi-tion in the middle plane of the membrane and radius R =1.5 nm; parameters: σ = 0.02nm^{-2}, I =100 mM, D^0=10^{-9}cm^2/s. B.1, B.2, and B.3 correspond to charge numbers α =1,2, and 3, respectively.

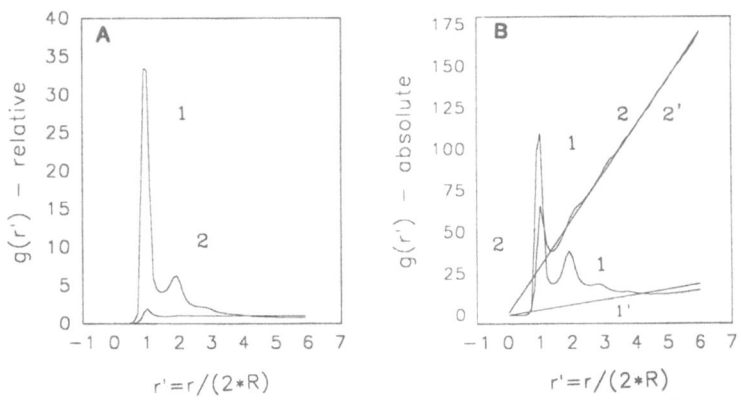

Figure 4. Radial Correlation Function of particle systems with charge position in the middle plane of the membrane (oppositely charged). Parameter: σ =0.02 nm^{-2}; I =100 mM; 1, R =0.5 nm, D^0=10^{-8} cm^2/s; 2, R =1.5 nm, D^0=10^{-9}cm^2/s. A) $g(r')$ with normalization; B) $g(r')$ without normalization; appropriate norm-alization curves are shown by lines 1' and 2', respectively.

We calculated the Born energy representing the main part of free energy if an ion is positioned into a dielectric medium (membrane). It was found that a very high energy barrier of several tens of $k_B T$ had to be overcome. Because this free energy is not drastically changed by aggregation of oppositely charged particles, charge locations within the membrane should only be possible for at least locally increased dielectric constants of the membrane and the simulated structuring might be of secondary importance in real systems.

4. Dependence of the Transport Rate of Ion Exchange Transport on the Transmembrane Potential of Electrostatically Coupled Transport Units

4.1. The model

Transport proteins with a density σ were located in a simulation box with periodic boundaries. The basic assumption was that these transport proteins possess a binding site in the middle plane of the membrane. This binding site carries $\alpha_0 \geq 0$ charges. A transport event is represented by a transition of this binding site from charge state α_0 to a state $\alpha_T < 0$. Also fractional values of the charge states were used to consider that in reality these charges might be partially screened. The transition rate was simulated as follows:

$$k_{\alpha_o \to \alpha_T} = k_o * \exp\left(-\frac{E_{\alpha_T} - E_{\alpha_o}}{k_B T}\right), \tag{3}$$

where k_0 is the transport rate in the absense of the field. The energy difference $E_{\alpha T} - E_{\alpha 0}$ was calculated by superposition of the charge movement in an applied transmembrane potential ($\pm \alpha_T \Phi_M / 2$) and the electric field caused by neighboring transport proteins. Once the charge state α_T is occupied, it is maintained for a certain time. In our simulation this transport time varied from the time it takes an ion to diffuse over the distance of the membrane thickness to the largest possible time corresponding to k_0. Afterwards the charge state α_0 was taken again. The times of binding and release of the transported ions were neglected. The binding site was only accessible sequentially. So the transport was electrically silent. The transport units were allowed to move according to their Brownian dynamics.

4.2. Results

The general results of the simulations of the model transport system was that the included lateral charge interactions reduced the dependence of the transport exchange rate on the transmembrane potential, as shown in Fig. 5. The effect was increased either by increasing the lateral electric energy by reducing the dielectric constant of the membrane, increasing the density of transport units or increasing the transport time. Simultaneously the absolute transport rate was influenced.

The reduced dependence of the exchange transport on the transmembrane potential was explained as follows. When a charged binding site in the middle plane of the membrane was assumed, the expected overall reduction of the transport rate due to the transmembrane potential would increase the lateral energy, due to the increase of the number of free charged binding sites. Therefore lateral interaction forces the system into a state where the occupation of binding sites with charges becomes higher. This compares to a situation where the distribution of binding sites on both surfaces of the membrane is more symmetric. The system approaches a situation where the transmembrane potential effect is reduced.

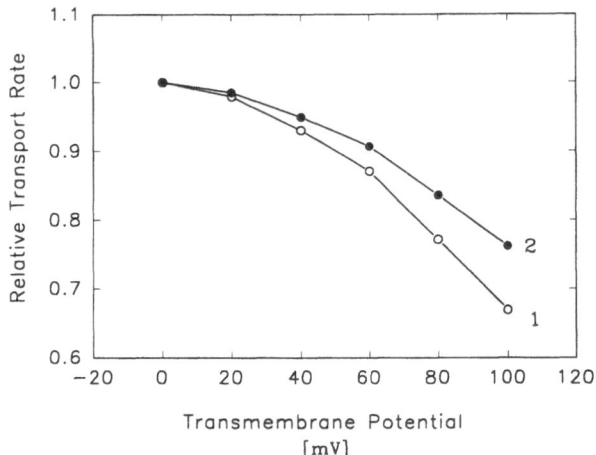

Figure 5. Dependence of the relative transport rate (transport rate at transmembrane potential $\Delta \Phi = x$ mV divided by the transport rate at $\Delta \Phi = 0$ mV) as a function of the transmembrane potential. Parameters: $\alpha_0 = +0.5$, $\alpha_T = -0.5$, diffusion coefficient of transport protein $D^0 = 10^{-9} cm^2/s$, $\sigma = 0.02$ nm^{-2}, $I = 100$ mM. 1, without lateral interactions (calculated analytically); 2, with lateral interactions of the transport units, transport time results from the one-dimensional diffusion of the transported ion with $D = 10^{-6} cm^2/s$ across the membrane.

This behavior could have consequences for the explanation of the transport characteristics of different real transport systems.

References

1. J.R. Abney and J.C. Owicki, Theories of protein-lipid and protein-protein interactions in membranes, *in*: "Progress in Protein-Lipid Interactions," A. Watts and J.J. DePont, eds, Elsevier, Amsterdam (1985).
2. M.J. Jennings, R.K. Schulz and M. Allen, Effects of membrane potential on electrically silent transport, *J. Gen. Physiol.* **15**:163-193 (1990).
3. V.B. Arakeljan, D. Walther and E. Donath, Electric potential distribution around discrete charges in a dielectric membrane - electrolyte solution system, *Coll. Polym. Sci.,* in press.
4. M. Lax, Classical noise. IV. Langevin methods, *Rev. Mol. Phys.* **38**:541 -566 (1966).
5. R. Zwanzig, Langevin theory of polymer dynamics in dilute solution, *Adv. Chem. Phys.* **15**:325-331 (1969).
6. D.L. Ermak, A computer simulation of charged particles in solution. I. Technique and equilibration properties, *J. Chem. Phys.* **62**:4189-4196 (1975).

EVOLUTIONARY STRATEGIES AND THE OPTIMIZATION OF CARRIER MODELS

J. Lenz and M. Höfer

Laboratory for Bioenergetics
Institute of Botany
The University of Bonn
Kirschallee 1, 5300 Bonn 1, Germany

1. Introduction

A very interesting question in the study of molecular systems concerns the principles of their design. Since the overall evolutionary fitness of an organism is, to a varying degree, a function of all kinds of catalytic proteins, selection pressure acts not only at the systemic level but also at the level of individual proteins. Consequently, present-day catalytic systems may be optimized with respect to one or several qualities. One approach to investigate such a kind of natural selection is to apply mathematical optimizing techniques to appropriate models of the systems under investigation[1-4]. In addition to our previous work[5] we tested different algorithms based on the theory of evolutionary strategies[6] with respect to the optimization of special qualities of the kinetic models for 2 particular mechanisms of a two-substrate symport, and for an active carrier mediated transport driven by dynamic asymmetry.

2. The Optimizing Procedure

Optimizing algorithms based on evolutionary strategies (ES) make use of principles of biological evolution. An algorithm of the (1,N)-ES type, for example, works as follows (Fig. 1): First, a quality Q is defined as a function of special parameters. Each step (generation) of the algorithm starts with a so-called parental set of parameters. N descendant sets are generated by altering the parameter values of the parental set randomly (mutation) by small amounts (mutation step sizes). Then, the quality Q is calculated for each descendant (expression of the phenotypes). The descendant set which corresponds to the best value of Q is selected as the parental set of the next generation (selection). We used also strategies of the (1+N)-ES type, which incorporate the parental set into the selection procedure. Introduction of variable mutation step sizes increased the effectiveness of the algorithms. We either defined the step sizes as a function of the generation number or optimized them in parallel to the parameters. After certain numbers of generations instabilities were induced by a sudden increase of the mutation step sizes or by allowing for descendants with worse Q-values to be selected. This made it possible to leave the optima already attained and thereby prevented the algorithms from being captured in suboptima.

3. Results

Fig. 2 shows the model of an A/B-symport with ordered binding. We assumed stationary conditions and the binding reactions to be in equilibrium. We maximized the unidirectional flux of A per symporter molecule J_A (with $A'=B'=1$, $A''=B''=0$) (Fig. 2 a) and

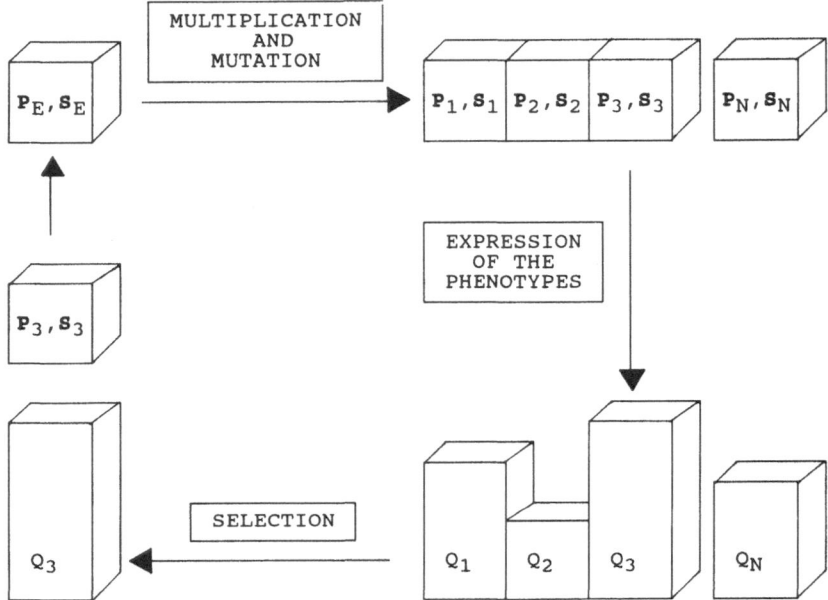

Figure 1. Schematic representation of a (1,N)-ES. **P** and **S** denote the set of parameters and mutation step sizes, respectively. Q is the quality function.

the accumulation ratio $R_A=A''/A'$ (with $B'=10$ and $B''=1$) (Fig. 2 b) as functions of the rate constants and of the dissociation constants. Two dissociation constants were calculated using the Wegscheider relations of the system. All parameter values were initially set to unity and their range of change restricted to the interval I=[1,10].

Fig. 3 shows J_A (a) and the mean value S (b) of the mutation step sizes of the individual parameters as functions of the number of generations. In this example, a (1,5)-ES with optimized mutation step sizes and three destabilizations was used. Initially, the mutation step sizes were set to unity. During each phase the algorithm selects small mutation step sizes. Algorithms with high values for N and those which do not incorporate the parental parameter set into the selection process turned out to be more effective. In this case, it was frequently observed that the selection towards small mutation step sizes got lost.

J_A reached a maximum for the dissociation constants approaching the lower limit, and the rate constants in both directions approaching the upper limit (Fig. 2 a). R_A reached a maximum for the rate constants of the transitions CA' $\leftrightarrow$ CA" and the dissociation

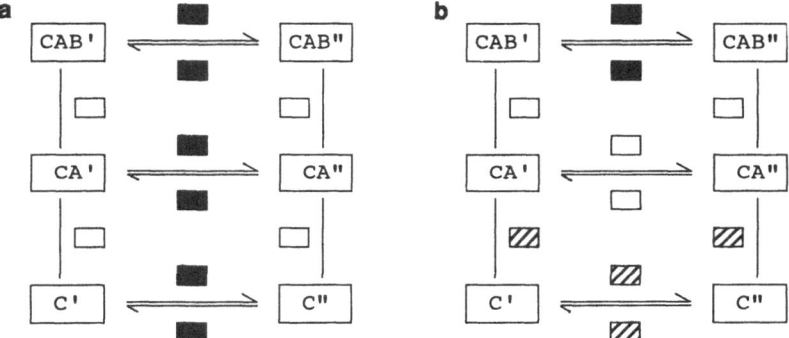

Figure 2. Optimization of the unidirectional flux per symporter molecule J_A (a) and the accumulation ratio R_A (b) for an A/B-symport with ordered binding. ' and " denote the two sides of a membrane, C the carrier protein and A and B the substrates. The rectangles show the optimal sets of the dissociation constants and the rate constants (black = upper limit, white = lower limit, hatched = no influence).

constants of the upper cycle approaching the lower limit and the rate constants of the transitions CAB' $\leftrightarrow$ CAB" approaching the upper limit (Fig. 2 b). The remaining parameters had no influence on R_A as indicated by their high standard deviations calculated from repeated optimizing procedures. It can easily be shown that there are two equations for the accumulation ratio each of which contains a different set of two rate constants and one dissociation constant of the upper cycle.

We obtained identical results on maximizing J_A and R_A for an A/B-symport with random binding, as well. If the net flux of A was maximized, the set of optimal parameters for both types of binding was dependent on the values of substrate concentrations.

Under the given conditions, all solutions were symmetric with respect to the two membrane sides; there is no need for any asymmetry in carrier conformation to obtain high accumulation ratios, even in the presence of a leak for substrate A. The leak must of course be small to yield a high accumulation ratio. It must be large to yield a high unidirectional flux.

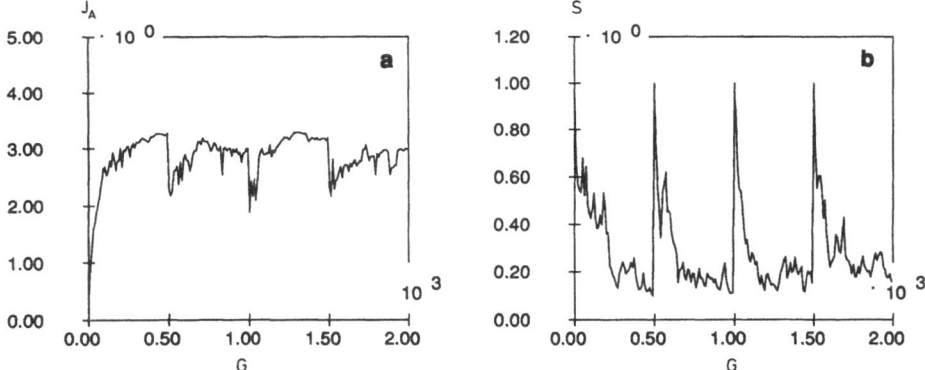

Figure 3. Optimization of the unidirectional flux J_A of A for the model of an A/B-symport with ordered binding using a [1,5]-ES with variable mutation step sizes and three destabilizations. The unidirectional flux J_A (a) and the mean value S of the mutation step sizes (b) are shown as functions of the number of generations.

Fig. 4 shows the model of a carrier for one substrate A, which also binds a non-transported ligand L on one membrane side. This system is capable of accumulating substrate A if the concentration of L starts to oscillate or fluctuate[7]. We maximized the net flux J_A (with $A'=A''=1$ M) as a function of the rate constants for different types of fluctuations of L (between 0 and 2 M) with a constant mean frequency of 1 s^{-1}. The rate constants were again initially set to unity, and their changes restricted to the interval I=[1,10]. One of the rate constants was calculated using the Wegscheider relation. Stationary conditions were not assumed. J_A was calculated by numerical integration of the set of differential equations describing the system.

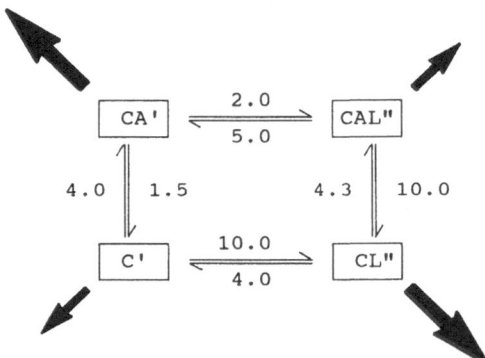

Figure 4. Optimization of an active transport of the substrate A due to fluctuation of the concentration of a ligand L. ' and " denote the two sides of a membrane. The numbers are the values of the rate constants of the optimal set. Thick arrows represent the stationary distribution of the corresponding carrier states.

The highest flux was observed for a dichotomous fluctuation of L. From the optimal set of rate constants we calculated the equilibrium distribution of the individual carrier states. CA' and CL'' were on the same high level, and C' and CAL'' on the same low level. The distribution was symmetric with respect to the diagonals of the kinetic scheme, but it did not represent the largest possible difference between the pairs CA', CL'' and C', CAL''. This may reflect the appearance of resonance-like phenomena.

4. Conclusion

The application of evolutionary strategies seems to be an efficient tool in optimizing kinetic models of catalytic systems. Due to the limited extent of investigations that we have done so far, the results are still preliminary. However, the appearance of symmetric solutions may have some meaning concerning the design of transport systems. We plan to extend our investigations to the optimization of other qualities and other transport systems under different substrate concentrations. Furthermore it will be interesting to have a look at the corresponding stationary concentrations of the individual carrier protein states.

Our results indicate that evolutionary strategies should be especially suited for the investigation of multiple optimal states of very complex dynamic systems containing several transport systems and for gaining insight into coevolution of coupled membrane transport processes.

References

1. B. Honig and W. D. Stein, Design principles for active transport systems, *J. theor. Biol.* **75**:299-305 (1978).
2. J. J. Burbaum, R. T. Raines, W. J. Albery and J. R. Knowles, Evolutionary optimization of the catalytic effectiveness of an enzyme, *Biochemistry* **28**:9293-9305 (1989).
3. R. Heinrich and E. Hoffmann, Kinetic parameters of enzymatic reactions in states of maximal activity; an evolutionary approach, *J. theor. Biol.* **151**:249-283 (1991).
4. R. Heinrich, S. Schuster and H.-G. Holzhütter, Mathematical analysis of enzymic reaction systems using optimization principles, *Eur. J. Biochem.* **201**:1-21 (1991).
5. J. Lenz and M. Höfer, Optimization of membrane transport models using evolutionary strategies, *in*: "Workshop on Yeast Transport and Energetics," Fundación Juan March, Madrid (1991).
6. I. Rechenberg. "Evolutionsstrategie," Friedrich Frommann, Stuttgart-Bad Cannstatt (1973).
7. Y.-d. Chen, Free energy transduction in membrane transport systems induced by externally imposed fluctuations in ligand concentrations, *Biochim. Biophys. Acta* **902**: 307-316 (1987).

II. MODELING OF CELL PROCESSES WITH APPLICATIONS TO BIOTECHNOLOGY AND MEDICINE

MATHEMATICAL MODEL
OF THE CREATINE KINASE REACTION
COUPLED TO ADENINE NUCLEOTIDE TRANSLOCATION
AND OXIDATIVE PHOSPHORYLATION

Mayis K. Aliev and Valdur A. Saks

Laboratory of Experimental Cardiac Pathology
and Laboratory of Bioenergetics
Cardiology Research Center
3rd Cherepkovskaya Str. 15A
121552 Moscow, Russia

1. Introduction

Detailed kinetic analysis of the creatine kinase (CK) reaction catalyzed by soluble enzymes has been given many years ago[1-3]. However, it has already been known since 1975 that the behavior of CK in mitochondria under conditions of oxidative phosphorylation is not governed by substrate concentration in the medium and soluble enzyme kinetics, but even the direction of the CK reaction may be different depending on the oxidative phosphorylation, which very significantly increases the rate of phosphocreatine production and decreases the rate of the reverse reaction of ATP formation[3].

The control of the mitochondrial CK reaction exerted by oxidative phosphorylation has been shown both kinetically[3-5] and thermodynamically[3,6]. All accumulated evidence was taken to show the functional coupling between mitochondrial CK and adenine nucleotide translocase: translocase directs ATP molecules directly to the active site of CK and simultaneously removes the reaction product, ADP[3-10]. Such close an interaction is based on the close proximity of the enzymes: creatine kinase is bound to the cardiolipin molecules closely surrounding adenine nucleotide translocase in the mitochondrial inner membrane, thus forming a transport protein-enzyme complex[11,12]. However, the functioning of this complex has not yet been quantitatively described in sufficient detail: only a short and rather general model has been published[13] in 1974.

It is the aim of this work to develop a mathematical model quantitatively describing the process of aerobic phosphocreatine synthesis coupled to oxidative phosphorylation in cardiac mitochondria.

Modern Trends in Biothermokinetics, Edited by
S. Schuster *et al.*, Plenum Press, New York, 1993

2. The Model

2.1. General considerations

The model is based on a probability approach to the description of the enzyme-enzyme interaction. In the model the mitochondria, which respire in a medium containing ATP, Cr and PCr, but not ADP, are considered as a three-component system (see Fig. 1), including:

1) oxidative phosphorylation (OP) reactions, which provide a stable ATP concentration in the mitochondrial matrix[14];
2) adenine nucleotide translocase (ANT), performs the exchange of matrix ATP for outside CK-supplied ADP when both substrates are simultaneously bound to translocase[15]; and
3) CK (shaded areas in Fig. 1), starts these reactions when activated by the substrates from the medium.

The specific feature of this system is a close proximity of CK and translocase molecules[11,12]. This results in a high probability of direct activation of translocase by CK-

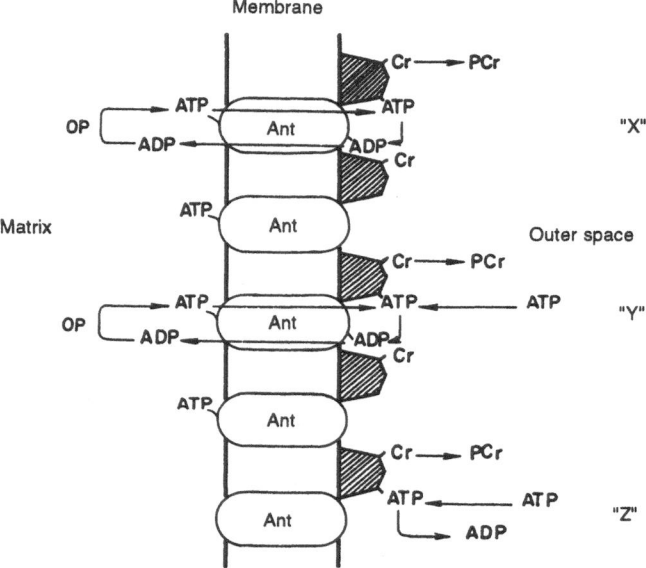

Figure 1. Schematic representation of three possible combinations of sources of ATP for mitochondrial phosphocreatine production. For further explanations see text.

derived ADP without its leak into the medium (concept of direct channeling, pathways "X" and "Y" in Fig. 1). In turn, the activated translocase directly provides, with the same high probability, creatine kinase with matrix-derived ATP. This "local" ATP (A_{loc}), when forming effective $E.A_{loc}.Cr$ complexes with CK (pathway "X" in Fig. 1), provides sustained PCr production from mitochondrial ATP. When CK cannot accept or retain locally supplied ATP molecules, they diffuse out into the solution (pathway "Y" in Fig. 1). In the pathway "Z", PCr is exclusively produced from ATP in the medium and the ADP formed is released into the medium, so this pathway is solely governed by substrate concentrations in the medium and the well defined soluble CK kinetics[2,3,13,16,17]. The enzyme functioning in states "X", "Y" and "Z" has been accounted for by the equations given in the following section.

2.2. CK activation by medium ATP in pathways "Z" and "Y"

The creatine kinase reaction mechanism is of rapid-equilibrium random binding BiBi type, according to Cleland's classification[2,6,17]. According to this mechanism the reaction rate in the forward direction is determined by interconversion of the central ternary complex E.Cr.A (enzyme-creatine-ATP) into the enzyme-product complex E.D.PCr (enzyme-ADP-phosphocreatine), with the rate constant k_{k1}. The equations for this reaction are given in several earlier papers[2,3,13]. In equilibrium or steady state the distribution of the enzyme between enzyme-substrate complexes and the free state can be expressed in terms of probablities:

$$P(E) = \frac{E}{E_{tot}} = \frac{1}{1 + \dfrac{Cr}{K_{icr}} + \dfrac{A}{K_{ia}} + \dfrac{A \cdot Cr}{K_{ia} \cdot K_{cr}} + \dfrac{PCr}{K_{icp}} + \dfrac{PCr \cdot A}{K_{icp} \cdot K_{Ia}}} = \frac{1}{Den} , \tag{1}$$

$$P(E.A) = \frac{EA}{E_{tot}} = \frac{A}{K_{ia} \cdot Den} , \tag{2}$$

$$P(E.Cr) = \frac{ECr}{E_{tot}} = \frac{Cr}{K_{icr} \cdot Den} , \tag{3}$$

$$P(E.A.Cr) = \frac{EACr}{E_{tot}} = \frac{A \cdot Cr}{K_{ia} \cdot K_{cr} \cdot Den} , \tag{4}$$

$$P(E.PCr) = \frac{EPCr}{E_{tot}} = \frac{PCr}{K_{icp} \cdot Den} , \tag{5}$$

$$P(E.PCr.A) = \frac{EPCrA}{E_{tot}} = \frac{PCr \cdot A}{K_{icp} \cdot K_{Ia} \cdot Den} . \tag{6}$$

In these equations, P designates the probability for the enzyme to subsist in the free state, E, or in the indicated enzyme-substrate complexes; E_{tot} denotes the total concentration of enzyme; K_{ia}, K_{icr}, and K_{icp} stand for the dissociation constants of ATP, Cr, and PCr, respectively, with respect to the primary enzyme-substrate complexes. K_{Ia} and K_{cr} are the dissociation constants of ATP and Cr, respectively, with respect to ternary complexes; and *Den* is an abbreviation for "denominator". The concentrations of substances are represented by italic symbols (e.g. *Cr* for the Cr concentration).

The rate of product (ADP and PCr) formation in the forward reaction is given by the following equation:

$$V_1 = E_{tot} \cdot P(E.A.Cr) \cdot k_{k1} . \tag{7}$$

2.3 The probability of translocase activation

The probability of translocase (T) activation, $P(T)_{ef}$, by CK-derived ADP in pathways "X", "Y" is expressed by the following equation:

$$P(T)_{ef} = P(CK)_{ef} \cdot P(T_{Dloc})_{out} \cdot P(T \cdot A)_{in} , \tag{8}$$

where $P(CK)_{ef}$ is the probability for CK to form catalytically effective complexes E.A.Cr to produce ADP when activated in pathways "X", "Y" and "Z"; $P(T_{Dloc})_{out}$ is the probability for local ADP to meet with T at the outer surface of the mitochondrial inner membrane; $P(T.A)_{in}$ is the probability for T to bind ATP on the inner, matrix side of the membrane.

2.4. The probability of CK activation

The probabilities of activation of CK by translocase-supplied mitochondrial ATP in the pathway "X" are expressed by the following equations:

$$P_1 = P(T)_{ef} \cdot P(CK_{Aloc}) \cdot P(E.Cr) , \tag{9}$$

$$P_2 = P(T)_{ef} \cdot P(CK_{Aloc}) \cdot P(E) \cdot Pc_1 , \tag{10}$$

$$Pc_1 = \frac{Cr \cdot K_{Cr+}}{Cr \cdot K_{Cr+} + k_{1-}} , \tag{10'}$$

$$P_3 = P(T)_{ef} \cdot P(CK_{Aloc}) \cdot P(E.PCr) \cdot Pc_1 \cdot Pc_2 , \tag{11}$$

$$Pc_2 = \frac{k_{4-}}{k_{4-} + k_{3-}} , \tag{11'}$$

where P_1, P_2, and P_3 designate the probabilities of E.A_{loc}.Cr complex formation from the complex E.Cr, free enzyme E and complex E.PCr, respectively; $P(CK_{Aloc})$ designates the probability for T-supplied ATP to meet with CK; Pc_1 (partitioning coefficient[18]) is the probability of transformation of the complex CK.A_{loc} to the complex E.A_{loc}.Cr; Pc_2 denotes the probability of transformation of the complex CK.PCr.A_{loc} to the complex CK.A_{loc}; K_{Cr+} is the diffusion-limited association rate constant for Cr; k_{1-} is the rate constant for ATP dissociation from the CK.A_{loc} complex; k_{3-} and k_{4-} are the rate constants for ATP and PCr dissociation, respectively, from the E.A_{loc}.PCr complex. A detailed description of these equations can be found in another paper[19].

Taken together, these probabilities were arranged in the final equations for the total probabilities of CK (eq. 12) and translocase (eq. 13) activation in the coupled reactions:

$$P(CK)_{ef} = P_1 + P_2 + P_3 + P(E.A.Cr)$$

$$= P(T)_{ef} \cdot P(CK_{Aloc}) \cdot \left\{ P(E.Cr) + \left[P(E) + P(E.PCr) \cdot P_{c2} \right] \cdot P_{c1} \right\} + P(E.A.Cr) , \tag{12}$$

$$P(T)_{ef} = \frac{P(E.A.Cr)}{\dfrac{1}{P(T_{Dloc})_{out} \cdot P(T.A)_{in}} - P(CK_{Aloc}) \cdot \left\{ P(E.Cr) + \left[P(E) + P(E.PCr) \cdot P_{c2} \right] \cdot P_{c1} \right\}} . \tag{13}$$

The expressions for the steady-state rates of functioning of translocase (eq. 14) and PCr production by mitochondrial CK (eq. 15) are

$$V_t = N \cdot k_{k1} \cdot P(T)_{ef} , \tag{14}$$

$$V_{PCr} = N \cdot k_{k1} \cdot P(CK)_{ef} , \tag{15}$$

where N denotes the content of CK in mitochondria[20].

2.5. Choice of the parameters for modeling

For the mitochondrial creatine kinase reaction in deenergized mitochondria all the data were taken from a paper by Jacobus and Saks[4]. They are indicated in Table 1, where $V_{1,max} = N \cdot k_{k1}$. $K_{Ia} = 11.25$ mM was calculated from the thermodynamic equation $K_{icp} \cdot K_{Ia} = K_{ia} \cdot K_{Icp}$.

$P(T_{Dloc})_{out}$ and $P(CK_{Aloc})$ were taken to be equal to 1.0 (concept of direct channeling). $P(T.A)_{in} = 0.9$ was found by a method of best approximation to the experimental data[19]. k_{1-} was calculated to be $1.5 \cdot 10^4$ s^{-1}, $K_{Cr+} = 4 \cdot 10^7$ M^{-1}s^{-1}, $k_{3-} = 2.25 \cdot 10^5$ s^{-1}, $k_{4-} = 9.6 \cdot 10^5$ s^{-1}.[19]

Table 1. Experimental[4] and simulated kinetic constants for rat heart mitochondrial creatine kinase as a function of oxidative phosphorylation (OP)[¶].

Elemental steps of CK reaction	Constants	Without OP		With OP	
		Experimental	Used in model	Experimental	Simulated
E.MgATP $\longleftrightarrow$ E + MgATP	K_{ia}, mM	0.75±0.06	0.75	0.29±0.04	0.15
E.Cr.MgATP $\longleftrightarrow$ E.Cr + MgATP	K_a, mM	0.15±0.01	0.15	0.014±0.005	0.015
E.Cr $\longleftrightarrow$ E + Cr	K_{icr}, mM	28.8±8.45	26.0	29.4±12.0	- -
E.MgATP.Cr $\longleftrightarrow$ E.MgATP + Cr	K_{cr}, mM	5.20±0.30	5.20	5.20±2.30	5.20
E.PCr $\longleftrightarrow$ E + PCr	K_{icp}, mM	1.60±0.20	1.60	1.40±0.20	0.91
E.MgATP.PCr $\longleftrightarrow$ E.MgATP + PCr	K_{Icp}, mM	20-50	24.0	20-50	24.0
	$V_{1,max}$, µmol ATP·min^{-1}·mg^{-1}	1.06±0.20	1.00	0.99±0.07	0.90

¶Oxygraph measurements were conducted at 30 °C in a medium containing 0.25 M sucrose, 10 mM HEPES, pH 7.4, 0.2 mM EDTA, 5 mM K$^+$ phosphate, 5 mM K$^+$ glutamate, 2 mM K$^+$ malate, 3.3 mM Mg(OAc)$_2$, 0.3 mM dithiothreitol, 1.0 mg/ml bovine serum albumin, 0.5-0.1 mg/ml of the rat heart mitochondrial protein. In respiring mitochondria the ATP/ADP translocation rates were calculated from oxygen consumption rates. In mitochondria, the respiration of which was completely inhibited by pretreatment with 10 µM rotenone and 5 µg/mg oligomycin, the creatine kinase reaction rates were determined spectrophotometrically, using a phosphoenolpyruvate (2 mM) - pyruvate kinase (2 IU/ml) system to regenerate exogenous ATP. Mean values and SD are given for five experiments[4]. Simulation parameters are indicated in Section 2.5.

3. Results

In cooperation with Dr. William Jacobus we systematically analyzed, in 1982, the mitochondrial creatine kinase reaction in the forward direction both in the absence of mitochondrial respiration and under conditions of coupling of this reaction to mitochondrial oxidative phosphorylation[4]. These data are especially well-suited for mathematical modeling since they give a quantitative description of the system in a wide range of the regimes of functioning. We have simulated all the experimental data of Jacobus and Saks[4], and in Table 1 the final results of the simulation (right column of the table) are compared with experimental data (left column).

The most interesting and important feature of the data in Table 1 is the theoretical confirmation that in the case of mitochondrial CK reaction coupled to oxidative phosphorylation the K_m value for MgATP (K_a) is decreased about ten times, while the apparent affinity of the E.MgATP complex to creatine ($1/K_{Cr}$) is not changed. This shows an apparent specific increase of the affinity of the system to MgATP. At the same time, if oxidative phosphorylation is operative, the calculated value of the apparent K_{ia}, i.e. the dissociation constant of MgATP from the E.MgATP complex, is lower (0.15 mM) than the corresponding K_{ia} from experiment (0.29 mM). The model describes the tendency of lowering this constant by oxidative phosphorylation[4] and confirms that the alteration of this constant by oxidative phosphorylation is much lower than the decrease in K_a under these conditions.

It is interesting to note that K_{icr}, i.e. the dissociation constant of Cr from the E.Cr complex, can, in the model, not be determined from the secondary slope plots if oxidative phosphorylation is operative (cf. Fig. 3B in Ref. 4), because the simulated points, although close to experimental ones, lie on the curved line[19]. The experimental points are extremely scattered in this case, 29.4±12.0 mM (Table 1). The dissociation constant of PCr from the dead-end complex E.MgATP.PCr, K_{Icp}, does not change if oxidative phosphorylation is operative, both in experiment and modeling. The small experimental decrease in the dissociation constant of PCr from the complex E.PCr, K_{icp}, under conditions of oxidative phosphorylation, is more prominent in the model (Table 1). As a whole, the present model may be regarded as satisfactory.

Fig. 2 presents the simulation data concerning the Cr dependence of the rates of PCr production (upper curve) and ATP/ADP translocation (lower curve) under conditions that approximate those *in vivo*: high MgATP concentrations (10 mM) and 20 mM Cr+PCr. Simulation parameters are those used for describing the data of Jacobus and Saks[4], and the MgADP concentration is taken to be zero. It follows from the data of Fig. 2 that coupled mitochondrial creatine kinase is able to exert strong control of oxidative phosphorylation under conditions that are close to those *in vivo*. The rates of ADP and PCr production by CK from medium ATP (pathway "Z") under these conditions is relatively low, as may be judged from the difference in the upper and lower curves.

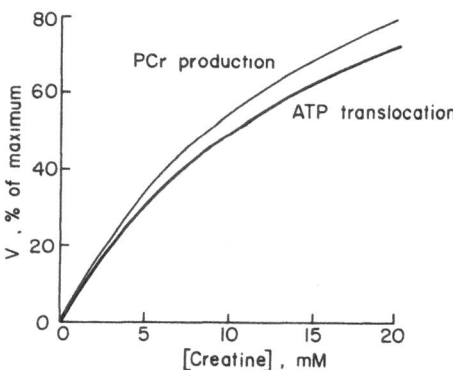

Figure 2. The modeling of the sole CK-control of mitochondrial ATP production (thick line) in a medium without added MgADP in the presence of 10 mM MgATP. *Cr + PCr* = const. = 20 mM. The thin line indicates the rate of PCr production. Modeling parameters as in Table 1.

4. Discussion

In this work we describe for the first time a quantitative model of the mitochondrial creatine kinase reaction coupled to the oxidative phosphorylation system in heart mitochondria. This model is based on a probability approach, and in addition to the conventional kinetic equations, it includes a description of the ATP transfer from the matrix by the adenine nucleotide translocase and its direct channeling to the active site of mitochondrial creatine kinase (CK_{mit}), which is located on the other side of the inner membrane.

The central problems of this approach are several assumptions concerning the values of probability factors. We have assumed that the probability of the binding of ATP translocated across the membrane by translocase is equal to one. That means that we have connected our model strictly to the concept of direct channeling of adenine nucleotides between creatine kinase and translocase. This is clearly different from the concept of dynamic compartmentation of adenine nucleotides in the intermembrane space due to the proposed impermeability of the outer membrane to adenine nucleotides[21].

Our proposition of the direct channeling is based on the experimental kinetic, thermodynamic and radioisotopic data[3-10], which shows that the phenomenon of the functional coupling between the creatine kinase reaction and oxidative phosphorylation is perfectly preserved in mitoplasts with destroyed outer membrane[6] and is lost in mitochondria in which the outer membrane is intact but the creatine kinase is released into the intermembrane space by KCl treatment[22]. It has also been shown by immunochemical methods for mitoplasts produced from heart mitochondria that CK_{mit} and translocase are structurally closely related to each other[12]. Furthermore, this concept is also directly related to a concept developed by Wallimann and coworkers[10,23-26], according to which mitochondrial creatine kinase forms octamers which are probably bound to the tetramers of the adenine nucleotide translocator and form one multienzyme complex, i.e. translocase-CK_{mit}-outer membrane pores. In this structure, ATP molecules transferred by translocase are inevitably directed to the active sites in the inner "channels" of the octamers of CK_{mit}, and, respectively, ADP also has a decreased diffusion distance from CK_{mit} to the translocase. Thus, there is rather strong and good functional and structural evidence for a high value of the probability of direct transfer of the ATP molecules between CK_{mit} and translocase.

The most important result of this modeling study is a predicted apparent decrease of the value of the dissociation constant for ATP from the ternary enzyme-substrate complex E.ATP.Cr, which is completely consistent with the experimental observations (see Table 1). This apparant decrease has been taken to show the recycling of adenine nucleotides in the tightly coupled system CK_{mit}-translocase-oxidative phosphorylation - an amplification effect resulting in multiple use of small numbers of ADP for ATP production and playing an important role in enhancing the regulatory signal in cardiac cells *in vivo*[3-10,27,28].

References

1. S.A. Kuby and E.A. Noltmann, Adenosine triphosphate-creatine transphosphorylase, *in*: "The Enzymes," Vol. 6, P.D. Boyer, H. Lardy and K. Myrback, eds, Academic Press, New York, p.515 (1962).
2. J.F. Morrison and E. James, The mechanism of the reaction catalyzed by adenosine triphosphate-creatine phosphotransferase, *Biochem. J.* **97**:37 (1965).
3. V.A. Saks, G.B. Chernousova, D.E. Gukovsky, V.N. Smirnov and E.I. Chazov, Studies of energy transport in heart cells. Mitochondrial isoenzyme of creatine phosphokinase: kinetic properties and regulatory action of Mg^{2+} ions, *Eur. J. Biochem.* **57**:273 (1975).
4. W.E. Jacobus and V.A. Saks, Creatine kinase of heart mitochondria: changes in its kinetic properties induced by coupling to oxidative phosphorylation, *Arch. Biochem. Biophys.* **219**:167 (1982).
5. V.A. Saks, V.V. Kupriyanov, G.V. Elizarova and W.E. Jacobus, Studies of energy transport in heart cells. The importance of creatine kinase localization for the coupling of mitochondrial phosphorylcreatine production to oxidative phosphorylation, *J. Biol. Chem.* **255**:755 (1980).
6. V.A. Saks, A.V. Kuznetsov, V.V. Kupriyanov, M.V. Miceli and W.E. Jacobus, Creatine kinase of rat heart mitochondria. The demonstration of functional coupling to oxidative phosphorylation in an inner membrane-matrix preparation, *J.Biol. Chem.* **260**:7757 (1985).
7. R.W. Moreadith and W.E. Jacobus, Creatine kinase of heart mitochondria. Functional coupling of ADP transfer to the adenosine nucleotide translocase, *J. Biol. Chem.* **257**:899 (1982).
8. R.L. Barbour, J. Ribaudo and S.H.P. Chan, Effect of creatine kinase activity on mitochondrial ADP/ATP transport. Evidence for a functional interaction. *J.Biol. Chem.* **259**:8246 (1984).
9. S.P. Bessman and P.J. Geiger, Transport of energy in muscle: the phosphocreatine shuttle, *Science* **211**: 448 (1981).

10. T. Walliman, M. Wyss, D. Brdiczka, K. Nicolay and H.M. Eppenberger, Intracellular compartmentation, structure and function of creatine kinase isoenzymes in tissues with high and fluctuating energy demands: the "phosphocreatine circuit" for cellular energy homeostasis, *Biochem. J.* **281**:21 (1992).

11. M. Muller, R. Moser, D. Cheneval and E. Carafoli, Cardiolipin is the membrane receptor for mitochondrial creatine phosphokinase, *J. Biol. Chem.* **260**:3839 (1985).

12. V.A. Saks, Z.A. Khuchua and A.V. Kuznetsov, Specific inhibition of ATP-ADP translocase in cardiac mitoplasts by antibodies against mitochondrial creatine kinase, *Biochim. Biophys. Acta* **891**:138 (1987).

13. V.A. Saks, N.V. Lipina, V.N. Smirnov and E.I. Chazov, Studies of energy transport in heart cells. The functional coupling between mitochondrial creatine phosphokinase and ATP-ADP translocase: kinetic evidence. *Arch. Biochem. Biophys.* **173**:34 (1976).

14. E.J. Davis and L. Lumeng, Relationships between the phosphorylation potentials generated by liver mitochondria and respiratory state under conditions of adenosine diphosphate control. *J. Biol. Chem.* **250**:2275 (1975).

15. P.V. Vignais, G. Brandolin, F. Boulay, P. Dalbon, M.R. Block and I. Gauche, Recent developments in the study of the conformational states and the nucleotide binding sites in the ADP/ATP carrier, in: "Anion Carriers of Mitochondrial Membranes," A. Azzi, K.A. Nalecz, M.J. Nalecz and L. Wojtczak, eds, Springer, Berlin, p. 133 (1989).

16. W.W. Cleland, Enzyme kinetics, *Ann. Rev. Biochem.* **36**:77 (1967).

17. G.L. Kenyon and G.H. Reed, Creatine kinase: structure-activity relationships, Adv. Enzymol. 54:367 (1983).

18. P.D. Boyer, L. de Meis, M.G.C. Carvalho and D.D. Hackney, Dynamic reversal of enzyme carboxyl group phosphorylation as the basis of the oxygen exchange catalysed by sarcoplasmic reticulum adenosine triphosphatase, *Biochemistry* **16**:136 (1977).

19. M.K. Aliev and V.A. Saks, Quantitative analysis of "phosphocreatine shuttle". I. A probability approach to the description of phosphocreatine production in the coupled creatine kinase - ATP/ADP translocase - oxidative phosphorylation reactions in heart mitochondria. *Biochim. Biophys. Acta (Bioenergetics)*, submitted.

20. A.V. Kuznetsov and V.A. Saks, Affinity modification of creatine kinase and ATP-ADP translocase in heart mitochondria: determination of their molar stoichiometry, *Biochem. Biophys. Res. Commun.* **134**:359 (1986).

21. F.N. Gellerich, M. Schlame, R. Bohnensack and W. Kunz, Dynamic compartmentation of adenine nucleotides in the mitochondrial intermembrane space of rat-heart mitochondria, *Biochim. Biophys. Acta* **890**:117 (1987).

22. A.V. Kuznetsov, Z.A. Khuchua, E.V. Vassil'eva, N.V. Medvedeva and V.A. Saks, Heart mitochondrial creatine kinase revisited: the outer mitochondrial membrane is not important for coupling of phosphocreatine production to oxidative phosphorylation, *Arch. Biochem. Biophys.* **268**:176 (1989).

23. J. Schlegel, B. Zurbriggen, G. Wegmann, M. Wyss, H.M. Eppenberger and T. Wallimann, Native mitochondrial creatine kinase forms octameric structures. I. Isolation of two interconvertible mitochondrial creatine kinase forms, dimeric and octameric mitochondrial creatine kinase: characterization, localization, and structure-function relationships. *J. Biol. Chem.* **263**:16942 (1988).

24. T. Schnyder, A. Engel, A. Lustig and T. Wallimann, Native mitochondrial creatine kinase forms octameric structures. II. Characterization of dimers and octamers by ultracentrifugation, direct mass measurements by scanning transmission electron microscopy, and image analysis of single mitochondrial creatine kinase octamers. *J. Biol. Chem.* **263**:16954 (1988).

25. V. Adams, W. Bosch, J. Schlegel, T. Wallimann and D. Brdiczka, Further characterization of contact sites from mitochondria of different tissues: topology of peripheral kinases, *Biochim. Biophys. Acta* **981**:213 (1989).

26. J. Schlegel, M. Wyss, H.M. Eppenberger and T. Wallimann, Functional studies with the octameric and dimeric form of mitochondrial creatine kinase. Differential pH-dependent association of the two oligomeric forms with the inner mitochondrial membrane, *J. Biol. Chem.* **265**:9221 (1990).

27. V.A. Saks, Y.O. Belikova and A.V. Kuznetsov, In vivo regulation of mitochondrial respiration in cardiomyocytes: specific restrictions for intracellular diffusion of ADP, *Biochim. Biophys. Acta* **1074**:302 (1991).

28. V.A. Saks, Y.O. Belikova, A.V. Kuznetsov, Z.A. Khuchua, T.H. Branishte, M.L. Semenovsky and V.G. Naumov, Phosphocreatine pathway for energy transport: ADP diffusion and cardiomyopathy, *Am. J. Physiol. Suppl.* **261**:30 (1991).

CRITICAL REGULATION OF MULTIENZYME COMPLEXES: A KINETIC MODEL FOR THE MAMMALIAN PYRUVATE DEHYDROGENASE COMPLEX

Boris N. Goldstein and Vitalii A. Selivanov

Institute of Theoretical and Experimental Biophysics
Academy of Sciences of Russia
142292 Pushchino, Moscow Region, Russia

1. Introduction

The pyruvate dehydrogenase complex (PDC) is one of the most intensively studied multienzyme complexes. The PDC plays a central role in intermediary metabolism, being a major site of regulation[1]. We construct a simple kinetic model, predicting the so-called 'critical' regulation[2] for mammalian PDC. The regulation of the critical type is characteristic of bistable systems having two alternative stable steady states[2] and may be observed as a steep change in metabolite levels.

The analysis showed that the critical regulation of mammalian PDC may be simply obtained due to the two different lipoyl-bearing components, assembled in its core. Two lipoyl-bearing proteins, E_2 and X, competing for the same substrate, both of them being potential acceptors of acetyl groups, may induce such a critical PDC regulation.

2. Results

The three catalytic components of the PDC, E_1, E_2, and E_3, contribute to sequential steps in the decarboxylation and dehydrogenation of pyruvate, leading to its conversion to acetyl-CoA. In the model, we consider only the reaction steps performed by the E_2 and protein X. The following designations are given to the reactants placed into the model:

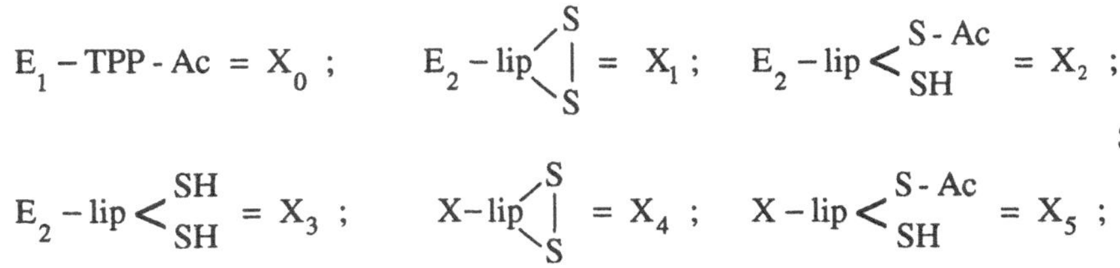

$$E_1 - TPP - Ac = X_0 \; ; \qquad E_2 - lip \underset{S}{\overset{S}{\diagdown}} = X_1 \; ; \qquad E_2 - lip \underset{SH}{\overset{S-Ac}{\diagdown}} = X_2 \; ;$$

Scheme 1

$$E_2 - lip \underset{SH}{\overset{SH}{\diagdown}} = X_3 \; ; \qquad X - lip \underset{S}{\overset{S}{\diagdown}} = X_4 \; ; \qquad X - lip \underset{SH}{\overset{S-Ac}{\diagdown}} = X_5 \; ;$$

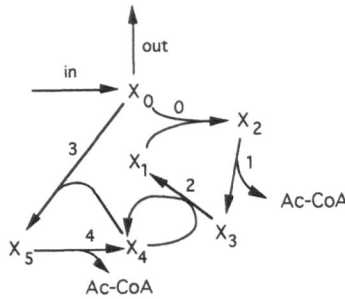

Figure 1. Reaction scheme of the mammalian pyruvate-dehydrogenase complex.

Accordingly, the complete scheme for the overall reaction catalyzed by the mammalian PDC can be represented as depicted in Fig. 1. This scheme is based on the simplifying assumption that the fraction of the reduced form of X_4 is negligibly small, so that only the oxidized form is represented.

The following biochemical properties of protein X and the E_2 are worth mentioning.

(1) The enzyme E_2 cannot transfer electrons directly to E_3.
(2) Protein X plays a central role in the transmission of reducing equivalents to E_3.
(3) Protein X can, similarly to E_2, be acetylated by E_1 and deacetylated by CoA.
(4) The number of molecules of protein X is one tenth of the number of the E_2 molecules.

The main purpose of this modeling is to determine the type of the kinetic behavior of the PDC induced by the above-mentioned properties of the protein X and E_2.

The system of ordinary differential equations describes processes shown by the scheme. The stability analysis of the system includes its linearization near a steady state. The characteristic polynomial for this linearized system reads

$$a_4 * \lambda^4 + a_3 * \lambda^3 + a_2 * \lambda^2 + a_1 * \lambda + a_0 = 0 . \tag{1}$$

The coefficients a_i in the characteristic polynomial were computed in dependence on the stationary concentrations, x_j, and stationary rates, v_j, with

$$v_j\left(x_i, x_k\right) = k_j * x_i * x_k \quad (i, j, k = 0, 1, ..., 4) , \tag{2}$$

where x_i denotes the concentration of X_i.

By using a computer program for searching the minimum points of multivariable functions, we found parameters giving rise to a negative a_0. If the free coefficient, a_0, is negative and no steady state exists on the boundary of the accessible region formed by the variables x_0, x_1, ..., then we necessarily have two stable steady states[2]. Taking all the derivatives to be equal to zero, we can find the steady-state concentration, x_0, in dependence on kinetic parameters. Let us fix all parameters but k_0, by choosing them in the bistability domain, and compute k_0 as a function of x_0. The resulting plot is shown in Fig. 2.

It is seen from Fig. 2 that for $3 < x_0 < 8$ the coefficient a_0 is positive, but k_0 is negative. Therefore, no detectable steady state exists for the middle part of the plot. There exist three different x_0 levels for $k_0 > 0.2$. The middle of these ($1.5 < x_0 < 3$) is unstable, as follows from the negative a_0 values corresponding to the middle x_0 branch. Thus, the two stable x_0 branches, shown in Fig. 2, correspond to the following x_0 values: $0 < x_0 < 1.5$ and $x_0 > 10$. Suppose that initially the system stays at the lower x_0 branch and the parameter k_0 decreases to about 0.1 (its critical value), then the system undergoes a steep 'critical' transition to the upper stable x_0 branch. If the parameter k_0 now increases, the system stays at the upper branch until changes in other parameters push it back to the lower branch.

Let us now consider the kinetic mechanism of the observed 'critical' switching between two stable alternative pyruvate levels. According to Fig. 2, a decrease in the parameter k_0 induces an increase in concentration x_0, which in turn stimulates step 3. As a result, protein X gets excluded from participation in the reaction cycle $X_1 \longrightarrow X_2 \longrightarrow X_3 \longrightarrow X_1$. Therefore, this reaction is progressively inhibited by the "critically" increased x_0 (pyruvate) concentration. The higher level is mainly determined not by the rate of the cycle $X_1 \longrightarrow X_2 \longrightarrow X_3 \longrightarrow X_1$, but by the rate of the competing, secondary reaction cycle $X_4 \longrightarrow X_5 \longrightarrow X_4$. If after that k_0 increases, the flux through the cycle $X_1 \longrightarrow X_2 \longrightarrow X_3 \longrightarrow X_1$ is not increased since protein X is excluded from this cycle. Thus, the higher pyruvate level is stable. Only external factors, such as a decrease in pyruvate concentration, can induce a reverse transition to the lower, stable pyruvate level.

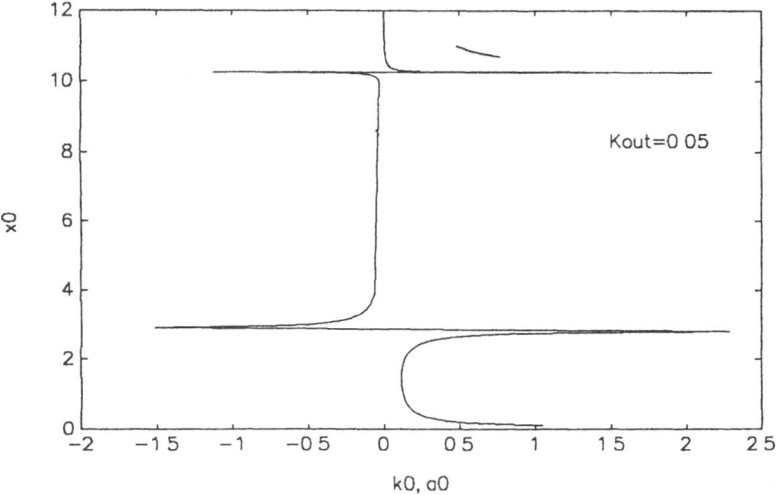

Figure 2. Plot of the results of a numerical simulation of the PDC model. The three different x_0 branches are shown by solid curves The three computed a_0 branches are represented by dotted curves, in dependence on x_0 The coefficient a_0 becomes negative if it corresponds to unstable x_0 values

The production rate of x_0 in glycolysis may be variable. This analysis shows that small changes in the input of x_0 may induce steep changes in the stationary level of x_0 . If the acetyl-CoA production by the PDC-catalyzed reactions would be changed like that of x_0 is, the flux through the tricarboxylic acid cycle would be destabilized, but this actually is not the case. Our model shows that the the acetyl-CoA production is only slightly dependent on the level of x_0, because there is a regulatory route in the PDC-catalyzed reactions, that conserves the stable acetyl-CoA production independently of the x_0 leaps, and protein X plays a dominant role in this regulation.

If the pyruvate concentration is low, the limiting step in the PDC reactions is the oxidative decarboxylation of pyruvate. Upon increase of the pyruvate concentration, the most part of protein X is acetylated and the electron transfer to NAD^+ becomes the bottleneck in the PDC reactions. Therefore, the acetyl-CoA production does not increase as the pyruvate level increases.

This investigation shows that protein X may be needed for stabilization of the tricarboxylic acid cycle independently of possible glycolysis variations.

The changes in the parameters k_0 and k_3 correspond to the changes in phosphorylation of pyruvate dehydrogenase (E_1). This is a well-known type of regulation of the PDC. Since phosphorylation needs ATP, fast phosphorylation/dephosphorylation does not save energy. Acetylation of protein X does not need energy of ATP and enables the PDC reactions to immediately respond to any changes in pyruvate input. Therefore, it may be supposed that the E_1 phosphorylation may serve as a long-term type of metabolic regulation, while the regulation by acetylation of protein X responds to fast changes in pyruvate level.

From the present model analysis, it can be seen that small parameter changes may cause sharp changes in pyruvate concentration. This may actually be the case for conditions under which glycolysis is reversed to gluconeogenesis.

References

1. M.S. Patel and T.E. Roche, Molecular biology and biochemistry of pyruvate dehyrdrogenase complexes, *FASEB J.* **4**:3224-3233 (1990).
2. B.N. Goldstein and A.N. Ivanova, Hormonal regulation of 6-phosphofructo-2-kinase/fructose-2,6-bis-phosphatase: kinetic models, *FEBS Lett.* **217**:212-215 (1987).
3. M. Rahmatullah, G.A. Radke, S.L. Powers-Greenwood, P.C. Andrews and T.E. Roche, Changes in the core of the mammalian pyruvate dehydrogenase complex upon selective removal of the lipoyl domain from the transacetylase component but not from the protein X component, *J. Biol. Chem.* **265**: 14512-14517 (1990).

THE ANALYSIS OF FLUX
IN SUBSTRATE CYCLES

David A. Fell

School of Biological and Molecular Sciences
Oxford Brookes University
Headington, Oxford OX3 0BP, United Kingdom

1. Introduction

Within the network that makes up the metabolism of a cell, there are to be found cyclic routes. These cycles can be of different types, but amongst them are the 'futile' or substrate cycles that are often defined as cyclic fluxes that lead to no net changes other than the dissipation of energy in the coupled reactions through which they are driven. It is presumed that *in vivo*, regulation of the enzyme(s) in opposite limbs of such cycles might limit the energy dissipation by unnecessary cycling, but this is an issue that must be determined by experiment. Often such experiments are carried out using isotopic labeling techniques in order to estimate the fluxes through individual reactions. Earlier work has looked at how these results might be analyzed in order to derive the cyclic flux[1,2]. My purpose in this paper is to show that once a network has a certain degree of complexity, knowing the individual reaction fluxes does not always allow an unambiguous estimate of the amount of cycling, and this problem appears to have been overlooked.

For reasons given previously[2], I define a pathway segment making a substrate cycle as having the following necessary features:

1. a cyclic route;
2. at least two independent fluxes, and
3. a potentially viable pathway structure if one limb is deleted.

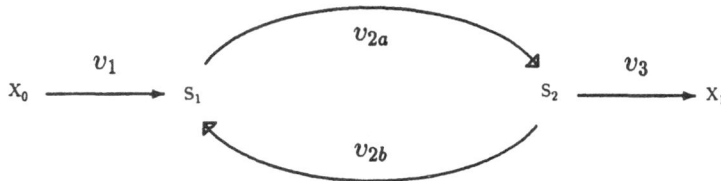

Figure 1. Minimum configuration for a substrate cycle corresponding to the definition in the text.

The simplest pathway possessing these features is shown in Fig. 1, with the coupling to the energy dissipating processes (such as phosphorylation and hydrolysis) omitted. In this case, there is no problem about resolving the pathway fluxes into a linear component flowing through the cycle and a dissipative cyclic flux. Blum and his coworkers, for example, have undertaken flux measurements in much larger metabolic networks, in both *Tetrahymena pyriformis* and liver cells, and in these networks, assigning the cyclic flux is much less obvious. The example I shall use here is a small segment taken by Leiser and Blum[1] from much more extensive measurements on hepatocytes, attributed apparently in error to Crawford and Blum, though the flux values are those of Fig. 2 of Rabkin and Blum[3]; this is shown in Fig. 2.

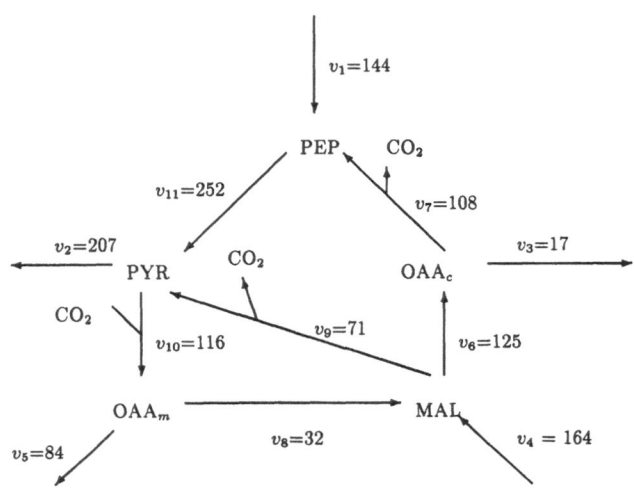

Figure 2. The pathway fragment analyzed by Leiser and Blum[1].
Their two cyclic modes were: a) $v_8 = v_9 = v_{10} = 116$; b) $v_6 = v_7 = -v_9 = v_{11} = 108$.

2. Methods for Assigning Cyclic Flux

Leiser and Blum[1] proposed that pathways should be decomposed into linear and cyclic fundamental modes that are linearly superimposed to give the observed flux distribution. The decomposition of the simple substrate cycle of Fig. 1 into a straight-through flux and a cyclic flux is straightforward and uncontroversial. In the case of Fig. 2, they proposed a decomposition into a non-futile mode consisting of a set of straight-through fluxes and two cyclic modes. One cycle involves pyruvate carboxylase, malate dehydrogenase and malic enzyme (reactions 8,9 and 10 in Fig. 2), and the other involves malate dehydrogenase, phospho*enol*pyruvate carboxykinase, malic enzyme in reverse and pyruvate kinase (reactions 6,7,9 and 11). However, they did not suggest a general procedure for deciding how many modes were needed for the analysis and how they were to be constructed.

I therefore suggested[2] that the modes could be identified with a set of vectors spanning the null space of the stoichiometry matrix of the pathway. This gives a definite criterion for the number of independent vectors that are needed, and allows automatic evaluation using a numerical algorithm if desired (see, e.g., Ref. 4). Vectors corresponding to cyclic pathways can be easily identified because they do not involve any of the reactions that connect to the external pools. The analysis can be illustrated with the simple substrate cycle as follows.

Any metabolic pathway at steady state satisfies the relationship:

$$\mathbf{N \cdot v = 0},$$ (1)

where $\mathbf{N}$ is the stoichiometry matrix of the pathway and $\mathbf{v}$ is a vector of rates. The stoichiometry matrix has n columns corresponding to the n reactions of the system, and m rows corresponding to variable metabolites (i.e. not the external metabolites) of the system. The entry in column i and row j indicates the change in the number of moles of metabolite j for a unit of the ith reaction. Eq. (1) states mathematically that at steady state, the metabolite concentrations are fixed because their rates of formation equal their rates of breakdown. Writing this equation for the system of Fig.1 gives:

$$\begin{bmatrix} 1 & -1 & 1 & 0 \\ 0 & 1 & -1 & -1 \end{bmatrix} \cdot \begin{bmatrix} v_1 \\ v_{2a} \\ v_{2b} \\ v_3 \end{bmatrix} = \begin{bmatrix} 0 \\ 0 \end{bmatrix}.$$ (2)

As pointed out by Reder[5,6], any observed set of velocities at steady state will be a linear combination of a set of vectors $\mathbf{K}$ referred to as the null space of the stoichiometry matrix. The number of vectors in the null space will be the number of reactions minus the number of independent metabolites (that is, the rank of the stoichiometry matrix, obtained by deleting rows that can be formed by simple combinations of other rows, as happens when there is moiety conservation, such as with the adenine of adenine nucleotides). In this example, a possible null space is:

$$\mathbf{K} = \begin{bmatrix} 1 & 0 \\ 1 & 1 \\ 0 & 1 \\ 1 & 0 \end{bmatrix},$$ (3)

and so for any possible set of steady state velocities, there must be two factors a and b such that:

$$\begin{bmatrix} 1 & 0 \\ 1 & 1 \\ 0 & 1 \\ 1 & 0 \end{bmatrix} \cdot \begin{bmatrix} a \\ b \end{bmatrix} = \begin{bmatrix} v_1 \\ v_{2a} \\ v_{2b} \\ v_3 \end{bmatrix}.$$ (4)

The solution of this equation for the linear flux, a and the cyclic flux, b, is trivial.

What happens when this procedure is applied to the second example? This has 11 reactions and 5 metabolites, so 6 vectors or modes will be needed to form the basis for a null space. The basis can be calculated by a computer program that parses the list of reactions in the pathway, forms the stoichiometry matrix and determines the null space (e.g. iMAP[7] or MetaCon[8,9]). In the general case, there is not a unique basis for a null space, and this is borne out by this scheme, where different null spaces can be obtained from the same program by the simple expedient of shuffling the order of rows in the stoichiometry matrix. Why should this pathway be more difficult to analyze than the minimal substrate cycle pathway? The complexity arises because there are 2 inputs and 3 outputs from the pathway; even if the pathway were reduced to a single intermediate connected to these 5 external pools, 4 vectors would be needed to form a basis, but there are 6 ways of connecting the inputs to the outputs with regard to direction, and 10 connections if the direction of flux is disregarded. Therefore, there are more potential elementary pathways than are needed to define a basis. It also follows that since the pathway requires 6 vectors, but a minimum of 4 are needed to describe the connections between the external pools, a maximum of 2 cyclic pathways can exist in the basis.

When the different bases for the null space are examined, a number of points emerge. Firstly, it is possible to find null spaces with 0, 1 or 2 cyclic pathways. However, some of these solutions contain negative values in the vectors. This corresponds to a reaction running in the reverse direction. Whether this is reasonable perhaps depends on the degree of irreversibility in the step involved; also, some of the steps shown are net balances of a number of grouped reactions. The solution that contained no cyclic elements involved a number of negative fluxes, some of which were greater than observed. Some of the solutions with 1 or 2 cyclic pathways also contained negative fluxes, but then so did the solution proposed by Leiser and Blum[1]. Apart from the two cyclic modes they had proposed, a solution with the outer cycle (v_6....v_8, v_{10} and v_{11}) was obtained. Two solutions were obtained where only 0 or positive elements appeared in the vectors obtained, which therefore corresponded to component pathways with the reactions aligned with the flux directions determined experimentally; one of these had 2 cyclic components and the other 1. However, when the weighting factors multiplying the vectors to give the steady state fluxes were calculated (cf. eq. 4), a negative weighting factor was needed for one of the cycles in the 2 cycle solution, equivalent to the whole cycle running in reverse. The solution with one cycle had positive weighting factors, and is shown in Fig. 3. The cycle is the pyruvate carboxylase - malic enzyme cycle selected by Leiser and Blum[1], but the rate ascribed to it is only 32 nmol/mg protein/20 min, compared with 116 attributed by them. The rest of the flux is attributed to linear flows between the inputs and outputs.

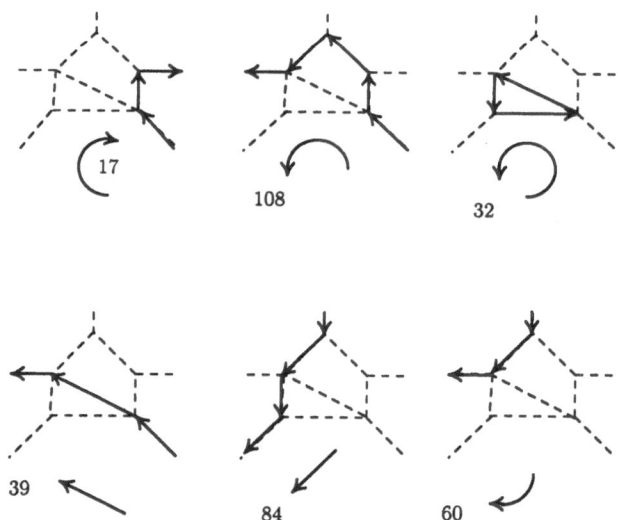

Figure 3. Component pathways corresponding to a set of non-negative vectors forming a basis for the null-space of the pathway in Fig. 2.
The numbers and arrows show the positive weighting factors giving the observed steady state fluxes.

3. Do Pathways and Cycles Objectively Exist?

This analysis demonstrates that once a metabolic network of moderate complexity is considered, it cannot simply be assumed that there is a unique way of decomposing it into component pathways, nor that the component pathways will necessarily correspond to those described in biochemistry textbooks. My earlier assumption that the component cyclic pathways could be assigned by examining the null space[2] is not a complete solution. The number of possible answers is reduced by rejecting those that involve negative fluxes, either in the null space vector or the weighting factors, but it is not obvious that this should be a criterion, given that some reactions are known to have bidirectional fluxes greater than the net flux, and that it is not intrinsically unreasonable that two oppositely directed pathways should cause a small resultant flux in a reaction where they coincide through 'destructive interference'. Where a number of different solutions are possible, it follows that different answers may be obtained for the energy cost attributable to substrate cycling.

Acknowledgement

This work was supported in part by project grant GR/E89117 from the Science and Engineering Research Council.

References

1. J. Leiser and J. J. Blum, On the analysis of substrate cycles in large metabolic systems, *Cell Biophys.* **11**:123-138 (1985).
2. D. A. Fell, Substrate cycles: Theoretical aspects of their role in metabolism, *Comments Theor. Biol.* **1**:341-357 (1990).
3. M. Rabkin and J. J. Blum, Quantitative analysis of intermediary metabolism in hepatocytes incubated in the presence and absence of glucagon with a substrate mixture containing glucose, ribose, fructose, alanine and acetate, *Biochem. J.* **255**:761-786 (1985).
4. D. E. Knuth. "Seminumerical Algorithms," Vol. 2 of "The Art of Computer Programming," 2nd edition, Addison-Wesley, Reading (1981).
5. C. Reder, Mimodrame mathématique sur les systèmes biochimiques, Technical Report 8608, U.E.R. de Mathématiques et d'Informatique, Université de Bordeaux I (1986).
6. C. Reder, Metabolic control theory: a structural approach, *J. Theor. Biol.* **135**:175-201 (1988).
7. H. M. Sauro and D. A. Fell, Analysers and simulators: a brief description of two computer programs, SCAMP and iMAP, *in*: "Biothermokinetics," H. V. Westerhoff, ed., Intercept, Andover, in press.
8. S. Thomas and D. A. Fell, *in*: "MetaCon User's Manual," Oxford Polytechnic, Oxford (1992).
9. S. Thomas and D. A. Fell, MetaCon - a computer program for the algebraic evaluation of control coefficients of metabolic networks, this volume.

ESTIMATION OF ENZYMATIC FLUX RATES FROM KINETIC ISOTOPE EXPERIMENTS

Hermann-Georg Holzhütter and Anke Schwendel

Institut für Biochemie
Medizinische Fakultät (Charité)
Humboldt-Universität zu Berlin
Hessische Str. 3-4, D-O-1040 Berlin, Germany

1. Introduction

The mathematical description of radioactive tracer experiments leads to a coupled system of differential equations for both the metabolite concentrations and the specific radioactivities. Computer simulations, i.e. numerical integration of these equations in combination with a fitting algorithm, enables one to estimate unknown kinetic parameters from the time courses of the specific radioactivities. In the present paper, the theoretical fundamentals of this approach are illustrated. As an example, we determine the flux rates in the adenine nucleotide metabolism of Ehrlich ascites tumor cells under stationary, spatially homogeneous metabolic conditions. The reaction scheme is shown in Fig.1.

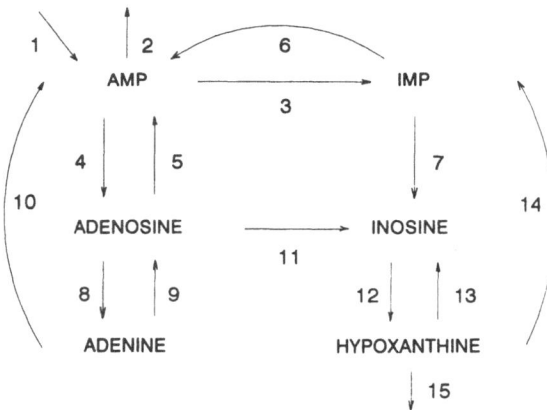

Figure 1. Reactions of purine metabolism of the Ehrlich ascites tumor cells considered in the model. The numbers indicate the flux rates.

Modern Trends in Biothermokinetics, Edited by
S. Schuster *et al.*, Plenum Press, New York, 1993

2. Basic Equations

The dependence of the metabolite concentrations x on time is governed by a system of ordinary differential equations

$$\frac{dx_i}{dt} = \sum_{j=1}^{m} c_{ij} v_j(x_1,...,x_n) , \quad i=1,...,n \tag{1}$$

with $\mathbf{C} = (c_{ij})$ giving the stoichiometric matrix, $\mathbf{V} = (v_j)$ indicating the vector of flux rates, i numbering the metabolites and j the reactions.

Applying eq. (1) to the system described in Fig.1, one gets:

$$dx_{amp}/dt = v_1 + v_5 + v_6 + v_{10} - v_2 - v_3 - v_4 \tag{2}$$

$$dx_{imp}/dt = v_3 + v_{14} - v_6 - v_7 \tag{3}$$

$$dx_{adenosine}/dt = v_4 + v_9 - v_5 - v_8 - v_{11} \tag{4}$$

$$dx_{inosine}/dt = v_7 + v_{11} + v_{13} - v_{12} \tag{5}$$

$$dx_{adenine}/dt = v_8 - v_9 - v_{10} \tag{6}$$

$$dx_{hypoxanthine}/dt = v_{12} - v_{13} - v_{14} - v_{15} . \tag{7}$$

The specific radioactivities are defined as

$$y_i = z\frac{\varepsilon_i}{x_i} , \quad i=1,...,n, \tag{8}$$

where z represents the number of radioactive counts per concentration unit and ε_i the labeled fraction of metabolite i.

The temporal evolution of the specific radioactivities are given by the following differential equations[1]

$$x_i\frac{dy_i}{dt} = \sum_{j=1}^{m} c_{ij}^+ v_j\left(\sum_{k=1}^{n} A_{ik}^j y_k - y_i\right), \tag{9}$$

$$c_{ij}^+ = \begin{cases} c_{ij} , & \text{if } c_{ij} \geq 0 \\ 0 , & \text{if } c_{ij} < 0 . \end{cases} \tag{9'}$$

Here A_{ik}^j denotes the probability for metabolite i to become labeled by metabolite k during reaction j.

For the system shown in Fig.1 we obtain the following equations

$$x_{amp} \, dy_{amp}/dt = -v_1y_1 + v_5(y_3 - y_1) + v_6(y_2 - y_1) + v_{10}(y_5 - y_1) \tag{10}$$

$$x_{imp} \, dy_{imp}/dt = v_3(y_1 - y_2) + v_{14}(y_6 - y_2) \tag{11}$$

$$x_{ado} \, dy_{ado}/dt = v_4(y_1 - y_3) + v_9(y_5 - y_1) \tag{12}$$

$$x_{ino} \, dy_{ino}/dt = v_7(y_2 - y_4) + v_{11}(y_3 - y_4) + v_{13}(y_6 - y_4) \tag{13}$$

$$x_{ade} \, dy_{ade}/dt = v_8(y_3 - y_5) \tag{14}$$

$$x_{hyp} \, dy_{hyp}/dt = v_{12}(y_4 - y_6) \, . \tag{15}$$

The coupled differential equation systems (1) and (9) govern the temporal evolution of the metabolic tracer system.

3. Metabolic Steady State

In the case that the system is in a metabolic steady state characterized by constant metabolite concentrations $\mathbf{X}^o = (x_i^o)$,

$$\frac{d\mathbf{X}^o}{dt} = \mathbf{C} \, \mathbf{V}(\mathbf{X}^o) = \mathbf{0} \, , \tag{16}$$

the differential equation system (1) becomes a linear equation system. Furthermore, eq. (9) can be written as

$$\frac{dy_i}{dt} = \sum_{j=1}^{m} \frac{c_{ij}^+ v_j^o}{x_i^o} \left(\sum_{k=1}^{n} A_{ik}^j y_k - y_i \right) , \tag{17}$$

where $v_j^o = v_j(\mathbf{X}^o)$.

If the stationary metabolite concentrations x_i^o are known, the flux rates $\mathbf{V}^o = (v_j^o)$ can be determined by fitting the solutions of equation system (17) to the observed time-courses of the specific radioactivities. Doing so one has to bear in mind that the flux rates v_j^o are not fully independent but related to each other due to the steady-state condition (16). Since the number of independent equations in (16) is given by the rank r of the stoichiometric matrix, one may choose $m-r$ independent flux rates, $\mathbf{V}^{ind} = (v_l^{ind})$, $l=1,...,m-r$. All flux rates can be expressed through these independed ones, i.e.

$$v_j^o = \sum_{l=1}^{m-r} f_{jl} v_l^{ind} , \tag{18}$$

where the f_{jl} are determined by eq. (16).

We have chosen v_1, v_3, v_5, v_8, v_9, v_{11}, v_{12}, v_{13}, v_{15} as independent flux rates to be determined by the fitting procedure, while

$$v_2 = v_1 - v_{15} \tag{19}$$

$$v_4 = v_5 + v_8 + v_{11} - v_9 \tag{20}$$

$$v_6 = v_3 + v_{11} - v_{15} \tag{21}$$

$$v_7 = v_{12} - v_{11} - v_{13} \tag{22}$$

$$v_{10} = v_8 - v_9 \tag{23}$$

$$v_{14} = v_{12} - v_{13} - v_{15} \, . \tag{24}$$

Inserting relation (18) into eq. (17) one obtains

$$\frac{dy_i}{dt} = \sum_{k=1}^{n}\sum_{j=1}^{m}\sum_{l=1}^{m-r}\frac{c_{ij}^{+}f_{jl}}{x_i^o}v_l^{ind}\left(A_{ik}^{j}-\delta_{ik}\right)y_k \,, \tag{25}$$

where δ_{ik} is the Kronecker delta.

4. Parameter Estimation

The software package SIMFIT[2] was used to estimate values of the unknown flux rates for the reaction scheme given in Fig.1. It fits the solutions of the equation system (25) to the observed time courses of the specific radioactivities[3]. Two examples of fitting are depicted in Fig. 2. The resulting flux rates are given in Tab.1.

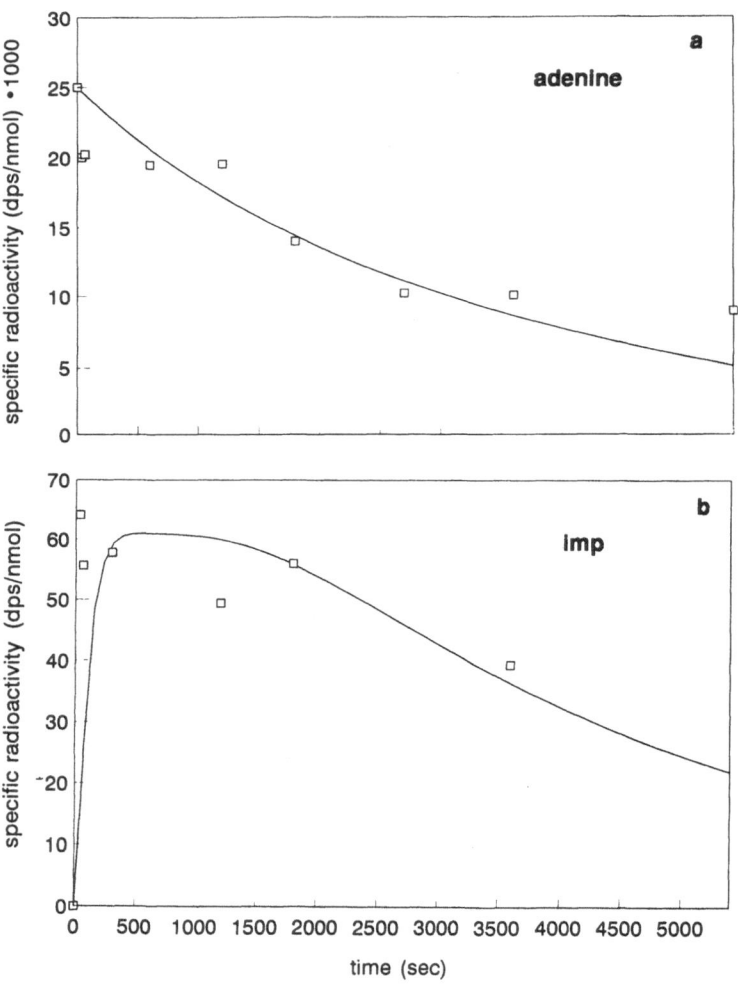

Figure 2. Time courses of the specific radioactivities of adenine (a) and inosine monophosphate (b), experimental points and fitted curves.

Table 1. Results of the fitting procedure.

Number of Flux	Flux rate [nmol/(l_{cells}* sec)] in the resting phase
1	7048
2	7046
3	8878
4	245
5	0
6	9107
7	948
8	23
9	7
10	15
11	228
12	1177
13	0
14	1175
15	3

References

1. R. Schuster, S. Schuster and H.-G. Holzhütter, Simplification of complex kinetic models used for the quantitative analysis of nuclear magnetic resonance or radioactive tracer studies, *J. Chem. Soc. Faraday Trans.* **88**:2837-2844 (1992).
2. H.-G. Holzhütter and A. Colosimo, SIMFIT: A microcomputer software-toolkit for modelistic studies in biochemistry, *Comp. Appl. Biosc.* **6**:23-28 (1990).
3. T. Grune, W.G. Siems, G. Gerber and R. Uhlig, Determination of the ultraviolet absorbance and radio-activity of purine compounds separated by high-performance liquid chromatography. Application to metabolic flux rate analysis, *J. Chromatogr.* **553**:193-199 (1991).

WHEN GOING BACKWARDS
MEANS PROGRESS:
ON THE SOLUTION OF
BIOCHEMICAL INVERSE PROBLEMS
USING ARTIFICIAL NEURAL NETWORKS

Douglas B. Kell, Chris L. Davey,
Royston Goodacre, and Herbert M. Sauro

Dept. of Biological Sciences
University of Wales
Aberystwyth, Dyfed SY23 3DA, U.K.

1. Introduction

Many (perhaps most) biochemical problems are actually "inverse" problems, or problems of system identification. In general, if we know the parameters of a system of interest, we can set up a model (or actual) experiment, run the model (or experiment), and observe the behavior or time evolution of the system. However, especially in a complex biological system, the things which are normally easiest to measure are the variables, not the parameters, and it is the variables which depend on the parameters, not *vice versa*.

In the case of metabolism, the usual parameters of interest are the enzymatic rate constants, which are difficult to measure accurately *in vitro*[1] and virtually impossible so to do *in vivo*. Yet to describe, understand, and simulate the system of interest we need knowledge of the parameters. In other words, we need somehow to go backwards from variables such as the steady-state fluxes and metabolite concentrations, which are relatively easy to measure, to the parameters, which are not.

A similar situation exists in biological spectroscopy. If we have chemical standards, whether pure or mixed, it is easy to obtain their spectra. The inverse problem then involves obtaining the concentrations of metabolites, or the overall spectral parameters, from the spectra observed in a complex system.

The purpose of this article is to describe our successful exploitation of artificial neural networks (ANNs) in the solution of such biochemical (and other) inverse problems. We apply the approach specifically to:

Modern Trends in Biothermokinetics, Edited by
S. Schuster *et al.*, Plenum Press, New York, 1993

(i)   model metabolic pathways studied by computer simulations,

(ii) parameter identification in biological dielectric spectroscopy, and

(iii) the extraction of chemical information from pyrolysis mass spectra of intact microbial cells.

2. Artificial Neural Networks

ANNs are collections of very simple "computational units" which can take a numerical input and transform it (usually via summation) into an output (see e.g. Refs 2-12). The inputs and outputs may be to and from the "external world" or to other units within the network. The way in which each unit transforms its input depends on the so-called "connection weight" (or "connection strength") and "bias" of the unit, which are modifiable. The output of each unit to another unit or to the external world then depends on both its strength and bias and on the weighted sum of all its inputs, which are transformed by a (normally) nonlinear weighting function refered to as its activation function.

The great power of neural networks stems from the fact that it is possible to present ("train") them with known inputs (and outputs) and provide some form of learning rule which may be used, iteratively, to modify the strengths and biases until the outputs of the network as a function of the inputs correspond to the desired ("true") outputs. The trained network may then be exposed to "unknown" inputs and will then provide its view of the "true" output(s).

A neural network therefore consists of at least three layers, representing the inputs and outputs and one or more so-called "hidden" layers. It is, in particular, the weights and biases of the interactions between inputs and outputs and the hidden layer(s) which reflect the underlying dynamics of the system of interest, even if its actual (physical) structure is not known. By training up a neural network with known data, then, it is possible to obtain outputs that can accurately predict things such as the (continuing) evolution of a time series, even if it is (deterministically) chaotic[13].

Other successful uses of neural networks include speech recognition, DNA sequence analysis, the correction of errors in optical astronomy, and the analysis of vapors by arrays of artificial sensors. One may also perhaps mention the successful use of simple neural nets in the analysis of chemical engineering systems[14].

3. Analysis of Metabolic Systems Using Neural Networks

As described in several other contributions in this volume, it is possible by computer simulation to determine steady-state variables such as fluxes and metabolite concentrations as a function of parameters such as the enzymatic rate constants and external metabolite concentrations. It is obviously then possible to change one or more of the parameters and to determine another set of associated variables, and so on. The idea is that having acquired related sets of parameters and variables, we would then be in a position to train neural networks *in which the (known) variables were the inputs and the parameters were the outputs*. When the nets had successfully learned to reflect the correct parameters when presented with the variables, we would have solved our problem. We could then present the net with "random" (experimental) variables and ask it for the parameters.

The correctness of the network's predictions could obviously be checked by running a simulation with the parameters provided by the network and seeing if they generated the variables used as the input to the net. The result would be that we *could* in fact obtain the (enzymatic) parameters of a metabolic network (*and hence the control coefficients and elasticities*) by measuring the variables alone. We have now implemented this strategy, as outlined in what follows.

For computational simplicity we concentrate here on a simple, three-step linear pathway, as shown in Scheme 1, with each enzyme (one substrate / one product) possessing reversible Michaelis-Menten kinetics, and with the steps having equilibrium constants of 567, 13.4 and 2.3 respectively.

$$X_0 \longleftrightarrow S_1 \longleftrightarrow S_2 \longleftrightarrow X_1 \qquad \text{Scheme 1}$$

The dataset was obtained by varying the V_{max} values of each enzyme and the first forward K_m value. All other K_ms were held constant. The concentration of S_1, S_2 and the steady-state flux were recorded for each parameter set. In addition, each parameter set was applied using three different concentrations (1, 10, 100) of the starting metabolite X_0 so that we could obtain three sets of variables for the same set of parameters (in an experimentally realizable fashion), and so aid the training process.

The parameters of the system were varied as follows. Each parameter was first assigned a uniform random number u between 0 and 2.0, which was then used to generate a non-skewed distribution spanning two decades using the formula, 10^u. The reason why the parameters were generated in this manner was so that the distribution of values would not be confined to the upper decades of the range and thus there would be a roughly equal number of random values between 1 and 10 as there would be between 10 and 100. In this way 500 random sets of K_m and V_{max} values were generated, from which the steady-state concentrations of S_1, S_2 and the fluxes were obtained. A 12-18-4 net was then trained (with the stochastic backpropagation algorithm) using the 12 values of X_0, S_1, S_2 and J as the input and the four varied parameters as the output.

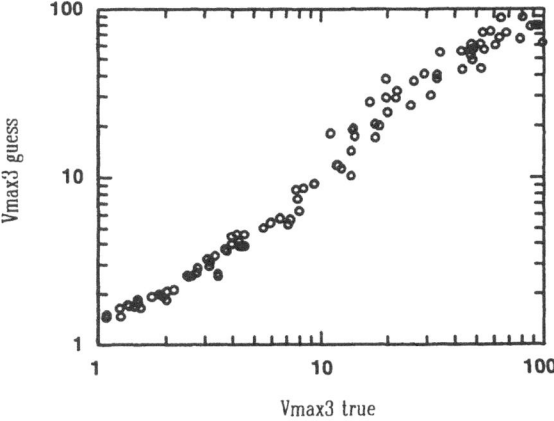

Figure 1. An Artificial Neural Network's estimate of unseen $V_{max,3}$ values.

After training to an RMS error of ca 0.08 (just under 4000 epochs), the net had successfully generalized, as illustrated in Fig. 1, which shows the network's estimate of the unknown values for $V_{max,3}$ against the "true" values. Other parameters had been learned to a more or less similar degree.

It is worth pointing out that it is only possible for ANNs to learn something that is actually learnable, so that if a variable is not significantly related to a particular parameter then the network will not learn it; this of course may be used to our advantage, since if the parameter of interest does not significantly control the variable it may be assumed that one is not interested in studying it anyway. Because updating of the network is based on the *overall* RMS error, "bad" relationships interfere with the learning of "good" ones; our experience to date suggests that any linear correlation coefficient (of logarithmic parameters) below some 0.4 will inhibit the net from converging.

4. Solution of the Inverse Problem in Biological Dielectric Spectroscopy

In biological dielectric spectroscopy, where dispersions are substantially broader than that expected from a purely Debye-like process, it is not always possible, due to technical limitations, to obtain data over a wide enough range of frequencies to encompass the entire dispersion(s) of interest. Similarly, because of the breadth of the dispersions, it is common to seek to characterize the dielectric behavior of interest by means of the Cole-Cole function[15]. Whilst it is possible to fit dielectric data to this equation using appropriate nonlinear least-squares methods[16], these methods are computationally rather demanding, and must be undergone, iteratively, for each set of data.

We have found[17] that it is possible to train an artificial neural network with small sets of dielectric data (permittivities measured at various frequencies) as the inputs and the attendant parameters of the Cole-Cole equation as the outputs. The trained net can then give an essentially instantaneous output of the limiting permittivities at frequencies that are both high and low with respect to the characteristic frequency, and thus of their difference, a parameter which, for the so-called β-dispersion, scales with the biomass content of cell suspensions[16].

5. Solution of the Inverse Problem in Pyrolysis Mass Spectrometry of Microbial Cells

Pyrolysis is the thermal degradation of a material in an inert atmosphere, and leads to the production of volatile fragments (pyrolysate) from non-volatile material such as microorganisms. Curie-point pyrolysis is a particularly reproducible and straightforward version of this technique, in which the sample, dried onto an appropriate metal is rapidly heated (0.6s is typical) to the Curie point of the metal, which may itself be chosen and is commonly 530°C. The pyrolysate may then be separated and analyzed in a mass spectrometer[18], and the combined technique is then known as Pyrolysis Mass Spectrometry or PyMS.

Conventionally (within microbiology and biotechnology), PyMS has been used as a taxonomic aid in the *identification* and *discrimination* of different microorganisms[19]. To this end, the reduction of the multivariate data (150 normalized values in the range m/z 51-200) generated by the PyMS system is normally carried out using Principal Components Analysis (PCA), a well-known technique for reducing the dimensionality of multivariate data whilst preserving most of the variance. Whilst PCA does not take account of any groupings in the data, neither does it require that the populations be normally distributed, *i.e.* it is a non-parametric method.

The closely related Canonical Variates Analysis technique then separates the samples into groups on the basis of the principal components and some *a priori* knowledge of the appropriate number of groupings. Provided that the data set contains "standards" (*i.e.* type or centro-strains) it is evident that one can establish the closeness of any unknown samples to a known organism, and thus effect the identification of the former. An excellent example of the discriminatory power of the approach is the demonstration[20] that one can use it to distinguish four strains of *Escherichia coli* which differ only in the presence or absence of a single plasmid.

We have found[21] that it is possible to train an ANN using the Pyrolysis Mass Spectra as the inputs and the known concentrations of target analytes in standards as the outputs. The trained net can then be tested with the Pyrolysis Mass Spectra of "unknowns", and then accurately outputs the concentration of the target analyte(s), in the case described here the concentration of tryptophan in the growth medium of indole-positive strains of *E. coli*.

We have also been able to effect a rapid distinction between extra virgin and adulterated olive oils using this approach[22]. It is obvious that this combination of PyMS and ANNs constitutes a powerful technology for the analysis of the concentration of appropriate substrates, metabolites and products in any biological process.

Acknowledgements

This work is supported by the Wellcome Trust and through SERC LINK schemes with Aber Instruments, FT Applikon, Horizon Instruments, ICI Biological Products and Neural Computer Sciences.

References

1. R.G. Duggleby, Analysis of biochemical data by nonlinear regression - is it a waste of time?, *Trends Biochem. Sci.* **16**:51-52 (1991).
2. D.E. Rumelhart, J.L. McClelland and the PDP Research Group. "Parallel Distributed Processing. Experiments in the Microstructure of Cognition," MIT Press, Cambridge (Mass.) (1986).
3. J.L. McClelland and D.E. Rumelhart. "Explorations in Parallel Distributed Processing; A Handbook of Models, Programs and Exercises," MIT Press, Cambridge (Mass.) (1988).
4. T. Kohonen. "Self-Organization and Associative Memory," 3rd Ed., Springer, Heidelberg (1989).
5. Y.-H. Pao. "Adaptive Pattern Recognition and Neural Networks," Addison-Wesley, Reading (Mass.) (1989).
6. P.D. Wasserman and R.M. Oetzel. "NeuralSource: the Bibliographic Guide to Artificial Neural Networks," Van Nostrand Reinhold, New York (1989).
7. P.D. Wasserman. "Neural Computing: Theory and Practice," Van Nostrand Reinhold, New York (1989).
8. R.C. Eberhart and R.W. Dobbins. "Neural Network PC Tools," Academic Press, London (1990).
9. P.K. Simpson. "Artificial Neural Systems," Pergamon Press, Oxford, (1990).
10. J.A. Freeman and D.M. Skapura. "Neural Networks," Addison-Wesley, Reading (Mass.) (1991).
11. J. Hertz, A. Krogh and R.G. Palmer. "Introduction to the Theory of Neural Computation," Addison-Wesley, Redwood City (1991).
12. J.M. Zurada. "Introduction to Artificial Neural Systems," West Publishing, St. Paul (1992).
13. D.M. Wolpert and R.C. Miall, Detecting chaos with neural networks, *Proc. Roy. Soc. Ser. B* **242**:82-86 (1990).
14. J.C. Hoskins and D.M. Himmelblau, Artificial neural network models of knowledge representation in chemical engineering, *Comput. Chem. Eng.* **12**:881-890 (1988).
15. R. Pethig and D.B. Kell, The passive electrical properties of biological systems: their significance in physiology, biophysics and biotechnology, *Phys. Med. Biol.* **32**:933-970 (1987).
16. C.L. Davey, H.M. Davey and D.B. Kell, On the dielectric properties of cell suspensions at high volume fractions, *Bioelectrochem. Bioenerg.* **28**:319-340 (1992).
17. D.B. Kell and C.L. Davey, On fitting dielectric spectra using artificial neural networks, *Bioelectrochem. Bioenerg.* **28**:425-434 (1992).
18. H.L.C. Meuzelaar, J. Haverkamp and F.D. Hileman. "Pyrolysis Mass Spectrometry of Recent and Fossil Biomaterials," Elsevier, Amsterdam (1982).
19. C.S. Gutteridge, Characterisation of microorganisms by pyrolysis mass spectrometry, *Meth. Microbiol.* **19**:227-272 (1987).

20. R. Goodacre and R.C.W. Berkeley, Detection of small genotypic changes in *Escherichia coli* by pyrolysis mass-spectrometry, *FEMS Microbiol. Lett.* **71**:133-138 (1990).

21. R. Goodacre and D.B. Kell, The rapid and quantitative analysis of indole production by bacteria, using pyrolysis mass spectrometry and neural networks, *J. Gen. Microbiol.*, in preparation.

22. R. Goodacre, D.B. Kell and G. Bianchi, Neural networks and olive oil, *Nature* **359**:594 (1992).

QUALITY ASSESSMENT
OF A METABOLIC MODEL
AND SYSTEMS ANALYSIS
OF CITRIC ACID PRODUCTION
BY *ASPERGILLUS NIGER*

Néstor Torres[1], Carlos Regalado[1],
Albert Sorribas[2] and Marta Cascante[3]

[1]Departamento de Bioquímica y Biología Molecular
   Facultad de Biología, Universidad de La Laguna
   38206-La Laguna, Tenerife, Canary Islands, Spain
[2]Departament de Bioquimica i Fisiologia
   Divisió de Ciències Experimentals i Matemàtiques
   Universitat de Barcelona
   Martí i Franquès, 1, 08071-Barcelona, Catalunya, Spain
[3]Departament de Ciències Mèdiques Bàsiques
   Universitat de Lleida
   Aviguda Rovira Roure, 44, 25006-Lleida, Catalunya, Spain

1. Introduction

Citric acid is a chemical of foremost importance in the food, cosmetic and chemical industry. Its annual production amounts up to 350,000 tons, with the carbohydrate fermentation by means of the filamentous fungus *Aspergillus niger* (ascomycetes group) being one of the leading processes for its production worldwide. A great deal of work has been done on the biochemistry of this process[1] and, consequently, information is available on most of its biochemical aspects. However, in spite of the amount of information on the system and its economic importance, no attempt has been made to explain its regulation on a quantitative basis.

In this context the aim of this work is, first to develop a mechanistic model where the information on many aspects of the process is integrated and, then, to translate the mechanistic picture so obtained into an S-system representation[2], which would allow a quantitative analysis of the control and regulation of citric acid production. Finally, we want to assess the quality of the model by carrying out a sensitivity analysis and studying the predictions of the model against perturbations of the basal steady state.

Modern Trends in Biothermokinetics, Edited by
S. Schuster *et al.*, Plenum Press, New York, 1993

2. Medium Requirements and Cellular Processes Involved in the Citric Acid Production by *Aspergillus niger*

The conidia suspensions are inoculated in a culture medium that contains sucrose as substrate (at saturating concentration), inorganic phosphate, Mg^{2+}, Ca^{2+} and Zn^{2+}. The culture medium contains no Mn^{2+}, which is determinant for the accumulation process. Once the conidia are inoculated, they exhibit a phase of growth, and from 96 to 288 hours, the system attains a steady state (idiophase, non-growing, fermentation stage) where citrate production is the only metabolic activity of quantitative importance[3].

Five sets of metabolic events are involved in the citric acid accumulation by *A. niger*:

(1) The metabolism of glucose, which involves a joint participation of the glycolytic and the pentose phosphate pathways. During idiophase the fractional involvement is 4:1. Citrate, which reaches concentrations of around 10 mM (average concentration in the mycelium) inhibits phosphofructokinase. However, NH_4^+, the concentration of which goes up to 3 mM, counteracts the inhibitory effects of citrate. Little difference has so far been detected in the mycelial contents of $F2,6P_2$, AMP and ATP under manganese-deficient and manganese-sufficient conditions.

(2) Anaplerotic pathways connected to the tricarboxylic acid cycle. The CO_2 formed from pyruvate upon its transformation to acetyl-CoA is used by pyruvate carboxylase (a cytoplasmatic enzyme) for the formation of oxaloacetate. The latter is converted into malate by means of cytoplasmatic malate dehydrogenase and then transported to the mitochondrion, thus acting as a countersubstrate of the tricarboxylic acid carrier.

(3) The tricarboxylic acid cycle. The glyoxylate shunt is almost inactive. A specific NADP-isocitrate dehydrogenase is inhibited by citrate and isocitrate. α-ketoglutarate dehydrogenase is inhibited owing to feedback loops from oxaloacetate and NADH. Succinate dehydrogenase is also inhibited by oxaloacetate.

(4) Transport processes. Transport of pyruvate, citrate, isocitrate and ketoglutarate through the mitochondria occurs by means of specific carriers. Glucose uptake into the cell has also been identified as a potential regulatory step.

(5) Oxygen consumption. Oxygen is at saturating concentrations, and there is an alternative respiratory system, sensitive to salicylhydroxamic acid that supports the standard respiratory chain in NADH reoxidation without concomitant ATP production.

3. The Model

All the facts mentioned above, together with the available data on metabolite pools, fluxes and enzyme levels at the considered steady state have been arranged in the mechanistic model shown in Fig. 1. Once the topological structure of the system has been constructed we can proceed to its mathematical representation. Here we have chosen the S-system representation developed by Savageau[2]. The first step in this process is to build up the mass balance equations.

3.1. Mass balance equations

In the following, we give the equations describing the dynamics of the system shown in Fig. 1.

$$\frac{dX_1}{dt} = V_{19,1} - (V_{1,2} + V_{1,0}),$$

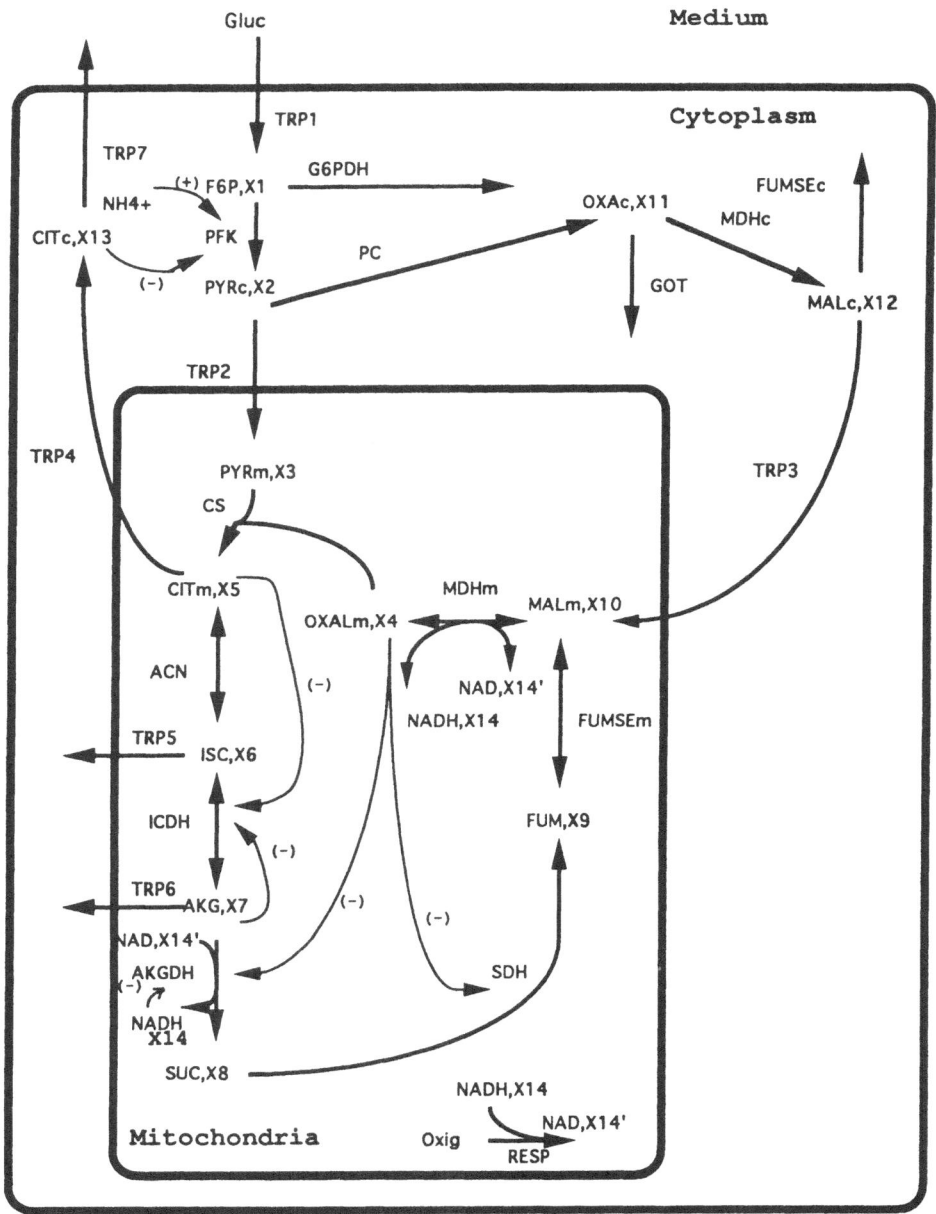

Figure 1. Schematic representation of the most important reactions in *Aspergillus niger*.
The independent variables are X_1, fructose 6-phosphate; X_2, cytosolic pyruvate; X_3, mitochondrial pyruvate; X_4, mitochondrial oxaloacetate; X_5, mitochondrial citrate; X_6, isocitrate; X_7, α-ketoglutarate; X_8, succinate; X_9, fumarate; X_{10}, mitochondrial malate; X_{11}, cytosolic oxaloacetate; X_{12}, cytosolic malate; X_{13}, cytosolic citrate; X_{14}, NADH; $X_{14'}$, NAD. See Table 1 for enzymatic and transport steps.

$$\frac{dX_2}{dt} = 2V_{1,2} - (V_{2,11} + V_{2,3}) \, ,$$

$$\frac{dX_3}{dt} = V_{2,3} - V_{3,5} \, ,$$

$$\frac{dX_4}{dt} = (V_{10,4} - V_{4,10}) - V_{3,5} \, ,$$

$$\frac{dX_5}{dt} = V_{3,5} - [(V_{5,6} + V_{6,5}) - V_{5,13}] \, ,$$

$$\frac{dX_6}{dt} = (V_{5,6} - V_{6,5}) - [V_{6,0} + (V_{6,7} - V_{7,6})] \, ,$$

$$\frac{dX_7}{dt} = (V_{6,7} - V_{7,6}) - (V_{7,0} + V_{7,8}) \, ,$$

$$\frac{dX_8}{dt} = V_{7,8} - (V_{8,9} - V_{9,8}) \, ,$$

$$\frac{dX_9}{dt} = (V_{8,9} - V_{9,8}) - (V_{9,10} - V_{10,9}) \, ,$$

$$\frac{dX_{10}}{dt} = [(V_{9,10} - V_{10,9}) + V_{12,10}] - (V_{10,4} - V_{4,10}) \, ,$$

$$\frac{dX_{11}}{dt} = V_{2,11} - (V_{11,12} + V_{11,0}) \, ,$$

$$\frac{dX_{12}}{dt} = V_{11,12} - (V_{12,10} + V_{12,0}) \, ,$$

$$\frac{dX_{13}}{dt} = V_{5,13} - V_{13,0} \, ,$$

$$\frac{dX_{14}}{dt} = [V_{7,8} + (V_{10,4} - V_{4,10})] - V_{15,14};$$

$$\frac{dX_i}{dt} = V_i^+ - V_i^- \, , \quad i = 1,\dots,14;$$

$$V_3^+ = V_4^+ = V_5^+ \, ,$$

$$V_8^- = V_9^+ \, ,$$

$$V_4^+ = V_{10}^- \, .$$

In these equations all the positive fluxes are grouped together, and so are all the negative ones. This is the so-called irreversible strategy. We can choose this kind of representation because the direction of flux through any reversible reaction is unchanged under the conditions examined in this model. In these equations, $V_{i,j}$ signifies the rate of interconversion of metabolite X_i into metabolite X_j; it also represents the flux through a reaction. The symbols V_i^+ and V_i^- represent the aggregate flux into and out of the pool X_i.

3.2. S-system representation

In this representation the elemental fluxes have been aggregated according to the pools they belong to. Each aggregated rate law, expressed in the power law form, is composed of power law functions, one for each variable that affects the rate of the process. The parameters $g_{i,j}$ and $h_{i,j}$ are called kinetic orders and have the same meaning as the elasticity coefficients as defined in Metabolic Control Theory[4,5] . The parameters α and β are rate constants.

The S-system representation is given by

$$\frac{dX_i}{dt} = \alpha_i \prod_{j=1}^{n+m} X_j^{g_{ij}} - \beta_i \prod_{j=1}^{n+m} X_j^{h_{ij}} \ ,$$

where

$$g_{ij} = \left(\frac{\partial V_i}{\partial X_j}\right)_{ij} \left(\frac{X_j}{V_i}\right)_o \ ,$$

$$h_{ij} = \left(\frac{\partial V_{-i}}{\partial X_j}\right)_{ij} \left(\frac{X_j}{V_{-i}}\right)_o \ ,$$

$$\alpha_i = V_{i0} \prod_j X_{j0}^{-g_{ij}} \ ,$$

$$\beta_i = V_{-i0} \prod_j X_{j0}^{-h_{ij}} \ .$$

For the reaction system considered in this paper, this representation specifies to:

$$\frac{dX_1}{dt} = \alpha_1 X_{17}^{g_{1,17}} X_{18}^{g_{1,18}} \qquad\qquad -\beta_1 X_1^{h_{1,1}} X_{13}^{h_{1,13}} X_{16}^{h_{1,16}} X_{19}^{h_{1,19}} X_{20}^{h_{1,20}} \ ,$$

$$\frac{dX_2}{dt} = \alpha_2 X_1^{g_{2,1}} X_{13}^{g_{2,13}} X_{16}^{g_{2,16}} X_{20}^{g_{2,20}} \qquad\qquad -\beta_2 X_2^{h_{2,2}} X_{21}^{h_{2,21}} X_{22}^{h_{2,22}} \ ,$$

$$\frac{dX_3}{dt} = \alpha_3 X_2^{g_{3,2}} X_{22}^{g_{3,22}} \qquad\qquad -\beta_3 X_3^{h_{3,3}} X_4^{h_{3,4}} X_{23}^{h_{3,23}} \ ,$$

$$\frac{dX_4}{dt} = \alpha_4 X_4^{g_{4,4}} X_{10}^{g_{4,10}} X_{14}^{g_{4,14}} X_{29}^{g_{4,29}} \qquad\qquad -\beta_4 X_3^{h_{4,3}} X_4^{h_{4,4}} X_{23}^{h_{4,23}} \ ,$$

$$\frac{dX_5}{dt} = \alpha_5 X_3^{g_{5,3}} X_4^{g_{5,4}} X_{23}^{g_{5,23}} \qquad\qquad -\beta_5 X_5^{h_{5,5}} X_6^{h_{5,6}} X_{24}^{h_{5,24}} X_{33}^{h_{5,33}} \ ,$$

$$\frac{dX_6}{dt} = \alpha_6 X_5^{g_{6,5}} X_6^{g_{6,6}} X_{24}^{g_{6,24}} \qquad -\beta_6 X_5^{h_{6,5}} X_6^{h_{6,6}} X_7^{h_{6,7}} X_{25}^{h_{6,25}} X_{35}^{h_{6,35}} \;,$$

$$\frac{dX_7}{dt} = \alpha_7 X_5^{g_{7,5}} X_6^{g_{7,6}} X_7^{g_{7,7}} X_{25}^{g_{7,25}} \qquad -\beta_7 X_4^{h_{7,4}} X_7^{h_{7,7}} X_{14}^{h_{7,14}} X_{26}^{h_{7,26}} X_{36}^{h_{7,36}} \;,$$

$$\frac{dX_8}{dt} = \alpha_8 X_4^{g_{8,4}} X_7^{g_{8,7}} X_{14}^{g_{8,14}} X_{26}^{g_{8,26}} \qquad -\beta_8 X_4^{h_{8,4}} X_8^{h_{8,8}} X_{27}^{h_{8,27}} \;,$$

$$\frac{dX_9}{dt} = \alpha_9 X_4^{g_{9,4}} X_8^{g_{9,8}} X_{27}^{g_{9,27}} \qquad -\beta_9 X_9^{h_{9,9}} X_{10}^{h_{9,10}} X_{28}^{h_{9,28}} \;,$$

$$\frac{dX_{10}}{dt} = \alpha_{10} X_9^{g_{10,9}} X_{10}^{g_{10,10}} X_{12}^{g_{10,12}} X_{28}^{g_{10,28}} X_{32}^{g_{10,32}} \qquad -\beta_{10} X_4^{h_{10,4}} X_{10}^{h_{10,10}} X_{14}^{h_{10,14}} X_{29}^{h_{10,29}}$$

$$\frac{dX_{11}}{dt} = \alpha_{11} X_2^{g_{11,2}} X_{21}^{g_{11,21}} \qquad -\beta_{11} X_{11}^{h_{11,11}} X_{30}^{h_{11,30}} X_{38}^{h_{11,38}} \;,$$

$$\frac{dX_{12}}{dt} = \alpha_{12} X_{11}^{g_{12,11}} X_{30}^{g_{11,30}} \qquad -\beta_{12} X_{12}^{h_{12,12}} X_{31}^{h_{12,31}} X_{32}^{h_{12,32}} \;,$$

$$\frac{dX_{13}}{dt} = \alpha_{13} X_5^{g_{13,5}} X_{33}^{g_{13,33}} \qquad -\beta_{13} X_{13}^{h_{13,13}} X_{34}^{h_{13,34}} \;,$$

$$\frac{dX_{14}}{dt} = \alpha_{14} X_4^{g_{14,4}} X_7^{g_{14,7}} X_{10}^{g_{14,10}} X_{14}^{g_{14,14}} X_{26}^{g_{14,26}} X_{29}^{g_{14,29}} \quad -\beta_{14} X_{14}^{h_{14,14}} X_{15}^{h_{14,15}} X_{37}^{h_{14,37}} \;.$$

The constraints are given by

$$X_{14} + X_{14'} = X_T \,, \quad X_T = \text{const.}\,,$$

$$X_{14'} = X_T - X_{14} \,,$$

$$X_{14'} = \gamma_{14} X_T^{f_{14',T}} X_{14}^{f_{14',14}} \;,$$

where

$$f_{14',T} = \frac{\partial X_{14'}}{\partial X_T} \cdot \frac{X_T}{X_{14}} = \frac{X_T}{X_{14}} \,,$$

$$f_{14',14} = \frac{\partial X_{14'}}{\partial X_{14}} \cdot \frac{X_{14}}{X_{14'}} = -\frac{X_{14}}{X_{14'}} \;.$$

The representation in the S-system form is constructed with respect to an operating point and is an exact representation of the system at this point. Moreover, it provides a good representation of the system in a local neighborhood of the operating point. In our model the NADH and NAD pools are not independent of each other. Thus the power law functions in terms of these variables cannot be simply absorbed into the rate constant parameters α and β as they can in simple cases. Because their sum is constant, we can represent this constraint in the power law formalism and use the resulting equation to eliminate $X_{14'}$. $f_{14',T}$ and $f_{14',14}$ are parameters analogous to the parameters α and β and are calculated in the same way (see above equations).

4. Characterization of the Steady State

Description of the steady state at the non-growing, 170 hours idiophase stage of citric acid fermentation implies the determination of metabolite pools, fluxes and kinetic orders. In our model we distinguish between the cytoplasmic and the mitochondrial pools. The metabolite concentrations are higher in the mitochondrial compartment owing to its smaller volume. We have calculated their values assuming that the mitochondria cover around 15% of the total cell volume and that the concentrations in the mitochondrial compartment are 5 times higher than in the cytosol. Regarding enzyme concentrations, only data referring to near-*in vivo* conditions have been considered. In some cases where data are not known, we have used data of related systems such as citric-acid producing yeasts. Fluxes were taken from the literature or calculated from the balance of fluxes and stoichiometric considerations. The flux profile so obtained accounts for the high yield of citric acid production (70-80% of the sugar is transformed into citric acid) obtained either industrially or in the laboratory during idiophase. Regarding the kinetic orders, although the main features of the kinetics of all the enzymes involved have been described, there is, however, a general lack of steady-state rate equations. Consequently, we have determined the kinetic orders mostly by using the information already available in the literature and by assuming Michaelis-Menten kinetics. This fact, rather than a shortcoming of the model is a characteristic of the method: rate equations, apart from being inaccurate in describing the kinetic behavior "*in vivo*", involve such a high number of kinetic constants often inaccessible to experimental determination; that makes this approach practically inappropriate.

5. Results

With the whole set of values of metabolite pools, kinetic orders, the parameters α and β and the independent variables of the system (substrates, NH_4^+ and enzyme concentrations) the system was analyzed by using the simulation package ESSYNS.[6] By this means we were able to carry out the stability and sensitivity analysis. The first requirement of a steady-state model to be biologically meaningful is that the predicted state is stable. This is guaranteed in our case since the real parts of all eigenvalues are negative. After that, our first concern was to see how the control of the citrate production flux was shared among the the different enzymes of the system.

The program ESSYNS gives us the so-called logarithmic gains which have the same definitions as the response and control coefficients in Metabolic Control Theory[4,5]. Table 1 shows how the particular enzyme-catalyzed and transport steps control this flux. As we can see in Table 1, among the whole set of 20 possible points of control only 4 play a significant role in the control of citric acid production: the influx of substrate, the effluxes of citrate out of the mitochondria and through the cell membrane and the step catalyzed by aconitase. Furthermore, one can see the quantitative importance of substrates and effectors on the flux of citrate production.

Table 1. Logarithmic gains.

Logarithmic gains of the enzymes and transport processes of the system towards the citric acid efflux.

Reaction	L (Citrate flux, X_i)
Substrate transport/hexokinase, TRP1	0.236
Glucose-6-phosphate dehydrogenase, G6P-DH	-0.048
Phosphofructokinase, PFK	$3.48 \cdot 10^{-3}$
Pyruvate carboxylase, PC	-0.0309
Pyruvate transport, TRP2	0.0227
Citrate synthase, CS	0
Aconitase, ACN	0.131
NADP-ISDH, ICDH	0.0208
α-Ketoglutarate dehydrogenase, AKG	$-6.32 \cdot 10^{-4}$
Succinate dehydrogenase, SUC	0
Fumarase (mitochondrial), FUMSEm	-0.0683
Malate dehydrogenase (mitochondrial), MALm	0
Malate dehydrogenase (cytosolic), MALc	$4.12 \cdot 10^{-3}$
Fumarase (cytosolic), FUMSEc	$-5.08 \cdot 10^{-3}$
Malate transport, TRP3	$4.31 \cdot 10^{-3}$
Citrate transport (mitochondrial), TRP4	0.519
Citrate transport (cytosolic), TRP7	0.298
Isocitrate transport, TRP5	-0.0486
α-Ketoglutarate transport, TRP6	-0.0352
Alternative respiratory system, RESP	0
Glutamate oxaloacetate transaminase, GOT	$-3.74 \cdot 10^{-3}$

Logarithmic gains of substrates and effectors towards the mitochondrial and cytosolic citric acid pools and the citric acid efflux.

Substrate or effector	L (Citrate Cyt, X_i)	L (Citrate Flux, X_i)
Oxygen	$-2.18 \cdot 10^{-4}$	$-7.63 \cdot 10^{-7}$
Ammonium	0.497	$2.24 \cdot 10^{-4}$
Glucose	0.0641	$1.74 \cdot 10^{-3}$

6. Quality Tests

The S-system representation allows us to carry out another kind of analysis that tests the quality of our model and its underlying assumptions and gives us an accurate diagnosis of the reliability of our system description. These quality tests are the sensitivity analysis, the stability and physical significance of the predicted steady states after perturbation of the reference steady state and the feasibility of the transient phase in the transition from the reference steady state towards the new ones.

6.1. Sensitivity analysis

In the context of the S-system representation sensitivity coefficients are defined as

$$S(X_i, g_{j,k}) = \frac{\partial X_i}{\partial g_{j,k}} \cdot \frac{g_{j,k}}{X_i} \quad \text{or}$$

$$S(X_i, h_{j,k}) = \frac{\partial X_i}{\partial h_{j,k}} \cdot \frac{h_{j,k}}{X_i} \quad .$$

These coefficients quantify the relative change in variable X_i in response to a change in the exponent $g_{j,k}$ or $h_{j,k}$.

$$S(X_i, \alpha_j) = \frac{\partial X_i}{\partial \alpha_j} \cdot \frac{\alpha_j}{X_i} \quad \text{or}$$

$$S(X_i, \beta_j) = \frac{\partial X_i}{\partial \beta_j} \cdot \frac{\beta_j}{X_i} \quad .$$

These coefficients quantify the relative change in variable X_i in response to a change in a rate constant α_j or β_j. High values of some of the sensitivities mean that a small perturbation in the parameter values (such as might occur in response to temperature changes, errors in transcription, etc.) would cause radical changes in concentrations and fluxes. Since real biological systems do not normally behave in this way, one can suspect that certain parts of the overall model are ill-determined. In our model the examination of the sensitivities of the system shows a fairly good situation: most of the values are low. Only a few ones have higher values of around 20-30. On this side the system behaves fairly well.

6.2. New steady-state predictions and transient phase behavior

The steady states that are reached when a perturbation is made in one of the relevant independent variables are unstable or, if stable, yield quite unrealistic values of the metabolite pools and fluxes (concentrations of 10^{+42} or 10^{-22} mM). As a consequence, the transient phase was extremely long, thus indicating the non-feasibility of the behavior of the system.

7. Conclusions

From these results we can conclude that the characterization of the steady state that we calculate using the published data available gives us a description of the control and regulation of the system that will eventually allow us to derive conclusions of physiological interest. However, once a stable steady state has been characterized, further analysis is necessary to assess the quality of our assumptions. Our model, in spite of having a fairly good behavior as far as sensitivity analysis is concerned, turns out to be ill-defined with respect to the other criteria.

8. Further Developments

Current developments of the model include a revised estimation of the metabolite pools and fluxes in order to reflect the effect of compartmentation more precisely and the re-examination of the enzyme activities and Michaelis-Menten kinetics. Eventually new interactions and enzymatic steps will be considered, such as the reactions that transform polyols into citric acid and the anaplerotic formation of oxaloacetate from aspartate by means of the glutamate transaminase reaction.

Acknowledgements

This work was supported by a "Beca para estancias en centros de investigación y universidades en el extranjero" awarded to Dr. Néstor Torres by the Consejería de Educación, Cultura y Deportes del Gobierno de Canarias and by a Research Project of the same institution (Subvención a proyectos de investigación, project number 91/109).

References

1. C.P. Kubicek and M. Rohr, Citric acid fermentation, *C.R.C. Crit. Rev. in Biotechn.* **3**:331-373 (1986).
2. M.A. Savageau. "Biochemical Systems Analysis: A study of function and design in molecular biology," Addison-Wesley, Reading (Mass.) (1976).
3. C.P. Kubicek, O. Zehentgruber and M. Rohr, An indirect method for studying the fine control of citric acid formation by *Aspergillus niger*, *Biotechn. Lett.* **1**:47-52 (1979).
4. H. Kacser and J.A. Burns, The control of flux, *Symp. Soc. Exp. Biol.* **27**:65-104 (1973).
5. R. Heinrich and T.A. Rapoport, A linear steady-state treatment of enzymatic chains. General properties, control and effector strength, *Eur. J. Biochem.* **42**:89-95 (1974).
6. E.O. Voit, D.H. Irvine and M.A. Savageau. "The User's Guide of ESSYNS," Medical University of South Carolina Press, Charleston (S.C.) (1989).

STEADY-STATE MEASUREMENTS AND IDENTIFIABILITY OF REGULATORY PATTERNS IN METABOLIC STUDIES

A. Sorribas[1] and M. Cascante[2]

[1]Departament de Ciències Mèdiques Bàsiques
Universitat de Lleida
Av.Rovira Roure 44, 25006-Lleida, Spain
[2]Departament de Bioquímica i Fisiologia
Universitat de Barcelona
Martí i Franqués 1, 08028-Barcelona, Spain

1. Introduction

A main objective in investigating a metabolic pathway is to obtain a definite idea of both its structure, in terms of material and information relationships, and its behavior. Traditionally, the schematic representation of a given pathway is based on previous information mainly obtained by experiments *in vitro*. Measurements on the intact system are then used for testing the resulting structure. This is justified by the idea that knowing the elements of a given pathway will lead us to a deep understanding of both its properties and its behavior under specified conditions. However, there is now increasing evidence that complete characterization of each element *in vitro* does not necessarily lead to a correct description of the whole system[1-4]. In practice, and especially if we consider the relevant regulatory signals within the system, it is often difficult to agree on a single scheme, and different possibilities arise. On one hand, the *in situ* conditions can make it some of the effects shown *in vitro* irrelevant. On the other hand, it is possible that some interactions not shown *in vitro* could be of interest *in situ*. Hence, we are dealing with a situation close to a *black box*, in the sense that the relationships within the system are not clearly established.

The limitations to using a reductionist approach make it necessary to focus on obtaining useful information in the intact system, and in developing appropriate tools for dealing with this information to build up an integrated description of the intact system. This will be especially important in trying to identify the structure of the target system. With this aim in mind, the definition of what is meant by *useful information* will largely depend on the tools we use for analyzing the experimental measurements obtained in metabolic systems. If we approach the problem with the whole system in mind, mathematical tools based on the so-called *power-law formalism* provide a convenient way of addressing this kind of question. Within this formalism, a number of different variant approaches have been defined. The two most widely used alternatives are Biochemical Systems Theory (BST) and Metabolic Control Analysis (MCA) (also referred to as Metabolic Control Theory). Although there are differences in nomenclature[5-16], both approaches deal basically with the same kind of experimental measurements and lead essentially to the same basic characterization of the system in a given steady-state (see Refs 8, 9, 15, 16 for detailed comparisons and discussion). However, when we consider the dynamic response of the system to a perturbation in an independent variable (external effector, kinetic parameter, the substrate of the pathway, etc.), the MCA approach does not provide appropriate tools. In this case, the *S-system representation* (see Ref. 8 and references therein) within BST should be used.

Because the S-system equations have the ability to combine both the steady-state characterization and the dynamic response of the system, they will be used as a theoretical framework for defining which kind of measurements are of interest for obtaining a precise image of the target system, especially in addressing the question of discriminating between alternative regulatory patterns in a given metabolic pathway. We will briefly discuss the limitations of the information contained in the steady-state measurements and how the measurement of the dynamic response can help in testing alternative patterns accounting for the regulatory interactions within a metabolic pathway.

2. Methods

We use the S-system representation within BST. This technique has been presented elsewhere (see Ref. 17 and references therein for an account of the state-of-the-art on S-system related methods) and its relationship with MCA has been discussed elsewhere[6-9,11,13,14]. The reader is referred to this literature for a detailed discussion of common concepts referred to under different nomenclature (see also Refs 15,16). Nevertheless, it is important to recall that the basic parameters in BST, i.e. kinetic orders, are *conceptually* equivalent to the elasticities defined in MCA. In this sense, g_{ij} and h_{ij} refer to the effect of variable X_j (whatever it represents: enzyme concentration, metabolite effector, V_m, internal metabolite) on the process of synthesis and degradation, respectively, of the dependent variable X_i. Hence, a given set of regulatory signals and chemical reactions will have a correspondence in a set of kinetic-orders. In MCA, these parameters are divided into several categories such as elasticities, and special elasticities.

The basic steady-state characterization in BST is obtained by means of the logarithmic gains, which correspond to the effect of a change in an independent variable (i.e. a variable that can be fixed externally) on a dependent variable (i.e. a variable whose value depends on the parameters and external conditions of the system). Logarithmic gains correspond to different concepts in MCA (Flux control coefficient, Concentration control coefficient, Response coefficient, and Parameter sensitivity). These correspondences have been widely discussed elsewhere (see references above).

3. Assessment of the Actual Regulatory Pattern

Once the flux of material through the system has been established, assessing the regulatory pattern for the system means specifying the set of kinetic-orders that correspond to the relevant regulatory signals, and obtaining numerical estimates for these parameters. In the general case in which we are not sure which signals are to be taken into account, we cannot apply the estimation procedures devised for a fixed set of parameters (see references in Ref.

8). In these situations, the solution procedure demands a strategy for discovering the relevant signals from measurements on the intact system.

We will address this question by showing that a given steady-state characterization is compatible with different regulatory patterns, and that we can obtain all the compatible patterns for a given situation. Then we will discuss which kind of experiment can be performed to discriminate between the different possibilities in order to identify the actual regulatory pattern.

4. Logarithmic Gains and the Regulatory Pattern

The basic characterization of a steady-state is achieved by means of the logarithmic gains of both fluxes and concentrations. Hence, and because the logarithmic gains are related to the kinetic orders by a simple equation, the measurement of the set of logarithmic gains seems to be a natural way to obtain the kinetic-order set. For a given dependent variable X_j $(i=1,..,n)$, the corresponding equations are[19]:

$$L(V_i,X_k) = g_{ir} + \sum_{j=1}^{n} g_{ij} \cdot L(X_j,X_r) , \quad r=(n+1),...,(n+m),$$ (1)

where m indicates the number of independent variables considered. To better appreciate the meaning of this equation, consider a system with $n=4$, $m=3$. Then for the synthesis of X_2, eq. (1) becomes:

$$\begin{pmatrix} L(V_2^+,X_5) \\ L(V_2^+,X_6) \\ L(V_2^+,X_7) \end{pmatrix} = \begin{pmatrix} L_{15} & L_{25} & L_{35} & L_{45} & 1 & 0 & 0 \\ L_{16} & L_{26} & L_{36} & L_{46} & 0 & 1 & 0 \\ L_{17} & L_{27} & L_{37} & L_{47} & 0 & 0 & 1 \end{pmatrix} \cdot \begin{pmatrix} g_{21} \\ g_{22} \\ g_{23} \\ g_{24} \\ g_{25} \\ g_{26} \\ g_{27} \end{pmatrix},$$ (2)

where $L_{ij}=L(X_i,X_j)$. Clearly, this equation, which represents the situation in which all the variables in the system have an effect on the process of synthesis of X_2, has no unique solution. Because there are three independent variables, the maximum number of variables we can include in eq. (2) with a kinetic order different from zero, is three (see, however, Ref. 20). Suppose we suspect that only X_1, X_2 and X_7 are considered to have an effect on V_2. In such a case, eq. (2) becomes:

$$\begin{pmatrix} L(V_2,X_5) \\ L(V_2,X_6) \\ L(V_2,X_7) \end{pmatrix} = \begin{pmatrix} L_{15} & L_{25} & 0 \\ L_{16} & L_{26} & 0 \\ L_{17} & L_{27} & 1 \end{pmatrix} \cdot \begin{pmatrix} g_{21} \\ g_{22} \\ g_{27} \end{pmatrix},$$ (3)

which has a unique solution provided the squared matrix has an inverse.

Consider now an alternative hypothesis that states that X_1, X_3 and X_7 are the variables involved in V_2. In this case, eq. (2) becomes:

$$\begin{pmatrix} L(V_2, X_5) \\ L(V_2, X_6) \\ L(V_2, X_7) \end{pmatrix} = \begin{pmatrix} L_{15} & L_{35} & 0 \\ L_{16} & L_{36} & 0 \\ L_{17} & L_{37} & 1 \end{pmatrix} \cdot \begin{pmatrix} g_{21} \\ g_{23} \\ g_{27} \end{pmatrix}, \tag{4}$$

which also has a unique solution. Following this rationale, and provided we have m independent variables and n dependent variables, there are $\binom{n+m}{m}$ possible *subsets* containing a maximum of m variables. For the example considered, with $n=4$ and $m=3$, this gives 35 possibilities for V_2^+. This should be considered the maximum number of possibilities because we rule out all those possibilities that do not agree with the flux of material. This means that if X_1 is the substrate for the synthesis of X_2, then g_{21} is a mandatory parameter for V_2^+. In consequence, all the cases in which X_2 does not appear should be rejected. In addition, some of the alternatives, given a set of measured logarithmic gains, could lead to non-meaningful sets of kinetic orders and would also be rejected. At the end, however, we will still end up with several possibilities that will need to be discriminated, in order to obtain the actual structure of the system (see Ref. 20 for numerical examples).

5. Discrimination between Alternative Patterns of Regulation

The results discussed above clarify the limitations of the steady-state characterization for identifying the actual structure of the regulatory signals. We have also shown that a single set of logarithmic gains can be compatible with a large number of different regulatory patterns. Hence, the question is: *How can we discriminate between the possibilities?*

If we focus on the kind of measurements on the intact system that can answer this question, we must turn to measurements that involve the dynamic response of the system. A measurement that can easily be related with the underlying kinetic-orders consists in evaluating the initial response of a given metabolite after a change in a variable of the system[18] (see also Ref. 5 for a related approach within MCA, although it applies to limited cases). This technique allows experimental estimation of the value of a_{ik} ($a_{ik}=g_{ik}-h_{ik}$) by following the progress curve of X_i after a perturbation of X_k (Fig.1). If s is the initial slope (initial rate of change) of this curve after perturbing X_k from its steady-state value (X_{k0}) to a

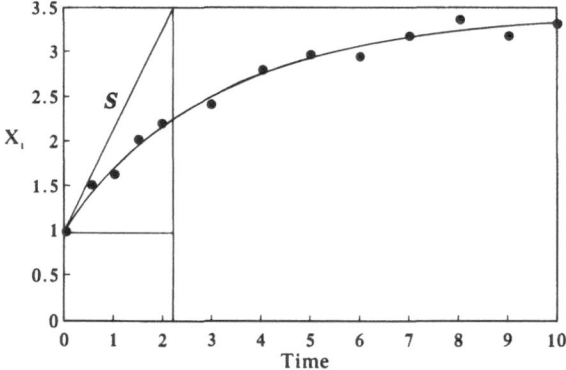

Figure 1. Experimental determination of a_{ik}. X_i is a dependent variable. At $t=0$, the variable X_k is increased. s indicates the initial rate of change of X_i after the perturbation. The solid lines indicate the computation of s as the slope of the progress curve at $t=0$. a_{ik} is computed as in eq. (6) (see text for explanation).

new value (X_{kp}), the value of a_{ik} can be estimated from this slope and from the steady-state values at the reference state of X_k (X_{k0}) and V_i (V_{i0}). That is (see Ref. 18 for a discussion of this result):

$$a_{ik} = \frac{s}{X_{kp} - X_{k0}} \cdot \frac{X_{k0}}{V_{i0}} \ . \tag{5}$$

Eq. (5) can be applied, in theory, to any X_i and X_k. However, in practice this procedure may be difficult to apply in some cases:

(1) When X_k is a dependent variable, and
(2) when X_k is an independent variable, but the required perturbation is difficult to perform experimentally.

These cases would limit the applicability of the method, and prevent its widespread application to solve for all the possible parameter values. Fortunately, in order to identify the regulatory pattern between the resulting alternative possibilities after the steady-state characterization, it is not necessary to estimate all the a_{ij} parameters experimentally. Only those a_{ij} parameters that are critical for discriminating between the competing alternative designs are necessary[20]. This can be better understood by an example. Consider a metabolic pathway with $n=4$ dependent variables, and $m=3$ independent variables. Consider also that after analyzing the steady-state characterization, we have found three competing alternatives for explaining V_1^+:

Alternative 1: $g_{11} = -0.12$, $g_{12} = -0.23$, $g_{16} = 0.34$;
Alternative 2: $g_{12} = -0.23$, $g_{14} = -0.20$, $g_{16} = 0.65$; $\qquad\qquad$ (6)
Alternative 3: $g_{12} = -0.20$, $g_{15} = -0.12$, $g_{16} = 0.97$.

In order to simplify the problem, consider that V_1^- includes neither X_5 nor X_6, which determines that $a_{15} = g_{15}$ and $a_{16} = g_{16}$. Then, instead of trying to measure each of the competing kinetic orders included in eq. (6), the best strategy is to measure the response of X_1 to a perturbation on X_6. Let us consider that X_6 has a steady-state value equal to 10, and that $V_1^+ = 2$. According to eq. (5), if we increase X_6 by a 25%, the expected initial rate of change in X_1 can be computed as:

$$s = \frac{a_{ik}}{X_{kp} - X_{k0}} \cdot \frac{X_{k0}}{V_{i0}} \ . \tag{7}$$

Hence, from the values of g_{16} in eq. (6), the *predicted* values of the initial slope are:

Alternative 1: $\hat{s} = 0.68$;
Alternative 2: $\hat{s} = 1.30$; $\qquad\qquad$ (8)
Alternative 3: $\hat{s} = 1.94$.

With this predicted behavior, a single experiment in which the value of this slope is measured (as in Fig.1) can be sufficient to discriminate between the three alternatives, and in turn to fix the values of the other kinetic orders involved. The information obtained after characterizing the steady-state, that is the set of compatible regulatory patterns, is fundamental for designing the best perturbation experiments leading to identification of the actual regulatory structure. This results in lowering the required experimental effort for characterizing the system.

6. Strategies for Identifying the Regulatory Pattern

If the aim is to study the regulatory pattern of a given metabolic pathway, we suggest the following strategy[20]:

1. Draw the scheme accounting for the flux of material through the system. Recognize the kinetic orders that must be included because they are related to a given metabolite by a flow of material. Collect information on the individual processes from experiments *in vitro*. Consider any reason that suggest a given regulatory interaction in the system.
2. Characterize the reference steady-state experimentally (measure the logarithmic gains for both metabolites and fluxes).
3. Determine all the compatible patterns of regulation for each V_i in the system (eq. 3).
4. Analyze the patterns obtained in order to rule out all cases that lead to absurd results. With the remaining alternatives, decide on the best perturbation experiments in order to discriminate between all the possibilities.
5. Perform the perturbation experiments. Rule out all the possibilities that do not match these experiments (eq. 8).
6. If there are still several undecidable alternatives, consider the information obtained *in vitro*. Consider also the possibility of using other techniques to measure a given kinetic order (see Ref. 18 for a discussion of the techniques available).

7. Discussion

In considering the use of mathematical models for understanding metabolic pathways, it is useful to remember that a given pathway is a complex system and that its behavior is more than the simple sum of its parts. Both the BST and MCA approaches have stressed the importance of considering the system as a whole in order to understand its properties and behavior. We have shown that this is particularly critical for identifying the underlying regulatory structure of a metabolic pathway. The strategy presented in this contribution should be considered a promising possibility to establish such a structure experimentally. Nowadays, the required measurements present some difficulties. However, the possibility of identifying the structure of the system, as shown in this paper, may act as stimulus to the search for new technical devices for obtaining the required data. Our theoretical results may help in encouraging an experimental approach based on measurements on the intact system.

Acknowledgements

The authors were supported by a Grant of the Comissió Interdepartamental de Recerca i Innovació Tecnològica (Programa de Química Fina) de la Generalitat de Catalunya.

References

1. F. Shira-ishi and M.A. Savageau, The tricarboxilic acid cycle in Dictyostelium discoideum I. Formulation of alternative kinetic representations, *J. Biol. Chem.* **267**:22912-22918 (1992).
2. F. Shira-ishi and M.A. Savageau, The tricarboxilic acid cycle in Dictyostelium discoideum II. Evaluation of model consistency and robustness, *J. Biol. Chem.* **267**:22919-22925 (1992).
3. F. Shira-ishi and M.A. Savageau, The tricarboxilic acid cycle in Dictyostelium discoideum III. Analysis of steady-state and dynamic behavior, *J. Biol. Chem.* **267**:22926-22933 (1992).
4. N. Torres, C. Regalado, A. Sorribas and M. Cascante, Quality assessment of a metabolic model and sys-

tems analysis of citric acid production by *Aspergillus niger*, this volume.

5. J. Delgado and J.C. Liao, Determination of Flux control coefficients from transient metabolite concentrations, *Biochem. J.* **282**:919-927 (1992).

6. M.A. Savageau, E.O. Voit and D.H. Irvine, Biochemical systems theory and metabolic control theory. 1. Fundamental similarities and differences, *Math. Biosci.* **86**:127-145 (1987).

7. M.A. Savageau, E.O. Voit and D.H. Irvine, Biochemical systems theory and metabolic control theory. 2. Flux oriented and metabolic control theories, *Math. Biosci.* **86**:147-169 (1987).

8. A. Sorribas and M.A. Savageau, A comparison of variant theories of intact biochemical systems. 1: Enzyme-enzyme interactions and biochemical systems theory, *Math. Biosci.* **94**:161-193 (1989).

9. A. Sorribas and M.A. Savageau, A comparison of variant theories of intact biochemical systems. 2: Flux oriented and metabolic control theories, *Math. Biosci.* **94**:195-238 (1989).

10. A. Sorribas and M.A. Savageau, Strategies for representing metabolic pathway within biochemical systems theory. Reversible pathways, *Math. Biosci.* **94**:239-269 (1989).

11. M. Cascante, R. Franco and E.I. Canela, Use of implicit methods from general sensitivity theory to develop a systemic approach to metabolic control. 1. Unbranched pathways, *Math. Biosci.* **94**:271-288 (1989).

12. M. Cascante, R. Franco and E.I. Canela, Use of implicit methods from general sensitivity theory to develop a systemic approach to metabolic control. 2. Complex systems, *Math. Biosci.* **94**:289-309 (1989).

13. M.A. Savageau, Biochemical systems theory: operational differences among variant representations and their significance, *J.Theor. Biol.* **151**:509-530 (1991).

14. M.A. Savageau, Dominance according to metabolic control analysis: major achievement or house of cards? *J.Theor. Biol.* **154**:131-136 (1992).

15. A. Sorribas, R. Curto and M. Cascante, Comparative characterization of the fermentation pathway of *Saccharomyzes cerevisiae* by using the Biochemical Systems Theory and the Metabolic Control Analysis: 1. Model definition and nomenclature, *Biochem. J.*, submitted.

16. A. Sorribas, R. Curto and M. Cascante, Comparative characterization of the fermentation pathway of *Saccharomyces cerevisiae* by using the Biochemical Systems Theory and the Metabolic Control Analysis: 2. Steady-state characterization and dynamics, *Biochem. J.*, submitted.

17. E.O. Voit (ed.). "Canonical Non-Linear Modelling: S-system Approach to Understanding Complexity," Van Nostrand Reindhold, New York (1991).

18. A. Sorribas, S. Samitier, E.I. Canela and M. Cascante, Metabolic pathway characterization from transient response data obtained *in situ*: Parameter estimation in S-system models, *J. Theor. Biol.*, in press.

19. M.A. Savageau, & Sorribas, A. Constraints among molecular and systemic properties: implications for physiological genetics, *J. Theor. Biol.* **141**:93-115 (1989).

20. A. Sorribas and M. Cascante, Structure identifiability in metabolic pathways: parameter estimation in models based on the power-law formalism, *Eur. J. Biochem.*, submitted.

A BIOCHEMICAL "NAND" GATE
AND ASSORTED CIRCUITS

Herbert M. Sauro

Penrodyn, Pontrhydygroes
Ystrad Meurig, Dyfed SY25 6DP, United Kingdom

1. Introduction

The study of metabolic cascades is often associated with the notion of 'switches', that is, mechanisms that can exist in one of two defined states, usually called 'ON' and 'OFF'. A number of workers have become interested in the possibility of designing digital and neural-like circuits based on the metabolic cascade[1-3]. In this short paper I would like to describe some metabolic digital 'devices' based on the cascade cycle. One of the most basic components in digital electronics is the NAND gate. This gate is important because almost any other logical device can be constructed from it; for example, NOR gates, clocks and flip-flops are easily built using particular configurations of NAND gates[4]. What follows will be a description of a basic metabolic NAND gate, a NOT gate, a novel metabolic `clock' based around the NOT gate and a metabolic flip-flop.

2. The Metabolic Cascade

The simplest metabolic cascade consists of two opposing reactions that interconvert two chemical species in a cyclic manner (see Fig. 1). Thus one reaction converts, say E_1 to E_2, and a second converts E_2 back to E_1, thereby forming a continuous cycle. Because there is no net loss of E moiety from the cycle, the total amount of moiety in the cycle, namely, $E_1 + E_2$, is constant. A second important characteristic of a cascade cycle is that each reaction must have a different thermodynamic equilibrium constant so that the cycle does not run to equilibrium. Clearly such a requirement implies the existence of different thermodynamic driving forces on each reaction. In real cascades this is usually achieved by driving one reaction with the conversion of ATP to ADP and the concomitant addition of phosphate and the reverse reaction by the hydrolysis and subsequent release of phosphate. A final requirement of the cascade, and one which greatly enhances its 'switching' characteristics, is the saturability of the reactions. If the reactions are first order over the whole range of cycle

Figure 1. The basic unit of the metabolic cascade.

moiety concentrations then no 'switching' can occur. The saturability of the steps is therefore essential[5]. In real cascades saturability is achieved by employing enzymes as the catalysts of the cycle reactions.

As a simple example, the time course simulation shown in Fig. 2, illustrates the basic 'switch' mechanism — a SCAMP listing of this model which includes all the values for the constants is given in Listing 1 (Appendix)[&]. In the first part of the simulation, the system is allowed to settle to its 'resting' state, which for argument's sake will be considered to be the 'OFF' position (concentration of E_1 is low). The convention that the cascade is 'OFF' is of course arbitrary because when one of the moieties is 'OFF' — i.e. low concentration — the other (by mass conservation) must be 'ON' — high concentration. At about time=20, the activity of the enzyme that catalyzes the conversion of E_2 to E_1 is increased and the cascade very rapidly switches over to its 'ON' position as indicated by the sudden change in concentration of the cycle moieties. At about time = 60, the activity of the same enzyme is brought back to its original level and the cascade rapidly returns to its 'OFF' position. These events can be repeated any number of times. This simple behavior immediately suggests a through gate, i.e. a logical gate that passes its input to its output (see Table 1). If instead of concentrating on the positively reacting moiety one considers the other moiety of the cycle, then the logical transformation is equivalent to the NOT gate so that when the enzyme activity is increased the concentration of moiety decreases, and *vice versa*, (see Table 1).

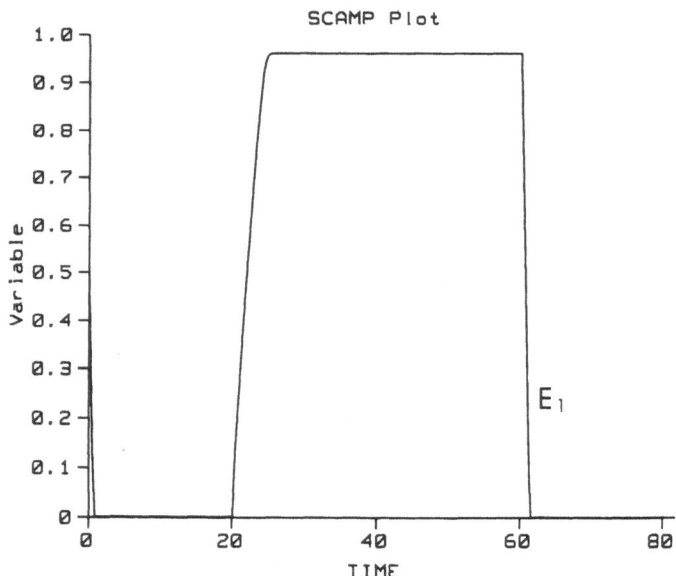

Figure 2. Simulation showing basic cascade switch response (ON at t=20, OFF at t=60), see also Listing 1.

[&]All simulations were performed using the simulation language SCAMP[6] on an Atari TT — SCAMP is also available for MS-DOS computers.

134

Table 1. Scheme showing 'through' and NOT logical transformations.

Input	Through gate	NOT gate
1	1	0
0	0	1

3. The NAND Gate and Ring Oscillator

If two effectors are used to change the rate of reaction instead of one, and at the same time the concentration of each applied effector is halved, then the cascade can emulate the NAND gate — or an AND gate if we observe the positively reacting moiety.

If two NOT gates are brought together so that the output of one goes to the input of the other, the net effect is to make a pass-through gate. If a third NOT gate is added then the net effect is to produce another NOT gate. An interesting effect occurs when the output of the third gate is connected to the input to the first gate (see Fig. 3a). If gate one moves from state

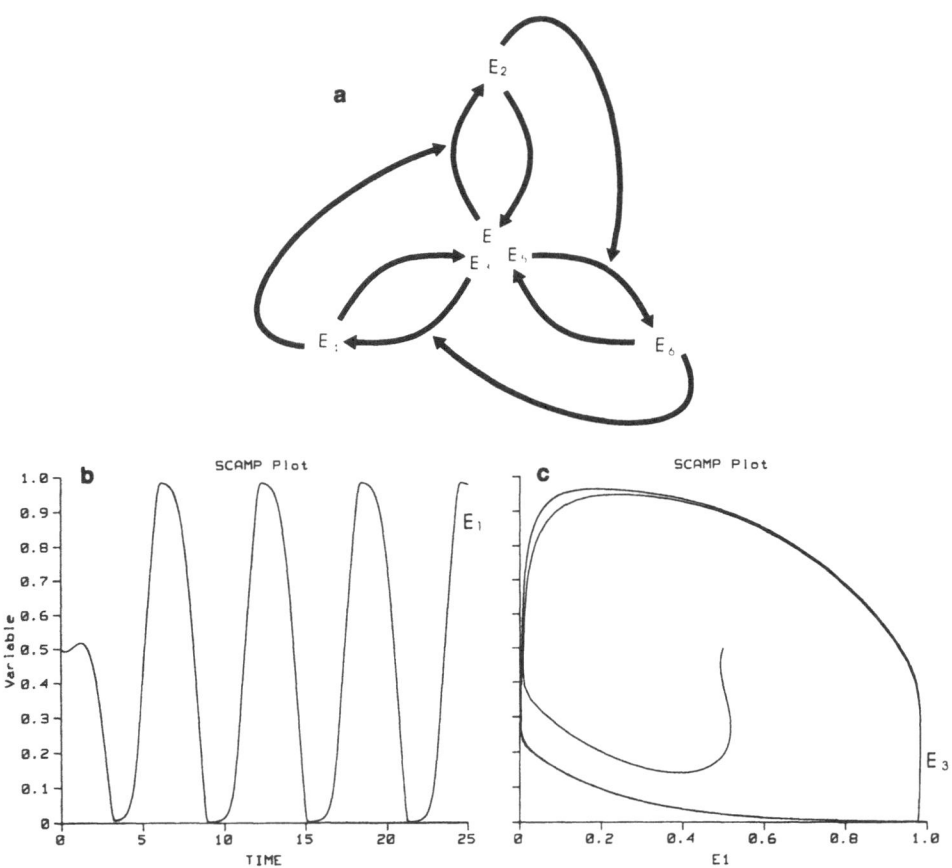

Figure 3. (a) Structure of the three NOT gate star oscillator. (b) Simulation of the NOT gate star oscillator showing evolution of the E_1 moiety. (c) Simulation of the NOT gate star oscillator showing phase plot of moiety E_1 vs. E_3.

Table 2. Scheme showing NAND logical transformation.

Input 1	Input 2	NAND gate
1	1	0
1	0	1
0	1	1
0	0	1

zero to state one then gate two moves from state one to state zero and the third gate moves from state zero to one. Since the third gate is connected to the first gate, the first gate acquires an input opposite to what it started with. As a result, the system begins to oscillate with each gate firing the next in a never ending sequence. Only an odd number of gates will oscillate, an even number will not. In addition, because each gate has the same kinetic characteristics (at least for the model presented here), the pulsing at each gate is 120 degrees out of phase from the next. A SCAMP listing for the ring oscillator is given in Listing 2 (Appendix), and a graph showing the pattern of oscillation for a single moiety is given in Figs 3b and 3c.

4. The Flip-flop and Binary Counter

One of the simplest logic blocks that can store information is the so-called RS flip-flop. (The RS is short-hand for set-reset[4]). This flip-flop has two inputs and two outputs (see Fig. 4). The two inputs are usually designated $\bar{S}$ and $\bar{R}$, the $\underline{S}$et input and $\underline{R}$eset input, respectively. The 'bar' above each input label refers to the fact that the actuating state is the OFF state. The $\bar{S}$ input line when held close to zero, causes a 1 bit to be stored in the flip-flop. The $\bar{R}$, on the other hand, when held close to zero causes a zero bit to be stored in the flip-flop. One problem with the simple RS flip-flop is that if both inputs are held close to zero then the condition is undefined. The two outputs, usually designated, Q and $\bar{Q}$, are the complement of each other and merely reflect the internal state of the flip-flop so that when the flip-flop has a 1 bit stored, the Q output is high and the $\bar{Q}$ output is low. Table 3 gives a summary of these input/output relationships.

Table 3. Scheme showing RS flip-flop states.

$\bar{S}$	$\bar{R}$	Q	$\bar{Q}$
1	1	unchanged	unchanged
1	0	1	0
0	1	1	1
0	0	disallowed	disallowed

Logically, the RS flip-flop can be readily built from two NAND gates, where one of the outputs of each NAND gate is made to go to the input of the other (see Fig. 4). Constructing an RS flip-flop from two cascades is straightforward and Fig. 5 shows a pathway scheme.

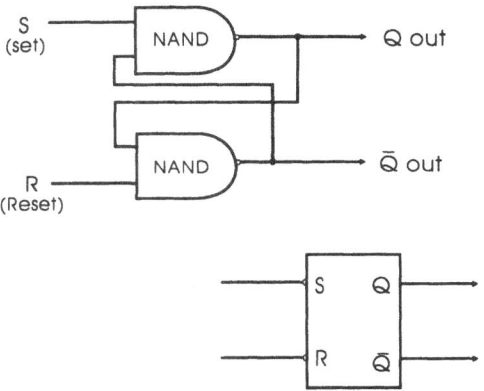

Figure 4. RS flip-flop schematic and internal NAND gate structure.

Two problems are associated with the RS flip-flop. The first problem, but perhaps a somewhat minor one, is the negative logic of the input lines. This means that in order to actuate a function an input line must be brought into a low state. The second problem relates to the way RS flip-flops are used to build other devices. Most operations in a digital device (especially the digital computer) occur synchronously, that is in a defined and regulated sequence. This is usually achieved by operating the system from a central clock where each

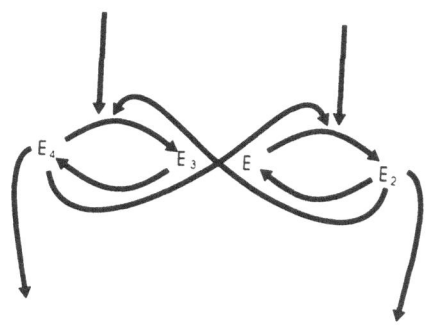

Figure 5. RS flip-flop built from two simple metabolic cascades.

device receives an appropriately timed pulse at the right moment. Unfortunately, the RS flip-flip has no facility to accept a clock line and so there is no way to synchronize its operations with other devices. The solution to the problem is to add two extra NAND gates coupled to an additional 'clock' input line (see Fig. 6). The idea is that even though the $\bar{S}$ and $\bar{R}$ may have valid inputs, only when the clock line is actuated does the flip-flop receive the information. In addition, because the new design places two NAND gates between the input lines and the RS flip-flop, the input lines become inverted so that 'normal' logic applies. With this type of flip-flop it is a trivial matter to construct other devices, for example a binary counter. The three basic elements of a digital computer have now been described, a clock to synchronize events, the NAND gate to perform both logical and arithmetic operations and the flip-flop for storage and retrieval of information. With these three elements, one could, in theory, build a metabolically based digital computer.

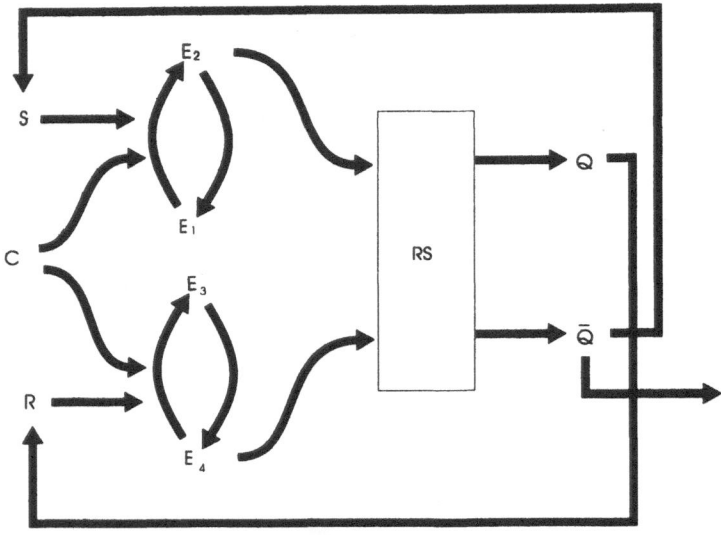

Figure 6. Clocked RS flip-flop built from four metabolic cascades.

References

1. M. Okamoto and K. Hayashi, Dynamic behaviour of cyclic enzyme systems, *J. theor. biol.* **104**:591-598 (1983).
2. M. Okamoto, T. Sakai and K. Hayashi, Biochemical switching device: how to turn on (off) the switch, *Biosystems* **21**:1-11 (1987)
3. A. Hjelmfelt, E.D. Weinberger and J. Ross, Chemical implementation of neural networks and Turing machines, *Proc. Natl. Acad. Sci. USA* **88**:10983-10987 (1991).
4. D. Lancaster. "TTL Cookbook," Howard W. Sons, Indianapolis (Indiana) (1974).
5. J.R. Small and D.A. Fell, Covalent modification and metabolic control analysis, *Eur. J. Biochem.* **191**:405-411 (1990) .
6. H.M. Sauro and D.A. Fell, SCAMP: A metabolic simulator and control analysis program, *Mathl. Comput. Model.* **15**:15-28 (1991).

Appendix: Listings for the Simulation Package "SCAMP"

Listing 1. The basic cascade switch described in SCAMP modeling script[6].

```
Title Model of Switchable Cascade Cycle;

Dec S1, S2; Simulate;

Reactions;

  S2 - S1;   D1*Input*S2/(KM1 + S2);
  S1 - S2;   D2*S1/(KM2 + S1);

eor;

Initialise;
```

```
    CSUM1 = 1.0;
    D1 = 1.4; D2 = 0.9;  # Vmax type parameters;

    Input = 0.1; S1 = 0.5; S2 = 0.5;

    KM1 = 0.01; KM2 = 0.01;  # Easily saturable steps;
ei;

# Switch the cascade 'on' and 'off' here;
monitor_sim
if TIME > 25 then Input = 0.8;
if TIME > 75 then Input = 0.05;
fend;

sim_points = 300;
Timeend = 100;
tolerance = 1E-3;

print_sim  TIME, S1 (results/2);
end
```

Listing 2. SCAMP script for a three NOT gate star oscillator.

```
Title NOT Gate Star Oscillator;

DEC E1, E2, E3, E4, E5, E6;

Simulate;

Reactions;

# NOT gate Unit 1;

  E2 - E1; D1*(Input+E6)*E2/(KM1 + E2);
  E1 - E2; D2*E1/(KM2 + E1);

# NOT gate Unit 2;

  E4 - E3; D1*E2*E4/(KM1 + E4);
  E3 - E4; D2*E3/(KM2 + E3);

# NOT gate Unit 3;

  E6 - E5; D1*E4*E6/(KM1 + E6);
  E5 - E6; D2*E5/(KM2 + E5);

eor;

Initialise;
  CSUM1 = 1.0; CSUM2 = 1.0; CSUM3 = 1.0;
  D1 = 1.4; D2 = 0.9;

  E1 = 0.5; E2 = 0.5;
  E3 = 0.5; E4 = 0.5;
  E5 = 0.5; E6 = 0.5;
  Input = 0.1;

  KM1 = 0.01;
  KM2 = 0.01;
```

```
ei;

sim_points = 300;
Timeend = 25;
tolerance = 1E-3;

print_sim  TIME, E1, E3, E5 (RES1/2);
end
```

_____

DYNAMIC MODELING OF CELL, SUBSTRATE, AND PRODUCT CONCENTRATIONS AND pH IN *E. COLI*-CONTAINING AGAR MEMBRANES

J. Lefebvre and J.C. Vincent

U.R.A. 500, C.N.R.S.
Université de Rouen
B.P. 118, 76134 Mont Saint Aignan Cedex, France

1. Introduction

Biotechnological processes using immobilized cells have been developed during recent years because of their advantages when compared to batch cultures. However, the complexity of the phenomena involved requires dynamic modeling of the systems, firstly, from a fundamental point of view, to gain a better understanding of the phenomena, and secondly, to better optimize and control the existing processes.

Two types of immobilized cell models can be distinguished[1]:

(i) structured models taking into account the modifications of the physiological state of the cells[2], and
(ii) unstructured models considering biomass as a single variable.

The new trend is towards dynamic modeling of the systems[3,4] instead of studying only their steady state. Modeling now tries to take into account a maximum of parameters which are related to both physiology (metabolism, cell death, inhibition...) and physico-chemistry (mass transfer)[5]. A few recent works show that the immobilized cell systems lead to a heterogeneous repartition of the cells inside the membrane, the cells reaching a maximum concentration in a layer near the membrane-solution interfaces[6-8].

In order to better understand the behavior of immobilized cell systems, we present here the unstructured dynamic modeling of *E. coli*-containing membrane systems. The modeling takes into account the cell metabolism characteristics studied in batch cultures, the diffusion parameters controlling the mass-transfer, and also the cell leakage at membrane-solution interfaces. pH was also studied and simulations were compared with the experimentals.

2. Material and Methods

Escherichia coli 1464402 was provided by the Le Havre municipal laboratory. Cells were cultivated at 37 °C in a growth medium described in Fig. 1. Cells were immobilized by suspending a calibrated cell inoculum in a 2.9% agar solution heated at 40 °C. The solution was then cast in order to obtain the desired area and thickness. Experiments were performed by plunging the gel slab containing the immobilized cells into the stirred growth medium. Finally, the gel slab was cut into 300 μm slices using a microtome.

In the external medium, glucose concentration was measured by using high performance liquid chromatography with a refractometric detector. Cell concentration was measured by spectrophotometry at 546 nm according to a calibrated curve. For immobilized cells, either whole membrane or gel slices were mechanically ground, prior to measurement of the cell concentration by spectrophotometry.

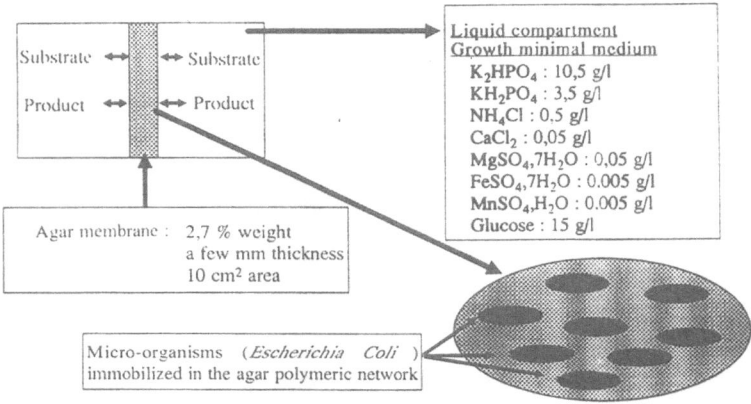

Figure 1. Immobilized cell system structure of the membrane and composition of the growth medium

3. Modeling

3.1. Characterization of the reaction and diffusion phenomena

The first step of the modeling was to define the characteristics of the cell metabolism. The experimental system concerns the anaerobic fermentation of glucose by *E. coli*. It has been already shown that this fermentation is characterized by a bioproduction of a few different weak acids (acetic, formic, lactic ...) responsible for the decrease in pH of the culture medium. For the sake of simplicity, we decided to represent this acid mixture by a single generic weak acid of $pK_a = 4.7$. In order to determine the physiological behavior of the batch culture and the associated kinetic parameters, we used an approach based on the graphical analysis of the substrate, cell and product kinetic curves. This analysis allowed us to define a simplified reaction scheme, which mainly shows two components: the first one coupled to growth and the second one uncoupled, both of them being inhibited by the weak acids produced.

Concerning the mass transfer, diffusion studies[9] in such immobilized cell membranes show that the diffusion coefficients were dependent on the biomass immobilized in the polymeric matrix.

3.2. Theory

Because of the geometry of the structure, a one-dimension (X axis) model for the coupling between diffusion, reaction and growth, was developed. We assumed that:

i) the cells in suspension and the immobilized cells have the same kinetic behavior;
ii) all the chemical species diffuse inside the membrane, but not the cells.

The time evolution of the system can be calculated from the following differential equations for the cell (B), substrate (S) and product (P) concentrations:

$$\frac{\partial B}{\partial t} = (\mu - K_d) \cdot B , \tag{1a}$$

$$\mu = \mu_m \frac{S}{S + K_m} \cdot \left(1 - \frac{P}{P_{i1}}\right), \tag{1b}$$

$$\frac{\partial S}{\partial t} = -\left[\frac{\mu}{Y_{xs}} \cdot B + \frac{r_p}{Y_{sp}} \cdot \left(1 - \frac{P}{P_{i2}}\right) \cdot B + m_s \cdot B\right] + D_s(B) \cdot \frac{\partial^2 S}{\partial x^2} , \tag{2}$$

$$\frac{\partial P}{\partial t} = \frac{\mu}{Y_{xp}} \cdot B + r_p \left(1 - \frac{P}{P_{i2}}\right) \cdot B + D_p(B) \cdot \frac{\partial^2 P}{\partial x^2} , \tag{3}$$

where K_d, m_s and μ are the death coefficient, maintenance coefficient and growth rate, respectively. r_p and μ_m denote the uncoupled bioproduction rate and specific growth rate, respectively. Y_{xs}, Y_{xp} and Y_{sp} stand for the yields, and P_{i1} and P_{i2} for the maximum inhibiting product concentrations (generic acid) (see Table 1), D_s and D_p the diffusion coefficients, which are expressed by[9]:

$$D_{s,p} = D_{s,p}^0 - k_{s,p} \cdot \log(B) . \tag{4}$$

Table 1. Parameters of the model.

μ_m (h^{-1})	1.30[*]	0.86[**]
P_{m1} (mol/l)	0.037	
Y_{xs} (cells/mol)	$18 \cdot 10^{13}$	
Y_{xp} (cells/mol)	$18 \cdot 10^{13}$	
Y_{sp}	0.35	
v_m (mol/h/cell)	$4.3 \cdot 10^{-15}$	
P_{m2} (mol/l)	0.08	
m_s (mol/h/cell)	$1.8 \cdot 10^{-16}$	
K_d (h^{-1})	0.011	
$D_s^0 = D_p^0$ (cm^{2}s^{-1})	$8.3 \cdot 10^{-6}$	
$K_s = k_p$ (cm^{2}s^{-1})	$0.27 \cdot 10^{-6}$	

[*]Cells in suspension. [**]Immobilized cells.

The modeling of the cell leakage was empirical and considered that only a fraction (approximately 10%) of the cells produced in the first layer can leak into the external medium. Fick's laws gave the concentrations of the chemical species in the external medium, according to:

$$\left[\frac{dZ}{dt}\right]_{\text{ext.med.}} = \frac{a}{V} \cdot D_z \left[\frac{\partial Z}{\partial x}\right]_{\text{membrane}} \tag{5}$$

with $Z \equiv S$ or P, a = membrane area (10 cm^2) and V = compartment volume (300ml).

In order to compare the experimental and simulated results, the relation between the weak acid produced, P, and pH, which is easy to measure experimentally in the external medium, was established by taking into account all the pH dependent species (generic acid, pH buffers):

$$P(t) = \left[H\right]_0^t + \frac{H_t + K_a}{K_a}\left[\frac{1}{H}\right]_0^t + \sum N_0 K_N \left[\frac{1}{H + K_N}\right]_0^t, \tag{6}$$

where N_0 and K_N are, respectively, the total concentration and the dissociation constant of each weak acid composing the pH buffer of the culture medium.

4. Results

Solving of the equation system (eqs 1 to 3) gave the time evolution of the concentrations of cells and of all the chemical species inside the membrane and in the external medium via eq. (5). We have discussed the influence of the initial cell concentration, membrane thickness and diffusion coefficients earlier[10]. The time evolution has also been analyzed in great detail in a previous paper[11]. Briefly, it was divided into three periods:

   i) in the first period diffusion is predominant and cell concentration is low,
   ii) in the second period cell concentration increases throughout the membrane and the reaction becomes predominant, and finally
   iii) the third period is characterized by strong diffusion constraints due to the high cell concentration and by a low reaction component due to the high concentration of the inhibitory product P.

Until now, it has been very difficult to compare these intramembrane concentration profiles with experiments. However, it is easier to compare the results concerning the concentrations in the external medium which are directly accessible by the experimental procedure. Fig. 2 shows the simulated time evolutions of the pH and of the cell and glucose concentrations in the external medium, together with the experimental points obtained with a 3 mm thick agar membrane. The good agreement between simulations and experiments required the adjustment of a unique parameter, the specific growth rate μ_m. This parameter thus seems to be lower under immobilized conditions (0.86 h^{-1}) than it is in suspension (1.30 h^{-1}). This decrease may be due to changes in the cell metabolism and/or to local pH effects.

Theoretically and experimentally, two kinds of results can be obtained for the membrane itself: the first one was the time evolution of the total biomass immobilized in the membrane. From the theoretical point of view, the total biomass immobilized was calculated by summing the intramembrane cell concentrations of each layer of the space discretization. Experimentally, several studies were worked out and stopped at different times. Membranes were then ground as indicated in the Material and Methods section and the cell concentration was measured. Fig. 3 shows the very good agreement between the simulated and experimental cell concentrations.

In fact, this time evolution of the total biomass does not reflect a homogeneous increase in the membrane cell concentration. Fig. 4 shows how the initial homogeneous cell concentration profile was progressively modified by the cell growth and diffusion constraints to finally show a highly heterogeneous profile. The U-shaped profile

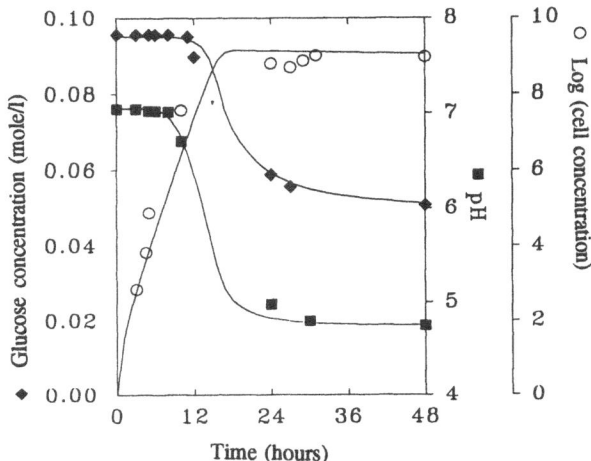

Figure 2. Kinetics of immobilized cells (membrane thickness = 3 mm): glucose and cell concentrations (cell leakage), and pH in the external liquid compartment as a function of time. Symbols represent experimental data and solid lines correspond to the simulated curves.

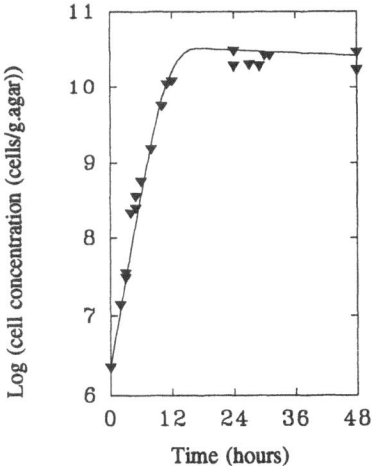

Figure 3. Kinetics of immobilized cells (membrane thickness = 3 mm): total biomass immobilized in the membrane (cells per g of agar gel) as a function of time. Symbols represent experimental data and solid lines correspond to the simulated curves.

characterizing the steady-state of the immobilized cell membrane has been compared with experiments: at the end of experiments, agar membranes were cut into slices approximately 300 μm thick, using a microtome. The layers were then ground and the cell concentration measured as above. Fig. 5 clearly shows the U-shaped steady state cell profiles obtained both theoretically and experimentally.

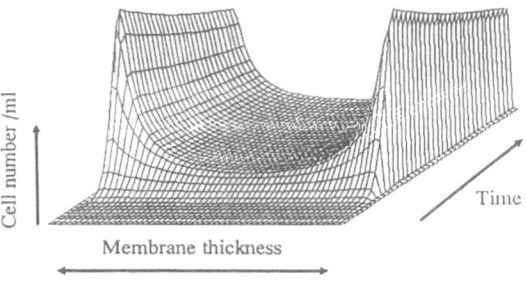

Figure 4. Temporal evolution of the spatial distribution of the immobilized cells. The intra-membrane profiles were obtained by simulation for a 3 mm membrane thickness and a final time of 48 hours.

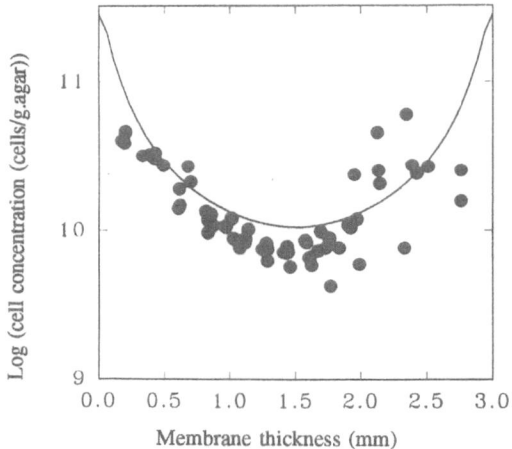

Figure 5. Steady-state profiles (time = 48 hours) of immobilized cells (cells per g of agar gel) inside the membrane (membrane thickness = 3 mm). The symbols represent experimental data and solid lines correspond to the simulated curves.

5. Conclusion

We designed a new model, based on the diffusion-reaction-growth couplings, for studying immobilized cell membranes. The reaction scheme and the metabolic parameters, determined from batch culture experiments, were used together with diffusion laws to compute the time evolution of the system and to compare it with experiments.

The solution of the time and space differential equation system gave us the kinetic evolution of the global concentrations throughout the membrane and in the external medium, especially for cell, substrate and product. pH was also studied. All the results were confronted with experimental data concerning agar membranes containing *Escherichia coli*. Theoretical and experimental results were in good agreement. The main results are:

i) a weaker maximal specific growth rate μ_m under immobilized conditions than under batch conditions;

ii) a steady-state characterized by a heterogeneous U-shaped distribution of the biomass inside the membrane.

In spite of the complexity of the bioproduction, it clearly appears that its representation by a simple generic acid is a good simplifying solution to describe the kinetic cell behavior; this holds as long as we are not concerned with metabolism. The main result is a lower apparent specific growth rate μ_m for the immobilized cells than for the free cells. We have also introduced an original method for the experimental study of the cell distribution inside the membrane. A theoretical and experimental confirmation of the heterogeneous U-shaped intramembrane cell distribution was obtained, in agreement with the very few reported studies. We are now interested in using this modeling for the study of composite structures eliminating cell leakage as well as the study of double flux reactors already designed and experimented on in our laboratory[12].

References

1. A.A. Esener, J.A. Roels and N.W.F. Kossen, Theory and applications of unstructured growth models: kinetic and energetic aspects, *Biotechnol. Bioeng.* **25**:2803-2841 (1983).
2. H.G. Monbouquette and D.F. Ollis, A structured model for immobilized cell kinetics, *Ann. N. Y. Acad. Sci.* **469**:230-244 (1986).
3. K. Nakasaki, T. Murai and T. Akiyama, Dynamic modeling of immobilized cell reactor: application of ethanol fermentation, *Biotechnol. Bioeng.* **24**:1749-1764 (1988).
4. C.D. De Gooijer, R.H. Wijffels and J. Tramper, Growth and substrate consumption of *Nitrobacter agilis* immobilized in Carrageenan: Part 1. Dynamic modeling. Part 2. Model evaluation, *Biotechnol. Bioeng.* **38**:224-240 (1991).
5. S.F. Karel, P.M. Salmon, P.S. Stewart and C.R. Robertson, Reaction and diffusion in immobilized cells: fact and fantazy, *in*: "Physiology of Immobilized Cells," J.A. De Bont, J. Visser, B. Mattiasson and J. Tramper, eds, Elsevier, Amsterdam, pp. 115-126 (1990).
6. C. Briasco, J.N. Barbotin and D. Thomas, Spatial distribution of viable cell concentration and plasmid stability in gel-immobilized recombinant *E. coli*, *in*: "Physiology of Immobilized Cells," J.A. De Bont, J. Visser, B. Mattiasson and J. Tramper, eds, Elsevier, Amsterdam, pp. 393-398 (1990).
7. L. De Baker, S. Devleminck, R. Willaert and G. Baron, Reaction and diffusion in a gel membrane. Reactor containing immobilized cells, *Biotechnol. Bioeng.* **40**:322-328 (1992).
8. P. De Taxis du Poët, Y. Arcand, R. Bernier, J.N. Barbotin and D. Thomas, Plasmid stability in immobilized and free recombinant *Escherichia coli*: importance of oxygen diffusion, *Appl. Environ. Microbiol.* **33**:1317-1323 (1987).
9. L. Mignot and G.A. Junter, Diffusion in immobilized-cell agar layers: influence of microbial burden and cell morphology on the diffusion coefficients of L-malic acid and glucose, *Appl. Microbiol. Biotechnol.* **32**:418-423 (1990).
10. J.C. Vincent and J. Lefebvre, Modelling of immobilized cell systems, *J. Mat. Sci., Mat. in Med.* **2**:234-237 (1991).
11. J. Lefebvre and J.C. Vincent, Dynamic simulations of cell-bearing membranes: modelling and optimization of bioreactors, *in*: "European Symposium on Computer Aided Process Engineering-2," Computers Chem. Engng., Vol. 17, Suppl., D. Depeyre, X. Joulia, B. Koehret and J.M. Le Lann, eds, Pergamon Press, Oxford, pp. 221-226 (1992).
12. D. Lemoine, Conception et expérimentation de bioréacteurs à microorganismes immobilisés adaptés à la dénitrification biologique de l'eau, Dr. Thesis, Université de Rouen (1991).

THE INTERPRETATION
OF FLOW CYTOMETRY PROFILES
IN TERMS OF GENE DOSAGE
AND GENE EXPRESSION

W.G. Bardsley[1] and E.K. Kyprianou[2]

[1]Department of Obstetrics and Gynaecology
  University of Manchester, St. Mary's Hospital
  Whitworth Park, Manchester M13 OJH, United Kingdom
[2]Department of Mathematics
  University of Manchester
  Manchester M13 9PL, United Kingdom

1. Introduction

It frequently happens that three flow cytometry histograms with fluorescence intensities X, Y, and Z are related by shifting and stretching under the following circumstances:

a) X is observed at one subsaturating concentration of lectin or antibody and Y is observed at another concentration with the same cell type;

b) X is observed at saturating concentration of ligand with one cell preparation and Y is observed at saturating concentration of the same ligand, but with another cell preparation;

c) X is observed at saturating concentration of ligand with one cell line, Y is observed at saturating concentration of ligand with the same cell line but with another ligand and Z is observed when both ligands are added together.

This paper briefly describes how the theory of ligand binding[1] can be used to develop a theory to interpret such shifts leading to a computer program CSAFIT[2,3] to analyze such effects quantitatively.

2. Definitions Required

The fluorescence intensities are random variables X, Y, Z with:

$$0 \le A = x_0 < x_1 < \cdots < x_n = B; \quad x_i = y_i = z_i; \quad i = 0, 1,..., n .$$

Modern Trends in Biothermokinetics, Edited by
S. Schuster *et al.*, Plenum Press, New York, 1993

The normalized fluorescence histograms are:

$$\sum_{i=1}^{n} (x_i - x_{i-1}) \, \Phi_X(x_i) = 1$$

and similar expressions for $\Phi_Y(y_i)$, $\Phi_Z(z_i)$. The unknown fluorescence probability density functions are:

$$f_X(x) \approx \Phi_X(x_i) , \quad x_{i-1} < x < x_i$$

and similar expressions for $f_Y(y)$ and $f_Z(z)$.

The problems are:

a) To calculate density functions for random variables X, Y, Z ;
b) to relate them to gene-dosage/gene-expression/epitope-structure;
c) to write software to quantitatively analyze X, Y, Z histograms.

3. The Probability of Just i Sites Being Occupied

The assumptions and theory necessary to calculate this are:

a) Each individual cell has n ligand binding sites.
b) The number of sites N is randomly distributed over the population of cell types.
c) The probability of a given fixed cell type having just n binding sites is

$$P(N=n) = f_N(n), \quad n \in 0, 1,..., n_{max} .$$

d) For each fixed cell type with just n sites there exists a binding polynomial $Q_n(L)$ in ligand activity L,

$$Q_n(L) = K_0 + K_1 L + K_2 L^2 + ... + K_n L^n , \quad K_0 = 1 .$$

e) The probability of observing just i sites occupied given one fixed cell type in thermodynamic equilibrium at ligand activity L is

$$P(I=i) = K_i L^i / Q_n(L) , \quad i \in 0, 1,..., n .$$

f) The sign of cooperativity depends on the sign of the Hessian $H(Q)$,

$$H(Q) = nQQ'' - (n-1)Q'^2 .$$

g) The binding polynomials are randomly distributed given N.
h) There exists a joint distribution function for the binding polynomials, i.e. binding constants K given N,

$$f_K(k \mid N=n) = f_n(k_1, k_2,..., k_n), \quad k \geq 0.$$

i) The probability of observing just i sites occupied given fixed N is given by

$$P(I=i \mid N=n) = \iint ... \int_{R_k} k_i L^i / Q_n(L) f_n(k_1, k_2,..., k_n) dk_1 dk_2 \cdots dk_n .$$

j) The probability of observing precisely i sites occupied is

$$P(I=i) = \sum_{n=0}^{\infty} P(I=i \mid N=n)\, P(N=n)\,.$$

k) The densities $f_N(n)$ and $f_n(k_1, k_2, ..., k_n)$ are not known, but useful expressions for interpreting flow cytometry profiles can be derived when all sites are identical, i.e.

$$Q_n(L) = (1 + KL)^n, \quad \forall\, K, n\,.$$

4. The Theory of Translation and Stretching

The expression for $f_Y(y)$ in terms of $f_X(x)$ assumes:

a) X and Y are dependent or independent random variables.
b) X and Y are sampled on the same restricted range $A \leq X \leq B$ and $A \leq Y \leq B$.
c) X is chosen as a reference and Y as a test random variable.
e) The distributions will be related if they are generated by common factors such as gene dosage or gene expression.
f) Then proportionate change in X leads to stretching of Y.
g) Also additive change in X leads to a translation of Y.

The probability density functions are then given by:

If $Y = \alpha X + \beta$ (in distribution), then $f_Y(y) = \dfrac{\gamma}{\alpha} f_X\left(\dfrac{y-\beta}{\alpha}\right)$,

where $\gamma^1 = \dfrac{1}{\alpha} \displaystyle\int_a^b f_X\left(\dfrac{u-\beta}{\alpha}\right) du$.

The limits of integration a and b are defined by:

$$a = \alpha A + \beta \ \text{ if } \ \alpha A + \beta > A \ (=A \text{ otherwise}),$$

$$b = \alpha B + \beta \ \text{ if } \ \alpha B + \beta < B \ (=B \text{ otherwise}),$$

$$\alpha \geq 0 \Rightarrow A \leq a < b \leq B\,.$$

5. The Interpretation of $\hat{\alpha}$ and $\hat{\beta}$

a) Geometric/model-free for all flow cytometry histograms:
   $\hat{\alpha}$ just measures the degree of stretching and $\hat{\beta}$ simply estimates the magnitude of any translation.
b) One antibody/one cell line where Y results from X by a change in ligand activity:
   $\hat{\alpha}$ and $\hat{\beta}$ depend on the distribution of the n sites and binding polynomials $Q_n(L)$. Under various assumptions $\hat{\alpha}(Q)$ and $\hat{\beta}(Q)$ can be inverted to give estimates for the binding constants K_i.
c) One antibody/one cell line where Y results from an alteration in gene dose/gene expression:

$\hat{\alpha}$ estimates the increase in gene expression proportional to the reference population gene expression while $\hat{\beta}$ measures the expectation of the random additional gene expression independent of the reference population gene expression.

d) Two antobodies/one cell line where

H_0: $Y = X$ (in distribution) can be rejected: the epitopes represent independent gene products.

e) Two antibodies/one cell line where X results from one antibody, Y results from another antibody and Z results from X and Y together in saturating amounts:

i) H_0: $X = Y$ (in distribution) can be rejected. The two antibodies bind to epitopes on two independent gene products and $f_Z(z)$ is the convolution of $f_X(x)$ with $f_Y(y)$.

ii) H_0: $X = Y$ (in distribution) cannot be rejected. $\hat{\alpha}$ and $\hat{\beta}$ from fitting $Z = \alpha X + \beta$ (in distribution) depend on λ. Epitopes are on the same gene product so $\hat{\alpha}(\lambda)$ and $\hat{\beta}(\lambda)$ can be calculated and λ estimated subject to assumptions.

P(both epitopes occupied) $= \lambda$,

P(one epitope occupied) $= 1-\lambda$,

$\lambda \approx 1 \implies$ epitopes are well separated,

$\lambda \approx 0 \implies$ epitopes are close together.

6. The SIMFIT Program CSAFIT

For performing the calculations explained above we developed the computer program CSAFIT, which is part of the package SIMFIT (see Ref. 4). The functions of CSAFIT are listed in Table 1.

Table 1. Flow diagram of the program CSAFIT.

Input the fluorescence histograms for X and Y Normalize to unit areas then perform Kolmogorov Smirnov test for statistically significant differences between X and Y
Fit least squares smoothing spline to X then normalize until best fit curve is positive and of unit area
Use normalized smoothing spline as density $f_X(x)$ Display graph of normalized best fit spline to X histogram Assess goodness of fit of spline to X histogram by comparing moments, run and sign tests and fractional differences in absolute areas
Estimate α and β by constrained optimization using $$f_Y(y) = \frac{1}{\alpha} f_X\left(\frac{y-\beta}{\alpha}\right) \Big/ \left\{ \int_B^A \frac{1}{\alpha} f_X\left(\frac{t-\beta}{\alpha}\right) dt \right\}$$
Display graph of normalized best fit curve to Y histogram Assess goodness of fit of theoretical curve to Y histogram by comparing moments, run and sign tests and fractional differences in absolute areas

References

1. W.G. Bardsley, R. Woolfson and J.-P. Mazat, Relationships between the magnitude of Hill plot slopes, apparent binding constants and factorability of binding polynomials and their Hessians, *J. theor. Biol.* **85**:247-284 (1987).
2. W.G. Bardsley, B.P. McMurray, A. Robson, S.D. D'Souza and G.M. Taylor, Analysis of gene-dosage effects on the expression of CD18 by trisomy 21 lymphoblastoid cell-lines using a statistical model to fit flow cytometry profiles, *Hum. Genet.* **86**:181-186 (1990).
3. W.G. Bardsley, A. Ross Wilson, E.K. Kyprianou and E.M. Melikhova, A statistical model and computer program to estimate association constants for the binding of fluorescent-labelled monoclonal antibodies to cell surface antigens and to interpret shifts in flow cytometry data resulting from alterations in gene expression, *J. Immun. Methods* **153**:235-247.
4. W.G. Bardsley, SIMFIT - A computer package for simulation, curve fitting and statistical analysis using life science models, this volume.

INVENTION OF A USEFUL
TRANSCRIPTIONAL REGULATION SYSTEM
USING MECHANISTIC MATHEMATIC MODELING

W. Chen[1], S.-B. Lee[2], P. Kallio[1] and J.E. Bailey[1]

Department of Chemical Engineering
California Institute of Technology
Pasadena, California 91125 USA
Present address:
[1]Institute of Biotechnology, ETH Hönggerberg, CH-8093 Zürich, Switzerland
[2]Pohang Institute of Science andTechnology, Pohang 790-330, Korea

1. Introduction

Regulation of gene expression is a process of central importance in cellular metabolism but also in disease, in development of differentiated organisms, in immunology, and many other life functions. Gene expression regulation is also paramount in the founding and still important process in the biotechnology industry, namely manufacture of a valuable protein based on overexpression of the corresponding cloned gene. The ideal qualitative profile of a typical batch process for producing such a protein involves an initial growth phase in which the cell density of recombinant cells is increased, followed by a production phase in which the desired protein is accumulated.

Realizing such a process is facilitated by availability of a regulated promoter which, after fusion to the product's gene, allows minimal transcription prior to some particular change in the process conditions (for example, addition of inducer or change in temperature) and affords extremely active transcription thereafter. Previously available regulated promoters for expression in *Escherichia coli,* which are the most diverse and well characterized set of such promoters, all suffer some disadvantages in preinduction or postinduction characteristics, some related to undesirable side effects of the culture manipulation required for induction.

Configurations of promoters, operators, and genes substantially more complicated than those presently used can be constructed using available DNA manipulation methods. Such constructions with the required checks for desired structure and function are, however, time-consuming. Mathematical simulations of the properties of various configurations of possible interest can in principle provide useful guidance for synthesis of useful new regulated promoters.

Modern Trends in Biothermokinetics, Edited by
S. Schuster *et al.*, Plenum Press, New York, 1993

Prior mathematical modeling studies have calculated the relative transcriptional activity of *lac* and *lambda* P_R promoters with and without inducer in different genetic backgrounds with results very consistent with experimental data for these systems[1-4]. The primary physical assumptions invoked in these models are

(i) transcription rate from a promoter-operator is proportional to the fraction of operator sites free of repressor,
(ii) the distribution of repressor among operator sites, solution, and complex with inducer is in thermodynamic equilibrium, and
(iii) the equilibrium configuration of pertinent sites and molecules can be calculated to reasonable approximation using classical (large-system) thermodynamics.

Support for the final assumption derives from prior computations showing typically small errors resulting from classical thermodynamic analysis relative to a more rigorous statistical thermodynamic description of the actual small-system situation (each cell in the population, in which the interactions actually occur, contains only a small number of interacting sites and molecules). This modeling framework has been articulated for transcriptional regulatory elements residing in the chromosome, multicopy plasmids, or a combination of the two.

These approaches, already shown to be useful in describing previously implemented genetic modifications in transcriptional control, have recently been applied to predict the characteristics of synthetic promoter-operator configurations chosen to be potentially useful in controling cloned gene expression in a batch production process. A superior transcription control design identified from these model calculations has been implemented in a new *E. coli* expression vector and evaluated in batch fermentations relative to a conventional construct.

2. Mathematical Modeling of Alternative Configurations for Transcriptional Regulation of a Cloned Gene

Applying the modeling approach summarized above, the transcriptional activity through the cloned gene (from the "product promoter") has been estimated for induced and uninduced conditions for several different configurations of promoters, operators, product gene, and repressor genes[5]. The configurations considered included cases already constructed and studied experimentally, such as constitutive synthesis of the repressor of the product promoter from a single chromosomal gene copy and from a gene contained in the same multicopy plasmid as the product promoter. Also considered were several configurations which had not been experimentally implemented. These include a cross-regulation system in which the product promoter and gene is fused in a synthetic operon to the gene encoding the repressor for a second promoter and in which the second promoter initiates transcription of the gene for repressor of the product promoter. The inducer in this case inactivates this repressor of the product promoter. The concepts motivating suggestion of this configuration are the prospect for extremely effective repression of product gene transcription preinduction due to high-level expression of product promoter repressor. Postinduction, the transcription of product gene is relaxed and, due to coordinated expression of repressor for the second promoter, a new steady state is achieved which involves a low level of product promoter repressor and concomitant high-level expression of product.

Correct predictions of the properties of such a complex system based on such qualitative reasoning is difficult and unreliable. Instead, the mathematical representation of the primary interactions in the system should be employed to predict its performance. The central purpose of these simulation calculations is to enable relatively effortless comparisons of the expected attributes of various regulation designs to select those to be implemented and tested in the laboratory.

The particular interactions and parameter values considered in these simulations came from previous validated models of individual promoter-operator systems. Specifically, the product promoter-operator was from the *E. coli lac* operon and the second promoter-operator was the phage *lambda* P_R promoter-multiple operator segment. Parameter values are given in prior publications. In contrast to the assumption of Lee and Bailey[1], the concentration of free cro repressor was here calculated considering the pertinent equilibria of cro-O_R operator binding.

These calculations show that the plasmid-based cross-regulation configuration provides the lowest transcription preinduction and the greatest transcription postinduction over a wide range of plasmid copy numbers. Extremely low preinduction activity can be a critically important quality when expressing cloned proteins which are particularly inhibitory to *E. coli* growth. The postinduction advantage of the cross-regulation system is larger at higher copy number.

Further model simulations considering variations in the model parameters indicate the following criteria should be fulfilled for successful implementation of the cross-regulation design:

(1) the product promoter should be strong,
(2) the second promoter should be stronger than the product promoter, and
(3) the second promoter should be less sensitive to its repressor than is the product promoter to its repressor.

Although these simulations involved particular promoter systems, their qualitative conclusions are expected to be generally applicable to other promoter-operator systems in any host.

3. Experimental Realization and Evaluation

An experimental program to construct an *E. coli* expression plasmid based on the cross-regulation concept and a companion plasmid providing constitutive transcription of product promoter repressor has been completed. Based on model suggestions of required properties of the regulatory components and on available genetic elements, the *tac* promoter was selected as the product promoter (with the lacI protein as corresponding repressor) and *lambda* P_L as the second promoter (cI repressor). Two DNA elements necessary for construction of the two required synthetic operons were synthesized by the polymerase chain reaction, with primer design to facilitate vector construction and to include needed translation start and transcription termination sequences. The PCR products were validated for corresponding biological activity in separate experiments. The product protein considered was the stable enzyme chloramphenicol acetyltransferase (CAT) which can be detected at very low levels and over a very broad range of activities using a radiolabeled substrate assay.

The cross-regulation and constitutive repressor synthesis vectors were each transformed into *E. coli* HB101. Both recombinant strains were cultivated in shake-flask cultures in LB medium at an initial pH of 7.0. Each was induced by addition of IPTG in late-exponential phase. The cross-regulation construct attained CAT specific activity levels approximately two-fold greater than those in the other culture. Moreover, as predicted by model simulation, CAT activity preinduction was much lower for the cross-regulation construct, a result compatible with SDS-PAGE analysis of soluble protein which showed much higher *lacI* levels preinduction for the cells with the cross-regulation vector.

4. Discussion

A mathematical model of multicopy promoter regulation was effective in predicting a useful novel regulatory network which has been successfully implemented and demonstrated experimentally. The ability to calculate the central characteristics of these transcriptional regulatory circuits correctly using a model with so few components and neglecting myriad related cellular processes is likely a consequence of the highly specific interactions involved in this system. Because of this, it functions with minimal coupling with the balance of cell metabolism.

Acknowledgement

This work was supported by the National Science Foundation and by the Advanced Industrial Concepts Division of the U.S. Department of Energy.

References

1. S. B. Lee and J. E. Bailey, A mathematical model for lambda dv plasmid replication: Analysis of wild-type plasmid, *Plasmid* **11**:151 (1984).
2. S. B. Lee and J. E. Bailey, A mathematical model for lambda dv plasmid replication: Analysis of copy num-ber mutants, *Plasmid* **11**:166 (1984).
3. S. B. Lee and J. E. Bailey, Genetically structured models for *lac* promoter-operator function in the chromosome and in multicopy plasmids: *lac* operator function," Biotechnol. Bioeng. **26**:1372 (1984).
4. S. B. Lee and J. E. Bailey, Genetically structured models for *lac* promoter-operator function in the chromosome and in multicopy plasmids: *lac* promoter function, *Biotechnol. Bioeng.* **26**:1383 (1984).
5. W. Chen, J. E. Bailey, and S. B. Lee, Molecular design of expression systems: Comparison of different repressor control configurations using molecular mechanism models, *Biotechnol. Bioeng.* **38**:679-687 (1991).

TUMOR PROGRESSION
AND CATASTROPHE THEORY

Henny Daams[1], Sonia Cortassa[2,3],
Miguel A. Aon[2,3] and Hans V. Westerhoff[2,4]

[1]Division of Tumor Biology
[2]Division of Molecular Biology
  The Netherlands Cancer Institute, Antoni van Leeuwenhoekhuis
  Plesmanlaan 121, 1066 CX Amsterdam, The Netherlands
[3]INSIBIO, Universidad Nacional de Tucumán
  Chacabuco 461, Tucumán, Argentina
[4]E.C. Slater Institute for Biochemical Research
  University of Amsterdam
  Plantage Muidergracht 12, NL-1018 TV Amsterdam, The Netherlands

1. Introduction

Progression is the development from bad to worse that is often observed in tumors. It can follow different courses, not only in different organs, but also in the same organ: sometimes there is a smooth, continuous progression from normal to malignant cells, sometimes suddenly a highly malignant tumor arises.

We first show that a qualitative catastrophe model can describe phenomena observed in tumor progression. Then we construct mathematical models to explain this on the basis of mechanisms from molecular biology. These indicate that, in addition to mutations, deregulation of feedback systems may play an important role in tumor progression.

Clinically, progression is very important when cancer cells acquire resistance against chemotherapy. Leslie Foulds[1] defined progression as an irreversible, qualitative change in tumors, and he identified the following rules:

Progression

- is independent in multiple tumors,
- is independent for each character in a tumor,
- is independent of growth,
- can be continuous as well as discontinuous,
- can follow several alternative pathways,
- does not always reach an endpoint in a lifetime.

Modern Trends in Biothermokinetics, Edited by
S. Schuster *et al.*, Plenum Press, New York, 1993

At the same time Berenblum and Shubik[2-5] discovered two phases in the genesis of skin cancer. First, a short initiation by a carcinogen changes normal cells into seemingly normal but potential tumor cells. Then, promotion ensues, caused by a substance that itself has no carcinogenic effect but converts the initiated cells into cancer cells.

Later, Foulds[6] combined these data into a scheme where malignancy is plotted against time (Fig. 1). After initiation, cells would develop an increasing carcinogenic potential, which may be revealed by the action of a tumor promoter. This is the start of overt progression. Regression sometimes occurs, but carcinogenic potency remains.

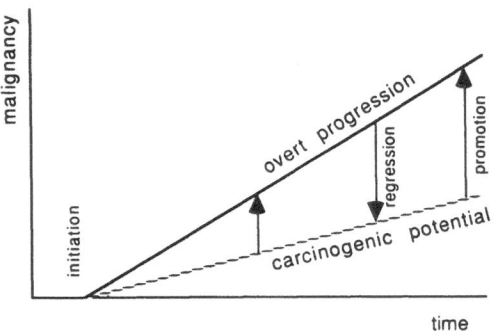

Figure 1. A generalized scheme of tumor progression (after Foulds[6]).

2. A Catastrophe Model: Foulds Folded

According to René Thom[7], seven elementary catastrophe models describe all discontinuous changes in phenomena that have four or less control variables. The scheme of Foulds[6] can be folded into the surface of a cusp catastrophe, one of the simpler catastrophe classes (Fig. 2). In this three-dimensional graph the vertical axis represents malignancy, and the two horizontal axes represent progression in time and promotion.

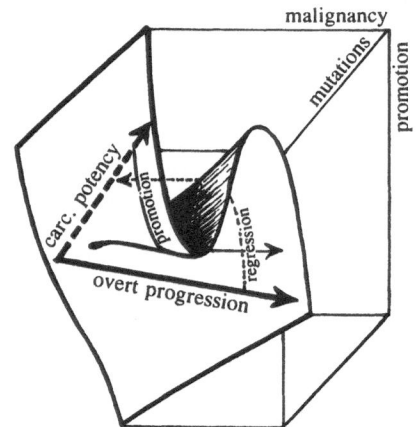

Figure 2. A catastrophe model of tumor progression. See description in text.

Our qualitative model shows all characteristics of tumor development. After initiation, the carcinogenic potential is increasing with time. Progression can follow alternative pathways, it can be overt and continuous, or hidden and discontinuous, depending on promotion, which is the bifurcation factor. When a considerable carcinogenic potential has developed in seemingly normal cells, a small increase of promotion can turn these into highly malignant cancer cells. The reverse is possible: *in vitro* sometimes regression can be induced. However, as it is proper to catastrophes, regression occurs at a different value of the promotion variable: we have hysteresis in this model.

3. Molecular Biology

How to relate this catastrophe description of tumorigenesis to molecular biology? Again, we have to discriminate between progression and promotion.

3.1. Progression: mutations

During steady progression, mutations in proto-oncogenes and tumor suppressor genes accumulate in many tumors[8]. It is generally assumed that this is mainly due to clonal selection: such mutations would give a cell an advantage in the competition for nutrients and oxygen, and it would expand as a clone[9].

In our model we may replace the successive steps in progression by their corresponding mutations. Initiation will be the first of these mutations.

3.2. Promotion: no mutations

Promotion by substances like phorbol esters induces tumors in cells that are already carrying one or more mutations. However, tumor promoters are not themselves mutagenic[8]. Moreover, their effect is reversible in the first phases of progression.

Some tumor promoters act by increasing the phosphorylation of protein kinases, probably the first step leading to the induction of cell division. It is supposed that in this way a mutated cell can expand clonally to a population of tumor cells.

3.3. Continuous or discontinuous

From the frequency of human cancer incidence with age it has been calculated that 5-6 steps are required for a diagnosable tumor[10]. Not always all these steps have been found in well studied tumor systems, nor the corresponding mutations.

In colon carcinoma a continuous series of steps leads to the development of carcinoma. Indeed, six genetic alterations could be demonstrated here, although not appearing in a strict order (Fig 3, after Fearon and Vogelstein[11]). Rarely a colon cancer arises *de novo* without precursor[12]. The most common form of breast cancer, on the contrary, shows a more discontinuous development, and premalignant phases are difficult to recognize[13]. Until now only four mutations of proto-oncogenes have been discovered in mammary tumors[14]. Here the catastrophe may prevail.

The contrasting behavior of tumor progression in these two organs may be due to a difference in cell division control. The epithelium of the colon renews itself continuously, while the mammary epithelium has short periods of intensive proliferation in puberty and pregnancy under the influence of hormones[15].

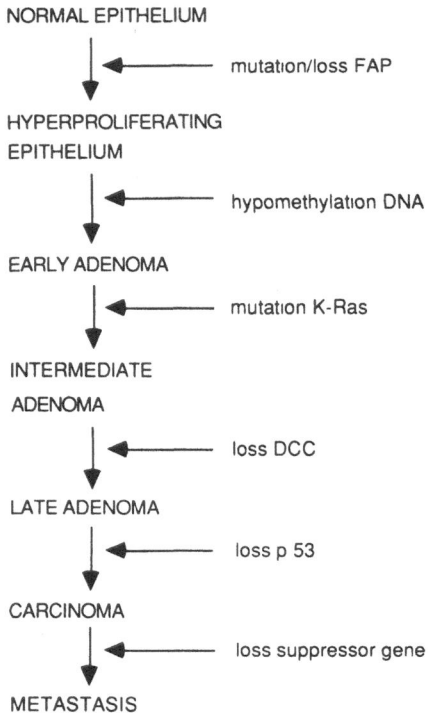

NORMAL EPITHELIUM

mutation/loss FAP

HYPERPROLIFERATING
EPITHELIUM

hypomethylation DNA

EARLY ADENOMA

mutation K-Ras

INTERMEDIATE
ADENOMA

loss DCC

LATE ADENOMA

loss p 53

CARCINOMA

loss suppressor gene

METASTASIS

Figure 3. A genetic model for colorectal tumorigenesis after Fearon and Vogelstein[11]. Abbreviations: FAP, familiar adenomatous polyposis; K-Ras, Kirsten-Ras proto-oncogene; DCC, deleted in colon carcinoma.

4. Quantitative Models

4.1. Kinetics

Catastrophe models give a phenomenological description and do not refer to actual mechanisms. To find the latter, we must identify the possible components and then show that their kinetic behavior corresponds to that of a catastrophe.

Here we use the fact that transcription factors are mostly active as dimers to construct the following mathematical models.

4.2. Model A: Two growth factors and a receptor for one

Here, we propose that proliferation is a direct function of the concentration of the receptor for a growth factor. We assume that upon binding of a growth factor a, the receptor R trimerizes which leads to degradation at a rate aR^3. A second growth factor b causes the receptor to dimerize and to stimulate the transcription of its own gene at a rate bR^2. Monomeric receptor is degraded at a rate cR, whereas the parameters d and e qualify the zero order degradation and the constitutive synthesis of the receptor, respectively (Fig.4). Therefore

$$\frac{\mathrm{d}R}{\mathrm{d}t} = -aR^3 + bR^2 - cR - d + e .$$

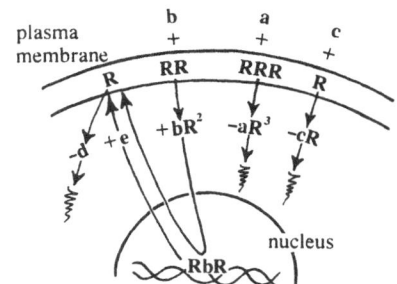

Figure 4. Schematic representation of model A: interaction of two growth factors, *a* and *b*, with receptor *R* for *a*. The dimeric complex of *R* activates the gene that produces *R*: positive feed back. Constitutive synthesis of *R* is represented by *e*, first-order degradation by *d*. In addition, breakdown of trimeric and monomeric *R* occurs with rate constants *a* and *b*, respectively.

The steady state requires that the overall concentration of *R* is constant, hence:

$$aR^3 - bR^2 + cR + d - e = 0.$$

We take *e* as the bifurcation parameter. Steady states are given by

$$e = aR^3 - bR^2 + cR - d.$$

This is the equation of the cusp catastrophe. By variation of the values for *c*, we get different sections through the steady-state surface. In Fig. 5, continuous lines indicate the stable steady states, broken lines the unstable ones. This is checked using:

$$\frac{\partial}{\partial R}\frac{dR}{dt} < 0 : \text{stable},$$

$$\frac{\partial}{\partial R}\frac{dR}{dt} > 0 : \text{unstable}.$$

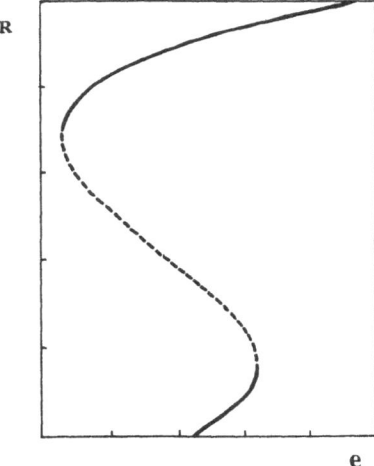

Figure 5. Variation of the steady-state receptor concentration *R* with the rate of its constitutive synthesis *e*, calculated for model A. Full lines refer to stable and dashed lines to unstable steady states.

4.3. Model B: One growth factor and a signal for its induction

This type of model is taken from Lewis et al.[16]. We suppose that growth factor G is produced by its gene under the influence of signal S. In addition, a dimer of G exerts a positive feedback on the formation of G. Further, G is degraded at a rate proportional to its concentration:

$$\frac{dG}{dt} = k_1 S - k_2 G + \frac{k_3 G^2}{K_m + G^2} \quad .$$

where k_1, k_2 and k_3 are reaction constants, K_m is the Michaelis constant and the last term is the Michaelis-Menten expression for the synthesis of G in terms of the concentration of the dimeric molecule.

At steady state:

$$\frac{dG}{dt} = k_1 K_m S - k_2 K_m G + k_1 G^2 S - k_2 G^3 + k_3 G^2 = 0 \, .$$

Replacing the reaction constants by one letter gives:

$$d\,G^3 - (c\,S + e)G^2 + b\,G - a\,S = 0 \, .$$

If S is constant, this is the equation for the cusp catastrophe. In a simple way it can be demonstrated[17] that for certain values of the signal S, the concentration of growth factor G remains on a stable, low level, while at higher values there are two steady states, one of them disappearing after a certain treshold level so that only a single stable high level of G remains (Fig. 6). The system is locked in this situation: it is impossible to return to a lower level.

This could well explain the continuing proliferation sometimes observed at high concentrations of a tumor promoter. It also shows that a transient signal can control discontinuous transitions[16] and it gives an explanation for the fact that sometimes only few mutations are found in a tumor: possibly there was a very brief expression of other genes that together constitute a "cancer cascade". Their signal can disappear completely, while their effect remains.

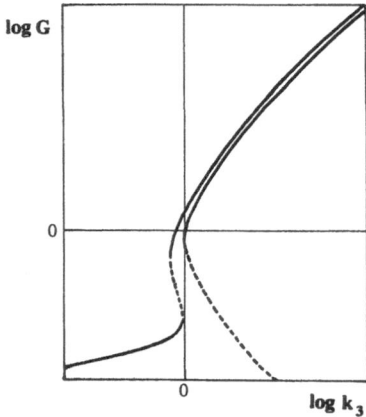

Figure 6. Concentration of growth factor G versus rate constant k_3 for the formation of its dimer G_2. Curves are given for two different values of the signal S.

5. Experimental Evidence

Some recent data in the literature match a catastrophe model:

Positive feedback has been demonstrated for a growth factor in carcinoma cell lines. These cells produce both transforming growth factor and its receptor, which combine to stimulate their own growth[18]. Positive feedback also exists in the activation of cyclins[19].

Hysteresis is suggested by the difference between pathways in tumor promotion and tumor regression. Tumor promotion by tetraphorbolic acid or nerve growth factor starts with tyrosine phosphorylation of MAP (mitogen-activated protein) kinases[20] while regression induced by retinoic acid is effected by the formation of a heterodimer of retinoic acid receptors which acts as a transcription factor[21].

Bifurcation may occur in the interaction between the products of c-*myc* and *bcl*-2 proto-oncogenes: "c-Myc can provide the first signal, leading either to apoptosis (programmed cell death) or to progression, and certain growth factors (for which Bcl-2 may substitute) may provide a second signal to inhibit apoptosis and allow c-Myc to drive cells into the cell cycle" (Ref. 22, p. 554, cf. also Ref. 23). In our model, the activation of the c-*myc* gene might be one of the steps on the progression axis, with the Bcl-2 product acting as a tumor promoter. Apoptosis then has to be considered as an extreme form of the normal cell differentiation that finally also results in cell death.

Transient signals of oncogenes can have a lasting effect, as predicted by the Lewis model: In normal cells stimulated to divide, a series of proto-oncogenes is successively switched on for a short time, in the end followed by mitosis[24]. The signal for activation of these genes is presumably passed from one gene to the next. A mouse mammary tumor cell line which is tumorigenic but nonmetastatic acquires metastatic potential when transfected with the activated Ha-ras gene. However, subclones that have lost this gene were found to be equally metastatic: ras acts as a "metastasis promoter"[25].

6. Conclusions

Deregulation of feedback systems that control cell proliferation, such as proposed here, can explain the mechanism of promotion, as well the lack of mutations in the progression of many tumors. The proto-oncogenes in this case could be activated only for a short period, giving a signal to the next in the cancer cascade. The rate of this mechanism can be higher than the normal mutation rate. It remains to be established experimentally to what extent catastrophes contribute to tumorigenesis.

References

1. L. Foulds, Tumor progression in mice: Growth and progression of spontaneous tumors, *Brit. J. Cancer* **3**:345-375 (1949).
2. I. Berenblum and P. Shubik, The role of croton oil applications, associated with a single painting of a carcinogen, in tumour induction of the mouse's skin, *Brit. J. Cancer* **1**:379-382 (1947).
3. I. Berenblum and P. Shubik, A new, quantitative approach to the study of the stages of chemical carcinogenesis, *Brit. J. Cancer* **1**:383-391 (1947).
4. I. Berenblum and P. Shubik, An experimental study of the initiating stage of carcinogenesis, and a re-examination of the somatic cell mutation theory of cancer, *Brit. J. Cancer* **3**:109-118 (1949).
5. I. Berenblum and P. Shubik, The persistence of latent tumor cells induced in the mouse's skin by a single application of 9:10-dimethyl-1:2-benzanthracene, *Brit. J. Cancer* **3**:384-386 (1949).
6. L. Foulds, "Neoplastic Development," vol. 1, Academic Press, London and New York, p. 78 (1969).
7. R. Thom, "Structural Stability and Morphogenesis," Benjamin, New York (1972), reprinted by Addison Wesley, New York, London and Amsterdam (1989).

8. T. Hunter, Cooperation between oncogenes, *Cell* **64**:249-270 (1991).

9. P. Nowell, Mechanisms of tumor progression, *Cancer Research* **46**:2203-2207 (1986).

10. R. Peto, F.J.C. Roe, P.N. Lee, L. Levy and J. Clack, Cancer and aging in mice and men, *Brit. J. Cancer* **32**:411-426 (1975).

11. E.R. Fearon and B. Vogelstein, A genetic model for colorectal tumorigenesis, *Cell* **61**:759-767 (1990).

12. J. Sadaway, M.I. Friedman, W.E. Katzin and G. Mendelsohn, Role of the transitional mucosa of the colon in differentiating primary adenocarcinoma from carcinomas metastatic to the colon, *Am. J. Surgical Pathol.* **15**:136-144 (1991).

13. S.P. Ethier and G.H. Heppner, Biology of breast cancer, *in*: "Breast Diseases," J.R. Harris, S. Hellmann, I. Craig Henderson and D.W. Kinne, eds, Lippincott Company, Philadelphia, pp. 135-146 (1987).

14. M.C. Sunderland and W.L. McGuire, Oncogenes as clinical prognostic indicators, *in*: "Regulatory Mechanisms in Breast Cancer," M.E. Lippmann and R.B. Dickson, eds, Kluwer Academic Publishers, Boston, Dordrecht and London, pp. 3-22 (1990).

15. J.H. Daams, A. Sonnenberg, T. Sakakura and J. Hilgers, Changes in antigen patterns during development of the mouse mammary gland: Implications for tumorigenesis, *in*: "Cellular and Molecular Biology of Mammalian Cancer," D. Medina, W. Kidwell, G. Heppner and E. Anderson, eds, Plenum Press, New York, London, Washington and Boston, pp. 1-8 (1987).

16. J. Lewis, J.M.W. Slack and L. Wolpert, Thresholds in development, *J. Theor. Biol.* **65**:579-590 (1977).

17. L. Edelstein-Keshet, "Mathematical Models in Biology," Random House, New York, pp. 283-287 (1988).

18. K. Stromberg, Th.J. Collins IV, A.W. Gordon, C.L. Jackson and G.R. Johnson, Transforming growth factor-alpha acts as an autocrine growth factor in ovarian carcinoma cell lines, *Canc. Res.* **52**:341-347 (1992).

19. I. Hoffmann, P.R. Clarke, M.J. Marcote, E. Karsenti and G. Draetta, Phosphorylation and activation of human cdc25-C by cdc2-cyclin B and its involvement in the self-amplification of MPF at mitosis, *EMBO J.* **12**:53-63 (1993).

20. S. M. Thomas, M. DeMarco, G. d'Arcangelo, S. Halegoua and J.S. Brugge, Ras is essential for nerve growth factor- and phorbol ester-induced tyrosine phosphorylation of MAP kinases, *Cell* **68**:1031-1040 (1992).

21. B. Durand, M. Saunders, P. Leroy, M. Leid and P. Chambon, All-trans and 9-cis retinoid acid induction of CRABPII transcription is mediated by RAR-RXR heterodimers bound to DR1 and DR2 repeated motifs, *Cell* **71**:73-85 (1992).

22. R.P. Bissonnette, F. Echeverri, A. Mahboubi and D.R. Green, Apoptic cell death induced by c-*myc* is inhibited by *bcl*-2, *Nature* **359**:552-554 (1992).

23. A. Fanidi, E.A. Harrington and G.I. Evan, Cooperative interaction between c-*myc* and *bcl*-2 proto-oncogenes, *Nature* **359**:554-556 (1992).

24. H. Herbst, S. Milani, D. Schuppan and H. Stein, Temporal and spatial patterns of proto-oncogene expression at early stages of toxic liver injury in the rat, *Lab. Invest.* **65**:324-333 (1991).

25. B. Schlatter and C.G. Waghorne, Persistence of Ha-ras-induced metastatic potential of SP1 mouse mammary tumors despite loss of the Ha-ras shuttle vector, *Proc.Natl.Acad.Sci. U.S.A.* **89**:9986-9990 (1992).

BIOCHEMICAL NEURON:
APPLICATION TO NEURAL NETWORK STUDIES
AND PRACTICAL IMPLEMENTATION OF DEVICE

Masahiro Okamoto, Yukihiro Maki and Tatsuya Sekiguchi

Department of Biochemical Engineering and Science
Kyushu Institute of Technology
Iizuka, Fukuoka 820, Japan

1. Introduction

In a series of papers[1-3] we showed by computer simulations that cyclic enzyme systems (basic switching elements) have the reliability of ON-OFF types of operation (McCulloch-Pitts' neuronic equation) capable of storing short-memory, and the applicability for a prototype of an artificial neuronic device. Constructing the integrated artificial neural network system being composed of this device, we shall show a physiological phenomenon termed "selective elimination of synapses"[4] generally produced as a result of a low-frequency train of electric stimuli to the synapses. Furthermore, for a practical implementation of the device, the analog circuit of the basic switching element was designed by using the PSpice (electronic circuit simulator) program.

2. Selective Elimination of Synapses

Fig. 1. represents a schematic circuit model of an artificial neuron, where, at time t, $Y_i(t)$ and $Z_j(t)$ are the excitatory and the inhibitory input signals, respectively. $A(t)$ denotes the output signal. w_i and w'_j show the synaptic efficiencies of Y_i and Z_j, respectively, and θ is the threshold value.

The mathematical model of this scheme can be written as follows:

$$\frac{dX_1}{dt} = \sum_{i=1}^{n} w_i Y_i - k_1 X_1 (1 - A) - k_3 X_1 , \tag{1a}$$

$$\frac{dX_3}{dt} = \sum_{j=1}^{m} w'_j Z_j + \theta - k_2 X_3 A - k_4 X_3 , \tag{1b}$$

Modern Trends in Biothermokinetics, Edited by
S. Schuster *et al.*, Plenum Press, New York, 1993

$$\frac{dA}{dt} = k_1 X_1 (1 - A) - k_2 X_3 A \ . \tag{1c}$$

In neurophysiology, a phenomenon called "selective elimination of synapses" was demonstrated. Since repeated stimulation with high frequency is the key factor to inhibit the selective elimination of synapses, we investigated the effect of the frequency of external stimuli on the dynamic response of $A_i(t)$ in a scheme of a branched chain with excitatory interaction (see Fig. 2). Each element in Fig. 2 is assumed to be composed of the basic unit shown in Fig. 1 and to transmit the output signal, $A_i(t)$, unidirectionally to the subsequent one.

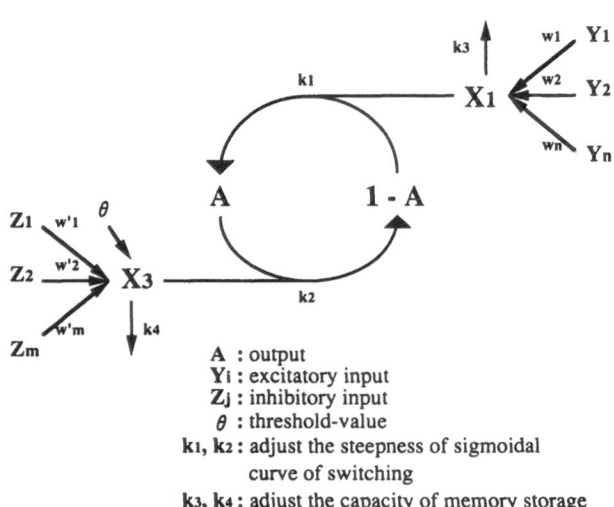

A : output
Y_i : excitatory input
Z_j : inhibitory input
θ : threshold-value
k_1, k_2 : adjust the steepness of sigmoidal
curve of switching
k_3, k_4 : adjust the capacity of memory storage

Figure 1. Schematic circuit model of an artificial neuron based on biochemical reactions.

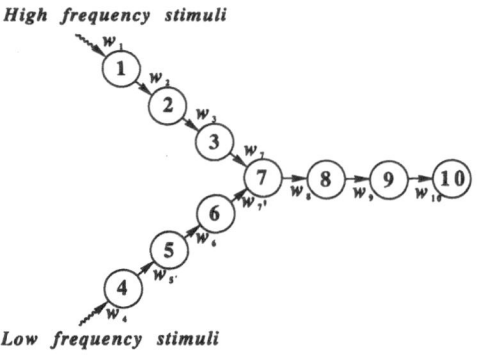

Figure 2. Branched network with excitatory scheme.

Fig. 3 shows the dynamic response of $A_i(t)$ in Fig. 2 under the condition that excitatory stimuli with high frequency are introduced into the first element and those with low frequency into the fourth.

Excitatory stimuli with high frequency introduced to the first element are amplified and are successively transmitted to the tenth element. Contrary to this, excitatory stimuli with low frequency introduced to the fourth element are attenuated during propagation, which

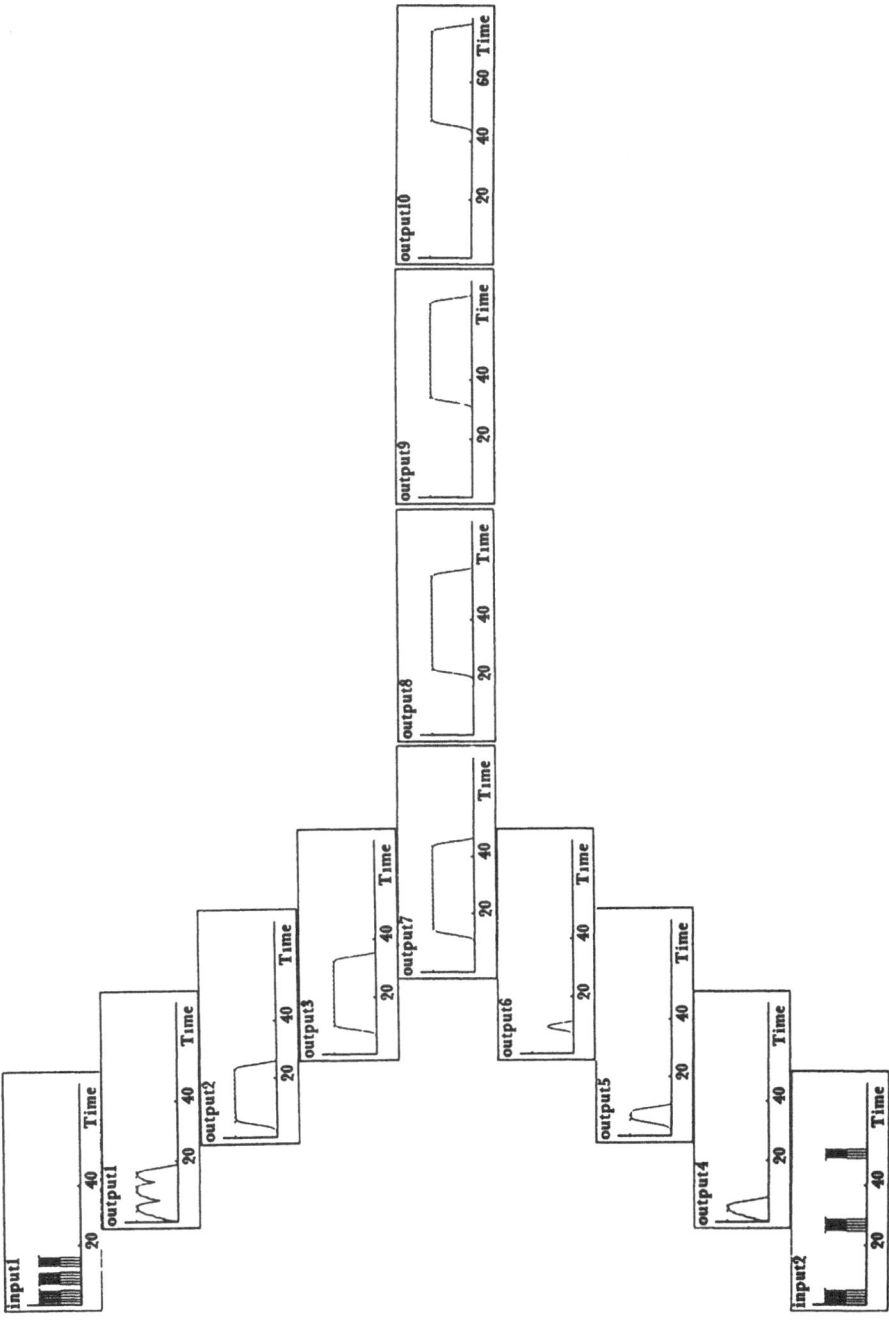

Figure 3. Dynamic response of output $A_i(t)$ in the scheme of Fig. 2. Notation: 'input1', excitatory stimuli with high frequency introduced to the 1-st element; 'input2', excitatory stimuli with low frequency introduced to the 4-th element; output i', output, $A_i(t)$, of the ith element. The maximum value of the ordinate (dashed line) is 1.0.

169

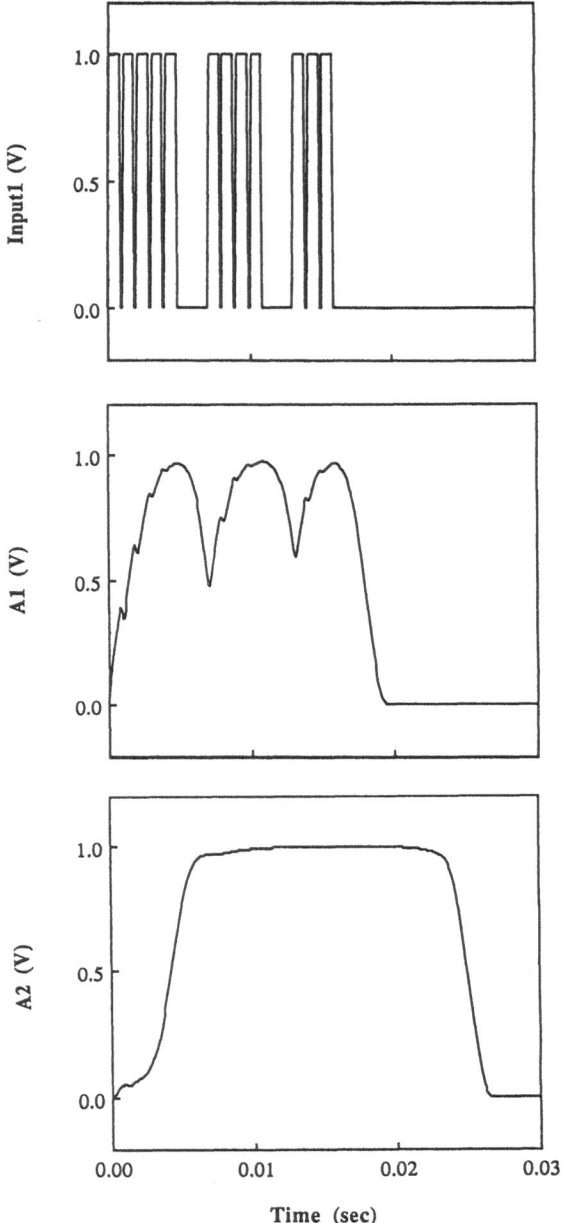

Figure 4. Results of a hardware simulation with PSpice. Input1, excited stimuli introduced to the 1-st element; A1, output $A_1(t)$; A2, output $A_2(t)$.

leads to a selective elimination of the synaptic connection between the sixth and the seventh element.

In the artificial neural network model, most rules for unsupervised learning are based on the proposal of Hebb[5]. Mathematical models of this Hebbian rule can be written as follows:

$$\tau_j \frac{du_j}{dt} = -u_j + \sum_{i=1}^{n} w_{ji} x_i \,, \tag{2a}$$

$$\tau'_j \frac{dw_{ji}}{dt} = -w_{ji} + c_j f(u_j) x_i \,, \tag{2b}$$

$$f(u_j) = \frac{\arctan\left(\dfrac{u_j - \theta}{\varepsilon}\right) + \dfrac{\pi}{2}}{\pi} \,, \tag{2c}$$

where, at time t, u_j represents the membrane potential of the jth element. w_{ji} is the synaptic weight of the transmission from the ith element to the jth. $f(u_j)$ and θ denote the Heaviside function and the threshold value of $f(u_j)$, respectively. The c_j, τ_j and τ'_j are parameters predominant for the control of learning. The parameter ε can regulate the steepness of the Heaviside function.

With the mathematical model (2), selective elimination of synapses could be observed under the same condition as in Fig. 3 (data not shown).

3. Artificial Neuronic Device

The conditions required for the artificial neuron can roughly be classified as follows:

(a) threshold-logic unit,
(b) multi-input and single-ouput systems,
(c) synaptic efficiencies or weights of input signals,
(d) integration of input signals,
(e) suitable selection of the threshold value, and
(f) plastic change in each synaptic efficiency[6].

The device proposed here (see Fig. 1) satisfies these requirements except for (f). By modifying the scheme shown in Fig. 1, we have recently designed an artificial neuronic device that satisfies all the above-mentioned requirements.

4. Practical Implementation of the Device

Our next step is the practical implementation of the device. Mimicking the switching algorithm of Fig. 1 based on a set of differential equations, we can design a unique IC-chip. As the first step of implementing the device, the analog circuit of Fig. 1 was designed by using the PSpice program (electronic circuit simulator).

According to eqs (1), this circuit supports multi-input (Y_i, Z_j), involving the following adjustable parameters: the steepness of the sigmoid curve of switching (k_1, k_2), the capacity of memory storage (k_3, k_4) and the threshold value (θ).

Fig. 4 shows the results of a hardware simulation with PSpice under the same conditions as in Fig. 3. Comparison of the results with those shown in Fig. 3 reveals that the dynamic responses of $A_1(t)$ and $A_2(t)$ completely coincide with each other.

Figure 5. Photograph of a "biochemical neuron" (cf. Fig. 1).

Based on the results of the PSpice program, we have implemented this electronic device as shown in Fig. 5. Integration of the device and the connection into a network will allow us to realize a hardware "neuro-simulator" in the near future.

Acknowledgements

This research is supported by the Nissan Science Foundation, Japan and by the Association for Progress of New Chemistry, Japan.

We thank Mr. Kei-ichi Yoshino at Kitakyushu College of Technology for his technical advice upon practical implementation of the electronic device.

References

1. M. Okamoto, T. Sakai, and K. Hayashi, Switching mechanism of a cyclic enzyme system: Role as a 'chemical diode', *BioSystems* **21**:1-11 (1987).
2. M. Okamoto, T. Sakai, K. Hayashi, Biochemical switching device realizing McCulloch-Pitts type equation, *Biol. Cybern.* **58**:295-299 (1988).
3. M. Okamoto, Development of a biochemical switching device: Mathematical model, *Biomed. Biochim. Acta* **49**:917-933 (1990).
4. Y. Kuroda, 'Tracing circuit' model for the memory process in human brain: roles of ATP and adenosine derivatives for dynamic change of synaptic connections, *Neurochem. Int.* **14**:309-319 (1989).
5. D.O. Hebb. "The organization of behavior", Wiley, New York (1949).
6. T. Kohonen, An introduction to neural computing, *Neural Networks* **1**:3-16 (1989).

EXAMPLE OF THE USE OF COMPUTERS TO STUDY THE VARIABILITY OF SEED DEVELOPMENT OF A PARASITIC WEED, *Striga hermonthica* (Del.) Benth. (SCROPHULARIACEAE), AFTER TREATMENT BY HERBICIDE

J. Paré

Université de Picardie Jules Verne
Faculté des Sciences
Laboratoire de Botanique et d'Embryologie Végétale
33 Rue Saint Leu, 80039 Amiens Cedex 1, France

1. Introduction

Striga hermonthica (Del.) Benth. (Scrophulariaceae), an enormously destructive weed is known as one of the greatest scourges of food crops production in Tropical Africa ; it interferes with the production of native African cereals[1,2].

Because of the formidable problem of the parasitic weed, we are now working with special interest for embryological research, to get a synthetic view of *Striga* development in natural conditions, then to compare the embryological development after treatment by herbicide, finally to evaluate the action of herbicide on seeds.

2. Material and Methods

The observations were carried out on *S. hermonthica*, collected in Mali from natural populations and after treatment by herbicide (2,4-D, 900g a.i./ha[-1]). Complete inflorescences were fixed with F.A.A., at the time they were collected. Standard light microscopy is the most usual way of examining seeds in routine embryology ; so, the classical paraffin method was used, to obtain serial sections (15 μm) stained with Heidenhain's iron alun hematoxylin. An average of 2000 seeds at various stages of development were examined. Axial sections of seeds were observed with an Olympus BHS microscope connected to a Panasonic WV-CD52 camera.

Pictures were digitized by a Compaq Deskpro 386 microcomputer with the Biocom Society mechanism using the Biocom 200 device. The digital image was displayed on a high resolution screen CD 233, BARC, and results were obtained by the morphometric method, on Imagenia software version 2. The different values obtained were then taken up again and graphically interpreted with Excel 2.2a software on an Apple Macintosh SE microcomputer.

Modern Trends in Biothermokinetics, Edited by
S. Schuster *et al.*, Plenum Press, New York, 1993

3. Outline of the Embryological Development

The vast production of seeds, every one of which is fertile, is one of the most important elements of the action of the parasite.

If numerous information exists concerning the phenologic stages of the biological cycle of *Striga*, no modern research describes intraseminal development in detail, from the fecondation until the maturity of seed.

By proceeding stage by stage, that is to say by describing the filiation of the embryonic cells from generation to generation, it is possible to determine the precise origin of all the blastomeres until their final differentiation. Thus it is possible to obtain, from precise embryogenic characters, information and analytic criteria, which are reliable and of highest interest.

Detailed embryogeny formed the subject of recent works: a summary of the essential stages of development of the embryo and seed allows us to describe a usual and unvariable embryogenesis, similar to those described for *Capsella*, with a precocity of differentiation of the embryonic tiers[3-5].

Table 1. Size of the mature embryo and of the endosperm cells in different areas of the mature seed; under natural conditions and after treatment. (C: total cell, V: vacuole, N: nucleus).

Development of the embryo	Total inner surface of the seed	Total surface of the embryo	Areas measured (μm^2)		
				Cells of endosperm	
			Micropylar zone (haustorium) (Mi)	Middle part around the embryo (Mid)	Chalazal zone (Cha)
Natural conditions	17400	3749.44	C 101.03 V 9.10 N 19.66	C 205.32 V 20.80 N 26.74	C 354.35 V 24.06 N 38.79
After treatment	18380	6873.2	C 104.95 V - N 22.06	C 310.22 V - N 32.07	C 573.64 V - N 33.09

The endosperm is *ab initio* cellular ; it develops although the zygote has not yet divided. The chalazal part is formed by two uni-nucleate cells which have dense cytoplasm contents. In the middle part of the seed, uni-nucleate cells lie parallel to one another. The tubular micropylar zone (haustorium) surrounding the suspensor, consists of 5 to 6 small cells, ending in a vesicle.

4. Results

We wish to present comparative results obtained on seeds with embryo at the young plantlet stage when all the parts of the future plant are formed and when the embryo is of axial symmetry. At this stage, embryos have generally escaped the action of the herbicide (see Tables 1, 2).

Table 2. Comparison of the size (in μm) of the total seed and of the seed-coat for different parts of the mature seed under natural conditions and after treatment.

	Part of the seed					
	Amount of thickening between the integument and the endosperm			Thickness of the seed coat (middle part of the seed)	Internal diameter of the seed (middle part)	Perimeter of the inner part of the seed
Development of the embryo	Micropylar zone	Middle zone	Chalazal zone			
Natural conditions	2.84	2.73	3.26	28.03	93.48	612
After treatment	2.84	2.35	3.99	27.89	118.54	544

The processes of development and the structure of the embryo and endosperm are not profoundly disturbed by treatment[4]. But quantitative analysis reveals hyperhydricity in most parts of seeds. The seeds after treatment are more rounded ; the general surface of seeds (in axial section) is about 10% more important. Nevertheless the importance of the embryo is generally double in the seed after treatment than in natural conditions.

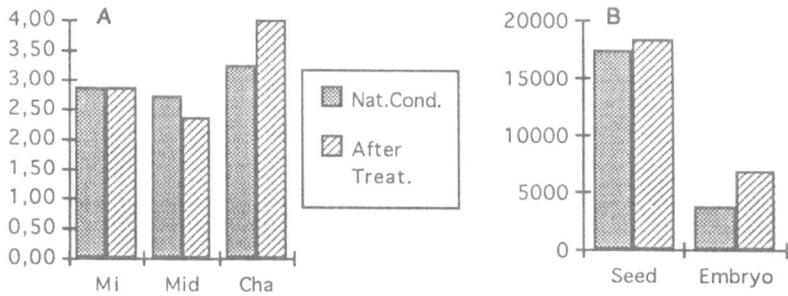

Figure 1. Thickening between the integument and the endosperm (A) and size of seed and embryo (B) under natural conditions and after treatment.

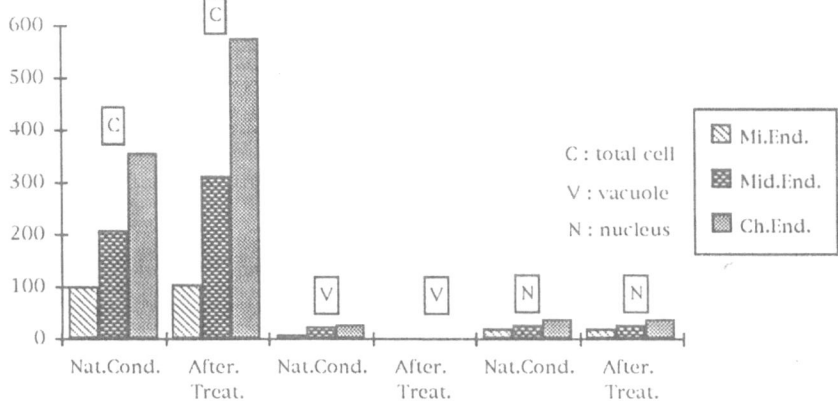

Figure 2. Comparison of endosperm cells in mature seed, in natural conditions and after treatment.

The endosperm cells of the micropylar haustorial zone do not seem to be under the influence of treatment. But these of the middle part of the seed, around the cotyledonary part of the embryo, and of the chalazal zone, appear modified. We can observe disconnected cells, with thick walls, granular cytoplasm (without large vacuoles), of untimely ageing aspect. The thickening between the integument and the endosperm, and the thickness of the seed coat, are quite identical in natural conditions and after treatment (see Figs 1 and 2).

5. Discussion and Conclusions

The major reason why we use computers in plant embryology is that we wish to analyze the shapes and functions of embryological organization, and to use basic mathematical and statistic structures in order to make a comparison between seed development in natural conditions and after treatment, also to study and specify the variability of seed development of a parasitic weed.

Embryogenesis in *S. hermonthica* is unvariable and typical of the family of Scrophulariaceae. The precocity of differentiation of the embryonic tiers guarantees the standard development of the embryo in spite of modifications of the environmental conditions after treatment by herbicide.

Due to the structure of the cellular endosperm, in natural conditions, the cell wall becomes cutinized at an early stage; the embryo must derive its nourishment through the micropylar haustorial zone. In mature seeds, the endosperm cells rounded and the wall thickened. We found the same aspect, of endosperm cells untimely ageing, inside the treated seeds, since the early stages of development of the proembryo. So, the endosperm seems to play the role of a filter, to protect the embryo. Under such conditions (in addition to the precocity of differentiation of the embryo, rapidity of biological cycle, capacity of germination, protection of seed by lignified testa and the cutinized outer-wall of the endosperm) it is easy to understand that *S. hermonthica* has the best chance to withstand treatments.

Considering the severity of the action of this parasitic weed, and the difficulty to eradicate it because of its high capacity of reproduction, the results obtained allow us to suggest the use of treatment before flowering, on the vegetative parts of the plant, in order to try to stop the formation of seeds. We can expect that this new data could be fruitful for future agronomical methods.

Acknowledgements

This research is supported by an E.E.C. grant (TS2 - 0236C (GDF)).

References

1. G. Sallé et A. Raynal-Roques, Le *Striga, La Recherche* **206**:44-52 (1989).
2. G. Sallé, B. Dembélé, A. Raynal-Roques et C. Tuquet, Le *Striga* : dégâts, biologie et lutte, *Afrique, Agriculture* **158**:41-43 (1990).
3. J. Paré, M. Bugnicourt, A. Raynal-Roques and G. Sallé, Embryological investigations on the development of the parasitic weed *Striga hermonthica* (Del.) Benth. Scrophulariaceae, *in*: "Proceedings of the Vth International Symposium of parasitic weeds," J.K. Ransom, L.J. Musselman, A.D. Worsham and C. Parker, eds, CIMMYT, Nairobi (Kenya), pp. 523-524 (1991).
4. J. Paré, A. Raynal-Roques and G. Sallé, *Striga hermonthica* (Del.) Benth., Scrophulariaceae, parasitic weed of food crops in tropical Africa. New embryological aspects, *in*: "Vth Conference of Plant Embryologists from Poland and Czecho-Slovakia," Bachotek, Torun (Poland), in press.
5. J. Paré and A. Raynal, The genus *Striga* (Scrophulariaceae): taxonomy, geographical distribution. Embryological results in *Striga hermonthica* (Del.) Benth. scourge of food crops in Tropical Africa with special reference to the action of herbicide on seed development, *Acta Biologica Cracoviensa*, in press.

III. METABOLIC CONTROL THEORY:
NEW DEVELOPMENTS AND APPLICATIONS

RESPONSES OF METABOLIC SYSTEMS TO LARGE CHANGES IN ENZYME ACTIVITIES

J. Rankin Small and Henrik Kacser

ICAPB, Crew Building
The University of Edinburgh
King's Buildings, Edinburgh, EH9 3JN, United Kingdom

1. Introduction

It is now possible in many organisms to use molecular biological tools to overexpress a gene coding for an enzyme. This overexpression can result in very large changes in enzyme activity. The effects of this overexpression on the metabolic system in which this enzyme acts are difficult to predict. Metabolic Control Analysis[1] may give us some indication of the likely effect (small or large), via the control coefficient of that enzyme to a particular flux or metabolite, but due to the infinitesimal nature of these coefficients, i.e.,

$$C_{E_i^o}^{J^o} = \left(\frac{\partial J}{\partial E_i} \right)^o \cdot \frac{E_i^o}{J^o} , \tag{1}$$

where $\left(\dfrac{\partial J}{\partial E_i} \right)^o$ is the slope of the J versus E relationship at (J^o, E_i^o), extrapolation to large changes is unlikely to be quantitatively accurate.

The problem to be considered can be illustrated by Fig. 1. Here we show an arbitrary relationship between flux and enzyme activity. We will consider that the information we have is limited to the initial enzyme activity, E_i^o, the initial flux, J^o at E_i^o and a measurement of the flux control coefficient at the initial point, $C_{E_i^o}^{J^o}$, (hence we know the slope $\partial J / \partial E$). We would like to know the effect on the flux, ΔJ, due to a large change in enzyme activity, ΔE_i. Recently we have considered this problem[2,3] and we will discuss some of the results here.

To facilitate the analysis of the problem we have defined the *Deviation index*, D as:

$$D_{E_i^r}^{J^r} = \left(\frac{\Delta J}{\Delta E_i} \right) \cdot \frac{E_i^r}{J^r} , \tag{2}$$

Modern Trends in Biothermokinetics, Edited by
S. Schuster *et al.*, Plenum Press, New York, 1993

where $\Delta J = J_r$-J_0, and $\Delta E_i^r = E_i^r - E_i^0$. Note that, in contrast to the control coefficient, the scaling factor chosen here is the ratio of enzyme activity to flux at the new and not the original point. In the analysis that follows we see that under certain conditions an exact algebraic relationship between the deviation index and the control coefficient can be established.

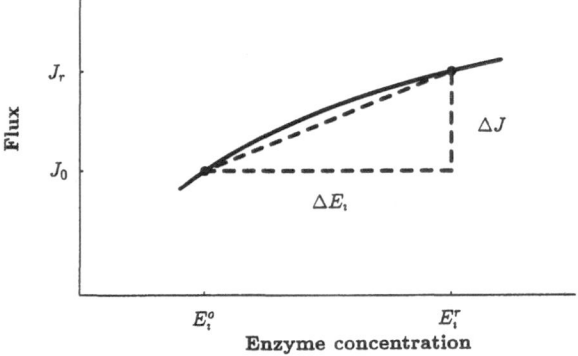

Figure 1. Flux-enzyme relationship.

2. Analysis

A general algebraic solution for the relationship between a metabolic flux and any particular enzyme is not possible to obtain. Hence, we have first considered a system where a solution can be obtained. We assume that all metabolites show linear (or quasi-linear) kinetics with respect to the reactions they are involved in, i.e., if the rate through a step is governed by the following non-linear mechanism,

$$v_i = \frac{V_{\text{max}_i} / K_{\text{m}_i} \cdot \left(S_i - S_j / K_{\text{eq}_{ij}} \right)}{1 + S_i / K_{\text{m}_i} + S_j / K_{\text{m}_j}} \quad ,$$

then it can be approximated by a linear form,

$$v_i = e_i \cdot \frac{\left(S_i - S_j / K_{\text{eq}_{ij}} \right)}{k_s} \quad , \tag{3}$$

where $e_i = [E_i].f(k_{\text{cat}_i}, K_{\text{m}_i}, K_{\text{m}_j}, ...)$ and k_s is the 'saturation function'. S_i, S_j are concentrations of the metabolites involved in the transformation. In the case when $S_i \ll K_{\text{m}_i}$ and $S_j \ll K_{\text{m}_j}$ (i.e. an 'unsaturated system'), the value for k_s is unity. Even if $k_s \neq 1$, the equation can still be treated as linear, provided there is no significant *change* in the saturation during the flux change considered, when k_s can be incorporated into e_i.

If the above assumptions are valid then the flux through an unbranched chain can be expressed algebraically as a function of all the enzyme activities. Using this expression it is possible to obtain further algebraic expressions for both control coefficients and deviation indices. Analysis of these expressions has shown[2] that the deviation index for an enzyme in an unbranched chain, for any change in activity r, is numerically equivalent to the flux control coefficient of that enzyme, determined at the initial point, i.e.,

$$D_{E_i^r}^{J^r} \equiv C_{E_i^o}^{J^o} \ . \tag{4}$$

This result can be generalized to any group of enzymes changed by the same factor or differing factors[2,3].

There are a number of applications using the above results. The identity between the flux control coefficient and the deviation index means that, from a single large change in activity r, we can obtain an estimate of the flux control coefficient of that enzyme. To illustrate this point we have considered the data of Salter and coworkers[4]. These authors were interested in aromatic amino acid catabolism in rat hepatocytes. They obtained measurements of tryptophan flux and the activity of tryptophan 2,3-dioxygenase under 8 different conditions. Table 1 shows a comparison between an estimate of the control coefficient for the basal (un-induced) state, $C_{\text{base}}^{J_{\text{trp}}}$, obtained by nonlinear curve-fitting to *all* 8 points and the deviation index, D, for a change in enzyme activity from the basal state to a hormonally induced state (i.e. obtained using only two of the 8 data points available). It is clear that the deviation index in this case is a very good estimate of the flux control coefficient.

Table 1. Determination of the deviation index for the basal tryptophan flux due to a change in activity of tryptophan 2,3-dioxygenase from a basal to an induced level. Standard deviations are shown in brackets.

r	J^r/J^o	D(s.d)	$C_{\text{base}}^{J_{\text{trp}}}$
7.37	3.0	0.77 (0.05)	0.78

We have also studied other experimental systems[2,3] and also found good agreement between values of deviation indices obtained from two data points only and flux control coefficients obtained from non-linear curve fitting to many points.

One of our aims was to be able to predict, using only a minimal amount of information, the effect of a large change in enzyme activity. The results of above indicate that, provided the assumptions are valid, we now have a means of estimating this effect from an estimate of the flux control coefficient value. We have defined an *Amplification factor*, f, for a change in flux J due to a large change in enzyme E_i by a factor r as follows:

$$f_{E_i^r}^J \equiv \frac{J^r}{J^o} \ . \tag{5}$$

Using the result of eq. (4) and the definition of a deviation index, it is possible to show that

$$f_{E_i^r}^J = \frac{1}{1 - C_{E_i^o}^{J^o} \cdot \dfrac{r-1}{r}} \tag{6}$$

This relationship is also valid for the effect of a change in a group of enzymes by the same factor r, if the single control coefficient, $C_{E_i^o}^{J^o}$, is replaced by the group control coefficient, which is equal in value to the sum of the individual control coefficients of the group, i.e.

$\sum_{i=j}^{m} \ C_{E_i^0}^{J^o}$. Fig. 2 shows the dependence of f on the value of this group control coefficient, at three different values of r. There are three major points that can be seen.

1) The size of the change in flux increases, in a non-linear manner, as the size of the control coefficient increases;

2) At control coefficient values less than 0.5, increases in activity above $r=10$ will have little additional effect compared to a 10-fold increase;

3) Only as the value of the control coefficient approaches unity will very large increases in flux occur. Indeed, even with a group control coefficient value of 0.5 (a significant amount of control), less than a twofold increase in flux would be expected after an increase in activity of the group by a factor equal to 20.

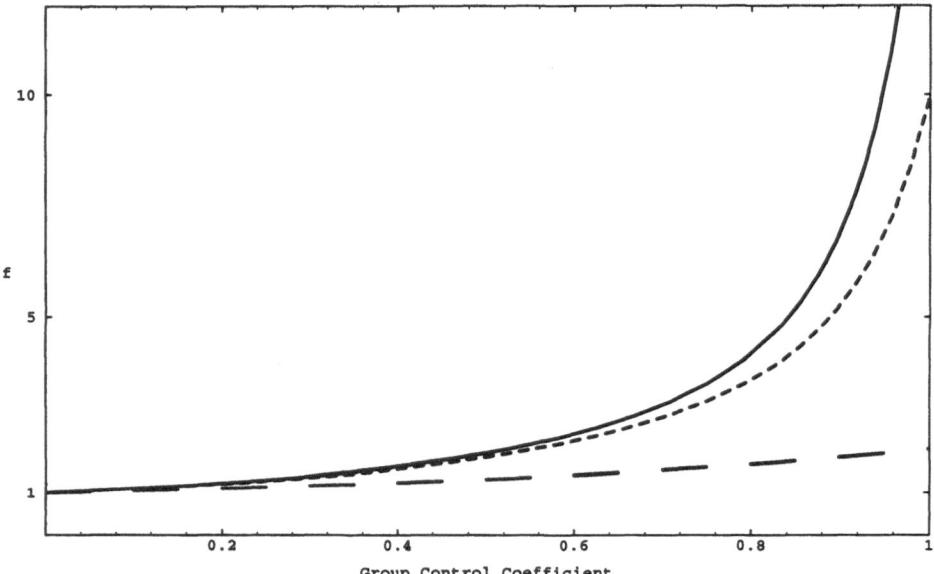

Figure 2. Dependence of f on C (group control coefficient) at three different r values for an enzyme in an unbranched chain. (— — —), $r = 2$; (– – – –), $r = 10$; (———), $r = 20$.

So far we have only considered unbranched chains. Branched metabolic pathways are, however, ubiquitous, and hence we have applied the same process to these systems as we have already shown for unbranched chains[3]. In branched systems we must distinguish between two types of responses. The first is where the enzyme modified and the flux monitored occur within the same branch of the pathway. In this case we obtain an identical relationship between the deviation index and the flux control coefficient as for the unbranched chain; e.g. if the activity of enzyme E_a, which occurs in branch A, is changed by a factor r and we monitor the change in flux J_A then the deviation index can be expressed as:

$$D_{E_a^r}^{J_A^r} \equiv C_{E_a^0}^{J_A^o} . \tag{7}$$

The applications discussed above in relationship to unbranched chains are therefore valid for this case in branched systems as well.

If, on the other hand, we monitor a flux through a branch other than the one in which the enzyme occurs, then we obtain a different type of response. If, for example, we monitor the flux through branch B (J_B), after changing enzyme E_a as above, we obtain the following expression for the deviation index:

$$D_{E_a^r}^{J_B^r} \equiv C_{E_a^o}^{J_B^o} \cdot \frac{1}{1 - \left(C_{E_a^o}^{J_A^o} - C_{E_a^o}^{J_B^o}\right) \cdot \frac{r-1}{r}} \cdot \tag{8}$$

Note that, in contrast to the case above and for the unbranched chain, this deviation index depends not only on the appropriate control coefficient, $C_{E_a^o}^{J_B^o}$, but also on $C_{E_a^o}^{J_A^o}$ (the control coefficient of enzyme E_a with respect to its 'own' branch flux), and the factor by which the enzyme was changed, r.

The form of this expression has a number of consequences when it comes to its use. To obtain an estimate of the value of the flux control coefficient $C_{E_a^o}^{J_B^o}$ from a single large change in enzyme E_a, two fluxes must be monitored, i.e.

$$C_{E_a^o}^{J_B^o} = \frac{f_a^{J_B} - 1}{f_a^{J_A} \cdot \frac{r_a - 1}{r_a}} \cdot \tag{9}$$

The amplification factor of flux J_B due to a change in activity of enzyme E_a is given by:

$$f_{E_a^r}^{J_b} = 1 + \frac{C_{E_a^o}^{J_B^o} \cdot \frac{r-1}{r}}{1 - C_{E_a^o}^{J_A^o} \cdot \frac{r-1}{r}} \cdot \tag{10}$$

Fig. 3 illustrates why it is important to know the values of both control coefficients when attempting to predict the value of the amplification factor in this sort of situation. Two important points can be seen in this figure. The first is that if $C_{E_a^o}^{J_A^o}$ is very small, then no matter what the value of $C_{E_a^o}^{J_B^o}$, the response of flux J_B to E_a will be small. Conversely, the second point is that as $C_{E_a^o}^{J_A^o}$ tends to unity (its maximum value), a large response of flux J_B can be obtained even if $C_{E_a^o}^{J_B^o}$ is small in value. For example, if $C_{E_a^o}^{J_B^o}$ has a value of 0.05 and $C_{E_a^o}^{J_A^o}$ has a value of 1 the amplification factor for flux J_B for a 20-fold increase in enzyme E_a would be 1.95. To put this value into perspective, it is a larger response in flux J_B than would be obtained from a change in enzyme E_b (an enzyme occurring in branch B) assuming that that enzyme had a flux control coefficient, $C_{E_b^o}^{J_B^o}$, equal to 0.5.

Similarly, relationships can be derived relating deviation indices and control coefficients with respect to other metabolic variables, e.g. metabolite concentrations.

All the above results are based on the assumption that linear kinetics can reasonably describe the behavior of the system. These, and other results, can be used to test whether the assumptions are admissible. We have devised a number of 'tests' for non-linearity, which are summarized below.

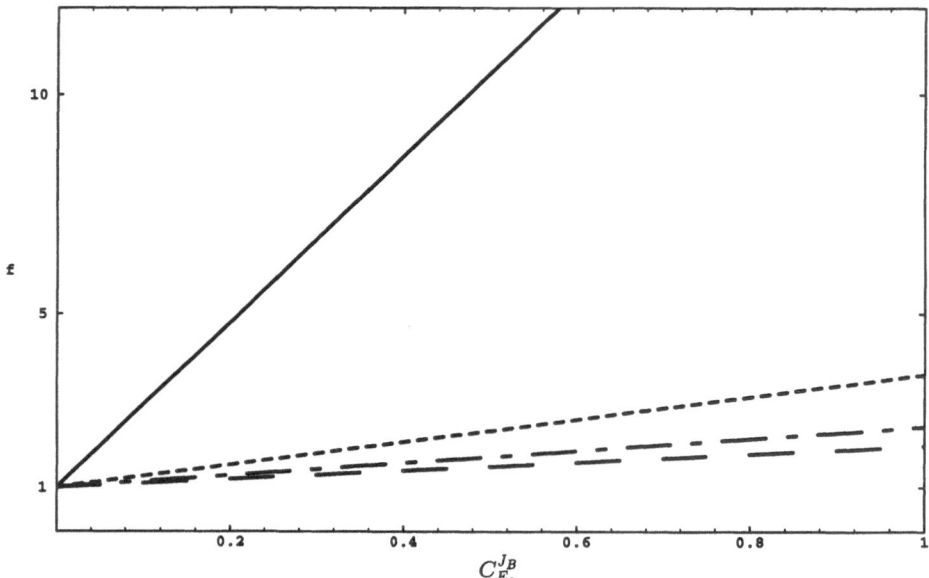

Figure 3. Dependence of f on $C_{E_a}^{J_B}$ and $C_{E_a}^{J_A}$ for a change in enzyme E_a with flux J_B monitored. (— — —), $C_{E_a}^{J_A} = 0$; (— – —), $C_{E_a}^{J_A} = 0.33$; (– – – –), $C_{E_a}^{J_A} = 0.67$; (———), $C_{E_a}^{J_A} = 1$. In all cases $r = 20$.

1) In an unbranched system, the flux deviation index is found to be numerically equal to the flux control coefficient with respect to the enzyme changed. Thus, the value of this deviation index is independent of the size of r. Hence, if two changes with different r's in the same enzyme are made, and the resulting values of the measured deviation indices are not significantly different, then the assumptions of kinetic linearity can be assumed to be valid for these changes. This particular test is also valid for branched systems when the flux monitored and the enzyme(s) changed are from the same branch of the system.

2) It can be shown[2,3] that the response of any flux or metabolite concentration, in both a branched and unbranched system, is related to the response coefficient of the pool in the following manner:

$$^*D_{X_0}^V \equiv R_{X_0}^V \; ,$$

where $^*D$ is defined as:

$$^*D_{X_0}^V = \frac{\Delta V}{\Delta X_0} \cdot \frac{X_0^o}{V^o}$$

and V is the particular variable monitored. Note that, in contrast to the 'normal' deviation index, D, the scaling factor in this $^*D$ is the variable/pool ratio measured at the *original* point. Since the value of the deviation index is independent of r then, again, two different r changes in the initial pool should give the same value if the assumption of linear kinetics is valid. Monitoring the response of more than one variable will provide extra checks on the validity of the assumptions.

184

3) If it is found that two different changes in external pool give different values for the resulting deviation indices then this may be due to saturation of the first step with the initial pool. Thus the test for non-linearity would fail even though all intermediate metabolites show linear kinetics with respect to the enzymes and the relationships derived for enzyme changes would still be valid. If the rate equation of the first step is given by:

$$v_1 = \frac{V_{max}^1 \cdot \left(X_0 - S_1 / K_{eq_1} \right)}{K_m^1 + X_0}$$

then, if three different r changes in the external pool are made and the values of the amplification factors and deviation indices are obtained, the following relationship between the values is true, provided all the internal metabolites show linear kinetics with respect to the enzymes:

$$\frac{{}^*D_2 - {}^*D_1}{f_1 - f_2} = \frac{{}^*D_3 - {}^*D_1}{f_1 - f_3} \ ,$$

where ${}^*D_i$ and f_i are the deviation index and amplification factor for the ith r change.

3. Conclusions

We have shown that with the assumption of linear kinetics it is possible to analyze the response of large changes in enzyme activities in terms of metabolic control analysis, and how this analysis can be used to test for the admissibility of these assumptions. A more in-depth discussion can be found in Refs 2,3. In the second of these papers we show the results of computer simulations which suggest that even if the kinetics is non-linear, predictions based on linear kinetics may not be drastically wrong.

References

1. D.A. Fell, Metabolic Control Analysis: a survey of its theoretical and experimental development, *Biochem. J.* **286**:313-330 (1992).
2. J.R. Small and H. Kacser, Responses of metabolic systems to large changes in enzyme activities and effectors. I. The linear treatment of unbranched chains, *Eur. J. Biochem.*, in press.
3. J.R. Small and H. Kacser, Responses of metabolic systems to large changes in enzyme activities and effectors. II. The linear treatment of branched pathways, *Eur. J. Biochem.*, in press.
4. M. Salter, R.G. Knowles and C.I. Pogson, Quantification of the importance of individual steps in the control of aromatic amino acid metabolism, *Biochem. J.* **234**:635-647 (1986).

AN EXTENSION OF METABOLIC CONTROL ANALYSIS
TO LARGER PARAMETER PERTURBATIONS
BY INCLUSION OF SECOND-ORDER EFFECTS

Thomas Höfer and Reinhart Heinrich

Humboldt-Universität zu Berlin
Fachbereich Biologie, Institut für Biophysik
Invalidenstr. 42, 1040 Berlin, Germany

1. General Theory

In the formalism of metabolic control analysis the changes in concentrations and fluxes have been related to the underlying parameter perturbations by a linear approximation, thus restricting the analysis to very small perturbations (for general definitions cf. Ref. 1). The response of the system variables to larger parameter perturbations can be described more accurately if in addition to the linear terms the second-order terms are considered:

$$\Delta \mathbf{Y} \cong \frac{\partial \mathbf{Y}}{\partial \mathbf{p}}(\mathbf{p})\Delta \mathbf{p} + \frac{1}{2} \sum_{\alpha,\beta} \frac{\partial^2 \mathbf{Y}}{\partial p_\alpha \partial p_\beta}(\mathbf{p})\Delta p_\alpha \Delta p_\beta . \tag{1}$$

$\mathbf{Y}$ stands both for the concentration vector $\mathbf{S} = (S_1,...,S_n)^T$ and the flux vector $\mathbf{J} = (J_1,..., J_r)^T$, $\mathbf{p}$ denotes the parameter vector $(p_1,..., p_m)^T$. Suppose $\mathbf{S}^o(\mathbf{p})$ to be a stable steady state of the reaction network for a given parameter vector $\mathbf{p}$. $\mathbf{S}^o(\mathbf{p})$ is the solution of the equation system

$$\mathbf{N} \, \mathbf{v}(\mathbf{S},\mathbf{p}) = \mathbf{0} , \tag{2}$$

where $\mathbf{N}$ and $\mathbf{v}(\mathbf{S},\mathbf{p})$ denote the stoichiometric matrix and the vector of reaction rates, $(v_1,...,v_r)^T$, respectively. We write the corresponding steady-state flux as $\mathbf{J} = \mathbf{v}(\mathbf{S}^o(\mathbf{p}),\mathbf{p})$.

The parameter change $\Delta \mathbf{p}$ leads to a perturbation of the stable steady state. The resulting concentration and flux changes are approximated by eq. (1). For the sake of simplicity it is assumed that the metabolic network has no conservation relations, i.e. $\text{rank}\mathbf{N} = n$. The derivatives in eq. (1) can be calculated by differentiation of eq. (2) with respect to $\mathbf{p}$. For the linear terms one obtains[2,3]

$$\frac{\partial \mathbf{S}}{\partial \mathbf{p}}(\mathbf{p}) = \mathbf{C}^S \frac{\partial \mathbf{v}}{\partial \mathbf{p}}(\mathbf{p}) , \quad \mathbf{C}^S = - \left(\mathbf{N} \frac{\partial \mathbf{v}}{\partial \mathbf{S}} \right)^{-1} \mathbf{N} , \tag{3}$$

Modern Trends in Biothermokinetics, Edited by
S. Schuster *et al.*, Plenum Press, New York, 1993

$$\frac{\partial \mathbf{J}}{\partial \mathbf{p}}(\mathbf{p}) = \mathbf{C}^{\mathbf{J}} \frac{\partial \mathbf{v}}{\partial \mathbf{p}}(\mathbf{p}), \quad \mathbf{C}^{\mathbf{J}} = \mathbf{I} - \frac{\partial \mathbf{v}}{\partial \mathbf{S}}\left(\mathbf{N}\frac{\partial \mathbf{v}}{\partial \mathbf{S}}\right)^{-1}\mathbf{N} = \mathbf{I} + \frac{\partial \mathbf{v}}{\partial \mathbf{S}}\mathbf{C}^{\mathbf{S}} \ . \tag{4}$$

$\mathbf{I}$ denotes the $r \times r$ identity matrix. The elements of $\mathbf{C}^{\mathbf{S}}$ and $\mathbf{C}^{\mathbf{J}}$ are the non-normalized concentration and flux control coefficients, respectively. Using eqs (3) and (4) one finds for the second-order terms

$$\frac{\partial^2 Y_a}{\partial p_\alpha \partial p_\beta} = \sum_{i,l,h=1}^{r} \sum_{j,k=1}^{n} C_{ai}^Y \frac{\partial^2 v_i}{\partial S_j \partial S_k} C_{jl}^S C_{kh}^S \frac{\partial v_l}{\partial p_\alpha}\frac{\partial v_h}{\partial p_\beta} + \sum_{i=1}^{r} C_{ai}^Y \frac{\partial^2 v_i}{\partial p_\alpha \partial p_\beta} +$$

$$+ \sum_{i,k=1}^{r} \sum_{j=1}^{n} C_{ai}^Y \left(\frac{\partial^2 v_i}{\partial S_j \partial p_\beta} C_{jk}^S \frac{\partial v_k}{\partial p_\alpha} + \frac{\partial^2 v_i}{\partial S_j \partial p_\alpha} C_{jk}^S \frac{\partial v_k}{\partial p_\beta} \right). \tag{5}$$

The general structure of the second-order approximation becomes more transparent with the following abbreviated notation:

$$\Delta \mathbf{Y} \cong \mathbf{C}^{\mathbf{Y}}\left[\delta_{\mathbf{p}}\mathbf{v} + \frac{1}{2}(\delta^2_{\mathbf{pp}}\mathbf{v} + 2\delta^2_{\mathbf{Sp}}\mathbf{v} + \delta^2_{\mathbf{SS}}\mathbf{v}) \right], \tag{6}$$

where the vectors $\delta_{\mathbf{p}}\mathbf{v}$, $\delta^2_{\mathbf{SS}}\mathbf{v}$, $\delta^2_{\mathbf{Sp}}\mathbf{v}$ and $\delta^2_{\mathbf{pp}}\mathbf{v}$ have the components

$$\delta_{\mathbf{p}}v_i = \sum_{\beta=1}^{m} \frac{\partial v_i}{\partial p_\beta}\Delta p_\beta, \quad \delta^2_{\mathbf{S,S}}v_i = \sum_{j,k=1}^{n} \frac{\partial^2 v_i}{\partial S_j \partial S_k}\delta S_j \delta S_k, \tag{7a,b}$$

$$\delta^2_{\mathbf{Sp}}v_i = \sum_{\beta=1}^{m}\sum_{j=1}^{n} \frac{\partial^2 v_i}{\partial S_j \partial p_\beta}\delta S_j \Delta p_\beta, \quad \delta^2_{\mathbf{pp}}v_i = \sum_{\alpha,\beta=1}^{m} \frac{\partial^2 v_i}{\partial p_\alpha \partial p_\beta}\Delta p_\alpha \Delta p_\beta \tag{7c,d}$$

with δS_j being the concentration change in the linear approximation.

In addition to the quantities of the linear theory (ε- and π-elasticities, concentration and flux control coefficients), calculation of the second-order terms requires knowledge of the following second derivatives:

(i) $\dfrac{\partial^2 v_i}{\partial S_j \partial S_k}$, "second-order ε-elasticities", (ii) $\dfrac{\partial^2 v_i}{\partial p_\alpha \partial p_\beta}$, "second-order π-elasticities", and

(iii) $\dfrac{\partial^2 v_i}{\partial S_j \partial p_\alpha}$, "mixed second-order ε-π-elasticities".

Owing to the occurrence of mixed second derivatives of the reaction rates with respect to concentrations and parameters, a general definition of parameter-independent second-order control coefficients is not possible. In contrast to the linear theory[2,3], the parameter perturbations can therefore not be replaced by the rate perturbations as independent variables. This is in line with the fact that the steady-state concentrations and fluxes are functions of the kinetic parameters but cannot be expressed in terms of the individual rates. Furthermore, the second-order terms contain, besides derivatives characterizing the influence of a single reaction, also mixed derivatives $\partial^2 Y_a / \partial p_\alpha \partial p_\beta$ where p_α and p_β may belong to different rate equations. Hence the effects of simultaneous perturbations of several rates are not simply approximated as the sum of the individual effects as in the case of the linear theory.

2. Linear Perturbation Parameters

If the perturbation parameters act specifically and linearly on individual reaction rates,

$$v_j = p_j w_j(S_1,...,S_n) ; \quad \frac{\partial v_i}{\partial p_j} \begin{cases} \neq 0 \text{ if } i = j \\ \equiv 0 \text{ if } i \neq j \end{cases} , \tag{8}$$

the following definition of second-order control coefficients is appropriate

$$D^Y_{a\alpha\beta} = \frac{1}{2} \frac{\partial^2 Y_a}{\partial p_\alpha \partial p_\beta} \left/ \frac{\partial v_\alpha}{\partial p_\alpha} \frac{\partial v_\beta}{\partial p_\beta} \right. . \tag{9}$$

These coefficients can be calculated from the control coefficients of the linear theory, the fluxes and the second-order ε-elasticities (as can be derived straightforwardly from eq. (6)[4]). Expansion (1) can now be written with the rate rather than the parameter perturbations as independent variables

$$\Delta Y_a = \sum_\alpha C^Y_{a\alpha} \Delta v_\alpha + \sum_{\alpha,\beta} D^Y_{a\alpha\beta} \Delta v_\alpha \Delta v_\beta . \tag{10}$$

The second-order control coefficients are symmetric, i.e. $D^Y_{a\alpha\beta} = D^Y_{a\beta\alpha}$. They satisfy the following summation relations

$$\sum_{\alpha,\beta} D^S_{a\alpha\beta} k_{\alpha\gamma} k_{\beta\delta} = 0 , \quad \sum_{\alpha,\beta} D^J_{a\alpha\beta} k_{\alpha\gamma} k_{\beta\delta} = 0 , \tag{11a,b}$$

where k_γ and k_δ are two vectors from kernel N ($k_\gamma \in$ kernel N, if $N k_\gamma = 0$). Eqs (11a,b) can be verified with the help of the generalized summation theorems of the linear theory[4].

3. Examples

We investigated the accuracy of both the linear and the second-order approximations in the case of finite parameter perturbations for flux control in two model systems, namely an unbranched metabolic chain and a skeleton model of glycolysis.

$$P_1 \underset{\longleftarrow}{\overset{v_1}{\longrightarrow}} S_1 \underset{\longleftarrow}{\overset{v_2}{\longrightarrow}} \cdots \underset{\longleftarrow}{\overset{v_n}{\longrightarrow}} S_n \underset{\longleftarrow}{\overset{v_{n+1}}{\longrightarrow}} P_2$$

For the reaction chain depicted above, we consider the case of reversible Michaelis-Menten rate laws

$$v_i = \left(V_i^+ \frac{S_{i-1}}{K_i^+} - V_i^- \frac{S_i}{K_i^-} \right) \left/ \left(1 + \frac{S_{i-1}}{K_i^+} + \frac{S_i}{K_i^-} \right) \right. . \tag{12}$$

In the case that all substrate concentrations are far below their respective K_m-values ($S_i \ll K_i^-$, $S_i \ll K_{i+1}^+$), eq. (13) can approximated by a linear expression which allows an analytical treatment. In particular the actual flux change for arbitrary rate perturbations can be expressed in terms of the linear flux control coefficients and the rate perturbations:

$$\frac{\Delta J}{J} = \sum_{\alpha} \frac{\Delta v_{\alpha} / v_{\alpha}}{\Delta v_{\alpha} / v_{\alpha} + 1} C_{\alpha}^{J} \Bigg/ \sum_{\alpha} \frac{1}{\Delta v_{\alpha} / v_{\alpha} + 1} C_{\alpha}^{J} \; . \tag{13}$$

The second-order flux control coefficients (7) take the simple form

$$D_{\alpha\beta}^{J} = C_{\alpha}^{J} (C_{\beta}^{J} - \delta_{\alpha\beta}) \; . \tag{14}$$

If only one reaction is perturbed, the relative errors, $\eta^{J}=\Delta J(\text{approximated})/\Delta J(\text{exact})-1$, read

$$\eta_{\text{lin}}^{J} = (1 - C_{\alpha}^{J}) \frac{\Delta v_{\alpha}}{v_{\alpha}} > 0 \; , \quad \eta_{\text{sec}}^{J} = -(1 - C_{\alpha}^{J})^{2} \left(\frac{\Delta v_{\alpha}}{v_{\alpha}} \right)^{2} < 0 \tag{15a,b}$$

for the linear and second-order approximations, respectively. The rate perturbations which lead to a certain admissible error $\tilde{\eta}$ of the linear and second-order approximations, respectively, are related as follows

$$\left(\Delta v_{\alpha} \right)_{\text{sec}} = \frac{1}{\sqrt{\tilde{\eta}}} \left(\Delta v_{\alpha} \right)_{\text{lin}} \; . \tag{16}$$

An illustration to the general situation of several simultaneously perturbed reaction rates is given in Fig.1A. To obtain analytical results with the non-linear expressions (15), the chain was reduced to two reactions. We investigated, in particular, the case of saturation of the first reaction with P_1 $(V^+_1 < V^+_2)$, which is illustrated in Fig.1B.

The results for the unbranched chain can be summarized as follows:

- The second-order approximation is more accurate than the linear one for a wide range of rate perturbations.
- Eventually the error of the second-order approximation becomes larger than that of the linear one, owing to the rapid divergence of the quadratic terms for large Δv.

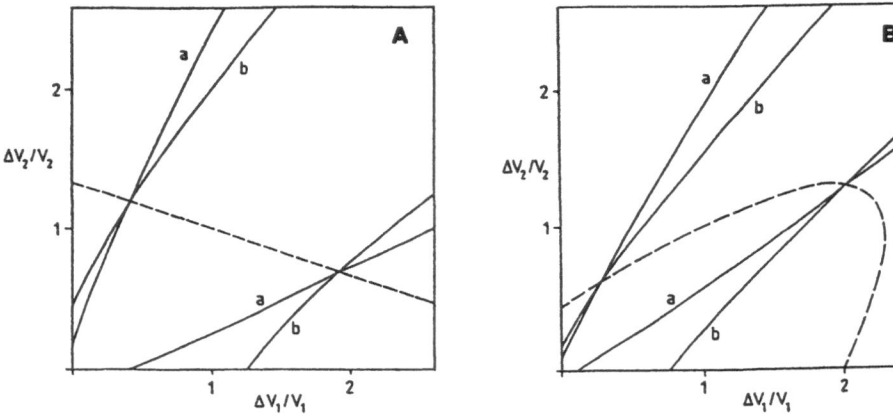

Figure 1. Boundaries of the regions where the relative errors of the linear and second-order approximations are below 10% (solid lines, curves a and b, respectively), and the boundary of the region where the second-order approximation is more accurate than the linear one (dashed line), for an unbranched chain of two reactions with the linear flux control coefficients $C^J_1=0.75$, $C^J_2=0.25$; (A) linear kinetics, (B) reversible Michaelis-Menten kinetics $(P_1/K^+_1= 80, P_2/K^-_2= 0, K^-_1= K^+_2=1, V^+_1=V^-_1 = 1, V^+_2=V^-_2=1.2)$.

- A large linear flux control coefficient of the perturbed reaction corresponds to small relative errors of both approximations. However, this behavior is much less pronounced in the saturated case: the accuracy of both approximations largely decreases if the perturbation of the maximal activity of the first reaction leads to an increase beyond the maximal activity of the second one (by which the latter becomes practically rate limiting). This can be seen by comparing the extension of the 10%-relative-error regions in Figs. 1A and B.

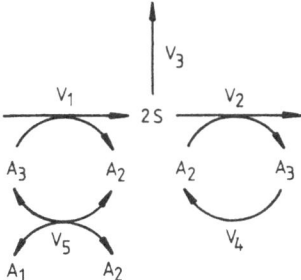

Figure 2. Simplified stoichiometric model of erythrocyte glycolysis. S - energy-rich metabolites of the middle section of glycolysis; A_1, A_2 and A_3 - adenine nucleotides AMP, ADP and ATP, respectively; $v_1 = k_1 A_3$ - ATP-consuming processes of the upper part of glycolysis; $v_2 = k_2 A_2 S$ - ATP-producing processes of the lower part of glycolysis; $v_3 = k_3 S$ - irreversible utilization of energy-rich intermediates for synthesis; $v_4 = k_4 A_3$ - non-glycolytic ATP-consuming processes; $v_5 = k_5 A_2^2 - k_{-5} A_1 A_3$ - adenylate kinase reaction.

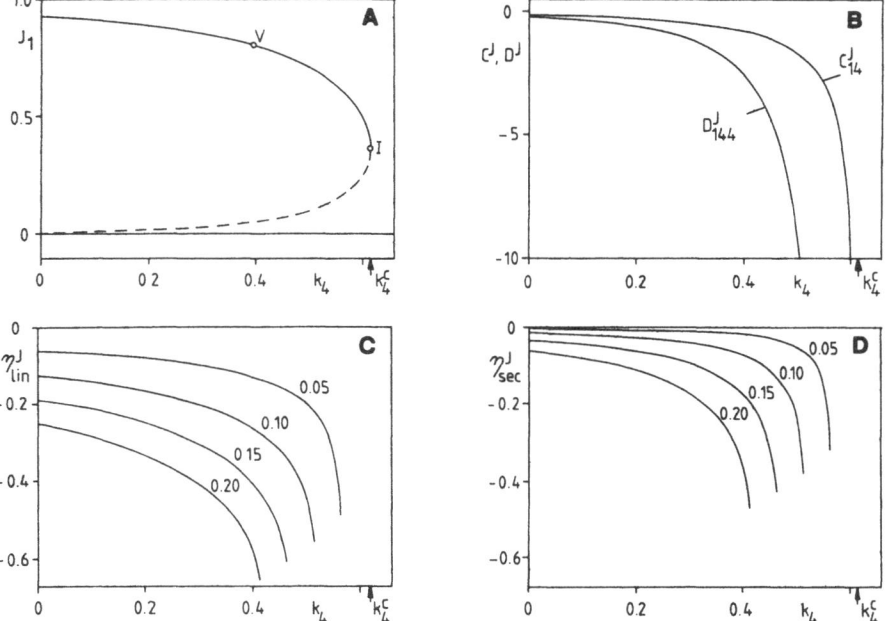

Figure 3. Stationary flux J_1 (A), linear and second-order flux control coefficients C^J_{14} and D^J_{144}, respectively (B), and the relative errors of the linear (C) and the second-order (D) approximations for various parameter perturbations Δk_4 as a function of the energetic load (k_4) for the model depicted in Fig. 2. In (A), dashed and solid lines correspond to unstable and stable steady states, respectively; I indicates the critical point. The numbers at the curves in (C) and (D) indicate the values of the parameter perturbations Δk_4. Kinetic parameters are chosen to reproduce qualitatively the flux distribution in the *in-vivo* state V ($v_3 = 0.3\ J_1$, $A_3 = 0.8\ A$): $k_1 = 1$, $k_3/k_2 = 0.06275$, $K_{AK} = 2$, $A = 1$).

As a more realistic example we consider the approximations for the control of the glycolytic flux by the "energetic load" in a simplified model of this pathway[5] (cf. Fig.2). In Fig. 3A the stationary solution for the glycolytic flux as a function of k_4 (measuring the energetic load) is shown. There is a bifurcation at a critical energetic load k_4^c where the number of steady states changes. The following results are obtained (Figs.3B-D):

- If the steady state of the system is not too close to the bifurcation point, both approximations are of reasonable accuracy. However, the error of the second-order approximation is distinctly smaller for the shown rate perturbations.
- As the steady state approaches the critical point, both approximations get less reliable but the second-order approximation is still more accurate. At the same time the range of admissible rate perturbations decreases, since a prediction of the flux change for a perturbation beyond the instability is not possible on the basis of a Taylor series (cf. Figs.3C,D).
- At the critical point itself, where a global change in the behavior of the system occurs, both the linear and the second-order flux control coefficients diverge (cf. Fig.3B).

4. Conclusion

We extended metabolic control analysis by including second-order effects. The analysis is focused on changes in stationary concentrations and fluxes after finite, but not too large parameter perturbations. To keep the resulting expressions as transparent as possible, only systems without conservation quantities are considered. Yet, the mathematical treatment allows inclusion of conservation relations in a similar manner as done for the linear theory[3].

The second-order approximation is essentially a nonlinear function of the parameter perturbations. Therefore it can retain two general properties of the perturbation problem which are lost by considering only linear terms. Firstly, the perturbations of the individual rates are not *a priori* the appropriate independent variables; and secondly, "cross-effects" which occur when more than one reaction is perturbed are taken into account (cf. Section 1).

The general concept of metabolic control analysis proved to be principally applicable to larger parameter perturbations. Compared to the linear theory better approximations can be achieved by extending the local characterization of the individual reactions in a given stationary state of a metabolic network to the second-order elasticity coefficients. This approach requires less effort than development of a kinetic model because no detailed information about the enzymatic rate laws is needed. On the other hand, the local approach of metabolic control analysis cannot replace kinetic modeling if one is interested in an integrated picture of the investigated processes including the prediction of the effects of large parameter perturbations.

References

1. S. Schuster and R. Heinrich, The definitions of metabolic control analysis revisited, *BioSystems* **27**:1-15 (1992).
2. R. Heinrich, S.M. Rapoport and T.A. Rapoport, Metabolic regulation and mathematical models, *Progr. Biophys. Mol. Biol.* **32**:1-82 (1977).
3. C. Reder, Metabolic control theory: A structural approach, *J. theor. Biol.* **135**:175-201 (1988).
4. T. Höfer and R. Heinrich, A second-order approach to metabolic control analysis, *J. theor. Biol.*, submitted.
5. E.E. Selkov, Stabilization of energy charge, generation of oscillations and multiple steady states, *Eur. J. Biochem.* **59**:151-157 (1975).

A CONTROL ANALYSIS
OF METABOLIC REGULATION

Jan-Hendrik S. Hofmeyr[1]* and Athel Cornish-Bowden[2]

[1]Department of Biochemistry, University of Stellenbosch
  Stellenbosch 7600, South Africa
[2]Laboratoire de Chimie Bactérienne
  Centre National de la Recherche Scientifique
  31 chemin Joseph-Aiguier, 13402 Marseille Cedex 20, France

1. Introduction

Why is it that many of the classical ideas of metabolic regulation have been so difficult to reconcile with metabolic control analysis? A major reason seems to be a difference in the questions asked by the two approaches. Analysis of metabolic control asks how changes in *parameters* affect variables[1,2] (for recent review see Ref. 3). Analysis of metabolic regulation asks how changes in *metabolite concentrations* or functions of metabolite concentrations affect variables[4]. Although at first sight there seems to be little difference between these questions, deeper study shows that there is an important difference. From a systems point of view parameters are constant quantities, and some metabolites that act as regulators, e.g. hormones, are often parameters of metabolic systems. Nevertheless, many of the metabolites classically regarded as regulators are themselves variables, so that the question of metabolic regulation often concerns the effect of one variable on another, e.g., of the concentration of a feedback inhibitor on the flux through its pool. The latter type of question has up to now been avoided in control analysis; here we will show how it can be answered.

If a study of metabolic regulation is a study of the effects of metabolites, then we must first distinguish between metabolites with regard to their systemic status. Metabolic systems contain external, internal and conserved metabolites. External metabolite concentrations are fixed: they are system parameters. Internal metabolite concentrations are steady-state output variables: they are fully determined by the system parameters. Conserved metabolites, which occur in moiety-conserved cycles, form part of a parameter, the sum of conserved metabolites: they are free to assume a steady-state value within this conservation constraint. Conserved metabolites therefore have dual status: they are part external, part internal.

In spite of the differences between the three classes of metabolites they all have two properties in common: first, the effects caused by changes in their concentrations can only

---

*To whom correspondence should be addressed.

be transmitted via the enzymes with which they interact directly; second, the net effect of a change in any metabolite concentration is always the sum of the effects which occur individually via the directly affected enzymes (note that, as required by control analysis, we consider small fractional changes). In this paper we show how these two properties can be described in terms of metabolic control analysis.

2. Regulatory Strengths of a Metabolite

Consider for any metabolite S, whether it is external, internal or conserved, a small change $\delta\ln s$ in its concentration. This change affects the rates of those reactions catalyzed by enzymes that have direct interaction with the metabolite, i.e., for which the metabolite is a substrate, product or effector. Now consider one of these enzymes, E_l. From the definition of elasticity coefficients it follows that the change in its rate, $\delta\ln v_l$, is given by $\varepsilon_s^{v_l}\cdot\delta\ln s$. In turn, the activity change $\delta\ln v_l$ causes a change $\delta\ln y$ in any steady-state variable, which, from the definition of control coefficients, is equal to $C_l^y\cdot\delta\ln v_l$. The effect $\delta\ln y$ is therefore obtained as $C_l^y\,\varepsilon_s^{v_l}\cdot\delta\ln s$. In the limit, where $\delta\to 0$, the effect that a change in s has on y via E_l can be quantified as

$$^{E_l}R_s^y = C_l^y\,\varepsilon_s^{v_l} = \frac{\partial\ln y}{\partial\ln v_l}\cdot\frac{\partial\ln v_l}{\partial\ln s}\,,\tag{1}$$

where $^{E_l}R_s^y$ is called the *regulatory strength* of s on y via E_l .[5] C_l^y can be regarded as the *regulatory capacity* of E_l and $\varepsilon_s^{v_l}$ as its *regulability* by s. The formulation in terms of partial derivatives is explicitly shown to remind ourselves that, in general, it is incorrect to regard the regulatory strength as a measure of $\partial\ln y/\partial\ln s$. Only when s is an external parameter that interacts with only one enzyme is this formulation correct, and here the regulatory strength corresponds to a response coefficient. Fig. 1 gives a qualitative picture of the regulatory strengths of external, internal and conserved metabolites in a hypothetical pathway constructed solely for illustrative purposes.

Eq. (1) is a generalization of the combined response property[1] which has, up to now, been used only to describe the effect of an external metabolite on steady-state variables. Here we have seen that it holds for any type of metabolite provided that its interaction with only one directly-connected enzyme is considered.

Whereas the above argument and definition is rigorous in terms of control analysis, the objection may be raised that it is impossible to isolate the individual effects of a metabolite perturbation; they all occur simultaneously. In a sense this is true, but there is nevertheless a simple way that does in principle allow the direct experimental measurement of an individual regulatory strength. All one needs to know is the elasticity coefficient $\varepsilon_s^{v_l}$. It is then always possible to translate the effect $\delta\ln v_l = \varepsilon_s^{v_l}\cdot\delta\ln s$ into an identical effect $\delta\ln v_l = \varepsilon_{e_l}^{v_l}\cdot\delta\ln e_l$ of a change in the concentration e_l of E_l. Using numbers makes the argument more transparent. Say a 1 % change in s causes a 0.4 % change in v_l. If we make the usual assumption that reaction rate is directly proportional to enzyme concentration so that $\varepsilon_{e_l}^{v_l}=1$, it follows that a 0.4 % change in e_l will cause the same 0.4 % change in v_l. Therefore the effect that we measure on the steady-state variable caused by a 0.4 % change in e_l is the same as that caused by a 1 % change in s if the effect were allowed to be transmitted only via E_l.

3. Net Regulatory Effects of Metabolites

In the previous section we showed how control analysis allows us to quantify the effect of a metabolite perturbation via a single enzyme in terms of a regulatory strength. It now remains to show how, for the different classes of metabolites, the regulatory strengths sum to give the net effect of a metabolite perturbation. Fig. 1 illustrates these net effects qualitatively.

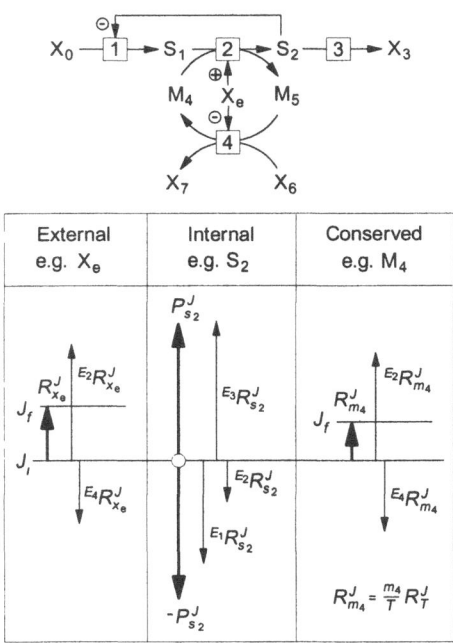

Figure 1. The effects of perturbations in the concentrations of the different classes of metabolites on the flux J through a hypothetical metabolic system. In the table a bold arrow represents either a response coefficient of an external and conserved metabolite, or the regulatory potential of an internal metabolite. Each thin arrow represents a regulatory strength. For an internal metabolite the regulatory potential is equally the sum of positive regulatory strengths or the sum of negative regulatory strengths. Whether a regulatory strength is positive or negative depends on the type of metabolite-enzyme interaction; here we regard the effect on the rate of substrate-enzyme interactions as positive, and of product-enzyme interactions as negative; the nature of other interactions are explicitly shown on the diagram. All the J-control coefficients are regarded as positive.

For a change in an external metabolite concentration x the net effect on variable y is given by a response coefficient[1]:

$$R_x^y = \frac{\partial \ln y}{\partial \ln x} = \sum_i {}^{E_i}R_x^y = \sum_i C_i^y \varepsilon_x^{v_i} \, , \tag{2}$$

where E_i is any enzyme with which X interacts directly.

For a change in an internal metabolite concentration s the net effect on another variable y is zero, according to the connectivity theorem of control analysis[1]:

$$\sum_i {}^{E_i}R_s^y = \sum_i C_i^y \varepsilon_s^{v_i} = 0 \, , \tag{3}$$

where E_i is any enzyme with which S interacts directly.

For a change in a conserved metabolite concentration m the net effect on another variable y is given by a response coefficient:

$$R_m^y = \frac{\partial \ln y}{\partial \ln m} = \sum_i {}^{E_i}R_m^y = \sum_i C_i^y \varepsilon_m^{v_i} \, , \tag{4}$$

where E_i is any enzyme with which M interacts directly. This response coefficient is related to the response coefficient of y defined with respect to T, the sum of conserved metabolite concentrations of which m forms part:

$$R_m^y = \frac{m}{T} \cdot R_T^y \, . \tag{5}$$

Ratios of conserved metabolites, on the other hand, have the same free variable status of internal metabolites, i.e., regulatory strengths in the connectivity equations of such ratios sum to zero. This issue will not be pursued here, but will be fully discussed in a future publication.

4. The Regulatory Potential of Internal Metabolites

We have seen how the regulatory strengths sum to the net effect of a change in a metabolite concentration on y. With external and conserved metabolites the net effect is quantified by a response coefficient, but with internal metabolites the net effect is zero. However, this means that some regulatory strengths of internal metabolites must be positive, while the others must be negative. The sum of positive regulatory strengths must cancel the sum of negative regulatory strengths. Either sum gives an indication of how sensitive the variable y is to a perturbation in the concentration of the metabolite, i.e., to what degree the system 'senses' the perturbation. We have termed the absolute value of this sum the *regulatory potential* of the metabolite[6]. In general, the definition of the regulatory potential of metabolite S is

$$P_S^y = \sum_{positive} {}^{E_i}R_S^y = - \sum_{negative} {}^{E_j}R_S^y$$

$$= \sum_{positive} C_i^y \varepsilon_S^{v_i} = - \sum_{negative} C_j^y \varepsilon_S^{v_j} \, , \tag{6}$$

where each E_i is an enzyme through which S exerts positive effects on y, while each E_j is an enzyme through which S exerts negative effects on y.

The regulatory potential can be used as a criterion for identifying those metabolites that have the greatest potential for regulating y. The example in Fig. 2 demonstrates this use. The steady state of the linear sequence of enzymes of the pathway in Fig. 1 was calculated, under the conditions described in the Figure legend, for a range of activities of E_3, which represents the demand on the end-product S_2. The system without a feedback loop was compared to the system in which S_2 inhibits E_1 with increasing Hill coefficient. In the absence of the feedback loop there is little difference between the regulatory potentials of S_1 and S_2 with respect to the flux, J, through the sequence, that of S_1 being marginally higher throughout. In the presence of the feedback loop the regulatory potential of S_1 decreases

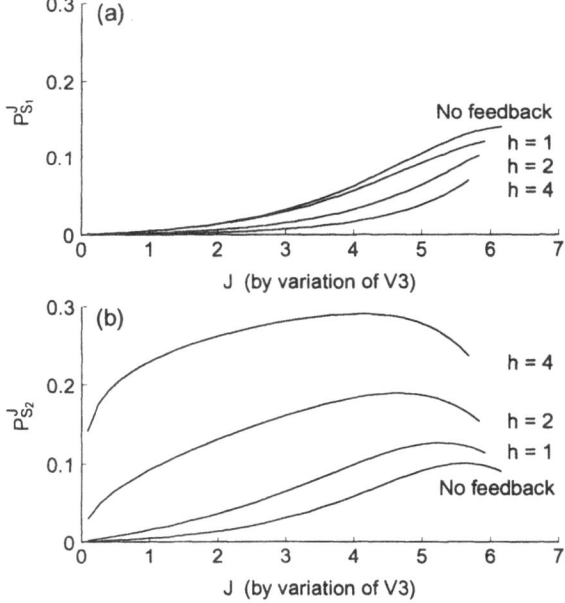

Figure 2. The regulatory potentials of (a) S_1 and (b) S_2 in the linear enzyme sequence of the metabolic scheme in Fig. 1. For this system $P_{s_1}^J$ is equal to the regulatory strength $^{E_2}R_{s_1}^J$, and $P_{s_2}^J$ is equal to $^{E_3}R_{s_2}^J$. Steady-state and control-analytic calculations were done with MetaModel[7]. The concentration of X_0 was fixed at 10 (arbitrary units). The rate equations were: reaction 1, $(10x_0-s_1)/(1+x_0+s_1+s_2^h)$; reaction 2, $(10s_1-s_2)/(1+s_1+s_2)$; reaction 3, $V_3s_2/(1+s_2)$. For the system without a feedback loop the term in s_2 was absent from the denominator of rate equation 1. V_3, the limiting rate of E_3, was varied from 0.1 to 10.1. To enable comparison between the systems, regulatory potentials are plotted against the flux. The P_s^J-axis scale is the same in both graphs so that the magnitudes of regulatory potentials of S_1 and S_2 can be compared directly.

while that of S_2 increases, the effect becoming more dramatic as the Hill coefficient of inhibition increases. The results show clearly how the feedback loop increases the sensitivity of the flux to changes in the end-product S_2. To our knowledge, no other quantitative measure is able to show this phenomenon as clearly as the regulatory potential.

In conclusion, with the development of the concepts of regulatory strength and regulatory potential we now have quantitative tools not only for assessing the validity of the classical ideas of metabolic regulation, but also for discovering new principles. We do not claim that these concepts are sufficient for answering all questions to do with regulation; additional concepts will have to be developed as the need for them arises.

Acknowledgements

We would like to thank Henrik Kacser, Daniel Kahn, Hans Westerhoff, Pedro Mendes, Marta Cascante and Albert Sorribas for discussion of the ideas presented in this paper, and especially Johann Rohwer for his contribution to the development of the concept of regulatory potential.

References

1. H. Kacser and J.A. Burns, The control of flux, *Symp. Soc. Exp. Biol.* **27**:65-104 (1973).
2. R. Heinrich and T.A. Rapoport, A linear steady-state treatment of enzymatic chains, *Eur. J. Biochem.* **42**:89-95 (1974).
3. D.A. Fell, Metabolic Control Analysis: a survey of its theoretical and experimental development, *Biochem. J.* **286**:313-330 (1992).
4. J.-H.S. Hofmeyr and A. Cornish-Bowden, Quantitative assessment of regulation in metabolic systems, *Eur. J. Biochem.* **200**:223-236 (1991).
5. D. Kahn and H.V. Westerhoff, The Regulatory Strength: How to be precise about regulation and homeostasis, *Acta Biotheor.*, in press.
6. J.-H.S. Hofmeyr and A. Cornish-Bowden, The Regulatory Potential: a quantitative criterion for identifying internal regulators, in: "Proc. First World Congress of Nonlinear Analysts," V. Lakshmikantham, ed., Walter de Gruyter, Berlin, in press.
7. A. Cornish-Bowden and J.-H.S. Hofmeyr, MetaModel: a program for modelling and control analysis of metabolic pathways on the IBM PC and compatibles, *Comp. Appl. Biosci.* **7**:89-93 (1991).

REGULATION AND HOMEOSTASIS IN METABOLIC CONTROL THEORY: INTERPLAY BETWEEN FLUCTUATIONS OF VARIABLES AND PARAMETER CHANGES

Daniel Kahn[1] and Hans V. Westerhoff[2,3]

[1]Laboratoire de Biologie Moléculaire, INRA-CNRS
 31326 Castanet-Tolosan, France
[2]Division of Molecular Biology
 Netherlands Cancer Institute,
 Plesmanlaan 121, 1066 CX Amsterdam, The Netherlands
[3]E.C. Slater Institute, University of Amsterdam
 Muidergracht 12, 1018 TV Amsterdam, The Netherlands

1. Introduction

The question of regulation is invariably a question of *sensitivity* of a system either to changes in its parameters or to fluctuations of its internal state. To investigate quantitative aspects of the regulation of metabolic systems, here we calculate the time-dependent relaxation of a system challenged with a *perturbation* of an enzyme activity (a *parameter*). Second we derive the time-dependent relaxation of the system challenged with the *fluctuation* of the concentration of an internal metabolite (a *variable*). We show that the *response* of a system to a transient fluctuation of a metabolite concentration, can be exactly mimicked by an appropriate perturbation of enzyme rates. This allows us to interpret the different terms of the system's response to a fluctuation of a metabolite concentration, in terms of additive effects exercised by this metabolite on enzyme rates. The magnitude of each of these effects on a systemic property, is measured by its *regulatory strength*, which is the product of an elasticity and a control coefficient.

2. The Time-dependent Response of a System to a Perturbation of the Parameters

The nomenclature used in this paper is defined in Table 1 (see Appendix), and is consistent with previously used nomenclature[1,2]. Consider a metabolic system at a stable steady state. Let x^0 be the vector of independent molarities, and v the vector of reaction rates (see Table 1). Steady state implies:

$$dx^0/dt = N^0 \, v = 0 \, . \tag{1}$$

First we will analyze the response of the system to a small change, δp, of the *parameters* at time 0. This change results in an immediate change of the rates, δv, which causes a small departure from the steady state. The system in turn will respond according to the following equations:

$$dx^0/dt = N^0 \, v \approx M\{x^0(t) - X_2^0\} \, , \tag{2}$$

$$dx^0/dt(0) = N^0 \, \delta v \, ,$$

$$x^0(0) = X_1^0$$

where $M = N^0 \, D \, L$ is the Jacobian matrix of the metabolic system[1], and X_1^0 and X_2^0 are the steady-state values of the metabolite vector, respectively before and after the parameter perturbation. Note that eq. 2 is valid only in the neighborhood of the steady state. This will be fulfilled because the system is stable. The system's evolution is of the form:

$$x^0(t) = x^0(0) + x_r^0(t) \, , \tag{3}$$

where $x_r^0(t)$ is defined as the *response* of the system to the parameter perturbation[3]. The solution of eqns (2) and (3) is :

$$x_r^0(t) = - \{I - (\exp M \, t)\} \, M^{-1} \, N^0 \, \delta v \, . \tag{4}$$

Because the system is stable, M must be negative definite. A consequence of these equations as $t \longrightarrow \infty$ is that x will tend to the new steady-state value X_2:

$$X_2 = X_1 - L \, M^{-1} \, N^0 \, \delta v \, , \tag{5}$$

and therefore

$$\delta X = - L \, M^{-1} \, N^0 \, \delta v \, , \tag{6}$$

which is valid for any small perturbation δv and

$$\Gamma = - L \, M^{-1} \, N^0 \, . \tag{7}$$

This expression for the concentration control matrix Γ is classical[1].

3. The Time-dependent Response of a System to a Fluctuation of Variables

Consider the same metabolic system at the same stable steady state. Now we will analyze the response of the system to a small *fluctuation* of the metabolite concentrations $\delta x = L \delta x^0$ at time 0. This change results in an immediate change of the rates δv which causes a small departure from the steady state. The system in turn will respond according to the following equations:

$$dx^0/dt = N^0 \, v \approx M\{x^0(t) - X_1^0\} \, , \tag{8}$$

$$dx^0/dt(0) = N^0 \, \delta v = N^0 \, D \, L \, \delta x^0 = M \, \delta x^0 \,, \tag{8'}$$

$$x^0(0) = X^0_1 + \delta x^0 \,,$$

where X^0_1 is the steady-state value of the metabolite vector at the given value of the parameters p. Eq. (8) is valid in the neighborhood of the steady state, where the system remains because of stability. The system's evolution is of the form:

$$x^0(t) = x^0(0) + x^0_r(t) = X^0_1 + \delta x^0 + x^0_r(t) \,, \tag{9}$$

where $x^0_r(t)$ is defined as the *response* of the system to the fluctuation. The solution of eqs (8) and (9) is:

$$x^0_r(t) = - \{I - (\exp M \, t \,)\} \, \delta x^0 \,. \tag{10}$$

Note that a consequence of these equations as $t \longrightarrow \infty$ is that x will tend to the previous steady-state value X_1, because M is negative definite. This expresses the system's stability with respect to internal fluctuations.

We will now show that the response $x^0_r(t)$ is identical to the response obtained if, instead of the fluctuation δx^0, we consider a perturbation of the *parameters* δp that would cause the same immediate changes in reaction velocities δv and is therefore defined by:

$$(\partial v/\partial p) \, \delta p = \delta v = D \, \delta x = D \, L \, \delta x^0 \,. \tag{11}$$

Indeed, in the previous section the solution $x^0_r(t)$ had the same form (compare eqs (4) and (10)), except for the multipliers, $-M^{-1}N^0\delta v$ and $-\delta x^0$ respectively. With the choice of parameter change given in eq. (11), however:

$$- M^{-1} N^0 \, \delta v = - M^{-1} N^0 \, D \, L \, \delta x^0 = - \delta x^0 \,,$$

which demonstrates that $x^0_r(t)$ is identical in both situations. This is illustrated in Fig. 1. The significance of this result is that we can rigorously mimic the response of the system to a fluctuation in metabolite concentrations, by considering equivalent perturbations of rates.

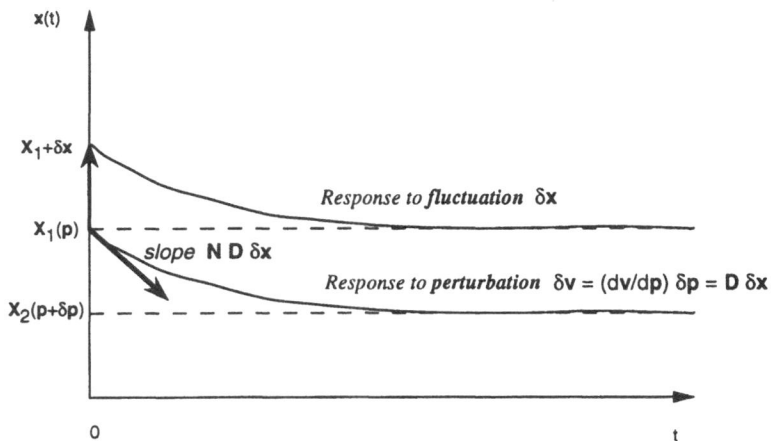

Figure 1. Identical responses to a fluctuation δx of the variables and to the equivalent parameter perturbation δp. Although the final values X_1 and X_2 are different, the responses $x_r(t) = x(t) - x(0)$ are identical.

Conversely, we can mimick the response to a parameter change $\delta\mathbf{p}$, resulting in a change of rates $\delta\mathbf{v}$, by considering the equivalent fluctuation:

$$\delta\mathbf{x}^0 = \mathbf{M}^{-1}\,\mathbf{N}^0\,\delta\mathbf{v}\;.$$

This shows how one can go back and forth between variables and parameters when analyzing the response of a system challenged at a stable steady state. Because it is usually simpler to interpret the response to a parameter change, we will now interpret the response to a fluctuation of variables in terms of the equivalent parameter perturbation.

4. The Concept of Regulatory Strength

The response of the system to a fluctuation of an *independent metabolite concentration* δx_j can now be analyzed as the response to the equivalent perturbation of parameters resulting in the following perturbation of the rates:

$$\delta v_k = D_{kj}\,\delta x_j\;.$$

The response of any independent concentration x_i when $t \longrightarrow \infty$ is:

$$x_{ir} = \sum_k \Gamma_{ik}\cdot D_{kj}\;\delta x_j = \begin{cases} -\delta x_j & \text{if } i = j \\ 0 & \text{otherwise.} \end{cases} \tag{12}$$

Each D_{kj} corresponds to a direct effect of metabolite j upon enzyme k. To describe the systemic effect of metabolite j on metabolite i *via* enzyme k, we define the corresponding *regulatory strength* as follows[4]:

$$^k R_{ij} = (x_j / x_i)\,\Gamma_{ik}\,D_{kj}\;. \tag{13}$$

Normalization of this coefficient is for practical purposes and is not essential. The regulatory strength quantifies the intensity of the system's response *via* a given interaction. Using flux control coefficients, one can define similar regulatory strengths expressing the regulation of a flux by a metabolite[4].

Classical connectivity relationships imply[5,6]:

$$\sum_k {}^k R_{ij} = \begin{cases} -1 & \text{if } i = j \\ 0 & \text{otherwise.} \end{cases} \tag{14}$$

The analysis above shows that relationships (12) and (14) can be interpreted in terms of the system's stability[3,6]: *the system's response to a fluctuation cancels the fluctuation, ultimately leaving other concentrations unchanged* (although they may change transiently).

Let us discuss first the implications of eq. (14) when $i{\neq}j$. If the sum contains two non-zero regulatory strengths, they must be of opposite magnitude: both regulatory strengths *counterbalance* each other, thus maintaining the concentration of metabolite i at steady-state. This can be regarded as a principle of action and reaction. Thus the regulatory strength quantifies the systemic action of metabolite j on metabolite i. If the sum contains several terms, each term quantifies one systemic effect, *via* enzyme k, of metabolite j on metabolite i. These regulatory strengths balance each other.

5. The Concept of Homeostatic Strength

We are now interested in analyzing the various interactions contributing to keep the concentration of a metabolite at a constant steady-state value, in response to a transient change. It is thus relevant to calculate the response of metabolite i to a fluctuation δx_i of its own molarity :

$$x_{ir} = \sum_k {}^k R_{ii} \, \delta x_i = -\delta x_i \ . \tag{15}$$

As above, this response can be seen as expressing the *reaction* of the system to action δx_i, which reaction is mediated by molecular interactions of metabolite i with several enzymes k. This allows to classify the regulatory effects according to their importance in stabilizing the system at a given steady state. Among the various terms in eq. (15), some terms may contribute more than others in counteracting the initial fluctuation. One can define *homeostatic strengths* by the following expressions[4]:

$$ {}^k H_{ii} = - {}^k R_{ii} \ . \tag{16}$$

The closer to 1 the homeostatic strength, the more homeostatic the corresponding regulatory route. Routes with smaller homeostatic strengths are less stabilizing. Negative homeostatic strengths refer to destabilizing effects. Regulatory routes with homeostatic strengths above 1 may be considered as hyperresponsive. They contribute to homeostasis in a system which would otherwise be destabilized by other interactions.

Appendix

Table 1. Nomenclature used in this paper.

$x = x(t,p)$	Molarity vector; a function of time t and of parameter vector p
$x^0 = x^0(t,p)$	Vector of *independent molarities*
$X = X(p)$	Molarity vector *at steady-state* (dx/dt = 0)
$v = v(x,p)$	Rate vector
$D = \partial v/\partial x$	*Elasticity matrix* (not normalized)
$dx/dt = N\ v$	N is the *stoichiometry matrix*
$N = L\ N^0$	N^0 is the *reduced stoichiometry matrix*,
	L is the *link matrix* ($dx = Ldx^0$)
$J = J(p)$	Flux vector *at steady-state*
$M = N^0\ D\ L$	Jacobian matrix : $M = \partial/\partial x\ (dx/dt)$
$\partial X/\partial p = \Gamma\ \partial v/\partial p$	Γ is the *concentration control matrix*
${}^k R_{ij} = (x_j/x_i)\ \Gamma_{ik}\ D_{kj}$	*Regulatory strength* of metabolite j on metabolite i *via* enzyme k
${}^k H_{ii} = - {}^k R_{ii}$	*Homeostatic strength* for metabolite i *via* enzyme k

Acknowledgements

This work was supported by the Institut National de la Recherche Agronomique and the Netherlands Organization for Scientific Research.

References

1. C. Reder, Metabolic control theory: a structural approach, *J. theor. Biol.* **135**:175-201 (1988).
2. D. Kahn and H.V. Westerhoff, Control theory of regulatory cascades, *J. theor. Biol.* **153**:255-285 (1991).
3. H.V. Westerhoff and K. van Dam. "Thermodynamics and Control of Biological Free-Energy Transduction," Elsevier, Amsterdam (1987).
4. D. Kahn and H.V. Westerhoff, The regulatory strength: how to be precise about regulation and homeostasis, *Acta Biotheoret.* **41**, in press.
5. H. Kacser and J.A. Burns, The control of flux, *Symp. Soc. Exp. Biol.* **27**:65-104 (1973).
6. H.V. Westerhoff and Y.D. Chen, How do enzyme activities control metabolic concentrations? An additional theorem in the theory of metabolic control, *Eur. J. Biochem.* **142**:425-430 (1984).

SUM OF THE FLUX CONTROL COEFFICIENTS: WHAT IS IT EQUAL TO IN DIFFERENT SYSTEMS?

Boris N. Kholodenko[1,2] and Hans V. Westerhoff[2]

[1]A.N. Belozersky Institute of Physico-Chemical Biology
  Moscow State University, Moscow 119899, Russia
[2]Division of Molecular Biology
  Netherlands Cancer Institute, H5
  Plesmanlaan 121, 1066 CX Amsterdam, The Netherlands

1. Introduction

Quantitative approaches have led to appreciable advances in understanding the control of cellular metabolism[1-3]. In the framework of this approach the contribution of any enzyme (E_i) to the control of the flux (J) is characterized by its control coefficient defined in terms of the fractional change $\partial J / J$ in the metabolic flux induced by the fractional change $\partial E_i / E_i$ in the enzyme concentration:

$$C_{E_i}^J = \frac{\partial J / J}{\partial E_i / E_i} = \frac{\partial \ln|J|}{\partial \ln E_i} . \tag{1}$$

In "simple" metabolic pathways simultaneous equal relative changes in enzyme concentrations lead to the same change of the pathway flux, i.e. the control exerted by all the enzymes adds up to unity[4,5]:

$$\sum_{i=1}^{n} C_{E_i}^J = 1 . \tag{2}$$

However, if a metabolic pathway is not "simple", i.e., if standard assumptions of the "classical" control theory do not hold true, the sum (2) may be unequal to unity[2,6,7]. In this paper we analyze the behavior of the sum of the flux control coefficients in different systems and demonstrate that measuring this sum can give a deeper insight into regulatory properties of the metabolic pathway studied.

Modern Trends in Biothermokinetics, Edited by
S. Schuster *et al.*, Plenum Press, New York, 1993

2. High Enzyme Concentrations and Moiety Conservation

The first evidence that the sum of the flux control coefficients could drop below unity was reported by Ottaway when modeling the tricarboxylic acid cycle at high enzyme concentrations[8]. Later, other researchers have shown that this is the case for systems with moiety-conserved cycles and enzyme concentrations comparable in magnitude to those of metabolites[9-11]. We here begin our analysis with considering such systems. The failure of the classical theory to treat them is related to the usual ignoring of the concentrations of enzyme-bound metabolites. Since the conserved moiety is, by definition, available in a fixed amount the metabolites sequestered by the enzymes cannot come from external pools.

Let T_i, T_i^{free} and T_i^{bound} be the ith moiety-conserved sum and the free and enzyme-bound parts of T_i, respectively:

$$\sum_j \mu_{ij}\left(X_j^{\text{free}} + X_j^{\text{bound}}\right) = T_i^{\text{free}} + T_i^{\text{bound}} = T_i , \qquad (3)$$

where X_j^{free} and X_j^{bound} are the free and enzyme-bound concentrations of metabolite X_j, μ_{ij} is the stoichiometric coefficient of metabolite X_j in the ith moiety conserved sum (T_i).

Let us suppose that, in order to measure the control coefficient of enzyme E_i, we vary the concentration of the latter. An increase of the concentration E_i leads to a transition of some of the metabolite molecules (X_j) from the free state (X_j^{free}) to the enzyme-bound state (X_j^{bound}). Therefore, although the total pool, T_i, remains unchanged, the available pool of free metabolites, T_i^{free}, which figures as a true parameter in the kinetic equations, does change.

As a consequence, the control coefficients relative to the enzyme concentration, $C_{E_i}^J$, will differ from those referring to a modulation of any other parameter, which affects the enzyme rate, v_i, but does not influence T_i^{bound} and T_i^{free} directly. One may choose, for example, the parameter k_{cat} (or V_{max}) in the following expression for the reaction rate, v_i:

$$v_i = k_i^{\text{cat}} \cdot E_i \cdot W_i (\mathbf{X}^{\text{free}}) , \qquad \mathbf{X} = X_1, ..., X_m . \qquad (4)$$

Here only W_i is a function of free metabolite concentrations. The control coefficients defined through a variation of any parameter, affecting only the rate, v_i, are usually termed the control coefficients with respect to the rate or enzyme activity and are denoted as $C_{v_i}^J$. In the case considered the coefficients $C_{E_i}^J$ will differ from $C_{v_i}^J$ by a term that includes the response coefficients, $R_{T_k}^J$, toward changes in the moiety-conserved sums, T_k (see Ref. 11):

$$C_{E_i}^J = C_{v_i}^J - \sum_k R_{T_k}^J \cdot \frac{[T_k E_i]}{T_k} , \qquad R_{T_k}^J = \frac{\partial \ln |J|}{\partial \ln T_k} . \qquad (5)$$

$[T_k E_i]$ denotes the part of sum T_k that is bound to E_i. Summing over all enzymes of the system and taking into account that the coefficients $C_{v_i}^J$ obey the classical summation theorem (i.e., their sum is equal to unity), we obtain the modified summation theorem for the system with high enzyme concentration and moiety-conserved cycles:

$$\sum_{i=1}^{n} C_{E_i}^{J} = 1 - \sum_{k} R_{T_k}^{J} \cdot \frac{T_k^{\text{bound}}}{T_k} \quad . \tag{6}$$

This equation implies that the sum of the flux control coefficients drops significantly below unity if at least one of the conserved sums (T_k) exerts a significant control on the flux and if the bound part of the conserved sum, T_k^{bound}, is comparable in magnitude with the free one. Indeed, it was demonstrated[11], that the flux control coefficients, $C_{E_i}^{J}$, and even their sum can take negative values in case when their analogs, $C_{v_i}^{J}$, are positive. In other words, in such systems an increase in the concentrations of all the enzymes (by the same factor) can lead to a drop rather than a rise in the flux; the enzymes "soup up" the metabolites.

It is instructive to consider here the experimental methods of determining the control coefficients, using inhibitors specific to a single enzyme[12]. By measuring the dependence of the flux on the concentration of an inhibitor (I) specific to the enzyme, E_i, one can determine the following coefficient:

$$^{\text{app}}C_{E_i}^{J} = \frac{\partial \ln |J| / \partial I}{\partial \ln |v_i| / \partial I} = \frac{v_i}{J} \cdot \frac{\partial J / \partial I}{\partial v_i / \partial I} \quad . \tag{7}$$

In simple pathways the result of such a measurement does not depend on the type of the inhibitor used, provided that the latter only affects the target enzyme. However, this is not true in our case. Affecting the system with a purely non-competitive inhibitor which changes k_i^{cat} only (i.e. $V_{\text{max},i}$), we determine the control coefficients with respect to the rate, $C_{v_i}^{J}$. Using an irreversible inhibitor that replaces metabolites at the active site, one finds the control coefficients with respect to the enzyme concentration, $C_{E_i}^{J}$. Therefore, contrary to the classical results, the control coefficients determined by titrating a system with an inhibitor may depend on the peculiarities of both the inhibitor and the system. Accordingly, we shall call these coefficients apparent control coefficients and denote them by a left-hand side superscript "app"[13]. They will be considered also for other types of non-classical enzyme systems.

3. Enzyme-enzyme Interactions

We will now study systems with enzyme-enzyme complexes. Various types of enzyme associations have been discussed in the literature[14,15]. In a control analysis study[16,17] quasi-irreversibly bound enzymes have been considered. In cases of reversible enzyme-enzyme associations, if a necessary condition for enzymes to form a complex is that one of them has already bound the common intermediate, then such an association is termed "dynamic" complex or dynamic channel. We may also have the so-called static channel when the enzyme complex exists independently of the presence of a common intermediate or, rather, longer than the mean passage time of metabolites through that part of the pathway. We consider here the simple example of a static channel which was analyzed by Sauro and Kacser[7] (see Fig. 1a).

This system involves two unbound enzymes E_1^f and E_2^f, which are assumed to be in equilibrium with their complex, Q. The upper route in Fig. 1a represents the path of the intermediate through the bulk phase and the lower route represents the channeling of the intermediate. It was shown that the sum of the control coefficients $C_{E_i}^J$ of the enzymes can differ from unity[7]:

$$C_{E_1}^J + C_{E_2}^J = 1 + \Delta .\qquad(8)$$

In the case of equimolar enzyme concentrations the difference, Δ, is positive if the complex has higher specific activity than the free enzyme combined, otherwise Δ is negative.

Consequently, if a direct transfer of the intermediate takes place, the sum of flux control coefficients ($C_{E_i}^J$) over all enzymes will generally differ from unity. The contrary is not, however, true. In fact, suppose that enzyme association leads only to a change in kinetics of the constituent enzymes, and channeling of the intermediate does not occur (see Fig. 1b). In this case there is only a bulk phase pool of the intermediate (X), and all reaction pathways include diffusional steps to (from) X. However, since the ratios between concentrations of the free enzymes and the complex changes with a change in the total enzyme concentration, also in this case, the sum of the flux control coefficients will differ from unity for this system (as well as for the system with channeling shown in Fig.1a).

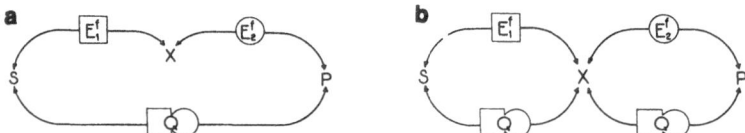

Figure 1. The system of two enzymes capable to form a complex (Q). (a) direct transfer of the intermediate ("channeling") is possible (the lower route through the complex Q). The upper route represents the usual reaction pathways trough the bulk phase, catalyzed by free enzymes, E_1^f and E_2^f. (b) Channeling is absent in the system, i.e. there is only a bulk phase pool of the intermediate (X), and all reaction pathways include diffusional steps to and from X.

Therefore, this sum may not be used as the sole criterion for discriminating between channeling and a free diffusion mechanism. In the particular case of equimolar concentrations of E_1 and E_2, Sauro and Kacser[7] proposed to distinguish between them by following the bulk phase intermediate X. When the total enzyme concentration, E_1+E_2, is varied, the concentration of X does change in the system shown in Fig. 1b (in which the direct transfer is absent), whereas it does not change in the case of channeling (Fig. 1a).

However, this criterion fails in systems with dynamic complexes. As mentioned above, in this case an enzyme complex is formed only if one of the enzymes has already bound the common intermediate (Fig. 2). The lower route in the Figure represents the dynamic channeling of the intermediate. It can be shown that even at equimolar concentrations of E_1 and E_2 the concentration of the bulk phase intermediate X will change upon a change in the total enzyme concentration[13].

Remarkably, the sum of the apparent control coefficients seemed to be useful for discrimination between channeling and a free-diffusion mechanism. According to the above definition the apparent control coefficients are obtained by titrating a system with specific inhibitors. For example, for an irreversible inhibitor one can derive from Eq. (7)[12]:

$$^{\text{app}}C_{E_i}^J = -\frac{I_{\text{end}}}{J} \cdot \frac{\partial J}{\partial I}\bigg|_{I=0}, \tag{9}$$

where I_{end} is the inhibitor concentration necessary to completely suppress the enzyme.

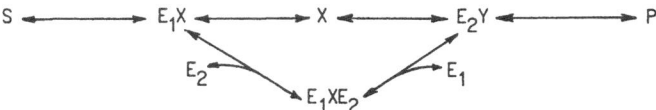

Figure 2. Dynamic channel. The complex E_1XE_2 is formed after binding of X to E_1. The upper route represents the reaction pathways through the bulk phase, catalyzed by free enzymes, which do not form a complex.

However, the result of such a measurement will strongly depend on whether inhibitors affect the enzyme-enzyme interactions. Here we assume that the inhibitor binding does not change the ability of the enzyme to form a complex with another enzyme. This may be so if a binding site for the inhibitor is distinct from a binding site for another enzyme.

An alternative approach for measuring the apparent control coefficients may be used for systems reconstituted from isolated enzymes. Instead of inhibitor titration an irreversible modification of the enzyme catalytic site should be performed (in a separate sample), such that the ability of the enzyme to associate with other enzymes remains unchanged. Different amounts of the modified enzyme (E_i^{mod}, which has lost the catalytic activity) should be then added to the system at the constant total concentration of the active and inactive enzyme, $E_i^{\text{mod}} + E_i = E_i^0$. It is very easy to transform formula (9) for the irreversible inhibitor in this case[13]:

$$^{\text{app}}C_{E_i}^J = -\frac{E_i^0}{J} \cdot \frac{\partial J}{\partial E_i^{\text{mod}}}\bigg|_{E_i^{\text{mod}}=0}. \tag{10}$$

It can be shown that the sum of the apparent control coefficients measured by using both irreversible and non-competitive inhibitors or enzyme modification appeared to be greater than unity for a system with channeling of intermediates[13]. For static channeling:

$$^{\text{app}}C_{E_1}^J + {}^{\text{app}}C_{E_2}^J = 1 + \frac{J_{\text{chan}}}{J}, \tag{11}$$

where J_{chan} is the channeled flux and J is the total flux. In fact, when all flux runs through the channel and the pathway consists of n enzymes, the sum of the apparent control coefficients should be equal to n (see Ref. 17). With a slight modification, eq. (11) also pertains to dynamic channels[18].

Remarkably, at the same time the sum of the apparent control coefficients proves to be equal to unity for a system with enzyme complexes but with a free diffusion mechanism of metabolite transfer (Fig.1b)[13]. Thus, an experimental discrimination between channeling and a free diffusion mechanism is available by measuring the sum of the apparent control coefficients.

4. Conclusions

We conclude that in cases of moiety conservation and high enzyme concentrations, or channeling the sum of the control coefficients is equal to unity only if it includes independent activities of all ("elemental") processes[13]. Moreover, a control coefficient, measured by inhibitor titration, is a property of both the inhibitor and the pathway. To express this explicitly we propose to call the latter "apparent control coefficient". Depending on inhibitor type, the sum of apparent control coefficients can allow one to recognize channeling.

References

1. H.V. Westerhoff and K. Van Dam. "Thermodynamics and Control of Biological Free-Energy Transduction," Elsevier, Amsterdam (1987).
2. B.N. Kholodenko. "Modern Theory of Metabolic Control," (in Russian), VINITI PRESS, Moscow (1991).
3. D.A. Fell, Metabolic control analysis: a survey of its theoretical and experimental development, *Biochem. J.* **286**:313-330 (1992).
4. H. Kacser and J.A. Burns, The control of flux, *Symp. Soc. Exp. Biol.* **27**:65-104 (1973).
5. R. Heinrich, S.M. Rapoport and T.A. Rapoport, Metabolic regulation and mathematical models, *Prog. Biophys. Mol. Biol.* **32**:1-83 (1977).
6. H. Kacser, H.M. Sauro and L. Acerenza, Enzyme-enzyme interactions and control analysis. The case of non-additivity: monomer-oligomer associations, *Eur. J. Biochem.* **187**:481-491 (1990).
7. H.M. Sauro and H. Kacser, Enzyme-enzyme interactions and control analysis. The case of non-independence: heterologous associations, *Eur. J. Biochem.* **187**:493-500 (1990).
8. J.H. Ottaway, Control points in the citric acid cycle, *Biochem. Soc. Trans.* **4**:371-376 (1976).
9. C. Reder, Metabolic control theory: a structural approach, *J. theor. Biol.* **135**:175-201 (1988).
10. D.A. Fell and H.M. Sauro, Metabolic control analysis. The effects of high enzyme concentrations, *Eur. J. Biochem.* **192**:183-187 (1990).
11. B.N. Kholodenko, A.E. Lyubarev and B.I. Kurganov, Control of metabolic flux in a system with high enzyme concentrations and moiety-conserved cycles, *Eur. J. Biochem.* **210**:147-153 (1992).
12. A.K. Groen, R.J.A. Wanders, H.V. Westerhoff, R. Van der Meer and J.M. Tager, Quantification of the contribution of various steps to the control of mitochondrial respiration, *J. Biol. Chem.* **257**:2754-2757 (1982).
13. B.N. Kholodenko, Control theory of "non-classical" enzyme systems and methods for the study of metabolic channelling, *Biokhimia* **58**, in press.
14. P.A. Srere and J. Ovádi, Enzyme-enzyme interactions and their metabolic role, *FEBS Lett.* **268**:360-364 (1989).
15. P. Mendes, D.B. Kell and H.V. Westerhoff, Channelling can decrease pool size, *Eur. J. Biochem.* **204**: 257-266 (1992).
16. H.V. Westerhoff, B.A. Melandri, G. Venturoli, G.F. Azzone and D.B. Kell, A minimal hypothesis for membrane-linked free energy transduction emphasizing the role of the independence of the protonic energy coupling modules, *Biochim. Biophys. Acta* **768**:257-292 (1984).
17. H.V. Westerhoff and D.B. Kell, A control theoretical analysis of inhibitor titration assays of metabolic channelling, *Comm. Molec. Cellul. Biophys.* **5**:57-107 (1988).
18. B.N. Kholodenko and H.V. Westerhoff, Metabolic channeling and control of the flux, *FEBS Lett.* **320**:71-74 (1993).

CONTROL ANALYSIS
OF METABOLIC CHANNELING

Pedro Mendes and Douglas B. Kell

Dept of Biological Sciences
University of Wales
Aberystwyth, Dyfed SY23 3DA, United Kingdom

1. Introduction

A growing body of experimental evidence supports the view that many consecutive enzymes in metabolic pathways form aggregates which transfer their common intermediate metabolite(s) directly, without its escape into a bulk phase[1]. These enzyme-enzyme interactions may be classified in terms of the lifetime of the multienzyme complexes: static channels persist over long periods of time and are destroyed only by processes as drastic as proteolysis; dynamic channels have a short lifetime and dissociate very easily[2]. Several examples of the former type have been found and in the case of tryptophan synthase, pictures of the multienzyme complex with the channeled intermediate bound in several positions have been reconstructed from X-ray diffraction patterns[3]. Due to their dissociable nature, channels of the latter type have not been detected in this way; evidence for their existence has therefore been gathered by other methods[4].

Recently, some controversy has surrounded the interpretation of apparently conflicting data from kinetic experiments concerning the direct transfer of NADH between dehydrogenases[5-9] and of 1,3-diphosphoglycerate between glyceraldehyde-phosphate dehydrogenase and phosphoglycerate kinase in vitro[10,11]. There are only a few methods[5,9,12] that can distinguish between an exclusively classical 'pool' mechanism and one with channeling for enzyme couples.

It is widely assumed that if metabolic channeling exists *in vivo* then it must somehow be beneficial to the cell (although it could simply be a consequence of other phenomena such as the high enzyme concentrations in vivo). A special issue of the Journal of Theoretical Biology was dedicated to this topic, with an introductory review by Ovádi[1], followed by several commentaries. Two of the most widely cited "advantages" were smaller free intermediate metabolite pools and higher fluxes than those that would exist were there no channelling. Small concentrations of intermediates should be advantageous since in the majority of cases these metabolites serve no other purpose than being a route from one chemical to another[13], so this should spare the limited solvent capacity[14]. High flux is very important in pathways that respond quickly to perturbations, such as energy metabolism in muscle cells.

Modern Trends in Biothermokinetics, Edited by
S. Schuster *et al.*, Plenum Press, New York, 1993

Keizer and Smolen[15] modeled NADH channeling among dehydrogenases, and found that additional methods can regulate the NADH/NAD$^+$ cycle in glycolysis if channeling of the pyridine nucleotide occurs. Cornish-Bowden[16] and Mendes et al.[17] have studied a model of dynamic channeling to enquire as to what extent this type of enzyme-metabolite-enzyme interaction would decrease the concentration of the free 'pool' intermediate metabolite. Whilst this concentration could be substantially decreased by increasing the relative flux through the channel[17], channeling could also be quite ineffective, and even serve to increase the pool concentration by a small amount[16].

In this work we use the program GEPASI[18] to study the steady-state properties of two models of dynamic channeling within the framework of metabolic control analysis. We concentrate in particular on the effects of a 'leak' and of the enzyme concentrations on the extent to which the channel reduces the concentration of the intermediate and increases the flux.

2. Metabolic Channeling Viewed as Two Alternative Catalytic Pathways

To be able to separate the effects of the classical 'pool' mechanism from the channeled one it is necessary to increase the detail by describing the enzymatic reactions at the level of their elementary chemical steps[16]. It also seemed interesting to find out what, if any, properties arise just due to the existence of two alternate routes ('pool' and channel).

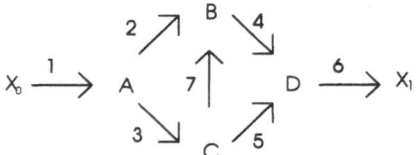

Figure 1. A simple model of dynamic channeling without enzymatic catalysis.

To achieve this goal we consider a very simple model in which there are no catalysts (enzymes). Such a model is depicted in Fig. 1, where the "external" metabolite X_0 is converted to X_1 (also external). In this case B represents the free metabolite, C the bound metabolite, and A and D transition species. Although the arrows representing the reactions point only in one direction (usually the net direction of flux) all steps are reversible and their rates obey normal chemical kinetics as in eq (1). Step 7 represents a 'leak' which interconverts the bound and free forms of the intermediate. It should be noted that this model is applicable not only to channeling but to any processes that have two alternative pathways between two points.

$$v_i = S \cdot k_i - P \cdot k_{-i} . \tag{1}$$

Special care must be taken when assigning values to the kinetic constants of steps 2, 3, 4, 5 and 7, as they must obey the relations in eqs (2) so as not to violate the principle of microscopic reversibility.

$$k_2 \cdot k_4 / k_{-2} \cdot k_{-4} = k_3 \cdot k_5 / k_{-3} \cdot k_{-5} , \tag{2a}$$

$$k_3 \cdot k_7 / k_{-3} \cdot k_{-7} = k_2 / k_{-2} . \tag{2b}$$

From eq. (2b) it can be concluded that the 'leak' is obviously not independent and interestingly, will only have an equilibrium constant greater than one if condition (3) is fulfilled.

$$k_2/k_{-2} > k_3/k_{-3} \Leftrightarrow K_2 > K_3 \, . \tag{3}$$

If the equilibrium constant of this step is smaller than one, then flux will be in the direction opposite to the arrow (effectively leaking from the pool into the channel). While the term 'leak' has always referred to a flow from the channel to the pool, one cannot state that a channel is really 'leaky' without information about the equilibrium constants of the alternative paths; the leak could in fact be a reverse leak! In such circumstances the 'leak' would effectively help in reducing the concentration of the free metabolite.

Two questions are now asked:

(i) how does the concentration B change with increasing flux through steps 3 and 5 (the channel), and

(ii) how does the total flux (J) change with the same increase of channel flux?

To find this out, all of the kinetic constants of these two steps are multiplied by a factor (p), keeping the equilibrium constant unchanged whilst allowing one simultaneously to change the 4 rate constants of these steps. Fig. 2 shows the effects on C^B_{channel}, and C^J_{channel} (respectively the concentration-control coefficient of the channel on B and the flux-control coeffcient of the channel). Both control coefficients are actually the sum of those of steps 3 and 5, thus describing how the channel limb controls the stated variables. The concentration-control coefficient is negative throughout the range of the factor p, reflecting

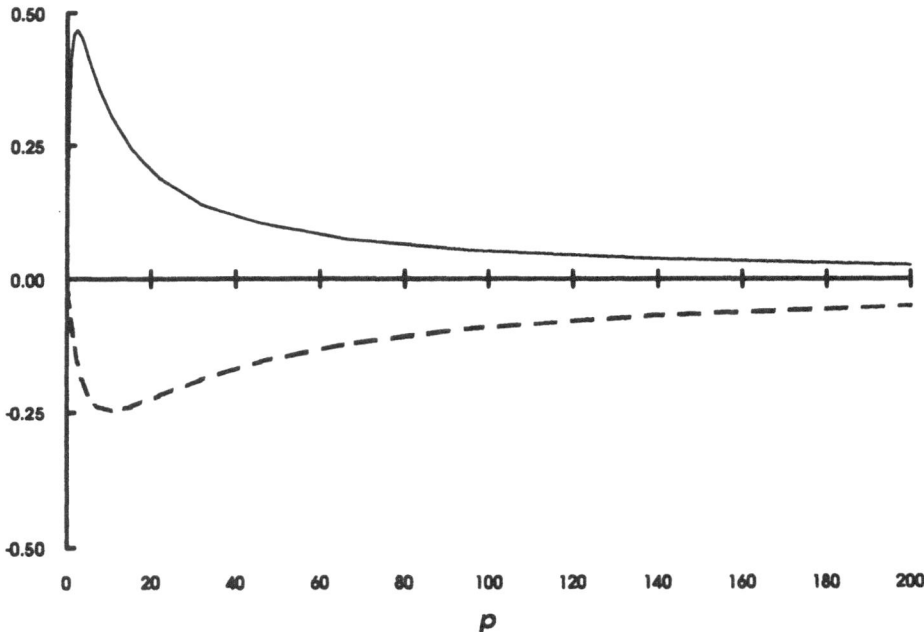

Figure 2. The effect of increasing flux through the channel limb on C^B_{channel} (broken line) and C^J_{channel} (continuous line).

the decrease in B with increased flux through the channel; the flux-control coefficient is always positive, reflecting a monotonic increase in J with increasing flux through the channel. It is interesting to note that both control coefficients have an inflection point at small values of p. The significance of this is that at these points the state variables (J and B) are more sensitive to infinitesimal perturbations than at any other point. It is also interesting that very low control coefficients are observed for the channel when the extent of channeling is either very low or very high.

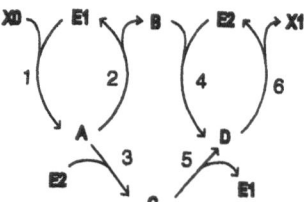

Figure 3. A dynamic channel with enzymatic catalysis.

3. Enzymes as Conserved Moieties

Now we extend the model of Fig. 1 to include catalysis by enzymes, as depicted in Fig. 3. Apart from the leak, all the other steps still convert the same species as before, now with the inclusion of E_1 and E_2, free forms of the enzymes. A and D are enzyme-metabolite complexes and C is the ternary enzyme-metabolite-enzyme complex (the channel). Analysis of the stoichiometry of this model by standard methods[18,19] reveals two moiety-conservation relations,

$$E_1 + A + C = E_{1,\text{tot}} ,\qquad(4a)$$

$$E_2 + D + C = E_{2,\text{tot}} .\qquad(4b)$$

These relations bring about saturation such that all internal metabolites except B, the free intermediate metabolite, have a maximum concentration. In some circumstances, increasing flux through the channel does not decrease B (Ref. 16); this is plausibly due to such saturation effects. However, in all our simulations J always increased.

Fig. 4 shows how the control of the total flux is distributed among the steps of the

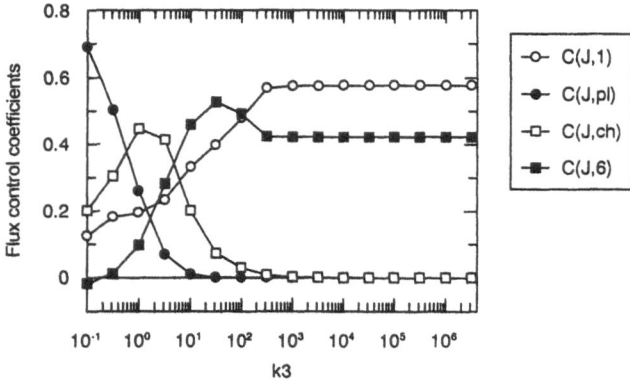

Figure 4. Effect of increasing channel flux on the flux-control distribution.

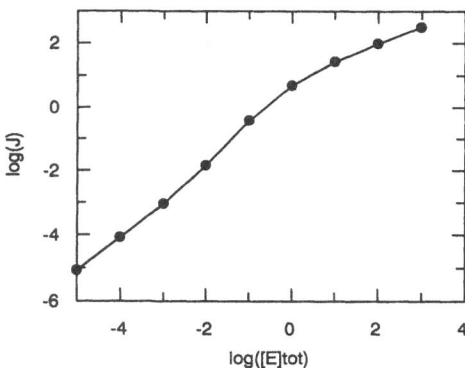

Figure 5. Effect of simultaneously increasing the total amount of enzymes 1 and 2 on the total flux.

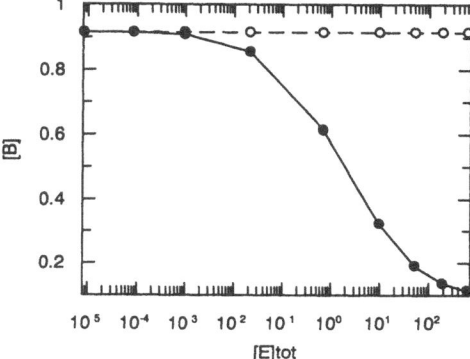

Figure 6. Effect of simultaneously increasing the total amount of enzymes 1 and 2 on B. The broken line corresponds to a pathway without channeling, the straight line to one with channeling.

pathway. Once again steps 3 and 5 were grouped (ch), as well as steps 2 and 4 (pl). As p increases (and therefore the proportion of flux through the channel) the control exerted by the pool steps decreases almost to zero. The control exerted by the channel steps increases until it reaches a maximum and then decreases to zero, because by then steps 1 and 6 share all the control, perhaps because their rates are limited by the concentrations of the enzyme moieties. B has a peculiar behavior, rising with increasing, but low, values of p then decreasing well below the initial level. The total flux increases sigmoidally, reaching a plateau at about the same value of p at which C_1^J and C_6^J become dominant (data not shown).

It is well known that in a linear multienzyme pathway where all intermediates are pools, increasing all the enzyme concentrations by the same amount will increase the flux proportionately without affecting the metabolite concentrations. Indeed this has been used to prove the summation theorems of control analysis[20,21]. However, changing all the enzyme concentrations in parallel when some intermediates are channeled will produce a deviation from a straight line of slope 1 in a plot of $\log J$ vs. $\log E$ (see Ref. 22). It was also noted that in some models of channeling the concentration of the free form of the channeled metabolite would change with an increase in enzyme concentration. In our model we increased the concentration of both enzymes by the same amounts to find that

(i) there is a modest deviation from a straight line in a plot of $\log J$ vs. $\log E$ (Fig. 5) and,

(ii) there is a large decrease in the steady-state concentration of B (Fig. 6).

At least for the sets of parameters used for our simulations, B_{ss} is better suited to reveal the existence of channeling than is J.

Acknowledgements

We thank the J.N.I.C.T., Portugal, for financial support and Herbert Sauro for helpful discussions.

References

1. J. Ovádi, Physiological significance of metabolic channeling. *J. Theoret. Biol.* **152**:1-22 (1991).
2. P. Friedrich, Dynamic compartmentation in soluble multienzyme systems, *in*: "Organized Multienzyme Systems. Catalytic Properties," G.R. Welch, ed., Academic Press, New York, pp. 141-176 (1985).
3. C.C. Hyde, S.A. Ahmed, E.A. Padlan, E.W. Miles and D.R. Davies, 3-Dimensional structure of the tryptophan synthase α-2-β-2 multienzyme complex from *Salmonella typhimurium, J. Biol. Chem.* **263**: 17857-17871 (1988).
4. T. Keleti, J. Ovádi and J. Batke, Kinetic and physicochemical analysis of enzyme complexes and their possible role in the control of metabolism, *Progr. Biophys. Molec. Biol.* **53**:105-152 (1989).
5. D.K. Srivastava and S.A. Bernhard, Mechanism of transfer of reduced nicotinamide adenine-dinucleotide among dehydrogenases, *Biochemistry* **24**:623-628 (1985).
6. B. Chock and H. Gutfreund, Reexamination of the kinetics of the transfer of NADH between its complexes with glycerol-3-phosphate dehydrogenase and with lactate-dehydrogenase, *Proc. Natl. Acad. Sci. U.S.A.* **85**:8870-8874 (1988).
7. D.K. Srivastava, P. Smolen, G.F. Betts, T. Fukushima, H.O. Spivey and S.A. Bernhard, Direct transfer of NADH between α-glycerol phosphate dehydrogenase and lactate- dehydrogenase - fact or misinterpretation, *Proc. Natl. Acad. Sci. U.S.A.* **86**:6464-6468 (1989).
8. X. Wu, H. Gutfreund, S. Lakatos and P.B. Chock, Substrate channeling in glycolysis - a phantom phenomenon, *Proc. Natl. Acad. Sci. U.S.A.* **88**:497-501 (1991).
9. T. Ushiroyama, T. Fukushima, J.D. Styre and H.O. Spivey, Substrate channelling of NADH in mitochondrial redox processes, *Curr. Top. Cell. Regul.* **33**:291-307 (1992).
10. J. Weber and S. Bernhard, Transfer of 1,3- diphosphoglycerate between glyceraldehyde-3-phosphate dehydrogenase and 3-phosphoglycerate kinase via an enzyme-substrate-enzyme complex, *Biochemistry* **21**:4189- 4194 (1982).
11. J. Kvassman and G. Pettersson, Mechanism of 1,3- bisphosphoglycerate transfer from phosphoglycerate kinase to glyceraldehyde-3-phosphate dehydrogenase, *Eur. J. Biochem.* **186**:265-272 (1989).
12. D.B. Kell and H.V. Westerhoff, Control analysis of multienzyme systems, *UCLA Symp. New Ser.* **134**:273-289 (1990).
13. P.A. Srere, Complexes of sequential metabolic enzymes, *Ann. Rev. Biochem.* **56**:89-124 (1987).
14. D.E. Atkinson, Limitation of metabolite concentrations and the conservation of solvent capacity in the living cell, *Curr. Top. Cell. Reg.* **1**:29-43 (1969).
15. J. Keizer and P. Smolen, Mechanisms of metabolite transfer between enzymes: diffusional versus direct transfer, *Curr. Top. Cell. Regul.* **33**:391-405 (1992).
16. A. Cornish-Bowden, Failure of channeling to maintain low concentrations of metabolic intermediates, *Eur. J. Biochem.* **195**:103-108 (1991).
17. P. Mendes, D.B. Kell and H.V. Westerhoff, Channelling can decrease pool size, *Eur. J. Biochem.* **204**:257-266 (1992).
18. P. Mendes, GEPASI: A user oriented metabolic simulator, this volume.
19. J.-H.S. Hofmeyr, H. Kacser and KJ. Van der Merwe, Metabolic control analysis of moiety-conserved cycles, *Eur. J. Biochem.* **155**:631-641 (1986).
20. C. Reder, Metabolic control-theory - a structural approach, *J. theoret. Biol.* **135**:175-201 (1988).
21. H. Kacser and J.A. Burns, The control of flux, *Symp. Soc. Exp. Biol.* **27**:65-104 (1973).
22. D.A. Fell and H.M. Sauro, Metabolic control and its analysis - additional relationships between elasticities and control coefficients, *Eur. J. Biochem.* **148**:555-561 (1985).
23. H.M. Sauro and H. Kacser, Enzyme-enzyme interactions and control analysis .2. The case of nonindependence - heterologous associations, *Eur J. Biochem.* **187**:493- 500 (1989).

PATH ANALYSIS
IN GENERAL CONTROL THEORY

Géza Meszéna

Population Biology Group
Department of Atomic Physics
Eötvös University
Puskin u. 5-7, H-1088 Budapest, Hungary

1. Introduction

In many instances in biological research the qualitative message of a mathematical model is more important than the quantitative results. Quantitative verification of a model is often impossible because of the experimental difficulties and the number of parameters involved.

The structure of matrix determinants offers a way to present mathematical results in an intuitive way. This is the so-called path (or loop) analysis. It is based on the cyclic decomposition of the permutation group involved in the definition of the determinant. Formulas involving calculation of a determinant (for instance eigenvalue equations or matrix inversions), can be expressed as a sum of terms, each of them corresponding to a path in a graph. Each path in turn corresponds to an easily understandable propagation of an effect. (The Feynman-graphs in quantum field theory and the King-Altmann-Hill diagrams for enzyme reactions are examples of other graphical methods in science.)

Levins and coworkers applied this technique to analyze stability of, and coevolution in, ecosystems[1,2]. Caswell's general theory of the complex life cycles is based on the same mathematical apparatus[3,4].

In the Population Biology Group, models were developed to predict how an evolutionary optimum is changed by a perturbation in the environment[5-7]. The emphasis was on qualitative predictions testable in field studies.

Metabolic Control Theory (MCT)[8-9] deals with a completely different subject: the regulation structure of metabolism. But it faces with the same difficulty: it is not feasible to collect all of the parameters involved in a description of a complex system. Hofmeyr[10] and Sen[11] developed graph methods to calculate control coefficients.

Our goal is to generalize these methods to find their common mathematical structure. A General Control Theory is presented for handling stable as well as optimal systems. Path analysis is reformulated in this framework. Application to MCT is presented as an example. (We will use the most general form of the Metabolic Control theory presented by Reder[9].)

Modern Trends in Biothermokinetics, Edited by
S. Schuster *et al.*, Plenum Press, New York, 1993

2. General Control Theory

Let us regard a system with internal variables σ and external parameters μ. The state of the system is a function of the parameters:

$$\sigma = \sigma(\mu) \tag{1}$$

determined implicitly by the equation system

$$w(\sigma,\mu) = 0 . \tag{2}$$

We suppose that eq. (2) unambiguously determines the function $\sigma(\mu)$, and w is differentiable with respect to both of its arguments. (This condition requires the equation $\dim(w) = \dim(\sigma)$ to hold. $\dim(\mu)$ may or may not be different.) Moreover, we suppose, that the matrix $\dfrac{\partial w}{\partial \sigma}$ is invertible.

One can easily express the derivatives of the variables with respect to the parameters:

$$\frac{\partial \sigma}{\partial \mu} = -\left(\frac{\partial w}{\partial \sigma}\right)^{-1} \cdot \frac{\partial w}{\partial \mu} . \tag{3}$$

This is the most general form of the Control Theory. It is rather abstract in this form, but this is the common root of the different applications. The matrix

$$E = \frac{\partial w}{\partial \sigma} \tag{4}$$

is equivalent to the matrix of elasticities, while the negative inverse of it,

$$C = - E^{-1} , \tag{5}$$

is the equivalent of the matrix of control coefficients.

In all of the cases below the matrix of elasticitiy coefficients will not only be invertible, but also negative definite.

3. Specific Cases

There are two different kind of systems, usually described by equations like eq. (2) with a negative definite elasticity matrix:

(A) stable systems; and
(B) optimal systems.

The connection between stable and optimal systems is not surprising: a stable system maximizes its Lyapunoff function, while an optimal state of a system is usually a stable fix-point of some optimization process.

3.1. Case (A). Stable systems

Let us regard a system with the dynamics

$$\frac{dx}{dt} = w(x,\mu) \; . \tag{6}$$

Eq. (2) is the condition for the state $x=\sigma$ to be a fix-point of the dynamics. If the elasticity matrix is negative definite, this fix-point is stable. For instance, we can investigate a stable equilibrium point of an ecosystem. Elasticity coefficients describe the effect of some species on the growth rate of others. Let the parameter μ_i describe any (natural, or artificial) new effect on growth rate (demographic parameters) of the ith species. Control coefficients describe the effect on the steady-state density of the members of the community.

More generally, elasticity coefficients are connected to the dynamical properties of the components of the system, while control coefficients describe the system (its steady state) as a whole.

In MCT the variables are the metabolite concentrations. (We restrict ourselves to the case of independent variables.) The rate of change of the concentrations are

$$\frac{dx}{dt} = w(x,\mu) = N \cdot v(x,\mu) \; , \tag{7}$$

where v is the vector of reaction rates, and N is the stoichiometric matrix. In MCT, usually the matrix

$$\varepsilon = \frac{\partial v}{\partial \sigma} \tag{8}$$

rather than $E=N \cdot \varepsilon$ is referred to as elasticity matrix. The stable steady state is controlled by the parameters in the following way:

$$\frac{\partial \sigma}{\partial \mu} = \Gamma \cdot \frac{\partial v}{\partial \mu} \; , \tag{9}$$

where

$$\Gamma = - \left(N \cdot \frac{\partial v}{\partial \sigma} \right)^{-1} \cdot N \tag{10}$$

is the usual form of the control matrix. This is just the formulation of MCT by Reder[9] for independent concentrations. But the terminology introduced above is more general, and makes possible to describe the behavior of the stable equilibrium of any system. Anyhow, elasticity coefficients describe the dynamics of the system, while the control coefficients describe the control exerted by the parameters on the stable fix point of the system.

3.2. Case (B). Optimal systems

Let us regard an optimization problem with

$$\lambda = \lambda(x,\mu) \tag{11}$$

as a function to maximize with respect to the variables **x**. With the appropriate conditions, the $\mathbf{x} = \sigma$ optimum point is determined implicitly by the equation

$$\mathbf{w}(\sigma,\mu) = \frac{\partial\lambda(\sigma,\mu)}{\partial\sigma} = \mathbf{0} . \tag{12}$$

This point is a maximum, if the elasticity matrix is negative definite.

One possible example is a population with fitness λ, in which selection has optimized the value of **x** to be equal to σ. Elasticity coefficients describe here the internal operation of the system, just like in case (A). Control coefficients describe the control exerted by external factors on the optimum.

4. The Path Analysis Theorem

Let us regard an $n \times n$ matrix **M** and the set of the positive integer numbers used to index the matrix elements:

$$H = \{1,2,3,...,n\} . \tag{13}$$

A series of the elements of this set will be called a 'path' under the condition that all of the elements can be used in the path at most once. For instance, the sequence '2,3,1,7,9' is a path if $n \geq 9$ (see Fig. 1). C_{ij} denotes the set of the possible paths beginning at i and ending at j. The example above is an element of $C_{2,9}$. The set C_{ii} has only one element, 'i'.

One can now derive the following theorem (T):

(T1) The inverse of the matrix **M** (if it exists) can be expressed in the following form:

$$(\mathbf{M}^{-1})_{ij} = - \sum_{c \in C_{ji}} \alpha_c \prod_{k=1}^{n_c-1} M_{c_{k+1}c_k} . \tag{14}$$

(T2) If the matrix **M** is negative definite, all of the α-coefficients are positive.
(The proof of a very similar theorem can be found in the Appendix of Ref. 1.)

Using theorem (T) one can express the solution of a system of linear equations:

$$\mathbf{M} \cdot \mathbf{x} = \mathbf{y} . \tag{15}$$

For simplicity's sake, let us suppose that only one element of **y**, say y_j is different from zero. (T1) leads to the result

$$x_i = -y_j \cdot \sum_{c \in C_{ji}} \alpha_c \prod_{k=1}^{n_c-1} M_{c_{k+1}c_k} . \tag{16}$$

This means that the effect of y_j is propagated to x_i along different paths. For instance, let $i=9$ and $j=2$. The path $\{2,3,1,7,9\}$ has a contribution

$$x_9 = -\alpha_{23179}\, y_2\, M_{32}\, M_{13}\, M_{71}\, M_{97} . \tag{17}$$

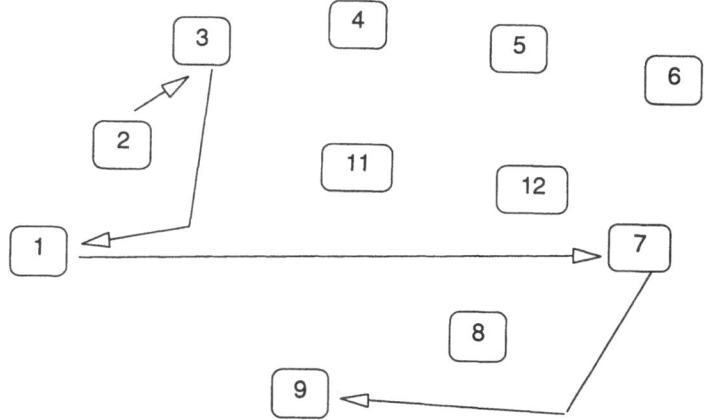

Figure 1. The path {2,3,1,7,9}.

If **M** is negative definite, and if we know or suppose that this path gives the dominant contribution to x_i, the sign of x_i is determined by y_j and by the matrix elements along the path independently of the matrix elements not involved in the path.

Use of this method is not recommended, if too many paths contribute. It is intended to give insight in cases where definite paths dominate the effect.

5. Application of the Theorem to Specific Cases

We will write down the equivalents of eq. (17) in different cases.

5.1. Case (A). Stable systems

Let us suppose for the sake of simplicity that the matrix $\dfrac{\partial \mathbf{w}}{\partial \mu}$ is equal to the unit matrix.

This means that a change of the parameter μ_i is nothing else than a change in the rate of production of the substance i caused by an external factor. According to theorem (T), a change in the steady-state value of the variable x_9 due to the parameter μ_2 through the path {2,3,1,7,9} is the following:

$$\frac{\partial \sigma_9}{\partial \mu_2} = \alpha_{23179} \cdot \frac{\partial w_3}{\partial \sigma_2} \cdot \frac{\partial w_1}{\partial \sigma_3} \cdot \frac{\partial w_7}{\partial \sigma_1} \cdot \frac{\partial w_9}{\partial \sigma_7} + \dots \tag{18}$$

where +... has been written to indicate that other paths might also have a contribution to the change. The meaning of this formula is very straightforward. μ_2 represents a change (an increase, if it is positive) in the rate of production of substance 2. Due to the increase in the amount of this substance, the rate of the production of substance 3 is modified, too. This effect is described by the first partial derivative in the r.h.s. of eq. (18). The modified amount of substance 3 modifies the rate of increase in substance 1, and so on. All of the partial derivatives correspond to one step in this path of effect propagation. The sign of change of σ_9 is simply the product of the signs of the steps, as should be expected. (α is positive because of the stability of the system.)

5.2. Case (B). Optimal systems

How is the optimal value σ_9 modified due to the change of the parameter μ? A path of effect propagation is the following:

$$\frac{\partial \sigma_9}{\partial \mu} = \alpha_{23179} \cdot \frac{\partial^2 \lambda}{\partial \mu \partial \sigma_2} \cdot \frac{\partial^2 \lambda}{\partial \sigma_2 \partial \sigma_3} \cdot \frac{\partial^2 \lambda}{\partial \sigma_3 \partial \sigma_1} \cdot \frac{\partial^2 \lambda}{\partial \sigma_1 \partial \sigma_7} \cdot \frac{\partial^2 \lambda}{\partial \sigma_7 \partial \sigma_9} +\ldots \qquad (19)$$

The second-order partial derivatives in the r.h.s. of this equation measure the effect of one variable on the optimal value of the other one. For instance, if $\dfrac{\partial^2 \lambda}{\partial \sigma_3 \partial \sigma_1} > 0$, an increase in σ_3 increases the advantage of a larger σ_1, so the optimal value of σ_1 is increased by an increase of σ_3. The meaning of eq. (19) is straightforward: a change in the external parameter changes the optimal value of the second variable. In turn, the change of the second variable changes the optimum of the third one, the change of third variable changes the optimum of the first one, and so on until the optimal value of σ_9 is adjusted. Positivity of the coefficient α ensures the possibility of a qualitative prediction.

6. Application to Metabolic Control

In MCT the variables are the metabolite concentrations. (We restrict ourselves to the case of independent variables. We follow Reder[9] in not using logarithmic derivatives.) The rate of change of the concentrations is given by eq. (7) with $\mathbf{v}$ and $\mathbf{N}$ being the vector of the reaction rates and the stoichiometric matrix, respectively. Path analysis is applicable. The only complication is that paths have to go not only through the metabolites, but also through the enzymes. Let μ_i be the activity of the ith enzyme. (So the matrix $\dfrac{\partial \mathbf{v}}{\partial \mu}$ is equal to the unit matrix.) A change in the steady-state value of the concentration x_9 due to the activity change of the enzyme X through the path $\{2,3,1,7,9\}$ is the following (concentration control coefficient):

$$\Gamma_{9X} = \frac{\partial \sigma_9}{\partial \mu_2} = \alpha_{23179} N_{2X} \frac{\partial v_{I}}{\partial \sigma_2} N_{3I} \frac{\partial v_{V}}{\partial \sigma_3} N_{1V} \frac{\partial v_{IX}}{\partial \sigma_1} N_{7IX} \frac{\partial v_{III}}{\partial \sigma_7} \cdot N_{9III} +\cdots \qquad (20)$$

where the Roman numbers are the indices of the enzymes involved in this path and +... indicates that other paths might also contribute to the change. The path contribution to the flux control coefficient is rather similar:

$$C_{III,X} = \frac{\partial v_{III}}{\partial \mu_2} = \alpha_{23179} N_{2X} \frac{\partial v_{I}}{\partial \sigma_2} N_{3I} \frac{\partial v_{V}}{\partial \sigma_3} N_{1V} \frac{\partial v_{IX}}{\partial \sigma_1} N_{7IX} \frac{\partial v_{III}}{\partial \sigma_7} +\cdots \qquad (21)$$

The formula makes possible to qualitatively assess the control exerted by enzyme X on substance 9 or on rate III through the given path, because the coefficient α is positive.

The meaning of these formulas is very straightforward. μ_X represents a change (an increase, to be definite) in the rate of production of substance 2. Due to the increase in the amount of this substance, the rate of production of substance 3 is modified, and so on (see Fig. 2).

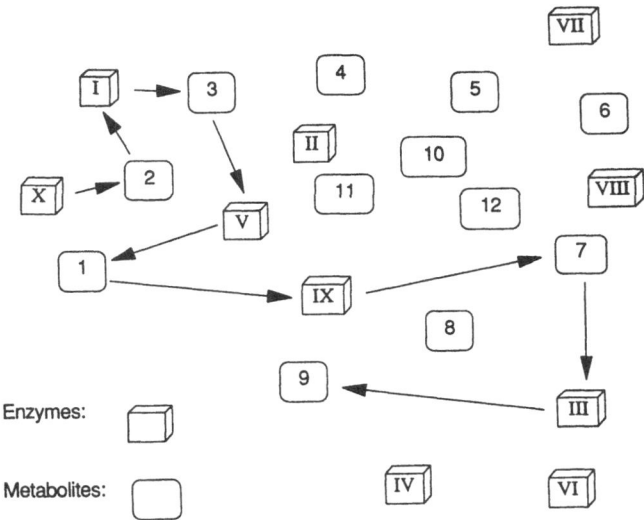

Figure 2. The contribution of the path $\{X,2,I,3,V,1,IX,7,III,9\}$ to the concentration control coefficient Γ_{9X}.

7. Summarizing Remarks

A generalization of Metabolic Control Theory is presented. It includes description of stable as well as optimal systems, and deals with the effect of an external change on the stable/optimal state. Path analysis is established in this framework. It gives a heuristic insight and establishes a connection between the mathematical treatment and qualitative expectations about the behavior of the system. Control coefficients can be expressed as a sum of terms, each of them representing a path of propagation of the effect through the system.

Acknowledgements

The author thanks Drs Liz Pásztor and Hans Westerhoff for introducing him into population biology and Metabolic Control Theory, respectively. Discussions with Drs Boris Kholodenko, Jan-Hendrik Hofmeyr, Hans Metz and Asok K. Sen were helpful. This work was partly supported by a grant OTKA 2222 and by the Peregrinatio foundation.

References

1. C. J. Puccia and R. Levins. "Qualitative Modeling of Complex Systems. An Introduction to Loop Analysis and Time Averaging," Harvard University Press, Cambridge (Mass.)/London (1985).
2. J. Roughgarden, The theory of coevolution, *in*: "Coevolution," E D. J. Futuyma and M. Slatkin, eds, Sinauer Associates, Suderland (Mass.) (1983).
3. H. Casewell, Optimal life histories and the maximization of reproductive value: a general theorem for complex life cycles, *Ecology* **63**:1218 (1982).
4. H. Casewell, Stable structures, reproductive value, and the life history analysis for populations with complex life histories, *Ecology* **63**:1223 (1982).
5. G. Meszéna and E. Pásztor, Population regulation and life-history strategies, *in*: "Proceeding in Nonlinear Science. Organizational Constraints on the Dynamics of Evolution," J. Maynard Smith and G. Vida, eds, Manchester University Press, Manchester and New York (1990).
6. Ê. Kisdi and G. Meszéna. "Density Dependent Life History Evolution in Fluctuating Environment," Lecture Notes in Biomathematics, Springer, Berlin, in press.

7. L. Pásztor, Unexploited dimensions of optimization life history theory, *in*: "Population Genetics and Evolution," G. de Jong, ed., Springer, Berlin (1986).

8. H. V. Westerhoff and K. van Dam, "Thermodynamics and Control of Biological Free-Energy Transduction," Elsevier, Amsterdam (1987).

9. C. Reder, Metabolic control theory: A structural approach, *J. theor Biol.* **135**:175-201 (1988).

10. J.-H. S. Hofmeyr, Control-pattern analysis of methabolic pathways. Flux and concentration control in linear pathways, *Eur. J. Biochem.* **186**:343-354 (1986).

11. A. K. Sen, Graph-theoretic analysis of control and regulation in complex metabolic pathways, this volume.

PHENOMENOLOGICAL KINETICS AND THE TOP-DOWN APPROACH TO METABOLIC CONTROL ANALYSIS

Guy C. Brown

Department of Biochemistry and Molecular Biology
University College London
Gower Street, London WC1E 6BT

1. Phenomenological Kinetics

In investigating the regulation of metabolic pathways one is often faced with a problem of the following type: some agent or change (e.g. hormone, drug, disease) has caused a change in a pathway flux, and you wish to know whether the agent has caused this change by altering the kinetics of one end of the pathway, or the other, or both. One approach to this problem is an analysis related to Cross-over analysis, which I shall call Phenomenological Kinetic Analysis. The pathway is conceptually divided into two parts around some intermediate metabolite (B in the scheme below) which can be measured in the system. The top end of the pathway (1 in the scheme) produces the intermediate (B), and the bottom end of the pathway (2 in the scheme) consumes the intermediate.

$$\text{Enzyme group:} \qquad 1 \qquad 2 \qquad\qquad \text{Scheme 1}$$
$$\text{Metabolite:} \qquad A \longrightarrow B \longrightarrow C$$

The response of the steady flux through the top and bottom ends of the pathway to changes in the level of the intermediate ([B]) are then measured. I will call these responses the phenomenological kinetics of the top and bottom ends of the pathway, and this is simply the dependence of flux on B measured in the system. The phenomenological kinetics of the bottom end of the pathway can be measured by altering the kinetics of the top end (e.g. by altering the pathway substrate concentration (A) or by adding a specific inhibitor of the top end 1), and measuring the dependence of the steady state flux on the level of the intermediate (B). Similarly the phenomenological kinetics of the top end of the pathway can be measured by altering the kinetics of the bottom end (e.g. by specifically inhibiting the bottom end). By plotting out the kinetics we get something like Fig. 1.

Modern Trends in Biothermokinetics, Edited by
S. Schuster *et al.*, Plenum Press, New York, 1993

2. Response to Agent X

We then add our agent (X) (not knowing whether it will act on the top or bottom end of the pathway), and measure the effect on the steady state flux and level of the intermediate (B), and plot this change (as in Fig. 1). If the agent (X) acts solely on the kinetics of the top end of the pathway then it does not change the phenomenological kinetics of the bottom end. Thus the new steady state will be the same as that caused by a specific effector of the top end; and when plotted as in Fig. 1 the new steady state (X) must lie on the J_2 curve.

Similarly if the agent (X) changes only the kinetics of the bottom end of the pathway, then the new steady state must lie on the J_1 curve when plotted as in Fig. 1. If the agent X acts on the kinetics of both the top and bottom ends of the pathway, then the new steady state must lie on neither the J_1 or J_2 curves (as in Fig. 1).

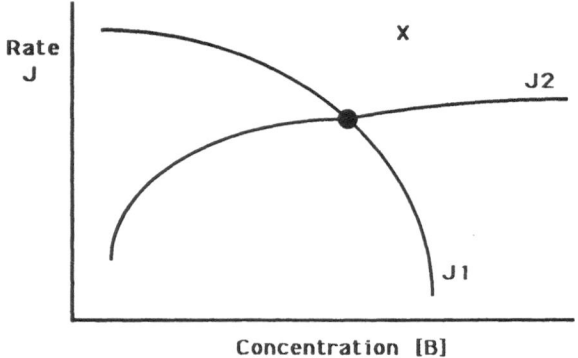

Figure 1. The phenomenological kinetics of the top (J_1) and bottom (J_2) ends of the pathway, i.e. the dependence of their fluxes on [B], measured by titrating the flux and [B] with specific effectors of either the bottom or top ends of the pathway. The effect of some other agent which stimulates both the top and bottom ends is also plotted as X.

This analysis can be repeated on other intermediates in the pathway, until the enzyme (or enzymes) whose kinetics are changed by the agent are specifically located. The analysis is easily extended to branched pathways. The main problem with this type of analysis is that their must be no kinetic interaction between the top and bottom ends of the pathway other than via the intermediate (B) (e.g. no feedback inhibition from the bottom end to the top via some other metabolite). Also the effectors (e.g. inhibitors) used to derive the phenomenological kinetics must be specific to either the top or bottom ends of the pathway.

This type of analysis has been used by: the author[1] to locate the site of action of vasopressin and extracellular ATP on respiration in isolated liver cells relative to the processes producing and using mitochondrial NADH; Nobes and coworkers[2] to investigate the affect of added fatty acids on the processes producing and using the mitochondrial membrane potential in isolated liver cells; Hafner and coworkers[3] to investigate which of the processes producing and using the mitochondrial proton motive force are changed by thyroid hormone in isolated mitochondria; and by Brand and coworkers[4] to locate the site of action of glucagon on the energy metabolism of isolated liver mitochondria.

3. Predictive Value of Phenomenological Kinetics

The phenomenological kinetics of pathway also have some predictive value. If the kinetics of the top or bottom ends of the pathway are changed by some known amount, then we can roughly predict what will be the new steady state flux for the whole pathway and the new level of the intermediate. The steady state is always given by the intercept of the two curves of the phenomenological kinetics (e.g. Fig. 1). A doubling of the activity of all the enzymes in the top end of the pathway will double the J_1 flux at any particular level of the intermediate (B), and thus the new steady state can be predicted from the phenomenological kinetics by superimposing the kinetic curves. The prediction is not accurate if the enzyme activities are not all changed in proportion. Accurate prediction of flux and metabolite changes when the kinetics of one or more pathway enzymes is changed requires knowledge of the phenomenological kinetics of the relevant parts of the pathway.

4. Top-down Control Analysis

The phenomenological kinetics can be used to derive the control coefficients of the top and bottom ends of the pathway. The connectivity theorem of metabolic control analysis has been used to show that the relative control coefficients are simply related to the relative elasticities of the top and bottom ends to the intermediate. The relative control coefficients are given by the relative slopes of the kinetics in the unperturbed state (i.e. the point of interception in Fig. 1). Since the control coefficients of the top and bottom ends must sum to unity, the absolute control coefficients of the two ends can also be derived. Repeating this analysis around other intermediates in the pathway allows the control coefficients of smaller and smaller groups of enzymes to be derived. This approach to deriving control coefficients is called the top-down approach to Metabolic Control Analysis[5]. This approach has been used by: Brand and coworkers[4] and Hafner and coworkers[6] to measure the control coefficients of the respiratory chain, proton leak and phosphorylation system over respiration and ATP synthesis in isolated mitochondria ; and the author[1] to measure similar coefficients in isolated liver cells.

5. Control Analysis of the Whole Body

The top-down approach can be adapted to analyze the control by body organs in animals over the fluxes of substances between the blood and organs. I will call the net flux (due to transport or metabolism) of a substance (e.g. metabolite, hormone, drug) into a particular organ or tissue the organ flux. Some organs will produce the substance and some use the substance. If the substance of interest reaches a steady state level in the blood then the net flux must be zero. The arterial concentration of the substance in the blood plasma acts like a kinetic intermediate between the organ fluxes. Organ fluxes of a substance may have a positive or negative elasticity to changes in arterial concentration of that substance. Using control analysis one can calculate the flux and (plasma) concentration control coefficients of the various organs from organs fluxes and elasticities.

There are several problems with such an analysis. (1) Kinetic interactions between organ fluxes other than via the substance of interest need to be taken into account. This is relatively simple for hormones when the hormone level is strictly related to the level of the substance of interest, but becomes difficult in more complex cases. (2) Analysis of the lungs and liver is difficult because their blood supply is not strictly arterial. (3) There must be a steady state level of the substance in the blood, and this must be reached relatively quickly to make analysis possible. (4) The practical determination of organ fluxes and elasticities is not easy.

References

1. G.C. Brown, P.L. Lakin-Thomas and M.D. Brand, Control of respiration and oxidative phosphorylation in isolated rat liver cells, *Eur. J. Biochem.* **198**:355-362 (1990).
2. C.D. Nobes, W.W. Hay and M.D. Brand, The mechanism of stimulation of respiration by fatty acids in isolated hepatocytes, *J. Biol. Chem.* **265**:12910-12915 (1990).
3. R.P. Hafner, G.C. Brown and M.D. Brand, Thyroid hormone control of state 3 respiration in isolated rat liver mitochondria, *Biochem. J.* **265**:731-734 (1990).
4. M.D. Brand, R.P. Hafner and M.D. Brand, Control of respiration in non-phosphorylating mitochondria is shared between the proton leak and respiratory chain, *Biochem. J.* **255**:535-539 (1988).
5. G.C. Brown, R.P. Hafner and M.D. Brand, A 'top-down' approach to the determination of control coefficients in metabolic control theory, *Eur. J. Biochem.* **188**:321-324 (1990).
6. R.P. Hafner, G.C. Brown and M.D. Brand, Analysis of the rate and protonmotive force in isolated mitochondria using the 'top-down' approach of metabolic control theory, *Eur. J. Biochem.* **188**:313-319 (1990).

A MODULAR APPROACH
TO THE DESCRIPTION OF THE CONTROL
OF CONNECTED METABOLIC SYSTEMS

Stefan Schuster[1,2], Daniel Kahn[2,3] and Hans V. Westerhoff[3,4]

[1]Université Bordeaux II, Dépt. de Biochimie Médicale
  146, rue Léo Saignat, F-33076 Bordeaux, France
[2]The Netherlands Cancer Institute, Dept. of Molecular Biology
  Plesmanlaan 121, NL-1066 CX Amsterdam, The Netherlands
[3]Laboratoire de Biologie Moleculaire, I.N.R.A-C.N.R.S.
  B.P. 27, F-31326 Castanet-Tolosan, France
[4]E.C. Slater Institute
  University of Amsterdam, The Netherlands

1. Introduction

A limitation of the existing Metabolic Control Analysis[1-4] was that it would discuss control only in terms of kinetic properties of the individual enzymes, and not in terms of control properties of functional metabolic subunits, such as cytosol and mitochondrion. A modular method was developed to analyze control in systems of unconnected modules in terms of the properties of the particular modules and the allosteric effects between them[5]. The latter approach is limited by the demand that there be no net flux between the modules (which can also occur in some connected systems[6]).

For the case of oxidative phosphorylation, a solution was devised which groups all the reactions into three parts: those connected with respiration, those connected with synthesis of extra-mitochondrial ATP and those that leak away $\Delta\mu_{H^+}$[3,7]. A similar procedure was developed to discuss control in branched metabolic pathways[8]. More recently, the approach has been developed more in detail and renamed "top-down approach"[9,10].

In the present paper we generalize overall control analysis to cases where modules have several degrees of freedom with respect to flux.

2. Modular Decomposition

We consider two types of modules,

type 1: subsystems for which we only wish to observe the reactions that link them with their surroundings, but not the internal details, type 2: subsystems subject to explicit observation.

We start with a decomposition into one module of type 1 and one module of type 2. The reactions can be classified into three groups: reactions internal to module 1 (subscript 1); reactions bridging the two modules (subscript b), and reactions pertinent to subsystem 2 only (subscript 2).

The system equations read

$$dX/dt = N \, v \tag{1}$$

with v and X denoting the vectors of reaction rates and of molarities, respectively. For multi-compartment systems, one should identify X with the vector of mole numbers rather than of molarities. Here, we shall assume that N has linearly independent rows only, that is that the system does not involve any conservation relations.

The modular decomposition can be formalized as follows,

$$N = \begin{pmatrix} N_{11} & N_{1b} & 0 \\ 0 & N_{2b} & N_{22} \end{pmatrix}, \quad X = \begin{pmatrix} X_1 \\ X_2 \end{pmatrix}, \quad v = \begin{pmatrix} v_1 \\ v_b \\ v_2 \end{pmatrix}. \tag{2a,b,c}$$

The matrix of non-normalized elasticities can be decomposed as:

$$D \underset{\text{def}}{=} \frac{\partial v}{\partial X} = \begin{pmatrix} D_{11} & D_{12} \\ D_{b1} & D_{b2} \\ D_{21} & D_{22} \end{pmatrix}. \tag{3}$$

We distinguish between fluxes, J, attained at a steady state of the whole system and fluxes, $^*J$, attained when module 1 is in a steady state on its own, with the molarities belonging to module 2 kept constant. We assume that the following overall elasticity coefficients can be measured,

$$^*D_{b2} = \partial\, ^*J_b/\partial X_2 , \quad ^*D_{22} = \partial\, ^*J_2/\partial X_2 . \tag{4a,b}$$

A case where these quantities are well defined is that in which module 1 is a fast subsystem. $^*D_{b2}$ and $^*D_{22}$ are determined by measuring $^*J_b$ and $^*J_2$ quickly after a perturbation in X_2, such that module 1 has reattained internal steady state whereas concentrations in module 2 are essentially unaltered, .

3. Reduction of Bridging Fluxes

Since the bridging fluxes are usually linearly dependent, it is appropriate to reduce them to a vector of independent fluxes, J_r, with the help of an appropriate matrix K,

$$J_b = K \cdot J_r . \tag{5}$$

Consider, for example, an unbranched reaction system as shown inScheme 1A.

$$P_1 \longrightarrow X_1 \longrightarrow X_2 \longrightarrow P_2 \qquad\qquad \text{Scheme 1A}$$

If we treat X_1 as module 1 and the first two reactions as bridging reactions, we can lump these reactions into one overall reaction. The reaction scheme can then be depicted as

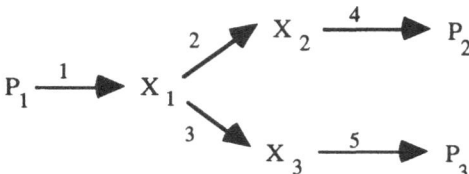

$$P_1 \xrightarrow[\text{Module 1}]{} X_2 \longrightarrow P_2 \qquad\qquad \text{Scheme 1B}$$

For this example, $\mathbf{K}$ reads $(1 \quad 1)^T$. In this special case, the aggregation method is essentially identical with the 'top-down' approach[9,10]. In this paper, we wish to include cases with more than one independent bridging flux, such as the system shown in Scheme 2A.

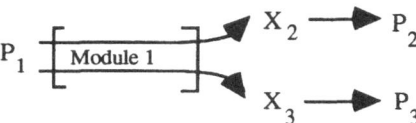

Scheme 2A

If X_1 is taken as module 1, it is linked with subsystem 2 by two independent steady-state fluxes. Accordingly, it is useful to redraw the scheme so that module 1 is a 'super-reaction' with two fluxes.

$$\text{Scheme 2B}$$

As in the traditional kinetic description of biochemical reactions, the rates are to be described as functions of parameters and of molarities external to the "super-reaction" occurring in module 1:

$$^*J_r = {}^*J_r(p,X_2) \, , \, ^*J_2 = {}^*J_2(p,X_2) \, . \tag{6a,b}$$

Let $^*\mathbf{D}_{r2}$ denote the matrix of overall elasticities pertaining to $^*\mathbf{J}_r$. This matrix is linked with $^*\mathbf{D}_{b2}$ by the same transformation rule as $\mathbf{J}_b$ is with $\mathbf{J}_r$.

4. Overall Control Coefficients

The matrices of non-normalized coefficients expressing the control exerted by module 2 can be defined as $(x = r, 2)$ (cf. Ref. 5),

$$\Gamma_{22} = \partial \mathbf{X}_2 / \partial \mathbf{p}_2 \, (\partial \mathbf{v}_2 / \partial \mathbf{p}_2)^{-1} \, , \tag{7}$$

$$\Phi_{x2} = \partial \mathbf{J}_x / \partial \mathbf{p}_2 \, (\partial \mathbf{v}_2 / \partial \mathbf{p}_2)^{-1} \, , \tag{8}$$

where $\mathbf{p}_2$ is a vector of parameters specificically influencing the reactions in module 2.

In a similar way, overall coefficients expressing the control exerted by the lumped bridging reactions may be defined (indicated by an asterisk),

$$^*\Gamma_{2r} = \partial \mathbf{X}_2 / \partial \mathbf{p}_r \, (\partial\, ^*\mathbf{J}_r / \partial \mathbf{p}_r)^{-1} \, , \tag{9}$$

$$^*\Phi_{2r} = \partial \mathbf{J}_x / \partial \mathbf{p}_r \, (\partial\, ^*\mathbf{J}_r / \partial \mathbf{p}_r)^{-1} \, , \tag{10}$$

with p_r being a vector of parameters influencing the rates $^*J_r$. These coefficients differ from the traditional control coefficients of the bridging reactions both conceptually and mathematically. The former express the control exerted by the lumped reactions of module 1 as a whole, whereas the latter refer to the control exerted by the particular bridging reactions. Mathematically, the two types of coefficients differ in the denominators, where the overall coefficients have a derivative $\partial^*J_r/\partial p_r$, which is in fact a matrix of *overall π-elasticities*. Note that, in modification to a basic idea of the 'top-down approach'[9], the overall control coefficients defined by eqs (9), (10) are not restricted to the situation that all enzymes belonging to one module or to one lumped reaction are changed by the same fractional amount.

The set of parameters p_r has to be "rich" enough to make the matrix (d^*J_r/dp_r) non-singular. It is worth noting that it is not, in general, possible to find parameters influencing the components of $^*J_r$ specifically.

5. Relations between Overall Coefficients

By eq. (5), we have the steady-state equation

$$N_{2b} K\, ^*J_r + N_{22}\, ^*J_2 = 0 . \tag{11}$$

Differentiation of this equation with respect to p yields

$$N_{2b}K\frac{\partial^* J_r}{\partial p} + N_{22}\frac{\partial^* J_2}{\partial p} + \Im^*\frac{\partial X_2}{\partial p} = 0 \tag{12}$$

with

$$\Im^* = N_{2b}\, K\, ^*D_{r2} + N_{22}\, ^*D_{22} . \tag{13}$$

The matrix $\Im^*$ is the Jacobian matrix of module 2 taking into account the presence of module 1. Eqs. (12) and (13) only contain quantities at the outside of system 1, i.e. quantities taken to be known.

Now we can calculate the concentration control coefficients defined in eqs (7,9)

$$^*\Gamma_{2r} = - (\Im^*)^{-1} N_{2b}\, K , \quad \Gamma_{22} = - (\Im^*)^{-1} N_{22} \tag{14a,b}$$

To obtain the flux control coefficients, one may differentiate the equations $J_x(p) = {}^*J_x(p,X_2)$ (x=r,2) with respect to p, to obtain

$$\frac{\partial J_x}{\partial p} = \frac{\partial^* J_x}{\partial p} + \frac{\partial^* J_x}{\partial X_2}\frac{\partial X_2}{\partial p} . \tag{15}$$

From this equation, we derive

$$^*\Phi_{rr} = I + {}^*D_{r2}\, ^*\Gamma_{2r} , \quad \Phi_{r2} = {}^*D_{r2}\, \Gamma_{22} , \tag{16a,b}$$

$$^*\Phi_{2r} = {}^*D_{22}\, ^*\Gamma_{2r} , \quad \Phi_{22} = I + {}^*D_{22}\, \Gamma_{22} . \tag{17a,b}$$

These results show that one is able to calculate the flux and concentration control related to module 2 even without knowing the internal details of module 1. The information of the inside of module 1 that is relevant for the whole network is apparently entirely represented by the overall elasticity coefficients ($^*D_{r2}$, $^*D_{22}$).

Using the equations obtained above, one can derive summation and connectivity relationships for the overall coefficients. For brevity's sake, we here give only the theorems for the normalized control coefficients:

$$(^*C^S_{2r}1 + C^S_{22}1) = 0 \, , \tag{18}$$

$$^*C^S_{2r} \, ^*\varepsilon_{r2} + C^S_{22} \, ^*\varepsilon_{22}) = -I, \tag{19}$$

$$(^*C^J_{xr}1 + C^J_{x2}1) = 1, \tag{20a}$$

$$^*C^J_{xr} \, ^*\varepsilon_{r2} + C^J_{x2} \, ^*\varepsilon_{22} = -I, \tag{20b}$$

where the symbols $^*C$ and $^*\varepsilon$ stand for normalized overall control coefficients and elasticities, respectively, and $1=(1\ 1\ ...\ 1\ 1)^T$.

An important situation is when the fluxes through all lumped bridging reactions are changed by the same fractional amount, α, $\delta \ln J_{r,k}=\alpha$ for any k (e.g. by changing the number of mitochondria in a suspension). In that case the control exerted by module 1 as a whole can be expressed by the sum of all normalized overall control coefficients pertaining to this module:

$$\delta \ln J_r = {}^*C^J_{rr} \, \alpha \, 1, \quad \delta \ln J_2 = {}^*C^J_{2r} \, \alpha \, 1, \quad \delta \ln X_2 = {}^*C^S_{2r} \, \alpha \, 1 \, . \tag{21a,b,c}$$

6. Two Modules with a Number of Metabolites in Between

We now consider the case where two modules of type 1 are connected by an arbitrary number of metabolites, which constitute a module of type 2, as shown in Fig. 1A. The two modules of type 1 can be regarded as the upper part (index u) and the lower part (index l) of a pathway. Note that in Sections 2-5, we considered a system which obtains from that shown in Fig. 1 by canceling module l.

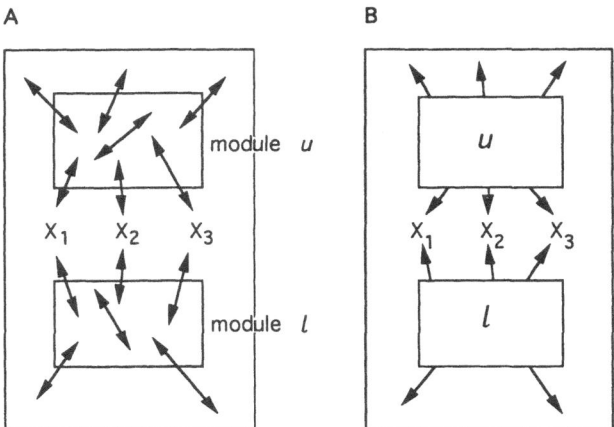

Figure 1. Schematic picture of a system with two modules of type 1, in detailed representation (A) and in a "black-box" representation (B).

We now define the overall elasticities

$$^*\mathbf{D}_{u2} = \partial\, ^*J_u/\partial\mathbf{X}_2\,, \quad ^*\mathbf{D}_{l2} = \partial\, ^*J_l/\partial\mathbf{X}_2\,. \tag{22a,b}$$

To simplify the notation, we suppose that the vectors $^*\mathbf{J}_u$ and $^*\mathbf{J}_l$ of bridging fluxes have already been reduced to contain independent fluxes only. We assume that no conservation relationship links $\mathbf{X}_2$, the upper and lower modules. Although the upper and lower modules are not directly connected stoichiometrically, they can be formally combined into one module of type 1. Thus, we write

$$\mathfrak{I}^* = \mathbf{N}_{2u}\,\mathbf{K}_u\, ^*\mathbf{D}_{u2} + \mathbf{N}_{2l}\,\mathbf{K}_l\, ^*\mathbf{D}_{l2}\,, \tag{23}$$

$$^*\Gamma_{2r} = (\,^*\Gamma_{2u}\quad ^*\Gamma_{2l}) = -\,(\mathfrak{I}^*)^{-1}\,(\,\mathbf{N}_{2u}\,\mathbf{K}_u \quad \mathbf{N}_{2l}\,\mathbf{K}_l\,) \tag{24}$$

$$^*\Phi_{rr} = \begin{pmatrix} ^*\Phi_{uu} & ^*\Phi_{ul} \\ ^*\Phi_{lu} & ^*\Phi_{ll} \end{pmatrix} = \begin{pmatrix} \mathbf{I} + ^*\mathbf{D}_{u2}\Gamma_{2u} & ^*\mathbf{D}_{u2}\Gamma_{2l} \\ ^*\mathbf{D}_{l2}\Gamma_{2u} & \mathbf{I} + ^*\mathbf{D}_{l2}\Gamma_{2l} \end{pmatrix} \tag{25}$$

Φ_{ul} and Φ_{lu} express the flux control exerted by the upper and lower modules upon each other.

7. Discussion

In traditional MCA, the global properties of metabolic networks are calculated in terms of the local properties of enzymes. If the latter are not completely known, it is more appropriate to start from "semi-global" features of certain subsystems (modules). In extension to earlier approaches[3,7,10], we have here allowed for the the black-box modules to have several independent "semi-global" rates, each of which can be considered to exert flux and concentration control.

As traditional MCA, modular control analysis can be based both on normalized and on non-normalized control coefficients (for comparison see Ref. 4).

For simplicity's sake, we have not considered systems with moiety conservation. In a more detailed analysis, we have shown that for such systems, modular decomposition has to be made in such a way that the modules are not linked by conservation relations[10].

Acknowledgements

This work was supported by The Netherlands Organization for Pure Research, the European Community (BRIDGE) and the Scientific Council of NATO.

References

1. H. Kacser and J.A. Burns, The control of flux, *Symp. Soc. Exp. Biol.* **27**:65-104 (1973).
2. R. Heinrich and T.A. Rapoport, A linear steady-state treatment of enzymatic chains, *Eur. J. Biochem.* **42**:89-95 (1974).
3. H.V. Westerhoff and K. van Dam, "Thermodynamics and Control of Biological Free-Energy Transduction," Elsevier, Amsterdam (1987).
4. S. Schuster and R. Heinrich, The definitions of metabolic control analysis revisited, *BioSystems* **27**:1-15 (1992).
5. D. Kahn and H.V. Westerhoff, Control theory of regulatory cascades, *J. theor. Biol.* **153**:255-285 (1991).

6. S. Schuster and R. Schuster, Decomposition of biochemical reaction systems according to flux control insusceptibility, *J. Chim. Phys.* **89**:1887-1910 (1992).

7. H.V. Westerhoff, A.K. Groen and R.J.A. Wanders, The thermodynamic basis for the partial control of oxidative phosphorylation by the adenine-nucleotide translocator, *Biochem. Soc. Trans.* **11**:90-91 (1983).

8. H. Kacser, The control of enzyme systems *in vivo*: Elasticity analysis of the steady state, *Biochem. Soc. Trans.* **11**:35-40 (1983).

9. G.C. Brown, R.P. Hafner and M.D. Brand, A 'top-down' approach to the determination of control coefficients in metabolic control theory, *Eur. J. Biochem.* **188**:321-325 (1990).

10. S. Schuster, D. Kahn and H.V. Westerhoff, Modular analysis of the control of complex metabolic pathways, *Biophys. Chem.*, submitted.

CONTROL ANALYSIS OF ISOENZYMIC PATHWAYS: APPLICATION OF DIRECTED GRAPHS AND ELECTRICAL ANALOGS

Asok K. Sen

Department of Mathematical Sciences
Purdue University School of Science
1125 East 38th Street, Indianapolis 46205-2810, Indiana, USA

1. Introduction

In many metabolic pathways a particular reaction step is catalyzed by different molecular forms of the same enzyme which are known as isoenzymes or simply isozymes. The isozymes are coded by different genes and differ in their amino acid composition. They may act as regulatory enzymes and control the metabolic fluxes and concentrations. An example is the metabolism of ethanol in horse liver by three isozymes of alcohol dehydrogenase[1]. Other examples include the branched biosynthetic pathway of the aromatic amino acids in *E. coli.* where the first enzyme, DAHP synthase, exists in the form of three isozymes each of which is inhibited by one of the three end products phenylalanine, tyrosine and tryptophan[2].

A quantitative analysis of the control exerted by the various isozymes in an isoenzymic pathway has been made[1,3]. These treatments are based on the Metabolic Control Theory of Kacser and Burns[4] and Heinrich and Rapoport[5]. Fell and Sauro[3] examined a linear pathway with two isozymes catalyzing the same reaction step. They used a matrix formulation of the governing equations to investigate the control distribution. Derr and Derr[1] generalized this matrix approach to determine the flux-control pattern among enzymes/isozymes in a linear pathway containing three or more isozymes.

In this paper we present a graph-theoretic approach and an electrical analog technique for analyzing the flux-control distribution in isoenzymic pathways. In particular, we describe a systematic procedure by which a directed graph or an analog circuit representing the control structure of an isoenzymic pathway can be constructed in a heuristic fashion directly from the reaction sequence of the pathway, without the necessity of writing down the governing equations for the flux control coefficients. From an inspection of the topology of the directed graph, expressions for the flux control coefficients of the various enzymes/isozymes are readily derived. The flux control coefficients are deduced from an electrical analog circuit with the application of Ohm's law. Our analysis is based on Metabolic Control Theory[4,5]. Directed graphs and electrical analogs have been used in our recent work[6-9] on control analysis of linear metabolic pathways.

Modern Trends in Biothermokinetics, Edited by
S. Schuster *et al.*, Plenum Press, New York, 1993

2. Application of Directed Graphs

2.1. Construction algorithm

As a paradigm of isoenzymic pathways we consider the following parallel pathway with three isozymes. Here X_0 is the initial substrate, P is the final product, and X_1 and X_2 are the intermediate metabolites. The first and last reactions are catalyzed by the enzymes E_1 and E_5, respectively, whereas the isozymes E_2, E_3 and E_4 catalyze the conversion of X_1 to X_2. A matrix formulation for control analysis of this pathway is given by Derr and Derr[1].

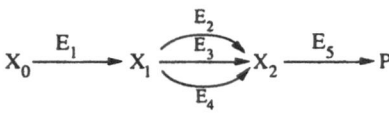

Figure 1. Reaction scheme of a pathway with three isoenzymes catalyzing the second reaction. (a) Usual representation, (b) representation with display of elasticities.

The construction of a directed graph is facilitated by redrawing the pathway with the elasticity coefficients of the enzymes/isozymes displayed as shown in Fig. 1. To preserve clarity the elasticity coefficients of the isozyme E_3 are not shown in this representation.

Our objective is to construct a directed graph for this pathway and evaluate the flux control coefficients of the various enzymes/isozymes. The flux control coefficients are defined relative to the flux through the enzyme E_1 or E_5. For the purpose of constructing a directed graph, we regard the very first enzyme E_1 as the *leader* enzyme and call any of the three isozymes, say E_2, the *primary* isozyme; the remaining isozymes, namely, E_3 and E_4 are referred to as *secondary* isozymes. The directed graph is drawn via the following steps.

Step 1. Consider the leader enzyme E_1, the primary isozyme E_2 and the enzyme E_5 to form a linear pathway and draw its directed graph (without a source node) in the manner described previously[6]. This consists of the nodes 1, 2 and 5 which are connected by the edges (1, 2), (2, 1), (2, 5) and (5, 1) with appropriate weights as shown in Fig. 2. To emphasize that node 2 represents the primary isozyme E_2, we have used the superscript P. An overbar on an elasticity coefficient indicates the magnitude of that elasticity coefficient which is negative.

Step 2. Draw the nodes 3 and 4 designating the secondary isozymes E_3 and E_4. From each of these nodes an edge of weight -1 is directed to the leader node 1.

Step 3. Consider each of the secondary isozymes, E_3 and E_4, which are contiguous to enzyme E_5 with X_2 as their adjoining metabolite. Accordingly, an edge is drawn *to* node 5 *from* each of the nodes 3 and 4 in the digraph. These edges, (3, 5) and (4, 5), are given the weights $\bar{\varepsilon}_2^3 / \varepsilon_2^5$ and $\bar{\varepsilon}_2^4 / \varepsilon_2^5$, which are respectively the ratios of the magnitudes of the elasticity coefficients of the isozymes E_3 and E_4 (with respect to X_2) to the elasticity coefficient of E_5 with respect to X_2.

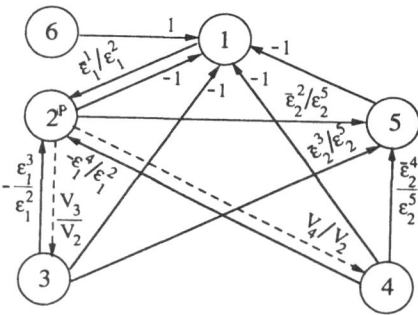

Figure 2. A directed graph for the control structure of the pathway shown in Fig. 1.

Step 4. Note that the isozymes E_2, E_3 and E_4 have a common substrate X_1. Their interactions are characterized in the directed graph in the following manner.

(i) An edge is drawn *from* each of the secondary isozyme nodes 3 and 4 *to* the primary isozyme node 2. The edge (3, 2) is given the weight $-\varepsilon_1^3 / \varepsilon_1^2$, which is the negative of the ratio of the elasticity coefficient of E_3 (with respect to X_1) to that of E_2 with respect to X_1. In choosing the negative sign, we have adopted the following convention: if both the elasticity coefficients in the weight point in the forward (i.e., downstream) direction in the representation (Fig. 1b), then a negative sign is attached to the ratio of the two elasticity coefficients. In a similar fashion, the edge (4, 2) is assigned the weight $-\varepsilon_1^4 / \varepsilon_1^2$.

(ii) An edge is directed *from* the primary isozyme node 2 *to* each of the secondary isozyme nodes 3 and 4 (shown by broken lines in Fig. 2). These edges, namely (2, 3) and (2, 4) are given the weights v_3/v_2 and v_4/v_2, respectively, where v_i with $i = 2, 3, 4$ denotes the velocity of the reaction catalyzed by the isozyme E_i.

Step 5. Draw a source node 6 and direct an edge from the source node to the leader node 1 carrying the weight unity.

2.2. Evaluation of the flux control coefficients

From Fig. 2 the flux control coefficient of an enzyme/isozyme E_i in the pathway shown in Fig. 1 can be derived with the aid of the formula[6]

$$C_i = \frac{1}{\Delta} \sum_k G_{ki} \Delta_{ki} \tag{1a}$$

where

$$\Delta = 1 - L_1 + L_2 - L_3 + \cdots . \tag{1b}$$

The quantity L_1 represents the sum of the weights of all first order cycles in Fig. 2, L_2 is the sum of the weights of all second order cycles, and so on; G_{ki} is the weight of the k-th directed path from the source node to node i, and Δ_{ki} is the sum of those terms in Δ (including the first term 1) which correspond to those cycles in Fig. 2 that do not touch the k-th directed path from the source node to node i. The weight of a directed path (or cycle) is the product of the weights on all the edges constituting the directed path (or cycle). Other relevant graph-theoretic definitions may be found in Ref. 6.

It is easy to see that there are eight first order cycles in Fig. 2; there are no higher order cycles in this Figure. Inserting the weights of these cycles into eq. (1b), we find

$$\Delta = 1 + \frac{\bar{\varepsilon}_1^1}{\varepsilon_1^2} + \frac{\bar{\varepsilon}_1^1}{\varepsilon_1^2}\frac{v_3}{v_2} + \frac{\bar{\varepsilon}_1^1}{\varepsilon_1^2}\frac{v_4}{v_2} + \frac{\bar{\varepsilon}_1^1\bar{\varepsilon}_2^2}{\varepsilon_1^2\varepsilon_2^5} + \frac{\bar{\varepsilon}_1^1\bar{\varepsilon}_2^3}{\varepsilon_1^2\varepsilon_2^5}\frac{v_3}{v_2} + \frac{\bar{\varepsilon}_1^1\bar{\varepsilon}_2^4}{\varepsilon_1^2\varepsilon_2^5}\frac{v_4}{v_2} + \frac{\varepsilon_1^3}{\varepsilon_1^2}\frac{v_3}{v_2} + \frac{\varepsilon_1^4}{\varepsilon_1^2}\frac{v_4}{v_2}.$$

(2)

Let us evaluate the flux control coefficient (C_1) of the enzyme E_1. To this end observe that in Fig. 2 there is only one directed path from the source node to node 1, with weight unity. In eq. (1a) we accordingly have $i=1$, $k=1$ and $G_{11}=1$. Observe also that there are two first order cycles that do not touch this directed path. Using the weights of these nontouching cycles from the expression (2) of Δ, including the first term 1, and the fact that $G_{11}=1$, we deduce from eq. (1a) that

$$C_1 = \frac{1}{\Delta}\left[1 + \frac{\varepsilon_1^3}{\varepsilon_1^2}\frac{v_3}{v_2} + \frac{\varepsilon_1^4}{\varepsilon_1^2}\frac{v_4}{v_2}\right].$$

(3)

The remaining flux control coefficients can be found from Fig. 2 in a similar manner by examining the directed paths from the source node to the respective nodes.

It should be emphasized that the numbering of the isozymes E_2, E_3 and E_4 in pathway (A) is completely arbitrary. In other words, in a given metabolic pathway the isozyme we wish to select as the *primary* isozyme can always be labeled as E_2, and a directed graph for the control structure may be constructed following the procedure described above.

2.3. The first reaction step is isoenzymic

In many metabolic pathways the very first reaction is catalyzed by isozymes. A case in point is the metabolism of ethanol in horse liver where three isozymes of alcohol dehydrogenase catalyze the first reaction step[1]. We examine below a pathway in which the first step is catalyzed by three isozymes and is followed by an enzymic reaction.

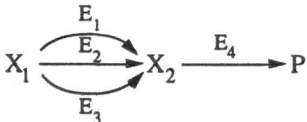

Figure 3. Reaction scheme of a pathway with three isoenzymes catalyzing the first reaction.

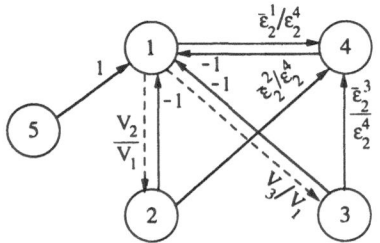

Figure 4. A directed graph for the control structure of the pathway shown in Fig. 3.

For control analysis of this pathway, a directed graph can be constructed by choosing any of the isozymes, say E_1, as the primary isozyme. The remaining isozymes, E_2 and E_3, are referred to as secondary isozymes. The primary isozyme is also regarded as the leader enzyme. The directed graph is portrayed in Fig. 4 from which the flux control coefficients can be easily determined.

3. Application of Electrical Analogs

We now describe how to construct an electrical analog circuit for the control structure of an isoenzymic pathway and evaluate the flux control coefficients of the various enzymes/isozymes. We reconsider the pathway shown in Fig. 1 for this purpose.

A voltage analog circuit for this pathway is presented in Fig. 5. To construct this circuit, label any of the isozymes, say, E_2 as the primary isozyme and refer to the remaining isozymes as secondary isozymes, and carry out the following steps.

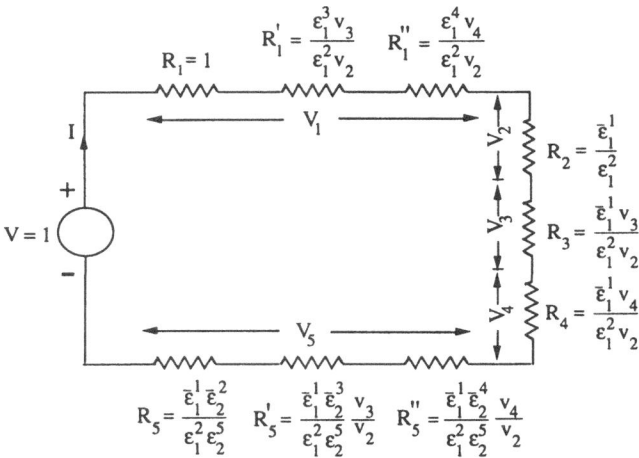

Figure 5. A voltage analog circuit for the control structure of the pathway shown in Fig. 1.

Step I. Consider the enzymes E_1, E_5 and the primary isozyme E_2 to form a linear pathway. Draw a voltage analog circuit for this linear pathway. It consists of the resistances R_1, R_2 and R_5 (see Fig. 5) whose magnitudes are assigned in the same manner as described previously[8].

Step II. For each secondary isozyme E_j ($j=3$, 4), introduce a resistance R_j in series in the voltage analog circuit. In other words, corresponding to the secondary isozymes E_3 and E_4, connect two resistances R_3 and R_4 as shown in Fig. 5. The magnitudes of these resistances are chosen respectively as $(v_3/v_2)R_2$ and $(v_4/v_2)R_2$. Recall that R_2 is the resistance associated with the primary isozyme E_2.

Step III. Note that the three isozymes have a common substrate X_1. Their interactions are incorporated into the voltage analog circuit of Fig. 5 as follows. For each secondary isozyme E_j ($j=3$, 4), an additional resistance is connected in series. The value of this resistance is the ratio of the magnitudes of the elasticity coefficients of E_j and the primary isozyme E_2 with respect to their common substrate X_1, multiplied by the ratio of the reaction velocities of the secondary isozyme E_j and the primary isozyme E_2. For the present case, the resistances R'_1 and R''_1 are connected as shown in Fig. 5, corresponding to the secondary isozymes E_3 and E_4, respectively.

Step IV. The three isozymes have a common product X_2. Their interactions are characterized into the voltage analog circuit of Fig. 5 in the following manner. For each secondary isozyme E_j ($j=3$, 4), connect a resistance in series. To select a value for this resistance, form the ratio ($\bar{\varepsilon}_2^j / \bar{\varepsilon}_2^2$) of the magnitudes of the elasticity coefficients of E_j and E_2 (the primary isozyme) with respect to their common product X_2, and multiply it by $(v_j/v_2)R_5$, v_2 and v_j being the reaction velocities of the primary isozyme E_2 and the secondary isozyme E_j, respectively. Corresponding to the secondary isozymes E_3 and E_4, we thus have the resistances R'_5 and R''_5 in the circuit of Fig. 5.

The flux control coefficient of the enzyme E_1 is given by the total voltage drop across the resistances R_1, R'_1 and R''_1 in this circuit. Similarly the flux control coefficient of the enzyme E_5 is equal to the total voltage drop across the resistances R_5, R'_5 and R''_5. The voltage drop across the resistances R_2, R_3 and R_4 designate the flux control coefficients of the isozymes E_2, E_3 and E_4, respectively. Using Ohm's law, it can be easily established that the flux control coefficients given by these voltage drops have the same expressions as those derived from the directed graph of Fig. 2. A current analog circuit for this pathway can be constructed by connecting in parallel nine resistances and a current source of strength unity. The magnitudes of these resistances are reciprocals of the corresponding resistances in the voltage analog circuit of Fig. 5.

A voltage or current analog circuit for the isoenzymic pathway shown in Fig. 3 in which the very first step is catalyzed by three isozymes can be designed in an analogous fashion. For the sake of brevity the details are not given here.

References

1. R. F. Derr and R. E. Derr, Control of flux by isozymes and alternate enzymes: application to horse liver alcohol dehydrogenase, *Biochem. Arch.* **3**:453 (1987).
2. H. E. Umbarger, Amino acid biosynthesis and its regulation, *Ann. Rev. Biochem.* **47**:533 (1978).
3. D. A. Fell and H. M. Sauro, Metabolic control and its analysis: additional relationships between elasticities and control coefficients, *Eur. J. Biochem.* **148**:555 (1985).
4. H. Kacser and J. A. Burns, The control of flux, *Symp. Soc. Exp. Biol.* **27**:65 (1973).
5. R. Heinrich and T. Rapoport, A linear steady state treatment of enzymatic chains, *Eur. J. Biochem.* **42**:89 (1974).
6. A. K. Sen, Metabolic control analyis: an application of signal flow graphs, *Biochem. J.* **269**:141 (1990).
7. A. K. Sen, A graph-theoretic analysis of metabolic regulation in linear pathways with multiple feedback loops and branched pathways, *Biochim. Biophys. Acta* **1059**:293 (1991).
8. A. K. Sen, Application of electrical analogs for control analysis of simple metabolic pathways, *Biochem. J.* **272**:65 (1990).
9. A. K. Sen, Application of electrical analogs to metabolic control analysis: linear pathways with multiple feedback, *Int. J. Biochem.* **22**:1295 (1990).

DERIVATION OF
A GENERAL MATRIX METHOD
FOR THE CALCULATION
OF CONTROL COEFFICIENTS

Magnus Ehlde and Guido Zacchi

Department of Chemical Engineering I
University of Lund
P.O. Box 124, S-221 00 Lund, Sweden

1. Introduction

As our knowledge about metabolism and metabolic pathways increases, it is becoming more important to be able to describe these processes quantitatively in mathematical terms. This is desirable if we are to be able to get a better understanding of the behavior and to make predictions of the metabolic systems or if a system is to be optimized. Since a metabolic network is a very complex system, a mathematical description must be simplified and limited to special cases; for instance the steady state of the system. Two such simplified and related tools have been developed: metabolic control analysis (MCA), developed by Kacser and Burns[1] and by Heinrich and Rapoport[2] in the early 1970s (for a review see Kell and Westerhoff[3]) and biochemical systems analysis (BSA), developed by Savageau[4-6] in the late 1960s. MCA and BSA are based on the same mathematics, but whereas MCA is a pure steady-state model, BSA deals, to some extent, with dynamic behavior.

A central part in MCA is the definition of the control coefficients and the elasticity coefficients. The control coefficients give quantitative values of the extent of control exerted by each enzyme in the pathway on the steady-state fluxes and the metabolite concentrations of the pathway. The control coefficients are systemic properties, i.e. their values are determined by the state of the whole pathway. The elasticity coefficients give values of the extent of the control exerted by the effectors of each enzyme, i.e. the substrates, products, inhibitors, allosteric effectors etc., on the rate of the enzyme. The elasticity coefficients are local properties, i.e. their values are determined only by the state of the specific enzyme and by its effectors. In MCA certain summation and connectivity theorems are formulated[1-3], which relate the control and elasticity coefficients. It is possible to study the parts of the system, i.e. the elasticity coefficients, and, using these theorems, gain information about the whole system, i.e. the control coefficients.

Although BSA and MCA are based on the same mathematics, in BSA the summation and connectivity theorems are not shown explicitly. It is possible, however, to use derivations, similar to those made in BSA, and arrive at relations which relate the elasticity coefficients to the control coefficients in the same way as the summation and connectivity theorems of MCA.

In this paper a general method for the calculation of the control coefficients from elasticity coefficients of pathways containing any number of branches and moiety-conserved cycles is presented. The derivation is based on general sensitivity analysis, which was used in the development of BSA, and is analogous to the derivations presented by Cascante and coworkers[7,8]. It is however generalized by the use of the stoichiometric matrix, following the guidelines given by Reder[9]. Since the method is general and based on matrix algebra, it is well suited for computer calculations. It has been incorporated into the computer program METSIM[10], running under MS Windows, which dynamically simulates and calculates the control coefficients of a user-defined metabolic pathway of any complexity.

2. Basic Relationships

Consider an arbitrary metabolic pathway, consisting of n enzymes, $E_1,...,E_n$, and m metabolites, $X_1,...,X_m$. If it is assumed that the pathway has r parameters, $P_1,...,P_r$, the state of the pathway is described by values of the metabolite concentrations, $x_1,...,x_m$, and values of the parameters, $p_1,...,p_r$. Enzyme activities, substrate and product concentrations, pH, temperature etc. are considered as parameters. At steady state the rate v_k of a certain enzyme k is equal to the steady-state flux J_k through that enzyme. This rate can be expressed as a function of the r parameter values and of the m metabolite concentrations:

$$J_k = v_{k,ss} = f(p_1, p_2,..., p_r, x_1, x_2,...,x_m).$$

This can be differentiated with respect to an enzyme activity e_j :

$$\frac{dJ_k}{de_j} = \frac{\partial v_k}{\partial p_1} \cdot \frac{dp_1}{de_j} +...+ \frac{\partial v_k}{\partial p_r} \cdot \frac{dp_r}{de_j} + \frac{\partial v_k}{\partial x_1} \cdot \frac{dx_1}{de_j} +...+ \frac{\partial v_k}{\partial x_m} \cdot \frac{dx_m}{de_j} . \tag{1}$$

If the parameters are not in any way coupled, i.e. a change in one parameter does not affect the others, it is noted that:

$$\frac{dp_i}{de_j} = \left\{ \begin{array}{l} 0 \; ; \; p_i \neq e_j \\ 1 \; ; \; p_i = e_j \end{array} \right. .$$

If the zero terms of eq. (1) are eliminated and the equation is multiplied by $e_j/J_k = e_j/v_k$ it can be rearranged to give:

$$\frac{dJ_k}{de_j} \cdot \frac{e_j}{J_k} = \frac{\partial v_k}{\partial e_j} \cdot \frac{e_j}{v_k} \cdot \frac{de_j}{de_j} + \sum_{i=1}^{m} \frac{\partial v_k}{\partial x_i} \cdot \frac{x_i}{v_k} \cdot \frac{dx_i}{de_j} \cdot \frac{e_j}{x_i} . \tag{2a}$$

In eq. (2a) the elasticity coefficients, the parameter elasticity coefficients, the flux control coefficients, and the concentration control coefficients can be recognized as:

$$\varepsilon_i^k = \frac{\partial v_k \cdot x_i}{\partial x_i \cdot v_k} , \qquad \pi_j^k = \frac{\partial v_k \cdot e_j}{\partial e_j \cdot v_k} , \qquad C^{J_k}_j = \frac{dJ_k \cdot e_j}{de_j \cdot J_k} , \qquad C^{X_i}_j = \frac{dx_i \cdot e_j}{de_j \cdot x_i} .$$

Eq. (2a) can then be written as:

$$C_j^{J_k} = \pi_j^k + \sum_{i=1}^{m} \varepsilon_i^k \cdot C_j^{X_i} . \tag{2b}$$

In vector form this is equal to:

$$C_j^{J_k} = \pi_j^k + \left(\varepsilon_1^k \quad \varepsilon_2^k \quad \cdots \quad \varepsilon_m^k \right) \cdot \begin{pmatrix} C_j^{X_1} \\ C_j^{X_2} \\ \cdots \\ C_j^{X_m} \end{pmatrix} . \tag{2c}$$

Extension to all enzymes in the pathway yields:

$$\left(C_1^{J_k} \cdots C_n^{J_k} \right) = \left(\pi_1^k \cdots \pi_n^k \right) + \left(\varepsilon_1^k \cdots \varepsilon_m^k \right) \cdot \begin{pmatrix} C_1^{X_1} & \cdots & C_n^{X_1} \\ \cdots & \cdot & \cdots \\ C_1^{X_m} & \cdots & C_n^{X_m} \end{pmatrix} .$$

Further extension to all steady-state enzyme fluxes yields:

$$\begin{pmatrix} C_1^{J_1} & \cdots & C_n^{J_1} \\ \cdots & & \cdots \\ C_1^{J_n} & \cdots & C_n^{J_n} \end{pmatrix} = \begin{pmatrix} \pi_1^1 & \cdots & \pi_n^1 \\ \cdots & & \cdots \\ \pi_1^n & \cdots & \pi_n^n \end{pmatrix} + \begin{pmatrix} \varepsilon_1^1 & \cdots & \varepsilon_m^1 \\ \cdots & & \cdots \\ \varepsilon_1^n & \cdots & \varepsilon_m^n \end{pmatrix} \cdot \begin{pmatrix} C_1^{X_1} & \cdots & C_n^{X_1} \\ \cdots & & \cdots \\ C_1^{X_m} & \cdots & C_n^{X_m} \end{pmatrix} . \tag{3a}$$

If the matrices in eq. (3a) are named it can be written as:

$$\mathbf{C^J = P + E \cdot C^X} . \tag{3b}$$

Eq. (3b) is a system of equations with $n \cdot n + m \cdot n$ unknowns (the control coefficients) and only $n \cdot n$ equations. More relations must thus be derived in order to solve it.

3. Choosing Independent Concentrations and Fluxes

The relations of the m metabolite concentrations and the n steady-state fluxes are defined in the stoichiometry matrix $\mathbf{N}$, which has the dimensions ($m \times n$), Each row corresponds to a metabolite (concentration) and each column corresponds to an enzyme (steady-state flux). The elements, n_{ij}, of $\mathbf{N}$ are defined as follows:

$n_{ij} = +z$,   if the reaction j produces z molecules of metabolite i,
$n_{ij} = -z$,   if the reaction j consumes z molecules of metabolite i,
$n_{ij} = 0$,   if the reaction j neither produces nor consumes metabolite i.

There may be constraints between some metabolite concentrations, due to moiety-conserved cycles in the pathway. In other words; all concentrations may not be independent and the dependent concentrations can be expressed in terms of the independent ones. Likewise, all steady-state fluxes may not be independent and the dependent fluxes can be expressed in terms of the independent ones. The pathway will thus contain a number of m_0 independent metabolite concentrations, less than or equal to m, and a number of n_0 independent steady-state enzyme fluxes, less than or equal to n. All this information is inherent in the stoichiometry matrix N. If the information can be extracted from N and incorporated into eq. (3b), it will be possible to solve the equation system.

As shown by Reder[9] the number of independent metabolite concentrations, m_0, is given by the rank of the stoichiometry matrix N and can be calculated by Gauss elimination. The independent rows of N will correspond to the independent concentrations. There are several combinations of concentrations (rows in N) that can be chosen as independent and the choice is arbitrary, but to simplify the derivation the metabolites are renumbered so that the independent rows are the m_0 first rows of N.

The column rank of a matrix is always equal to its row rank and there are thus m_0 independent columns of N. The m_0 independent columns will correspond to the dependent fluxes and the $n-m_0=n_0$ dependent columns will correspond to the independent fluxes. The fluxes (columns of N) are renumbered so that the independent columns are the m_0 last columns of N. Four new matrices are then constructed from N. The first consists of the m_0 independent (first) rows of N and is denoted N_R. The second consists of the m_0 independent (last) columns of N and is denoted N_C. The third consists of the m_0 independent (first) rows and the m_0 independent (last) columns of N and is denoted N_{RC}. The fourth consists of the m_0 independent (first) rows and the n_0 dependent (first) columns of N and is denoted N_0. The matrices are illustrated below.

$$N = \begin{pmatrix} n_{11} & \cdots & n_{1n} \\ \cdots & \cdots & \cdots \\ n_{m1} & \cdots & n_{mn} \end{pmatrix}, \qquad N_R = \begin{pmatrix} n_{11} & \cdots & n_{1n} \\ \cdots & \cdots & \cdots \\ n_{m_01} & \cdots & n_{m_0n} \end{pmatrix},$$
$$(m \times n) \qquad\qquad\qquad (m_0 \times n)$$

$$N_C = \begin{pmatrix} n_{1(n_0+1)} \cdots n_{1n} \\ \cdots \cdots\cdots \\ n_{m(n_0+1)} \cdots n_{mn} \end{pmatrix}, \quad N_{RC} = \begin{pmatrix} n_{1(n_0+1)} \cdots n_{1n} \\ \cdots \cdots \cdots \\ n_{m_0(n_0+1)} \cdots n_{m_0n} \end{pmatrix}, \quad N_0 = \begin{pmatrix} n_{11} \cdots n_{1n_0} \\ \cdots \cdots \cdots \\ n_{m_01} \cdots n_{m_0n_0} \end{pmatrix}.$$
$$(m \times m_0) \qquad\qquad (m_0 \times m_0) \qquad\qquad (m_0 \times n_0)$$

Note that the matrix N_{RC} is square, since $n_0=n-m_0$, and always invertible, since all rows (and columns) are independent.

4. Relations for the Concentration Control Coefficients

According to Reder[9] the stoichiometry matrix N can be decomposed as:

$$N = L^X \cdot N_R. \tag{4}$$

L^X (which is of dimension $m \times m_0$) is here called the concentration link matrix. If the dependent columns of N and N_R are deleted, L^X can be calculated as:

$$L^X = N_C \cdot N_{RC}^{-1}. \tag{5}$$

Note that if the pathway does not contain any moiety-conserved cycles, N_R will be equal to N and L^X will be the (m x m) identity matrix.

It can be shown[9] that:

$$d(\mathbf{X} - \mathbf{L}^X \cdot \mathbf{X_R}) = \mathbf{0}, \quad \text{or}$$

$$d\mathbf{X} = \mathbf{L}^X \cdot d\mathbf{X_R}, \qquad (6)$$

where $\mathbf{X}$ is an m-vector containing all m metabolites and $\mathbf{X_R}$ is an m_0-vector containing the m_0 independent metabolites. This equation specifies the conservation relationships of the metabolite concentrations, i.e. how the dependent concentrations can be expressed in terms of the independent ones. Using this equation a new equation can be derived, which specifies how the concentration control coefficients for the dependent metabolites can be expressed in terms of the concentration control coefficients for the independent metabolites. The derivation of this is as follows.

Row k of eq. (6) can be written as:

$$dx_k = \sum_{i=1}^{m_0} l_{ki} \cdot dx_i \ ,$$

where l_{ki} is an element of $\mathbf{L}^X$. If the differentiation is performed with respect to the enzyme activity e_j and the equation is multiplied by e_j/x_k the result is:

$$\frac{dx_k}{de_j} \cdot \frac{e_j}{x_k} = \frac{e_j}{x_k} \cdot \frac{e_j}{x_k} \cdot \sum_{i=1}^{m_0} l_{ki} \cdot \frac{dx_i}{de_j} \ ,$$

Rearranging and recognition of terms yields:

$$C_j^{X_k} = \sum_{i=1}^{m_0} l_{ki} \cdot \frac{x_i}{x_k} \cdot C_j^{X_i}, \quad k = 1,2,...,\text{m}; \quad j = 1,2,...,\text{n} \ .$$

If a new matrix, denoted $\mathbf{L}_F^X$, is constructed from $\mathbf{L}^X$, by multiplying each element l_{ki} of $\mathbf{L}^X$ by x_i/x_k, the equation can be written as:

$$\mathbf{C}^X = \mathbf{L}_F^X \cdot \mathbf{C}_R^X \qquad (7)$$

$\mathbf{C}^X$ contains all concentration control coefficients of the pathway and $\mathbf{C}_R^X$ contains the concentration control coefficients for the independent metabolites.

5. Relations for the Flux Control Coefficients

The following relation is true for any pathway[9]:

$$\mathbf{N_R} \cdot \mathbf{J} = \mathbf{0} \ . \qquad (8)$$

$\mathbf{J}$ is an n-vector containing the steady-state flux through each enzyme. Eq. (8) can be rearranged in order to obtain the $n-n_0=m_0$ dependent fluxes $\mathbf{J_0}$ as a function of the n_0 independent fluxes $\mathbf{J_R}$. Using the matrices defined above, eq. (8) can be rearranged as:

$$\mathbf{N}_0 \cdot \mathbf{J}_R + \mathbf{N}_{RC} \cdot \mathbf{J}_0 = \mathbf{0} , \quad \text{or}$$

$$\mathbf{J}_0 = -\mathbf{N}_{RC}^{-1} \cdot \mathbf{N}_0 \cdot \mathbf{J}_R .$$

The product $-\mathbf{N}_{RC}^{-1} \cdot \mathbf{N}_0$ will have the dimensions (n-n_0 x n_0). A new matrix is constructed from this product and an (n_0 x n_0) identity matrix as shown below. The resulting matrix is denoted $\mathbf{L}^J$ and has the dimensions (n x n_0):

$$\mathbf{L}^J = \begin{pmatrix} \mathbf{I}_{n_0} \\ -\mathbf{N}_{RC}^{-1} \cdot \mathbf{N}_0 \end{pmatrix}. \tag{9}$$

All n fluxes, $\mathbf{J}$, can now be expressed as a function of the n_0 independent fluxes $\mathbf{J}_R$:

$$\mathbf{J} = \mathbf{L}^J \cdot \mathbf{J}_R . \tag{10}$$

This equation can be differentiated as $d\mathbf{J} = \mathbf{L}^J \cdot d\mathbf{J}_R$. Exactly the same derivation can then be made as when eq. (7) was derived from eq. (6). The result is:

$$\mathbf{C}^J = \mathbf{L}_F^J \cdot \mathbf{C}_R^J . \tag{11}$$

$\mathbf{C}^J$ contains all the flux control coefficients and $\mathbf{C}_R^J$ contains the flux control coefficients for the independent fluxes. $\mathbf{L}_F^J$ is the result when each element l_{ki} of $\mathbf{L}^J$ is multiplied by J_i/J_k.

6. Results

Eqs (7) and (11) can now be inserted into eq. (3b) and the result is:

$$\mathbf{L}_F^J \cdot \mathbf{C}_R^J - \mathbf{E} \cdot \mathbf{L}_F^X \cdot \mathbf{C}_R^X = \mathbf{P} . \tag{12}$$

The number of unknowns has now been reduced to $n_0 \cdot n + m_0 \cdot n = n \cdot n$ and it is possible to solve the equation system.

Eq. (12) can be rearranged as

$$\left(\mathbf{L}_F^J - \mathbf{E} \cdot \mathbf{L}_F^X \right) \cdot \begin{pmatrix} \mathbf{C}_R^J \\ \mathbf{C}_R^X \end{pmatrix} = \mathbf{P}$$

and

$$\begin{pmatrix} \mathbf{C}_R^J \\ \mathbf{C}_R^X \end{pmatrix} = \left(\mathbf{L}_F^J - \mathbf{E} \cdot \mathbf{L}_F^X \right)^{-1} \cdot \mathbf{P} . \tag{13}$$

The control coefficients for the independent fluxes and concentrations can thus be analytically calculated using eq. (13). If required, all control coefficients can then be calculated using eq. (7) and eq. (11).

The general procedure for the calculation of the control coefficients can thus be summarized as follows:

- Construct the stoichiometric matrix $\mathbf{N}$.
- Calculate the rank of the matrix $\mathbf{N}$.
- Choose m_0 independent rows and m_0 independent columns of $\mathbf{N}$ and construct the matrices $\mathbf{N_C}$, $\mathbf{N_{RC}}$ and $\mathbf{N_0}$.
- Construct matrix $\mathbf{L_F^X}$ from eq. (5) and the metabolite concentrations.
- Construct matrix $\mathbf{L_F^J}$ from eq. (9) and the steady-state fluxes.
- Calculate the control coefficients for the independent fluxes and metabolites using eq. (13).
- If required, calculate all control coefficients using eqs (7) and (11).

The procedure is illustrated by the following simple example where numerical values from a computer simulation have been used.

7. Example

Consider the metabolic pathway shown in Fig. 1. The pathway contains the metabolites X_1 to X_5 ($m = 5$) and the enzymes E_1 to E_5 ($n = 5$). The stoichiometry matrix $\mathbf{N}$ will be:

$$\mathbf{N} = \begin{pmatrix} 1 & -1 & 0 & -1 & 0 \\ 0 & 1 & -1 & 0 & 0 \\ 0 & 0 & 0 & 1 & -1 \\ 0 & 0 & 0 & -1 & 1 \\ 0 & 0 & 0 & 1 & -1 \end{pmatrix}.$$

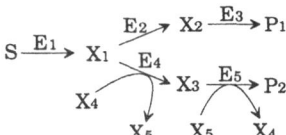

Figure 1. Reaction scheme of an exemplifying metabolic pathway.

The rank of the matrix $\mathbf{N}$ is 3, which means that the pathway contains 3 independent metabolites and 2 independent fluxes, i.e. $m_0 = 3$ and $n_0 = n - m_0 = 2$. If the three first rows and the three last columns are chosen as independent, the metabolites and enzymes do not have to be renumbered and the following matrices can be constructed:

$$\mathbf{N_C} = \begin{pmatrix} 0 & -1 & 0 \\ -1 & 0 & 0 \\ 0 & 1 & -1 \\ 0 & -1 & 1 \\ 0 & 1 & -1 \end{pmatrix},$$

$$\mathbf{N_{RC}} = \begin{pmatrix} 0 & -1 & 0 \\ -1 & 0 & 0 \\ 0 & 1 & -1 \end{pmatrix},$$

$$\mathbf{N_0} = \begin{pmatrix} 1 & -1 \\ 0 & 1 \\ 0 & 0 \end{pmatrix}.$$

Using eq. (5) the matrix $\mathbf{L^X}$ can be constructed as:

$$\mathbf{L^X} = \mathbf{N_C} \cdot \mathbf{N_{RC}}^{-1} = \begin{pmatrix} 0 & -1 & 0 \\ -1 & 0 & 0 \\ 0 & 1 & -1 \\ 0 & -1 & 1 \\ 0 & 1 & -1 \end{pmatrix} \cdot \begin{pmatrix} 0 & -1 & 0 \\ -1 & 0 & 0 \\ -1 & 0 & -1 \end{pmatrix} = \begin{pmatrix} 1 & 0 & 0 \\ 0 & 1 & 0 \\ 0 & 0 & 1 \\ 0 & 0 & -1 \\ 0 & 0 & 1 \end{pmatrix}.$$

The matrix $\mathbf{L^J}$ can be constructed from eq. (9) as:

$$\mathbf{L^J} = \begin{pmatrix} \mathbf{I}_{n_0} \\ -\mathbf{N_{RC}^{-1}} \cdot \mathbf{N_0} \end{pmatrix} = \begin{pmatrix} 1 & 0 \\ 0 & 1 \\ 0 & 1 \\ 1 & -1 \\ 1 & -1 \end{pmatrix}.$$

A simulation of the pathway with the rate equations shown in the Appendix yields the following metabolite concentrations, fluxes, and elasticity coefficients at steady state:

$$x_1 = 0.0769, \quad x_2 = 0.0532, \quad x_3 = 0.1643, \quad x_4 = 0.1357, \quad x_5 = 0.0643,$$

$$J_1 = 0.5408, \quad J_2 = J_3 = 0.4555, \quad J_4 = J_5 = 0.0853,$$

$$\mathbf{E} = \begin{pmatrix} -0.658 & 0 & 0 & 0 & 0 \\ 1.460 & -0.553 & 0 & 0 & 0 \\ 0 & 1.053 & 0 & 0 & 0 \\ 0.929 & 0 & 0 & 0.880 & 0 \\ 0 & 0 & 0.859 & 0 & 0.940 \end{pmatrix}.$$

The matrices $\mathbf{L_F^X}$ and $\mathbf{L_F^J}$ can now be constructed by multiplying each element l_{ij} of $\mathbf{L^X}$ by x_j/x_i and each element l_{ij} of $\mathbf{L^J}$ by J_j/J_i:

$$\mathbf{L_F^X} = \begin{pmatrix} 1 & 0 & 0 \\ 0 & 1 & 0 \\ 0 & 0 & 1 \\ 0 & 0 & -1.211 \\ 0 & 0 & 2.555 \end{pmatrix},$$

$$\mathbf{L_F^J} = \begin{pmatrix} 1 & 0 \\ 0 & 1 \\ 0 & 1 \\ 6.340 & -5.340 \\ 6.340 & -5.340 \end{pmatrix}.$$

For the metabolic pathway in this example the rate equations are homogeneous with respect to the enzyme activities so the matrix $\mathbf{P}$ will be an (n x n) identity matrix.

The following matrix multiplication can now be performed:

$$-\mathbf{E}\cdot\mathbf{L_F^X} = \begin{pmatrix} 0.658 & 0 & 0 \\ -1.460 & 0.553 & 0 \\ 0 & -1.053 & 0 \\ -0.929 & 0 & 1.066 \\ 0 & 0 & -3.260 \end{pmatrix},$$

and the control coefficients are obtained from eq. (13) as:

$$\begin{pmatrix} \mathbf{C_R^J} \\ \mathbf{C_R^X} \end{pmatrix} = \left(\mathbf{L_F^J} \quad -\mathbf{E}\cdot\mathbf{L_F^X} \right)^{-1} \cdot \mathbf{I}_n$$

$$= \begin{pmatrix} 1 & 0 & 0.658 & 0 & 0 \\ 0 & 1 & -1.460 & 0.553 & 0 \\ 0 & 1 & 0 & -1.053 & 0 \\ 6.340 & -5.340 & -0.929 & 0 & 1.066 \\ 6.340 & -5.340 & 0 & 0 & -3.260 \end{pmatrix}^{-1}.$$

This results in:

$$\begin{pmatrix} \mathbf{C_R^J} \\ \mathbf{C_R^X} \end{pmatrix} = \begin{pmatrix} 0.582 & 0.231 & 0.121 & 0.050 & 0.016 \\ 0.608 & 0.320 & 0.168 & -0.072 & -0.024 \\ 0.635 & -0.351 & -0.184 & -0.075 & -0.025 \\ 0.577 & 0.304 & -0.790 & -0.069 & -0.022 \\ 0.136 & -0.075 & -0.040 & 0.215 & -0.236 \end{pmatrix}.$$

If eqs (7) and (11) are used to calculate all the control coefficients the result is:

$$\mathbf{C^J} = \begin{pmatrix} 0.582 & 0.231 & 0.121 & 0.050 & 0.016 \\ 0.608 & 0.320 & 0.168 & -0.072 & -0.024 \\ 0.608 & 0.320 & 0.168 & -0.072 & -0.024 \\ 0.444 & -0.245 & -0.129 & 0.701 & 0.229 \\ 0.444 & -0.245 & -0.129 & 0.701 & 0.229 \end{pmatrix},$$

$$\mathbf{C^X} = \begin{pmatrix} 0.635 & -0.351 & -0.184 & -0.075 & -0.025 \\ 0.577 & 0.304 & -0.790 & -0.069 & -0.022 \\ 0.136 & -0.075 & -0.040 & 0.215 & -0.236 \\ -0.165 & 0.091 & 0.048 & -0.260 & 0.286 \\ 0.348 & -0.192 & -0.101 & 0.549 & -0.604 \end{pmatrix}.$$

References

1. H. Kacser, J. A. Burns, The control of flux, *Symp. Soc. Exp. Biol.* **27**:65-104 (1973).
2. R. Heinrich, T. A. Rapoport, A linear steady-state treatment of enzymatic chains: general properties, control and effector strength, *Eur. J. Biochem.* **42**:89-95 (1974).
3. D. B. Kell, H. V. Westerhoff, Metabolic control theory: its role in microbiology and biotechnology, *Microbiol. Rev.* **39**:305-320 (1986).
4. M. A. Savageau, Biochemical systems analysis. I. Some mathematical properties of the rate law for the component enzymatic reactions, *J. theor. Biol.* **25**:365-369 (1969).

5. M. A. Savageau, Biochemical systems analysis. II. The steady-state solutions for an *n*-pool system using a power law approximation, *J. theor. Biol.* **25**:365-369 (1969).

6. M. A. Savageau, Biochemical systems analysis. III. Dynamic solutions using a power law approximation, *J. theor. Biol.* **25**:365-369 (1969).

7. M. Cascante, R. Franco, E. I. Canela, Use of implicit methods from general sensitivity theory to develop a systematic approach to metabolic control. 1. Unbranched pathways, *Math. Biosci.* **94**:271-288 (1989).

8. M. Cascante, R. Franco, E. I. Canela, Use of implicit methods from general sensitivity theory to develop a systematic approach to metabolic control. 2. Complex systems, *Math. Biosci.* **94**:289-309 (1989).

9. C. Reder, Metabolic control theory: A structural approach, *J. Theor. Biol.* **135**:175-201 (1988).

10. M. Ehlde, G. Zacchi, METSIM - A user-friendly computer program for dynamic simulations of user-defined metabolic pathways, to be published.

Appendix: Rate Equations for the Example Pathway

Enzyme E_1: $v_1 = v_{max} \cdot \dfrac{s - x_1/k_{eq}}{k_s + s + x_1/k_{eq}}$, $\quad v_{max} = 10$, $k_s = 1$, $k_{eq} = 2$, $s = 0.1$;

Enzyme E_2: $v_2 = v_{max} \cdot \dfrac{x_1 - x_2/k_{eq}}{k_s + x_1 + x_2/k_{eq}}$, $\quad v_{max} = 10$, $k_s = 1$, $k_{eq} = 2$;

Enzyme E_3: $v_3 = v_{max} \cdot \dfrac{x_2 - p_1/k_{eq}}{k_s + x_2 + p_1/k_{eq}}$, $\quad v_{max} = 10$, $k_s = 1$, $k_{eq} = 2$, $p_1 = 0.01$;

Enzyme E_4: $v_4 = v_{max} \cdot \dfrac{x_1 \cdot x_4}{k_a \cdot k_b + k_a \cdot x_1 + k_b \cdot x_4 + x_1 \cdot x_4}$, $\quad v_{max} = 10$, $k_a = 1$, $k_b = 1$;

Enzyme E_5: $v_5 = v_{max} \cdot \dfrac{x_3 \cdot x_5}{k_a \cdot k_b + k_a \cdot x_3 + k_b \cdot x_5 + x_3 \cdot x_5}$, $\quad v_{max} = 10$, $k_a = 1$, $k_b = 1$.

INTERNAL REGULATION.
THE C·E = I = E·C SQUARE-MATRIX METHOD
ILLUSTRATED FOR A SIMPLE CASE
OF A COMPLEX PATHWAY

Anton A. van der Gugten[1] and Hans V. Westerhoff[2,3]

[1]Division Tumor Biology
[2]Division Molecular Biology
  Netherlands Cancer Institute, H5
  Plesmanlaan 121, NL-1066 CX Amsterdam, The Netherlands
[3]E.C. Slater Institute, University of Amsterdam
  Plantage Muidergracht 12, NL-1018 TV Amsterdam, The Netherlands

1. Introduction

Metabolic control analysis (for review see Ref. 1) allows insight in the control properties of complicated enzymatic pathways. It also attempts to clarify the regulatory processes that keep these pathways at their desired operation point[2,3]. It is based on the concept that enzymes show elasticity for the various metabolites in a pathway, or, from another point of view, that metabolites show regulability for the various enzymes, which is expressed in terms of elasticity coefficients. Elasticity is inversely related with the enzymatic control of maintaining the steady state as well as of keeping the concentration of the metabolites constant. The result of this internal control is that any intrusion into either steady state or metabolite concentration is nullified, so, that the uniqueness of steady states is conserved.

To have insight in the control of a *metabolic pathway*, however, is something else than understanding the regulation of *cellular physiology*. Even by knowing all characteristics of the enzymes involved, the sheer number of elements would greatly hamper both overview and computerized compilation of the details. Yet, an important aspect of regulation is its repetitiveness. This suggests that several metabolic processes together may constitute a functional unit ("module") with a global elasticity (the explicit resultant of the elasticity coefficients of the enzymes involved), and that the entire system can be understood in terms of control interactions between several such modules[2,4,5]. We intend to demonstrate control and its global aspects in simple terms. Here the emphasis will lie on the square-matrix method that allows concentration- and flux-control coefficients to be directly calculated from characteristics such as involvement of enzymes in a pathway (yes=1, no=0) and their elasticity coefficients regarding the metabolites.

Modern Trends in Biothermokinetics, Edited by
S. Schuster *et al.*, Plenum Press, New York, 1993

2. Simple Pathways

Fig. 1 shows two examples of simple pathways. Fig.1a describes Pathway I, i.e. the conversion by enzyme e_1 of a substrate S_1 into metabolite X and the subsequent turnover of X by enzyme e_2 into the product P_1. Similarly, Fig.1b describes Pathway II. Flux J represents the flux at steady state through pathway I, whereas flux j represents the flux through pathway II. Figs 1a,b each show two important square matrices. The explicitized elements of matrix **E** represent characteristics of the appropriate enzymes regarding involvement in the pathway (first column) and elasticity towards the metabolites in the pathway (elasticity coefficients, second column). The control matrix $\mathbf{C} = \mathbf{E}^{-1}$ is displayed on the right. The elements of **C** represent the flux-control coefficients (first, i.e. upper row) and the concentration-control coefficients (with respect to control on metabolite concentration; lower row) of the enzymes

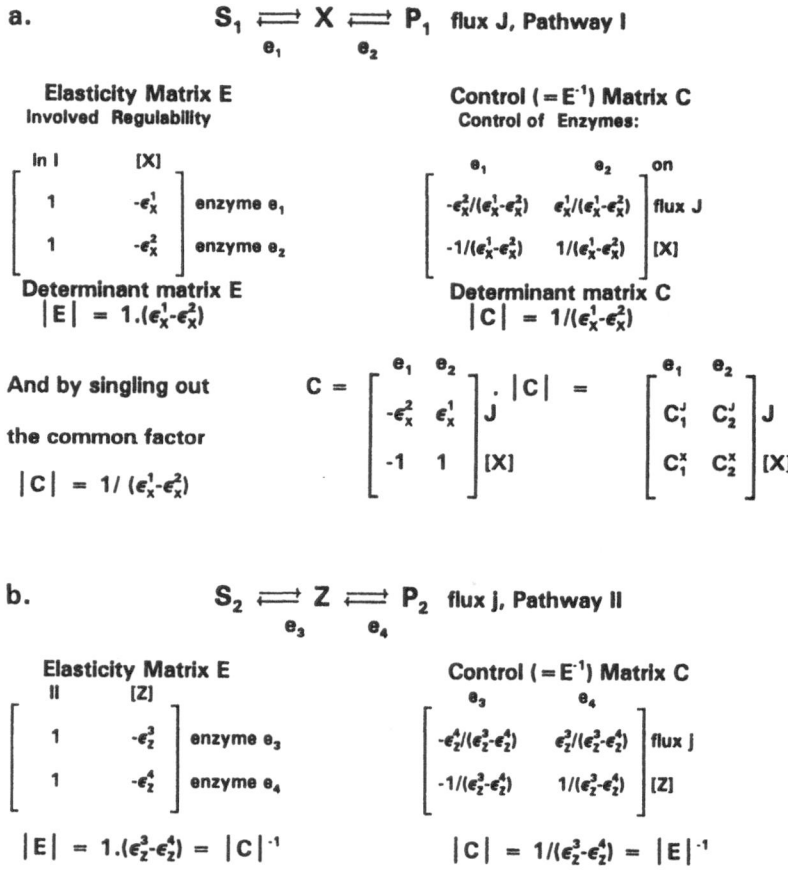

Figure 1. Two simple pathways, I in (a) and II in (b). The substrate in either pathway is, via one metabolite, straightforwardly turned over into its endproduct (P_1 and P_2, respectively), present at a fixed concentration. Elasticity matrices (**E**), arranging enzyme involvement in the pathway (yes=1, no=0) and elasticity coefficients regarding the respective metabolite (X in pathway I; Z in metabolite II), are on the left. Control matrices (**C**), arranging flux-control coefficients (elements on first row) and concentration-control coefficients (second row) of the enzymes involved, are shown on the right. Matrix **C** is the inverse of matrix **E**, provided **E** is non-singular; the discriminating factor is the determinant of **E** (=|E|), which must differ from zero. Fig. 1a further shows that the factor |C| common to all elements of matrix **C** may be singled out.

involved in the pathway. From matrix C a factor common to all its elements (i.e. $|C|$, the determinant of matrix C) may be singled out, as shown, and placed outside the matrix as a general multiplication factor.

2.1. The elasticity matrix E, the control (= E⁻¹) matrix C and their determinants

The control matrix C is the inverse of elasticity matrix E, provided that E can be inverted[6]. The deciding factor is the determinant of matrix E, the value of which must not equal zero. For, if it does, then the determinant $|C|$ which is equal to $|E|^{-1}$ would take the insolvable value of $1/0$. With respect to pathway I, $|E|$ takes the value of $\varepsilon_X^1 - \varepsilon_X^2$ which is unequal to zero provided $\varepsilon_X^1 \neq \varepsilon_X^2$. Likewise and with respect to pathway II, $|E|$ is unequal zero provided $\varepsilon_Z^3 \neq \varepsilon_Z^4$. We conjecture that these conditions are only met if the steady states considered, are stable, i.e., that only then matrix E is invertible (non-singular) and matrix C exists,

$$C = E^{-1} \text{ and } E = C^{-1} . \tag{1}$$

2.2. The identity matrix I demonstrating internal regulation; Summation and connectivity theorems

As the following applies to pathway I and II in exactly the same manner, we use pathway I as an example for making the points clear. The product of any matrix A and its inverse A^{-1} equals the identity matrix I and thus[7]:

$$E^{-1} \cdot E = C \cdot E = I = E \cdot C = E \cdot E^{-1}, \tag{2a}$$

or written in the form of matrices:

$$\begin{pmatrix} C_1^J & C_2^J \\ C_1^X & C_2^X \end{pmatrix} \cdot \begin{pmatrix} 1 & -\varepsilon_X^1 \\ 1 & -\varepsilon_X^2 \end{pmatrix} =$$

$$\begin{pmatrix} 1 = C_1^J + C_2^J = (-\varepsilon_X^2 + \varepsilon_X^1)|C| = |E| \cdot |C| & 0 = -(C_1^J \varepsilon_X^1 + C_2^J \varepsilon_X^2) = (\varepsilon_X^2 \varepsilon_X^1 - \varepsilon_X^1 \varepsilon_X^2) \cdot |C| \\ 0 = C_1^X + C_2^X = (-1+1) \cdot |C| & 1 = -(C_1^X \varepsilon_X^1 + C_2^X \varepsilon_X^2) = (\varepsilon_X^1 - \varepsilon_X^2) \cdot |C| = |E| \cdot |C| \end{pmatrix},$$

$$C \cdot E = I . \tag{2b}$$

We find that the four elements of I represent:

-*Upper left*: The flux-control summation theorem as applied to pathway I; *i.e.* the summation of the flux-control coefficients regarding flux J in pathway I equals 1.

-*Lower left*: The concentration-control summation theorem; *i.e.* the summation of the concentration-control coefficients with respect to metabolite X in pathway I equals 0.

-*Lower right*: The concentration-control connectivity theorem; *i.e.* the sum of the coefficients indicating the regulatory strength[3] of a given metabolite regarding the concentration of itself equals -1. With respect to pathway I this refers to the effect that metabolite X, *via* enzymes e_1 and e_2, has in stabilizing its own concentration (*i.e.* homeostatic strength[3]).

-*Upper right*: The flux-control connectivity theorem; *i.e.* the summation of the coefficients indicating the regulatory strength of a given metabolite on a flux equals 0. With respect to pathway I it refers to the effect of fluctuations of the metabolite X (via e_1 and e_2) on flux J.

The two elements of the right-hand column of $\mathbf{I} = \mathbf{C} \cdot \mathbf{E}$ then indicate that the pathway's response to a fluctuation cancels the fluctuation, ultimately leaving concentration(s) and flux unchanged. They also show the conditions for this regulation.

Matrix $\mathbf{I}$ may also present its elements in quite different terms (cf. Ref. 8).

$$\begin{pmatrix} 1 & -\varepsilon_X^1 \\ 1 & -\varepsilon_X^2 \end{pmatrix} \cdot \begin{pmatrix} C_1^J & C_2^J \\ C_1^X & C_2^X \end{pmatrix} =$$

$$\begin{pmatrix} 1 = C_1^J - \varepsilon_X^1 C_1^X = (-\varepsilon_X^2 + \varepsilon_X^1) \cdot |\mathbf{C}| = |\mathbf{E}| \cdot |\mathbf{C}| & 0 = C_2^J - \varepsilon_X^1 C_2^X = (\varepsilon_X^1 - \varepsilon_X^1) \cdot |\mathbf{C}| \\ 0 = C_1^J - \varepsilon_X^2 C_1^X = (-\varepsilon_X^2 + \varepsilon_X^2) \cdot |\mathbf{C}| & 1 = C_2^J - \varepsilon_X^2 C_2^X = (\varepsilon_X^1 - \varepsilon_X^2) \cdot |\mathbf{C}| = |\mathbf{E}| \cdot |\mathbf{C}| \end{pmatrix},$$

$$\mathbf{E} \cdot \mathbf{C} = \mathbf{I}. \tag{2c}$$

Interestingly, the interpretation of the basis of the control properties is different in this formulation; regulation may be looked at from a dual point of view. For instance the lower left element refers to the phenomenon that enzyme e_1 can control the flux through e_2 only via changes in the concentration of metabolite X. The approach of (2b) has the advantage of indicating the relation between the coefficients for flux-control and concentration-directed regulatory strength.

Note that since the inverse of any matrix is unique, there is no other solution to $\mathbf{C}$ than the one given, i.e. $\mathbf{C} = \mathbf{E}^{-1}$.

3. Complex Pathways

Fig.2a shows a "complex" system of pathways. Although looking very much the same as pathways I and II in Fig.1, there is the difference that in either pathway the metabolite is affecting enzymes in the other pathway.

3.1. The matrix E, its determinant |E| and the matrix C = E⁻¹

The elasticity matrix describing this situation is twice as large, *i.e.* a (4x4) instead of a (2x2) square matrix, since $\mathbf{E}$ contains all elements that play a role in regulating both pathways. It is subdivided into four submatrices indicated as $\mathbf{E}_{11}$ (containing the elements on the first set of rows and the first set of columns, upper left), $\mathbf{E}_{12}$ (first set of rows, second set of columns, upper right), $\mathbf{E}_{21}$ and $\mathbf{E}_{22}$.

Characteristics regarding pathway I *sensu stricto* are assembled in $\mathbf{E}_{11}$ and those of pathway II in $\mathbf{E}_{22}$. These submatrices look exactly the same as the corresponding elasticity matrices in Fig.1. Due to the cross-effects, there are some additional elasticity coefficients of enzymes e_1 and e_2 regarding Z and of e_3 and e_4 regarding X. Submatrices $\mathbf{E}_{12}$ and $\mathbf{E}_{21}$ contain zero elements to testify to the non-involvement of e_1 and e_2 in Pathway II and of e_3 and e_4 in Pathway I, respectively. They also contain the appropriate elasticity coefficients. Fig.2a further shows control matrix $\mathbf{C}$ in terms of flux- and concentration-control coefficients. The submatrices $\mathbf{C}_{11}$ and $\mathbf{C}_{22}$ in these terms describe the internal control in pathways I and II. Submatrix $\mathbf{C}_{12}$ describes the control that e_3 and e_4 (*via* Z) will exert on flux J and on the concentration of X in pathway I. Submatrix $\mathbf{C}_{21}$ describes the reverse, i.e.the control that, via X, enzymes e_1 and e_2 will exert on flux j and the concentration of Z in pathway II. Again $\mathbf{C}$ can only exist if $|\mathbf{E}| \neq 0$.

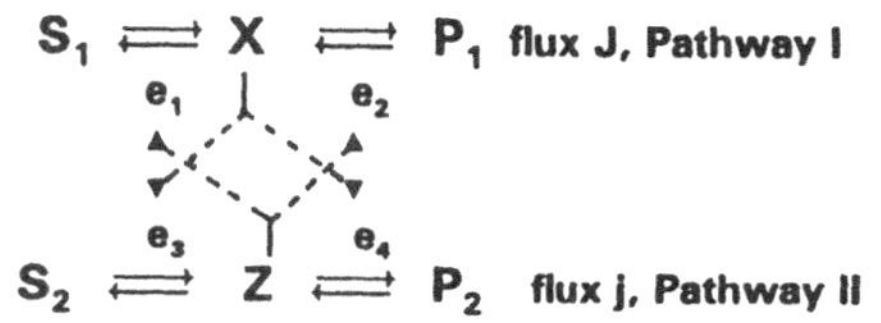

Elasticity Matrix E — Inv. I, Reg. [X], Inv. II, Reg. [Z]; rows I: e1, e2; II: e3, e4

Control (= E^{-1}) Matrix C — Control of enzymes e_1, e_2, e_3, e_4; rows I: flux J, [X]; II: flux j, [Z]

a.

$$E = \left[\begin{array}{cc|cc} 1 & -\epsilon^1_X & 0 & -\epsilon^1_Z \\ 1 & -\epsilon^2_X & 0 & -\epsilon^2_Z \\ \hline 0 & -\epsilon^3_X & 1 & -\epsilon^3_Z \\ 0 & -\epsilon^4_X & 1 & -\epsilon^4_Z \end{array}\right] \quad \begin{array}{cc} E_{11} & E_{12} \\ E_{21} & E_{22} \end{array}$$

$$C = \left[\begin{array}{cc|cc} C^J_1 & C^J_2 & C^J_3 & C^J_4 \\ C^X_1 & C^X_2 & C^X_3 & C^X_4 \\ \hline C^j_1 & C^j_2 & C^j_3 & C^j_4 \\ C^Z_1 & C^Z_2 & C^Z_3 & C^Z_4 \end{array}\right] \quad \begin{array}{cc} C_{11} & C_{12} \\ C_{21} & C_{22} \end{array}$$

Determinant $|E| = (\epsilon^1_X-\epsilon^2_X).(\epsilon^3_Z-\epsilon^4_Z) - (\epsilon^1_Z-\epsilon^2_Z).(\epsilon^3_X-\epsilon^4_X)$

Determinant $|C| = 1 / [(\epsilon^1_X-\epsilon^2_X).(\epsilon^3_Z-\epsilon^4_Z) - (\epsilon^1_Z-\epsilon^2_Z).(\epsilon^3_X-\epsilon^4_X)]$

b.

$$C = |C| . \left[\begin{array}{cc|cc} -(\epsilon^3_Z-\epsilon^4_Z)\epsilon^2_X+(\epsilon^3_X-\epsilon^4_X)\epsilon^2_Z & +(\epsilon^3_Z-\epsilon^4_Z)\epsilon^1_X-(\epsilon^3_X-\epsilon^4_X)\epsilon^1_Z & +(\epsilon^1_Z-\epsilon^2_Z)\epsilon^1_X-(\epsilon^1_X-\epsilon^2_X)\epsilon^1_Z & -(\epsilon^1_Z-\epsilon^2_Z)\epsilon^1_X+(\epsilon^1_X-\epsilon^2_X)\epsilon^1_Z \\ -(\epsilon^3_Z-\epsilon^4_Z) & \epsilon^3_Z-\epsilon^4_Z & \epsilon^1_Z-\epsilon^2_Z & -(\epsilon^1_Z-\epsilon^2_Z) \\ \hline -(\epsilon^3_Z-\epsilon^4_Z)\epsilon^3_X+(\epsilon^3_X-\epsilon^4_X)\epsilon^3_Z & +(\epsilon^3_Z-\epsilon^4_Z)\epsilon^2_X-(\epsilon^3_X-\epsilon^4_X)\epsilon^3_Z & -(\epsilon^1_X-\epsilon^2_X)\epsilon^4_Z+(\epsilon^1_Z-\epsilon^2_Z)\epsilon^4_X & +(\epsilon^1_X-\epsilon^2_X)\epsilon^3_Z-(\epsilon^1_Z-\epsilon^2_Z)\epsilon^3_X \\ \epsilon^3_X-\epsilon^4_X & -(\epsilon^3_X-\epsilon^4_X) & -(\epsilon^1_X-\epsilon^2_X) & \epsilon^1_X-\epsilon^2_X \end{array}\right] \quad \begin{array}{cc} C_{11} & C_{12} \\ C_{21} & C_{22} \end{array}$$

Figure 2. A "complex" system consisting of two *interacting* pathways I and II. In this case the metabolite in either pathway is affecting both enzymes in the other (dashed arrows!).
a) In addition to the characteristics regarding pathways I and II *sensu stricto* as pointed out in submatrices $\mathbf{E}_{11}$ and $\mathbf{E}_{22}$, matrix **E** also shows the cross-dependent character of the pathways through the cross elasticities $\varepsilon^1{}_Z$, $\varepsilon^2{}_Z$, $\varepsilon^3{}_X$, $\varepsilon^4{}_X$. It also reflects the non-involvement (=0) of e_1 and e_2 in pathway II (cf. submatrix $\mathbf{E}_{12}$) and of e_3 and e_4 in pathway I (cf. $\mathbf{E}_{21}$). Next to **E**, the control matrix **C** (=$\mathbf{E}^{-1}$) is shown in terms of flux-control and concentration-control coefficients. Control exerted by by e_3 and e_4 (via Z) on flux J and concentration X in pathway I, is shown in $\mathbf{C}_{12}$, and control exerted by by e_1 and e_2 (via X) on flux j and Z in pathway II, is shown in $\mathbf{C}_{21}$.
b) The same control matrix **C**, yet in terms of elasticity coefficients. Note that all values indicated should be multiplied by the determinant |C|.

The determinant |**E**| is complicated (Fig.2a). Two of its factors are recognizable, i.e $\varepsilon^1_X - \varepsilon^2_X$ = |$\mathbf{E}_{11}$| and $\varepsilon^3_Z - \varepsilon^4_Z$ = |$\mathbf{E}_{22}$|, identical to the determinants shown in Fig.1. These factors are unequal to zero if pathways I and II are in themselves controlled and the steady states considered are stable or metastable. Then the product of these factors is likewise unequal to zero. The remaining factors, i.e. $\varepsilon^1_Z - \varepsilon^2_Z$ and $\varepsilon^3_X - \varepsilon^4_X$ are the apparent determinants of $\mathbf{E}_{12}$ and $\mathbf{E}_{21}$, respectively (which will be indicated in italics, *i.e.* |$\boldsymbol{E}_{12}$| and |$\boldsymbol{E}_{21}$|). The determinant |**E**| therefore is unequal zero if $(\varepsilon^1_X - \varepsilon^2_X)(\varepsilon^3_Z - \varepsilon^4_Z) \neq (\varepsilon^1_Z - \varepsilon^2_Z)(\varepsilon^3_X - \varepsilon^4_X)$, *i.e.* |$\mathbf{E}_{11}$|·|$\mathbf{E}_{22}$|≠|$\boldsymbol{E}_{12}$|·|$\boldsymbol{E}_{21}$|. Given the assumption that pathways I and II reach steady state, the influence of X on e_3 and e_4 will affect the concentration of Z which in turn affects the enzymes e_1 and e_2; *i.e.* these mutual intrusions must balance to allow the steady states to remain stable. Then $|E|$ is unequal to zero, $|C| = 1/|E|$ is solvable and the control matrix $C = E^{-1}$ exists. Fig.2b further shows matrix C in terms of elasticity coefficients. At first sight matrix C looks confusingly complicated. However, by confronting C with the theorems of control analysis matters will be facilitated, as demonstrated below.

Pairwise and apart from sign the control coefficients of enzymes e_3 and e_4 (cf. C_{12} in Fig.2b) with respect to flux J as well as with respect to the concentration of metabolite X in pathway I are identical. The same is noted with respect to the control coefficients of enzymes e_1 and e_2 (cf. C_{21}) for flux j as well as for concentration of metabolite Z in pathway II.

$$C_3^J = -C_4^J = \left[\left(\varepsilon_Z^1 - \varepsilon_Z^2 \right) \cdot \varepsilon_X^1 - \left(\varepsilon_X^1 - \varepsilon_X^2 \right) \cdot \varepsilon_Z^1 \right] \cdot |C|$$

$$= |E_{12}| \cdot |C| \cdot \varepsilon_X^1 - |E_{11}| \cdot |C| \cdot \varepsilon_Z^1 , \qquad (3a)$$

$$C_3^X = -C_4^X = \left(\varepsilon_Z^1 - \varepsilon_Z^2 \right) \cdot |C| = |E_{12}| \cdot |C| , \qquad (3b)$$

$$C_1^j = -C_2^j = \left[-\left(\varepsilon_Z^3 - \varepsilon_Z^4 \right) \cdot \varepsilon_X^3 + \left(\varepsilon_X^3 - \varepsilon_X^4 \right) \cdot \varepsilon_Z^3 \right] \cdot |C|$$

$$= |E_{21}| \cdot |C| \cdot \varepsilon_Z^3 - |E_{22}| \cdot |C| \cdot \varepsilon_X^3 , \qquad (3c)$$

$$C_1^Z = -C_2^Z = \left(\varepsilon_X^3 - \varepsilon_X^4 \right) \cdot |C| = -|E_{21}| \cdot |C| . \qquad (3d)$$

This suggests that whatever the control a given enzyme in a certain pathway (*e.g.* e_3 in pathway II) has on the flux through another pathway (flux J in pathway I), this control is counterbalanced by the former pathway's other enzyme(s) (*i.e.* e_4 in pathway II)[8]. Moreover, by comparing expressions (3b,d) with (4a,b) it shows that control over the metabolite concentration in the other pathway is organized similarly to control over the metabolite concentration, but not the flux, in the pathway itself.

$$C_1^X = -C_2^X = -\left(\varepsilon_Z^3 - \varepsilon_Z^4 \right) \cdot |C| = -|E_{22}| \cdot |C| , \qquad (4a)$$

$$C_3^Z = -C_4^Z = -\left(\varepsilon_X^1 - \varepsilon_X^2 \right) \cdot |C| = -|E_{11}| \cdot |C| . \qquad (4b)$$

From another point of view, internal (within the proper pathway) regulation and external (out of the proper pathway) regulation of metabolites by enzymes are similar.

3.2. The regulation of control: The identity matrix $I = C \cdot E = E \cdot C$

Fig. 3 demonstrates the regulation of control in complex pathways in a similar way as did expressions (2b) and (2c) for simple pathways. Fig.3a ($I=C\cdot E$) describes the summation and connectivity theorems in their usual form and equations (3a-d, 4a-b) can be readily read from it. Fig.3b ($I=E\cdot C$) shows relations between (enzymatic) flux-control coefficients and (metabolite) regulatory strength.

3.2.1. Summation theorem for concentration control. From equations (4a), (3b), and (4b), (3d), respectively, it follows that:

$$C_1^X + C_2^X + C_3^X + C_4^X = C_1^X + C_2^X = 0\,, \tag{5a}$$

$$C_1^Z + C_2^Z + C_3^Z + C_4^Z = C_3^Z + C_4^Z = 0\,, \tag{5b}$$

indicating indeed that if some enzymes act to decrease the concentration of metabolites (X or Z) others act to increase the same.

3.2.2. Summation theorem for control on flux. From eqs (3a) and (3c), respectively, it follows that:

$$C_1^J + C_2^J + C_3^J + C_4^J = C_1^J + C_2^J\,, \tag{6a}$$

$$C_1^j + C_2^j + C_3^j + C_4^j = C_3^j + C_4^j\,, \tag{6b}$$

$$C_1^J + C_2^J = \left[-\left(\varepsilon_Z^3 - \varepsilon_Z^4\right)\varepsilon_X^2 + \left(\varepsilon_X^3 - \varepsilon_X^4\right)\varepsilon_Z^2 + \left(\varepsilon_Z^3 - \varepsilon_Z^4\right)\varepsilon_X^1 - \left(\varepsilon_X^3 - \varepsilon_X^4\right)\varepsilon_Z^1\right] \cdot |\mathbf{C}|\,,$$

$$= [-|\mathbf{E}_{22}| \cdot \varepsilon_X^2 + |\mathbf{E}_{21}| \cdot \varepsilon_Z^2 + |\mathbf{E}_{22}| \cdot \varepsilon_X^1 - |\mathbf{E}_{21}| \cdot \varepsilon_Z^1] \cdot |\mathbf{C}|\,,$$

$$= [|\mathbf{E}_{22}| \cdot |\mathbf{E}_{11}| - |\mathbf{E}_{21}| \cdot |\mathbf{E}_{12}|] \cdot |\mathbf{C}| = |\mathbf{E}| \cdot |\mathbf{C}| = 1\,, \tag{7a}$$

a.

$$\mathbf{C}\,.\,\mathbf{E} = \mathbf{I} \quad \begin{array}{cccc|c} \text{Summation Theorem} & \text{Connectivity Theorem} & \text{Summation Theorem} & \text{Connectivity Theorem} & \text{for control} \\ 1 = C_1^J + C_2^J & 0 = -(C_1^J\epsilon_X^1 + C_2^J\epsilon_X^2 + C_3^J\epsilon_X^3 + C_4^J\epsilon_X^4) & 0 = C_3^J + C_4^J & 0 = -(C_1^J\epsilon_Z^1 + C_2^J\epsilon_Z^2 + C_3^J\epsilon_Z^3 + C_4^J\epsilon_Z^4) & \text{on } J \\ 0 = C_1^X + C_2^X & 1 = -(C_1^X\epsilon_X^1 + C_2^X\epsilon_X^2 + C_3^X\epsilon_X^3 + C_4^X\epsilon_X^4) & 0 = C_3^X + C_4^X & 0 = -(C_1^X\epsilon_Z^1 + C_2^X\epsilon_Z^2 + C_3^X\epsilon_Z^3 + C_4^X\epsilon_Z^4) & [X] \\ \hdashline 0 = C_1^j + C_2^j & 0 = -(C_1^j\epsilon_X^1 + C_2^j\epsilon_X^2 + C_3^j\epsilon_X^3 + C_4^j\epsilon_X^4) & 1 = C_3^j + C_4^j & 0 = -(C_1^j\epsilon_Z^1 + C_2^j\epsilon_Z^2 + C_3^j\epsilon_Z^3 + C_4^j\epsilon_Z^4) & \text{on } j \\ 0 = C_1^Z + C_2^Z & 0 = -(C_1^Z\epsilon_X^1 + C_2^Z\epsilon_X^2 + C_3^Z\epsilon_X^3 + C_4^Z\epsilon_X^4) & 0 = C_3^Z + C_4^Z & 1 = -(C_1^Z\epsilon_Z^1 + C_2^Z\epsilon_Z^2 + C_3^Z\epsilon_Z^3 + C_4^Z\epsilon_Z^4) & [Z] \end{array}$$

b.

$$\mathbf{E}\,.\,\mathbf{C} = \mathbf{I} \quad \begin{array}{cccc|c} 1 = C_1^J - (C_1^X\epsilon_X^1 + C_1^Z\epsilon_Z^1) & 0 = C_2^J - (C_2^X\epsilon_X^1 + C_2^Z\epsilon_Z^1) & 0 = C_3^J - (C_3^X\epsilon_X^1 + C_3^Z\epsilon_Z^1) & 0 = C_4^J - (C_4^X\epsilon_X^1 + C_4^Z\epsilon_Z^1) & \text{on } J \\ 0 = C_1^J - (C_1^X\epsilon_X^2 + C_1^Z\epsilon_Z^2) & 1 = C_2^J - (C_2^X\epsilon_X^2 + C_2^Z\epsilon_Z^2) & 0 = C_3^J - (C_3^X\epsilon_X^2 + C_3^Z\epsilon_Z^2) & 0 = C_4^J - (C_4^X\epsilon_X^2 + C_4^Z\epsilon_Z^2) & [X] \\ \hdashline 0 = C_1^j - (C_1^X\epsilon_X^3 + C_1^Z\epsilon_Z^3) & 0 = C_2^j - (C_2^X\epsilon_X^3 + C_2^Z\epsilon_Z^3) & 1 = C_3^j - (C_3^X\epsilon_X^3 + C_3^Z\epsilon_Z^3) & 0 = C_4^j - (C_4^X\epsilon_X^3 + C_4^Z\epsilon_Z^3) & \text{on } j \\ 0 = C_1^j - (C_1^X\epsilon_X^4 + C_1^Z\epsilon_Z^4) & 0 = C_2^j - (C_2^X\epsilon_X^4 + C_2^Z\epsilon_Z^4) & 0 = C_3^j - (C_3^X\epsilon_X^4 + C_3^Z\epsilon_Z^4) & 1 = C_4^j - (C_4^X\epsilon_X^4 + C_4^Z\epsilon_Z^4) & [Z] \end{array}$$

Figure 3. The regulation of control in complex pathways as represented by identity matrix **I**. Matrix **I**, as a whole, then expresses: regulability (elasticity) times control (or *vice versa*) equals regulation of control. The terms in which the elements of **I** are expressed, depending on the way of multiplying, are not the same.
a) Regulation as expressed by the product **C**·**E**. Each element of **I** represents an aspect of either the Summation or the Connectivity Theroems in their usual form, i.e. summation of flux-control coefficients, of concentration-control coefficients, of metabolite-on-flux regulatory strengths, and of metabolite regulatory strengths.
b) Regulation as expressed by the product **E**·**C**, i.e. in the form of the connection between *enzymatic* flux-control coefficients and *metabolite* regulatory strength[3].

$$C_3^j + C_4^j = \left[-\left(\varepsilon_X^1 - \varepsilon_X^2 \right) \varepsilon_Z^4 + \left(\varepsilon_Z^1 - \varepsilon_Z^2 \right) \varepsilon_X^4 + \left(\varepsilon_X^1 - \varepsilon_X^2 \right) \varepsilon_Z^3 - \left(\varepsilon_Z^1 - \varepsilon_Z^2 \right) \varepsilon_X^3 \right] \cdot |C| \; ,$$

$$= \left[-|E_{11}| \cdot \varepsilon_Z^4 + |E_{12}| \cdot \varepsilon_X^4 + |E_{11}| \cdot \varepsilon_Z^3 - |E_{12}| \cdot \varepsilon_X^3 \right. \; ,$$

$$= \left[|E_{11}| \cdot |E_{22}| - |E_{12}| \cdot |E_{21}| \right] \cdot |C| = |E| \cdot |C| = 1 \; . \tag{7b}$$

Thus the flux-control coefficients within a pathway continue to meet a summation theorem even when an unconnected (none of its enzymes being involved in another pathway, nor the reverse) pathway is embedded in a longer system!

3.2.3. Concentration control connectivity theorem; regulatory strength on metabolite concentrations.

The flux-control coefficients in complex pathways can be expressed in terms of metabolite regulatory strength by applying eq. (7a) with respect to C_1^J and C_2^J and eq. (7b) with respect to C_3^j and C_4^j,

$$-\left[\left(\varepsilon_Z^3 - \varepsilon_Z^4 \right) \varepsilon_X^2 - \left(\varepsilon_X^3 - \varepsilon_X^4 \right) \varepsilon_Z^2 \right] \cdot |C| = -\left(C_2^X \varepsilon_X^2 + C_2^Z \varepsilon_Z^2 \right) = C_1^J \tag{8a}$$

and likewise with indices 1 and 2 interchanged;

$$-\left[\left(\varepsilon_Z^1 - \varepsilon_Z^2 \right) \varepsilon_X^4 - \left(\varepsilon_X^1 - \varepsilon_X^2 \right) \varepsilon_Z^4 \right] \cdot |C| = -\left(C_4^X \varepsilon_X^4 + C_4^Z \varepsilon_Z^4 \right) = C_3^j \tag{8b}$$

and likewise with indices 3 and 4 interchanged.

The equations indicate that the control on flux by an enzyme in a given pathway is directly related with the regulatory strength (defined as $C_i^X \varepsilon_X^i$, see Ref. 3) of all metabolites (*i.e.* both X *and* Z from pathway I *and* II) affecting the remaining enzyme(s) of that pathway (*i.e.* e_2 in I). The equations (8a,b), in concordance with Fig.3a, also demonstrate the cross-dependence of the pathways I and II, i.e. the concentration-control connectivity theorem regarding X is met by adding the first factors (eq. 8c), and regarding Z, by adding the second factors (eq. 8g). Due to eqs (4a), (3b) respectively (3d), (4b), the solutions to the homeostatic strengths of X and Z are:

$$C_1^X \varepsilon_X^1 + C_2^X \varepsilon_X^2 + C_3^X \varepsilon_X^3 + C_4^X \varepsilon_X^4 = C_1^X \left(\varepsilon_X^1 - \varepsilon_X^2 \right) + C_3^X \left(\varepsilon_X^3 - \varepsilon_X^4 \right) \; , \tag{8c}$$

which directly corresponds to

$$-\left(\varepsilon_Z^3 - \varepsilon_Z^4 \right) \cdot |C| \cdot \left(\varepsilon_X^1 - \varepsilon_X^2 \right) + \left(\varepsilon_Z^1 - \varepsilon_Z^2 \right) \cdot |C| \cdot \left(\varepsilon_X^3 - \varepsilon_X^4 \right) = -|E| \cdot |C| = -1 \; , \tag{8d}$$

which proves that

$$C_1^X \varepsilon_X^1 + C_2^X \varepsilon_X^2 = \left(\varepsilon_Z^3 - \varepsilon_Z^4 \right) C_3^Z = C_3^Z \varepsilon_Z^3 + C_4^Z \varepsilon_Z^4 = -|E_{11}| \cdot |E_{22}| \cdot |C| \; , \tag{8e}$$

and by analogy,

$$C_3^X \varepsilon_X^3 + C_4^X \varepsilon_X^4 = C_1^Z \varepsilon_Z^1 + C_2^Z \varepsilon_Z^2 = |E_{12}| \cdot |E_{21}| \cdot |C| \tag{8f}$$

and

$$C_1^Z \varepsilon_Z^1 + C_2^Z \varepsilon_Z^2 + C_3^Z \varepsilon_Z^3 + C_4^Z \varepsilon_Z^4 = -|\mathbf{E}| \cdot |\mathbf{C}| = -1 \,. \tag{8g}$$

These equations show that the internal homeostatic strength $(C_1^X \varepsilon_X^1 + C_2^X \varepsilon_X^2)$ of metabolite X in pathway I, via enzymes e_1 and e_2 of pathway I, matches the internal homeostatic strength $(C_3^Z \varepsilon_Z^3 + C_4^Z \varepsilon_Z^4)$ of metabolite Z in pathway II (eq. 8e), via enzymes e_3 and e_4 of pathway II; they are directly related with the first part, $|\mathbf{E}_{11}| \cdot |\mathbf{E}_{22}|$, of the determinant of matrix $\mathbf{E}$. The external homeostatic strengths of both X, $(C_3^X \varepsilon_X^3 + C_4^X \varepsilon_X^4)$, *and* Z, $(C_1^Z \varepsilon_Z^1 + C_2^Z \varepsilon_Z^2)$, via the enzymes of the other pathway (II and I, respectively), are equal to the right-hand side of eq. (8f), and are directly related to the second part of the determinant of matrix $\mathbf{E}$. The precondition for the system's steady state to remain stable is, however, $|\mathbf{E}_{11}| \cdot |\mathbf{E}_{22}| \neq |\mathbf{E}_{12}| \cdot |\mathbf{E}_{21}|$. Thus the system will cease to exist if the internal homeostatic strength via the enzymes of the pathway proper equals the external homeostatic strength via enzymes of the other pathway. Total homeostatic strength of X via all appropriate enzymes is completely expressible in terms of homeostatic strength of Z, and *vice versa*, indicating that homeostatic strengths of X and Z are completely coupled.

3.2.4. Regulatory strength on flux of the pathway proper. By multiplying the flux-control coefficients (eqs 8a,b) with the proper elasticity coefficients, it is relatively easy to demonstrate that the sum of the metabolite-on-flux regulatory strength equals zero. Due to eqs (4a), (3b) respectively (3d), (4b), the solutions to the homeostatic strengths of X and Z on fluxes J (pathway I) and j (pathway II) are:

$$C_1^J \varepsilon_X^1 = -C_2^X \varepsilon_X^2 \varepsilon_X^1 + C_1^Z \varepsilon_Z^2 \varepsilon_X^1 \,, \tag{9a}$$

$$C_2^J \varepsilon_X^2 = C_2^X \varepsilon_X^1 \varepsilon_X^2 - C_1^Z \varepsilon_Z^1 \varepsilon_X^2 \,, \tag{9b}$$

$$C_1^J \varepsilon_X^1 + C_2^J \varepsilon_X^2 = C_1^Z \left(\varepsilon_Z^2 \varepsilon_X^1 - \varepsilon_Z^1 \varepsilon_X^2 \right) = \left(\varepsilon_X^3 - \varepsilon_X^4 \right) \cdot |\mathbf{C}| \cdot \left(\varepsilon_Z^2 \varepsilon_X^1 - \varepsilon_Z^1 \varepsilon_X^2 \right)$$

$$= \left(\varepsilon_X^3 - \varepsilon_X^4 \right) \; C_4^J \; (!) \tag{9c}$$

and since $C_4^J = -C_3^J$, we have

$$C_1^J \varepsilon_X^1 + C_2^J \varepsilon_X^2 = -\left(C_3^J \varepsilon_X^3 + C_4^J \varepsilon_X^4 \right) \tag{9d}$$

and, consequently,

$$C_1^J \varepsilon_X^1 + C_2^J \varepsilon_X^2 + C_3^J \varepsilon_X^3 + C_4^J \varepsilon_X^4 = 0 \,. \tag{9e}$$

i.e. a value that can be calculated from elements of submatrices $\mathbf{C}_{11}$ and $\mathbf{E}_{11}$ only. Similarly:

$$C_1^j \varepsilon_Z^1 + C_2^j \varepsilon_Z^2 + C_3^j \varepsilon_Z^3 + C_4^j \varepsilon_Z^4 = -\left(C_3^j \varepsilon_Z^3 + C_4^j \varepsilon_Z^4 \right) + C_3^j \varepsilon_Z^3 + C_4^j \varepsilon_Z^4 = 0 \,. \tag{9f}$$

3.2.5. Regulatory strength on the flux through other pathways. Likewise (deduced similarly to eqs (9)):

$$C_1^J \varepsilon_Z^1 + C_2^J \varepsilon_Z^2 + C_3^J \varepsilon_Z^3 + C_4^J \varepsilon_Z^4 = 0 \,, \tag{10a}$$

$$C_1^j \varepsilon_X^1 + C_2^j \varepsilon_X^2 + C_3^j \varepsilon_X^3 + C_4^j \varepsilon_X^4 = 0 \,. \qquad\qquad (10b)$$

All of these equations confirm the corresponding lines in Fig.3a.

Eqs (9d,e) illustrate that the presence of pathway II (Fig.2) affects the homeostatic regulation of pathway I. The internal regulation of X on flux J within pathway I no longer suffices to maintain homeostasis. The deficit in internal regulatory strength is completely compensated for by the additional regulatory strength of X upon J through pathway II. Eq. (10a) demonstrates that the regulatory strength Z has on flux J through enzymes e_1 and e_2 in pathway I is completely nullified by the regulatory strength Z has on that same flux J through enzymes e_3 and e_4 in pathway II.

4. Summarizing Remarks

The internal regulation that is responsible for steady states to remain stable is similarly organized in simple and in complex metabolic networks. Mathematically, this regulation, repetitive as it should be, may be expressed in the form of an identity matrix **I**, the elements of which form a compilation of all accepted theorems regarding control analysis. We show that the product **C·E** offers the elements of **I** in terms of summation and connectivity theorems for flux- and concentration-control. The reverse product **E·C** offers the elements of **I** in the form of all relations existing between (enzymatic) flux-control coefficients and (metabolite) regulatory strength. Matrix **C** is a compilation of all flux- and concentration-control coefficients of enzymes pertinent to the network. The matrix **E** shows a limit wherein regulation can exist, in that its determinant must differ from zero. We demonstrate that external (in the sense of a regulatory loop through metabolism outside the pathway itself) regulation of flux and concentration of metabolites in a given pathway is very similar to additional internal regulation (positive/negative feedback). This offers the opportunity to greatly reduce the number of details necessary to understand the essential aspects of the regulation of complex pathways.

References

1. D.A. Fell, Metabolic control analysis: a survey of its theoretical and experimental development, *Biochem. J.* **286**:313 - 330.(1992).
2. H.V.Westerhoff and K. van Dam, "Thermodynamics and Control of Biological Free-Energy Transduction," Elsevier, Amsterdam (1987).
3. D. Kahn and H.V. Westerhoff, Regulation and homeostasis in Metabolic Control Theory: Interplay between fluctuations of variables and parameter changes, this volume.
4. D. Kahn and H.V. Westerhoff, Control theory of regulatory cascades, *J.theor.Biol.* **153**:255-285 (1991).
5. S. Schuster, D. Kahn and H.V. Westerhoff, A modular approach to the description of the control of connected metabolic systems'.
6. H.V. Westerhoff and D.B. Kell, Matrix maethod for determining steps most rate-limiting to metabolic fluxes in biotechnological processes, *Biotechn. Bioeng.* **30**:101-107 (1987).
7. F. Ayres, "Theory and Problems of Matrices," Schaum Publishing, New York (1962).
8. B. Crabtree and E.A. Newsholme, The derivation and interpretation of control coefficients, *Biochem. J.* **247**:113 -120 (1987).
9. H.V. Westerhoff, J.G. Koster, M. van Workum and K.E. Rudd, On the control of gene expression, *in*: "Control of Metabolic Processes," A. Cornish-Bowden and M.L. Cárdenas, eds, Plenum Press, New York, pp. 399-412 (1990).

MULTIPLICITY OF CONTROL

Hans V. Westerhoff[1,2], Peter Ruhdal Jensen[1], Anton A. van der Gugten[1],
Daniel Kahn[1,3], Boris N. Kholodenko[2,4], Stefan Schuster[5],
Nienke Oldenburg[1], Karel van Dam[2] and Wally C. van Heeswijk[1,2]

[1]Netherlands Cancer Institute
 Plesmanlaan 121, NL-1066 CX Amsterdam, The Netherlands
[2]E.C. Slater Institute
 University of Amsterdam, 1018 TV Amsterdam, The Netherlands
[3]Laboratoire de Biologie Moleculaire, I.N.R.A-C.N.R.S.
 31326 Castanet-Tolosan, France
[4]A.N. Belozersky Institute of Physico-Chemical Biology
 Moscow State University, Moscow 119899, Russia
[5]Dépt. de Biochimie Medicale, Universite Bordeaux II
 33076 Bordeaux, France

1. Introduction

The living cell is controlled by a multitude of regulatory mechanisms. In addition to the difficulty of approaching the molecular processes of life, it is this sheer complexity that makes biology one of the most challenging topics for the physical chemist. We are attempting to deal with complexity by reducing it along the lines indicated by the organization of cell physiology itself. With respect to control analysis, this implies that we attempt to discern the parts of the cell that function as units in interactions with other such units in the cell. We have identified such units in covalent cascades of enzymes modifying one another, in the regulation of enzyme concentrations through gene expression, as well as in the processes generating and the processes consuming the proton gradient in oxidative phosphorylation.

2. Modules with a Single Connecting Flow

An important experimental demonstration of the usefulness of metabolic control analysis has been its application to mitochondrial oxidative phosphorylation. First the distribution of control over the various processes was determined[1] and this solved ongoing debates about which is the rate-limiting step in that process (answer: there is no single such step). One of the surprising aspects of that study was that cytochrome oxidase, which catalyzes the irreversible reaction of oxidative phosphorylation, did not exert all of the control on mitochondrial respiration.

Modern Trends in Biothermokinetics, Edited by
S. Schuster *et al.*, Plenum Press, New York, 1993

As demonstrated by the flux-control connectivity theorem[2], the distribution of flux control over enzymes is *not* determined by the displacement of the corresponding reaction from equilibrium, but by the relative elasticity coefficients of all the enzymes. This pointed a way to understanding why cytochrome oxidase exerted less control than intuitively expected.

However, mitochondrial oxidative phosphorylation has many component reactions; the complete solution of the flux control coefficients in terms of all elasticity coefficients is tedious and any such solution would not be very useful because too many of the fundamental elasticity coefficients are unknown. This dilemma was resolved by viewing oxidative phosphorylation as subdivided into three subsystems, i.e., input (comprising uptake of respiratory substrate, respiration and export of respiratory product), leak (representing proton permeability of the inner mitochondrial membrane), and output (representing transport of adenine nucleotides and phosphate, as well as the phosphorylation of intramitochondrial ADP by the mitochondrial H+-ATPase) (cf. Fig. 1)[3-5]. "Overall" elasticity coefficients were defined as the logarithmic dependence of the flux through a subsystem on the electrochemical potential difference for protons ($\Delta\tilde{\mu}_H$), under conditions where the internal steady state of that subsystem is reattained, but all variables outside that subsystem are not allowed to change. To quantify the control exerted by subsystems as such, "overall" flux-control coefficients were defined for each subsystem as the effect on the steady-state flux through the entire system of simultaneously increasing all activities in that subsystem by the same factor leaving all activities in all other subsystems unaltered. With this, the following basis for the control of respiration by the respiratory subsystem was derived:

$$^*C_i = 1 \, / \, \{1 + (-^*\varepsilon_H^i/^*\varepsilon_H^o)\} \tag{1}$$

(where, for simplicity we have left out the proton leak; the complete expression is found elsewhere[4,5]). "i" and "o" refer to input and output subsystem respectively (cf. Fig. 1). The fact that *overall* elasticities and *overall* control coefficients are referred to here is indicated by the asterisks. Eq. (1) states that the control of the cytochrome oxidase on respiration ($^*C_i$) should increase as the absolute value of the elasticity of the input subsystem (i.e., respiration;

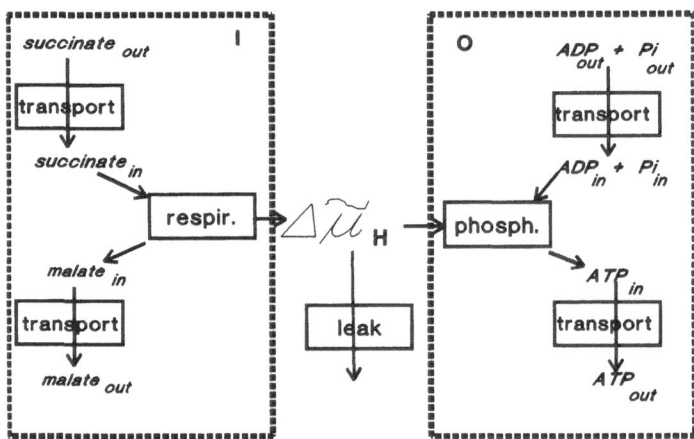

Figure 1. Overall control analysis of oxidative phosphorylation; an example of control involving modules with flow in between. The module "i" comprises transport and oxidation of respiratory substrate, followed by transport of the product. Module "o" involves transport of adenine nucleotides and phosphate as well as phosphorylation of ADP. The two modules are connected by flow through the electrochemical potential difference for protons, which may partly leak away.

$-^*\varepsilon_H^i$ is usually positive) decreases relative to the elasticity of the output (ATP synthesizing) subsystem, both elasticities taken with respect to the proton gradient.

The factor $(-^*\varepsilon_H^i/^*\varepsilon_H^o)$ was determined experimentally and found to equal 0.4,[6] explaining, according to the above equation that the control by the respiratory subsystem should amount to only 0.7, hence that cytochrome oxidase should not be a sole rate-limiting step (control coefficient of 1). Indeed, cytochrome oxidase plus the succinate translocator exhibited a total control close to 0.6.[1]

This *overall* approach was also useful when Wanders and colleagues[7] analyzed what determines the control exerted by the enzyme *consuming* the ATP, on oxidative phosphorylation. Treating the mitochondria as one subsystem and the ATP consuming enzyme (hexokinase or creatine kinase) as the second subsystem, they showed experimentally that the control exerted by the consuming enzyme depended on the ratio between (i) the overall elasticity coefficient of the mitochondria with respect to extramitochondrial ATP concentration, and (ii) the elasticity coefficient of the ATP consuming enzyme. Later, this type of overall approach to oxidative phosphorylation was renamed "top down" approach and elaborated somewhat further[8].

The overall elasticity coefficient of a subsystem, corresponds to a control coefficient by an external substrate if the subsystem were viewed as a system in isolation. This implies that the control of the total system may be analyzed by first analyzing (in terms of metabolic control analysis) the control within each subsystem and then, by treating the subsystems as (enzyme-like) units, analyzing the total system in terms of the interactions between its subsystems. Since this procedure can be repeated, this generates a method to analyze control in quite large systems. The complexity of such an approach is smaller than if the overall system were treated at once in terms of all its molecular components. And, such a strategy recognizes the important aspect of *multiplicity of control:* in many biological systems fluxes are not only controlled at one level (e.g., by product inhibition of an enzyme), but also at another level (such as by inhibition of communication between an organelle and the cytosol, or by altered gene expression).

3. Modules Connected by More Than a Single Flux

For the above strategy to be effective, it should work for many types of subsystem of a cell. The treatment of mitochondrial oxidative phosphorylation sketched above, was limited to just a single variable ($\Delta\tilde{\mu}_H$). Only recently, we have been able to eliminate this limitation[8]. Essentially this is done by allowing the intermediate ($\Delta\tilde{\mu}_H$ above) to be replaced by a vector of (independent) metabolite concentrations or the flux between the modules by a vector of independent fluxes. Subsystem 1 in Fig. 2 has two *overall* elasticity coefficients, one with respect to the concentration of X and one with respect to the concentration of Y. And, these coefficients themselves are vectors, i.e., consist of two components, one representing the dependence of flux J_{1X} and one representing the dependence of the flux J_{1Y} on [X], or [Y].

Importantly, the role of the control exerted by the subsystem 1 on a flux in the total system can no longer be expressed in terms of a single *overall* flux control coefficient defining the effect of a simultaneous increase of all processes in subsystem 1 (which has the effect that J_{1X} and J_{1Y} change by the same factor). For, the rates J_{1X} and J_{1Y} can change independently. In addition therefore, one may define an *overall* control coefficient by subsystem 1 that indicates the effect of a simultaneous change in all processes in subsystem 1 *such that J_{1X} increases by the same factor as J_{1Y} decreases.* For those interested in free-energy transduction, the latter coefficient could be used to assess the relevance of the mitochondrial P/O ratio for other cellular processes. Connectivity and summation theorems have been derived for this type of modular analysis[9] along the lines of the control theory set out by Reder[10].

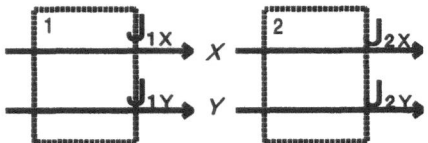

Figure 2. Overall control analysis with more than one flow or more than one metabolite connecting the modules.

4. Unconnected but Dependent Modules

Before the limitation to a single flux between subsystems was removed, the modular philosophy was extended to a very important set of subsystems in cell biology, i.e., those of intermediary metabolism versus gene expression[11]. In this case, the definition of the subsystems was limited to cases where they do not exchange matter, but only affect each other by regulatory effects of substances within one module on rates in another module. Cascades of enzymes modifying one another covalently also obey this criterion. The control of the entire system can be analyzed in terms of the control within and the elasticities between the modules[12]. In a parallel paper in this book, this modular method is elaborated in terms of the $\mathbf{C \cdot E = I}$ matrix method[13].

5. Zero Flux Control

In some actual cases the two processes constituting a single module are actually catalyzed by (two forms of) the same enzyme. The process in which glutamine synthetase is adenylylated is catalyzed by the same enzyme (adenylyl transferase) as the process in which the AMP group is hydrolysed from the enzyme. Modular control theory has derived that the sum of the control coefficients of processes 1 and 2 on the *metabolic* flux through glutamine synthetase should be zero. As a consequence, the enzyme adenylyl transferase should exert zero flux control on glutamine synthesis. Van Heeswijk and colleagues further analyze this system in a parallel paper[14].

Zero flux control has many interesting control aspects. It has been used as a tool to decompose biochemical reaction systems into modules that are functionally disconnected[15]. In a parallel paper Jensen and colleagues[16] discuss their observation that in *E. coli* the H+-ATPase exerts zero control on growth. This case is not related to the case of glutamine synthetase however.

6. Unconnected but Dependent Modules

The regulation of glutamine synthetase is an example where enzymes and processes do not map onto one another. More in general in such cases the flux-control summation theorem can be distinguished into two separate ones, one dealing with the sum of the controls by the process activities, and one dealing with the sum of the controls by the enzyme concentrations[17]. Fig. 3 shows another such case, that of a group-transfer pathway. Here in each process, two enzymes participate. In this system the sum of the flux control coefficients over all process activities is 1, whereas the sum over all enzyme concentrations plus the initial and final carriers of the transferred group is 2 (K. van Dam, J. van der Vlag, B.N. Kholodenko and H.V. Westerhoff, in preparation).

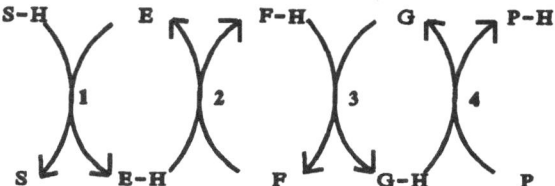

Figure 3. A group transfer system. In this example the group "H" is transferred from a substance (reductor) S to a substance (oxidator) P. The enzymes E, F, and G are catalysts of the transfer by being carriers of the group "H". The transfer processes are numbered 1 through 4. Flux control over the processes amounts to 1 when summed over the processes. When summed over the enzymes plus S and P, it amounts to 2.

7. Chemostat Control Analysis

A special case of modular control occurs where man and microbe meet, such as in the chemostat. Here one may profitably distinguish between a "bio" module and a "chemo" module. With respect to Fig. 4 this means that the dashed circle "2" corresponds to the "bio" module: the living cell, whereas reaction 1 corresponds to the chemostat. In a chemostat, the influx of substrate is *in*sensitive to the substrate concentration in the vessel. Using that usually the concentration of substrate in the feeding reservoir is much higher than in the chemostat, one finds (Westerhoff *et al.*, in preparation):

$$
{}^{*}\mathbf{C} = \begin{bmatrix} C_1^J & {}^{*}C_2^J \\ C_1^S & {}^{*}C_2^S \end{bmatrix} = {}^{*}\mathbf{E}^{-1} = \begin{bmatrix} {}^{*}e_S^2 & 0 \\ 1 & -1 \end{bmatrix} / \{ {}^{*}\varepsilon_S^2 \} . \tag{2}
$$

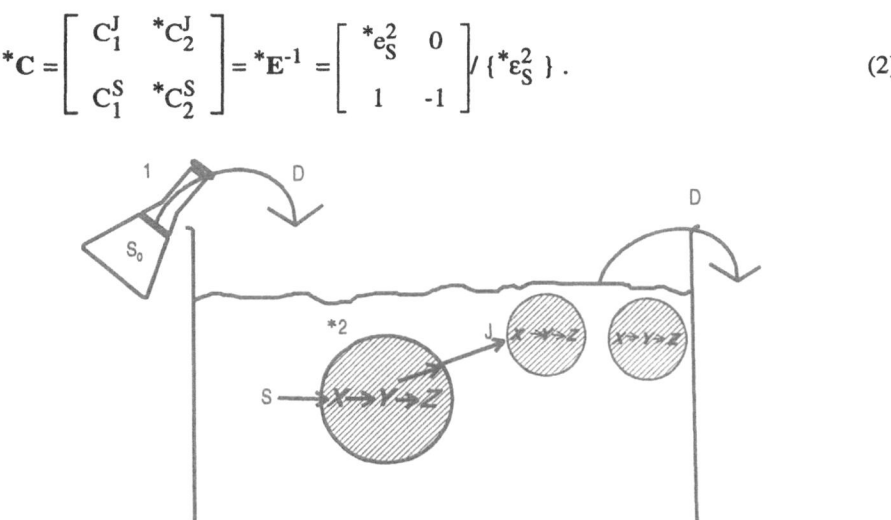

Figure 4. Control analysis of chemostats. In a chemostat, medium containing growth limiting substrate S at a concentration S_0 is pumped into a vessel at a rate D. The vessel contains cells and overflows at a rate D. The cells grow (divide) at a rate J, which at steady state equals D. Here the cells are treated as module "2".

This equation reflects that in a chemostat only the pump controls the growth rate (C_1^J=1 and ${}^{*}C_2^J$=0). It also indicates how one may use the chemostat to determine the control of growth rate by the substrate concentration, when under conditions of fixed substrate concentration. For, this property is identical to ${}^{*}\varepsilon_S^2$, which equals $1/C_1^S$ (cf. eq. (2)). C_1^S is the control exerted by the pump rate on the concentration of the growth limiting substrate (S), hence directly observable in the chemostat[18]. Clearly this approach allows one to add to the biological system a *non* biological system (the chemostat), and then use a modular type of control analysis to determine biological properties in the hybrid system.

8. Concluding remarks

We have indicated that control in physiological systems is multiple. Control Analysis is a suitable tool to disentangle such control especially if any one of the modular control analysis methods is employed.

Acknowledgements

This work was supported in part by the Netherlands Organization for Scientific Research (NWO). We thank Dr. A. de Waal for help with the manuscript.

References

1. A.K. Groen, R. van der Meer, H.V. Westerhoff, R.J.A. Wanders, T.P.M. Akerboom and J.M. Tager, Control of metabolic fluxes, *in*: "Metabolic Compartmentation," H. Sies, ed., pp. 9-37, Academic Press, New York (1982).
2. H. Kacser and J.A. Burns, The control of flux, *Symp. Soc. Exp. Biol.* **27**:65-104 (1973).
3. H.V. Westerhoff, A.K. Groen and R.J.A. Wanders, The thermodynamic basis for the partial control of oxidative phosphorylation by the adenine-nucleotide translocator, *Biochem. Soc. Trans.* **11**:90-91 (1983).
4. H.V. Westerhoff, P.J.A.M. Plomp, A.K. Groen and R.J.A. Wanders, Thermodynamics of the control of metabolism, *Cell Biophys.* **10**:239-267 (1987a).
5. H.V. Westerhoff and K. van Dam, "Thermodynamics and Control of Biological Free Energy Transduction," Elsevier, Amsterdam (1987).
6. H.V. Westerhoff, P.J.A.M. Plomp, A.K. Groen, R.J.A. Wanders, J.A. Bode and K. van Dam, On the origin of the limited control of mitochondrial respiration by the adenine nucleotide translocator, *Arch. Biochem. Biophys.* **257**:154-169.
7. R.J.A. Wanders, A.K. Groen, C.W.T. van Roermund and J.M. Tager, Factors determining the relative contribution of the adenine-nucleotide translocator and the ADP-regenerating system to the control of oxidative phosphorylation in isolated rat-liver mitochondria, *Eur. J. Biochem.* **142**:417-424 (1984).
8. R.P. Hafner, G.C. Brown and M.D. Brand, Analysis of the control of respiration rate, phosphorylation rate, proton leak rate and protonmotive force in isolated mitochondria using the 'top-down' approach of metabolic control theory, *Eur. J. Biochem.* **188**:313-319 (1990).
9. S. Schuster, D. Kahn and H.V. Westerhoff, A modular approach to the description of the control of connected metabolic systems, this volume.
10. C. Reder, Metabolic control theory: a structural approach, *J. theor. Biol.* **135**:175-202 (1988).
11. H.V. Westerhoff, J.G. Koster, M. van Workum and K.E. Rudd, On the control of gene expression, *in*: "Control of Metabolic Processes," A. Cornish-Bowden and M.-L. Cardenas, eds., pp. 399 - 412, Plenum, New York (1990).
12. D. Kahn and H.V. Westerhoff, Control theory of regulatory cascades, *J. theor. Biol.* **53**:255-285 (1991).
13. A.A. van der Gugten and H.V. Westerhoff, Internal regulation. The $C \cdot E = I = E \cdot C$ square-matrix method illustrated for a simple case of a complex pathway, this volume.
14. W.C. van Heeswijk, H.V. Westerhoff and D. Kahn, Cascade control of ammonia assimilation, this volume.
15. R. Schuster and S. Schuster, Decomposition of biochemical reaction systems according to flux control insusceptibility, *J. Chim. Phys.* **89**:1887-1910 (1992).
16. P.R. Jensen, N. Oldenburg, B. Petra, O. Michelsen and H.V. Westerhoff, Modulation of cellular energy state and DNA supercoiling, this volume.
17. B.N. Kholodenko and H.V. Westerhoff, Sum of the flux control coefficients: What is it equal to in different systems?, this volume.
18. H.V. Westerhoff, W. van Heeswijk, D. Kahn and D.B. Kell, Quantitative approaches to the analysis of the control and regulation of microbial metabolism, *Anth. Leeuwenh.* **60**:193-208 (1991).

CONTROL OF
THREONINE BIOSYNTHESIS
IN *E. COLI*

B. Rais[1], R. Heinrich[2],
M. Malgat[1] and J.P. Mazat[1]

[1]Université Bordeaux II
 G.E.S.B.I.
 146 Rue Léo Saignat, 33076 Bordeaux Cedex, France
[2]Humboldt-Universität zu Berlin
 Institut für Biophysik
 Invalidenstr. 42, 1040 Berlin, Germany

1. Introduction

In order to increase the production of amino acids by bacterial strains, one usually employs amplification of the genes coding for the enzymes of the pathway leading to the corresponding amino acid. Frequently, however, an increase in the gene copy number does not entail an increase in the production of amino acid. In order to better understand this phenomenon, it is useful to determine the flux control coefficient of this step in the strain to be amplified[1,2].

We perform this determination for the pathway of threonine biosynthesis for the wild type *E. coli* strain (see Fig. 1).

We have also modeled the threonine pathway in order to compare the theoretical effects of the threonine operon amplification with our experimental results.

2. Materials and Methods

2.1. Cells

The wild type *E. coli* K-12, carrying a desensitized aspartokinase I-homoserine dehydrogenase I for threonine, was grown in a minimal salt medium at 37 °C with 0.5 % glucose as carbon source.

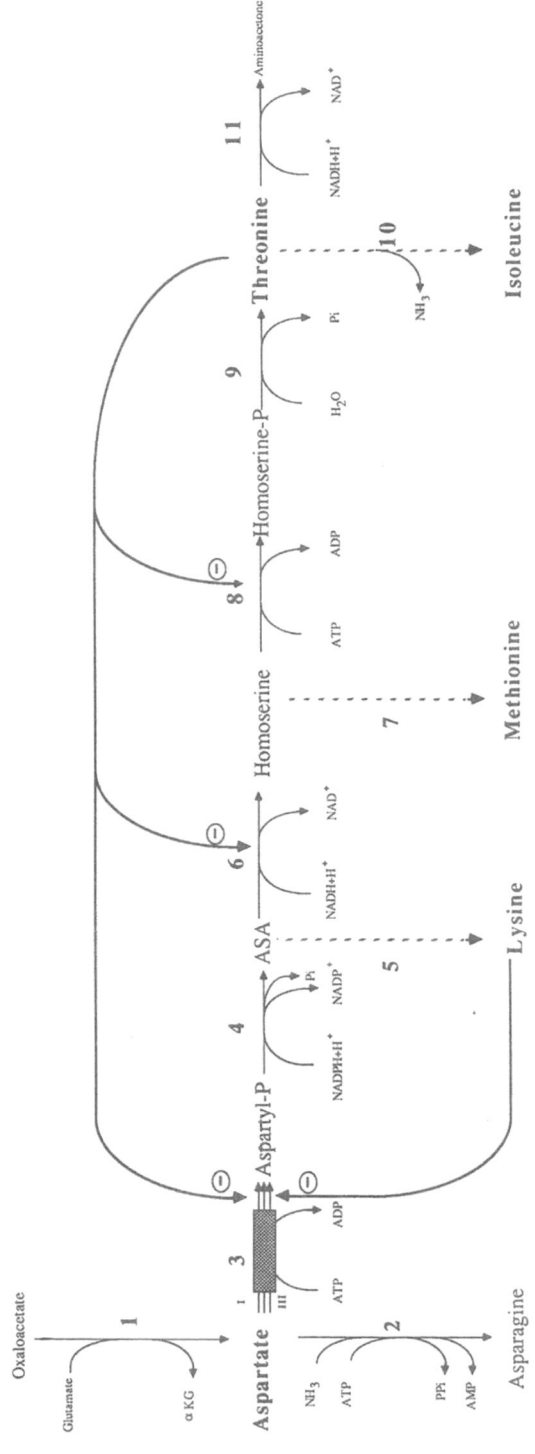

ENZYMES :

1 : ASPARTATE AMINOTRANSFERASE.

2 : ASPARAGINE SYNTHASE.

3 : ASPARTOKINASES I, II, III.

4 : ASA DEHYDROGENASE.

5 : Biosynthetic pathway of Lysine.

6 : HOMOSERINE DEHYDROGENASE.

7 : ACETYL-HOMOSERINE TRANSACETYLASE.

8 : HOMOSERINE KINASE.

9 : THREONINE SYNTASE.

10 : THREONINE DEAMINASE.

11 : THREONINE DEHYDROGENASE.

Figure 1. Biosynthetic pathway of threonine.

2.2. Buffers

The breaking buffer contained 20 mM potassium phosphate, pH 7.2, 2 mM Mg tritriplex, 1mM dithiothreitol, 0.15 M KCl, 1 mM lysine and 1 mM threonine.

All the enzymes activities were measured with the same assay buffer AT (for aspartate-threonine), which contained 23 mM Tes (pH 7.5), 114 mM KCl and 6 mM $MgCl_2$, supplemented with amino acids as described by Szczesiul and Wampler[3].

2.3. Metabolic intermediates

DL-aspartic ß-semialdehyde was prepared by ozonolysis of allylglycine as described by Szczesiul and Wampler[3].

2.4. Enzyme assays

Aspartokinase activity was measured by two methods: the first is the coupled assay which was described by Wampler and Westhead[4]. In the second method, the reaction mixture contained, in a volume of 1 ml, 23 mM Tes, pH 7.5, 114 mM KCl and 6 mM $MgCl_2$, 20 mM ATP, 0.8 M hydroxylamine and 40 mM aspartic acid. The aspartokinase activity was then determined by measuring the amount of aspartohydroxamate formed per minute and per mg of proteines.

The other activities were measured as described by Wampler[3].

3. Results

3.1. Enzymatic activities of the threonine pathway

The first step in the study of the whole flux from aspartate to threonine was to measure the activity of the individual steps under the same conditions, i.e. in particular in the same buffer. This lead us to adopt the buffer AT to record all the enzymatic activities which are summarized in Table 1.

Table 1. Enzymatic activities of the individual steps of threonine pathway in nmole/min/mg protein (n, number of experiments).

Enzyme	Activity	
Aspartokinase	120.41	(n=4)
Aspartokinase + threonine	124.41	(n=4)
Aspartokinase + lysine	94.46	(n=4)
ASA dehydrogenase	275.71	(n=4)
Homoserinedehydrogenase	1099.80	(n=4)
Homoserine kinase	103.78	(n=3)
Threonine deaminase	0.173	(n=5)
Threonine dehydrogenase	0	(n=5)

3.2. Flux of threonine production from aspartate and control coefficient of the first step

Fig. 2 shows an example of measurement of the flux of threonine production from aspartate. Part (a) is the endogenous oxidation of NADPH in the absence of aspartate. Part (b) mainly represents the flux of threonine synthesis (after addition of aspartate) and part (c) the inhibition of the latter flux by lysine, due to aspartokinase III inhibition.

Fig. 3a represents the percentage of flux inhibition by lysine due to inhibiton of aspartokinase III, which is represented in Fig. 3b. The slopes at zero concentration of lysine allow us to determine the control coefficient of the first step for the whole flux. This control coefficient is equal to 0.072.

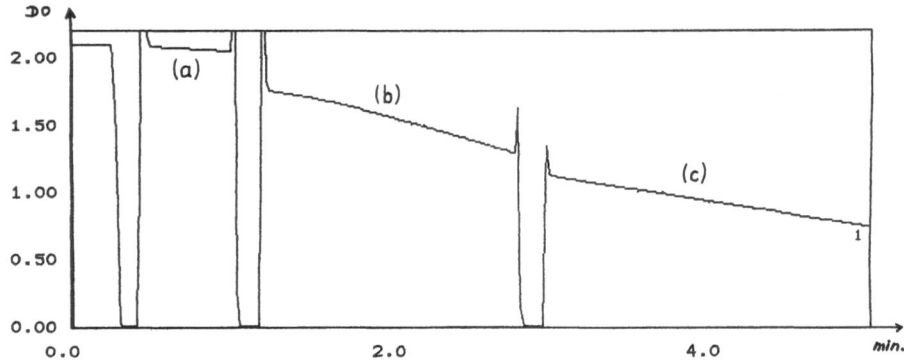

Figure 2 . The rate of threonine synthesis under different conditions. Endogenous activity (a), in the presence of aspartate (b), in the presence of aspartate and lysine (c).

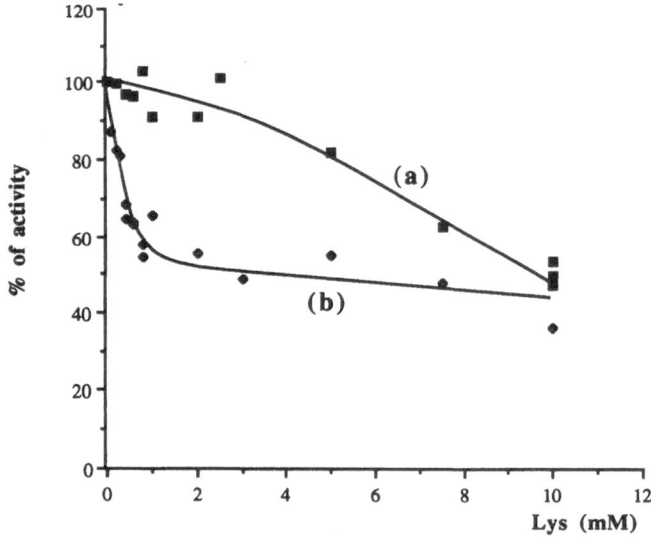

Figure 3. Inhibition of threonine synthesis flux by lysine (a). Inhibition of the aspartokinase activity by lysine (b).

3.3. Modeling of the pathway

In order to get a deeper insight into this amplification problem, we performed a modeling of the pathway flux at steady state, with very simple equations describing the enzymatic avtivity of each step (see Fig. 4). This model allows us to simulate the threonine production as a function of the copy number of threonine operon (see Fig. 5). The model allows also to calculate the control coefficient of each step as shown in Table 2.

$$V1 = \text{constante} \qquad V3 = \frac{k3 \cdot A \cdot X1}{1+(X6/q3)^{n1}} \qquad \text{Inhibition of AK I by Threonine}$$

$$V2 = k2 \cdot X1$$

$$V4 = k4 \cdot X2$$

$$V5 = k5 \cdot X3 \qquad V6 = \frac{k6 \cdot A \cdot X3}{1+(X6/q6)^{n2}} \qquad \text{Inhibition of HDH by Threonine}$$

$$V7 = k7 \cdot X4$$

$$V9 = k9 \cdot A \cdot X5 \qquad V8 = \frac{k8 \cdot A \cdot X4}{1+ X6/q8} \qquad \text{Inhibition of HSK by Threonine}$$

$$V10 = k10 \cdot X6$$

$$V11 = k11 \cdot X7 \qquad V12 = \frac{k12 \cdot X1}{1+(X7/q12)^{n3}} \qquad \text{Inhibition of AK III by Lysine}$$

Figure 4. Kinetic equations of the reactions (the numbers are those of the steps represented in Fig. 1). Kinetic parameters : V_1=4, k_2=0.5, $k_{3.I}$ =2.66, $k_{3.III}$ =1.5, k_4=15, k_5=10, k_6=69, k_7=5, k_8=11.7, k_9=4, k_{10}=1, k_{11}=2, $n_{1.I}$ =4, $n_{1.III}$ =2, n_2=4.
Inhibition constants:
Model Thr-8: $q_{3.I}$ = 0.8 mM, $q_{3.III}$ = 0.5 mM, q_6= 0,8 mM, q_8= 0.8 5mM;
Model Thr-9: $q_{3.I}$ = 10 mM, $q_{3.III}$ = 0.5 mM, q_6= 10 mM, q_8= 0.8 5mM;
Model Thr-10: $q_{3.I}$ = 10 mM, $q_{3.III}$ =0.5 mM, q_6= 10 mM, q_8= 10mM.

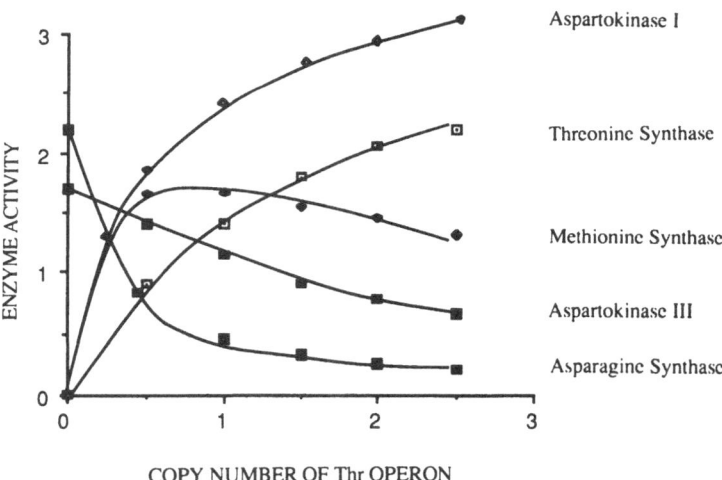

Figure 5. Variation of the fluxes (enzymatic activities) as a function of threonine operon amplification. Model Thr-9 with inhibition of AK III by lysine and inhibition of HSK by threonine.

Table 2. Flux control coefficients for threonine production.[§]

C_i	Thre-8	Thre-9	Thre-10
1-AATF	0.316	0.739	0.930
2-Asn-syn	-0.080	-0.083	-0.105
3-AK-I	0.040	0.057	0.072
3-AK-III	0.040	0.027	0.033
4-ASADH	0.000	0.000	0.000
5-Lys-syn	-0.149	-0.102	-0.128
6-HSDH	0.149	-0.102	-0.128
7-Met-syn	-0.172	-0.399	-0.318
8-HSK	0.172	0.399	0.318
9-Thr-syn	0.000	0.000	0.000
10-TDA	0.643	0.252	0.058
11-Lys-degr	0.041	0.009	0.011
$\Sigma_i C_i$	1.000	1.000	1.000

[§]Thr-8: model with inhibition of AK I, HSDH, HSK by threonine and AK III by lysine.
Thr-9: model with inhibition of HSK by threonine and AK III by lysine only.
Thr-10: model with inhibition of AK III by lysine only.

4. Discussion

The experimental determination of the control coefficient of aspartokinase activity shows good agreement with the prediction of the various models. In all the theoretical cases, this control coefficient is low, as experimentally measured.

This study will be carried on by measurement of the control coefficient of the other steps of the threonine pathway either by using the inhibitor method or by using the known kinetic properties of the enzyme and by measuring then the intermediate concentrations *in vivo* (giving access to the elasticity coefficients).

References

1. H. Kacser and J. A. Burns, The control of flux, *Symp. Soc. Exp. Biol.* **27**:65-104 (1973).
2. R. Heinrich and T. A. Rapoport, A linear steady-state treatment of enzymatic chains. General properties, control and effector strength, *Eur. J. Biochem.* **42**:89-95 (1974).
3. M. Szczesiul and D.E. Wampler, Regulation of a metabolic system *in vitro*: Synthesis of threonine from aspartic acid, *Biochemistry* **5**:2236-2244 (1976).
4. D.E. Wampler and E.W. Westhead, Two aspartokinases from *Escherichia coli*. Nature of the inhibition and molecular changes accompanying reversible inactivation, *Biochemistry* **7**:1661-1671 (1968)

CONTROL COEFFICIENTS
IN CELLULAR CATION HOMEOSTASIS:
A MODEL ANALYSIS

Jan Siegenbeek van Heukelom

Dept. of Experimental Zoology
Sect. Cell Biophysics
University of Amsterdam
Kruislaan 320, 1098 SM Amsterdam, The Netherlands

1. Introduction

The electrical component of the electrochemical gradients across cell membranes is the membrane potential V_m. Electrophysiologically it is defined as the potential in a cell with respect to the surrounding medium; more often then not it is defined experimentally as the change in potential measured with a microelectrode on impaling a cell. This microelectrode consists of a ultramicropipette ($\phi_{tip} \leq 0.5$ μm) filled with 3 M KCl, or another well-conducting solution, as a salt bridge[*]. Using such electrodes it was measured in mouse skeletal muscle fibres that V_m hyperpolarizes till about -95 mV when K_o is reduced moderately (≥ 1 mM). When the reduction is larger, V_m depolarizes to a steady state value of about - 50 mV.[2] Apparently membrane components controling V_m switch from one steady state to another and a singularity is observed in the plot $V_m(K_o)$. This mechanism was described in detail[2] and observed in many cell types, particularly in heart and muscle.

2. Description of the Model

Ion homeostasis is normally described as a "pump-leak" system, where the Na/K-pump forms the pump and the leak is formed by passive back-flux of potassium and sodium. The Na/K-pump, the ($Na^+ + K^+$)-ATPase, imports $2K^+$ ions and exports $3 Na^+$ ions at the expense of the hydrolysis of one ATP molecule. The main potassium leak in heart and muscle cells comprises the so-called Inward Potassium Rectifier (IKR). The molecular substrate of this

---

[*]Contrary to such electrodes, in another frequently used technique a patch clamp electrode ($\phi_{tip} > 0.5$ μm) in "whole cell configuration"[1] "breaks" open the cell to the electrode interior and thereby dialyzes the cytosol, thus disturbing the cellular homeostasis.

Modern Trends in Biothermokinetics, Edited by
S. Schuster *et al.*, Plenum Press, New York, 1993

rectifier is a membrane channel[3,4] that opens when approximately $V_m \leq E_K$ (E_K= potassium equilibrium potential). In the normal resting state, $V_m = V_{rest}$, the IKR is open and forms the major cationic conductive pathway through the membrane, even though V_m is slightly less negative then E_K (as can be deduced from eq. 1). The Na/K-pump is the major metabolite energized primary active cation membrane transporter in the membrane[§].

Changing K_o will influence the performances of both cation pathways leading to a new steady state value V_m. When the new steady state is obtained, the Kirchhoff current condition for all flows can be used to derive the well-known Goldman-Hodgkin-Katz equation[5], that relates V_m to the permeabilities (P_X) and concentrations of K^+, Na^+ and Cl^- inside (i) and outside (o) the cell:

$$V_m = \frac{RT}{F} \ln \frac{P_K[K]_o + P_{Na}[Na]_o + P_{Cl}[Cl]_i}{P_K[K]_i + P_{Na}[Na]_i + P_{Cl}[Cl]_o} . \tag{1}$$

However, the behavior of both components, IKR and Na/K-pump, depends strongly on K_o and V_m, therefore weakening the descriptive value of the equation considerably. Different equations, describing the same process, also demonstrate logarithmic dependences and do not simplify the outcome. As fluxes through the Na/K-pump can be related to the kinetics of its reaction scheme[5] the fluxes of the pump depend on K_o and Na_i: the simplest description commonly used is:

$$J_{pump} = \frac{J_{max}}{\left(1 + \frac{K_{mK}}{[K]_o}\right)^2 \left(1 + \frac{K_{mNa}}{[Na]_i}\right)^3} . \tag{2}$$

Equally, IKR depends in a rather complex way on a number of parameters and variables:

- intrinsic open channel conductances[&] depend in a square-root manner on K_o,[3]
- the open-close probability responds like a Boltzmann function to V_m, whereas the sensing of the field (V_h) also depends on E_K[£],

$$P_K = \frac{P_{Kmax}}{\sqrt{K_o}\left\{1 + \exp\frac{V_m - V_h}{v}\right\}} + 4P_{Na} \tag{3}$$

- the simplest assumption for V_h is $V_h = E_K$, but experimental data suggest that this dependence is more complex and varies from one type of cell to the other. In the model the equation V_h is modulated around E_K being lower when K_o is high ($\Delta V_{h(K=high)}$= -23 mV) and higher when K_o is low ($\Delta V_{h(K=low)}$= 30 mV) and $V_h = E_K$ at K_{co}:

$$V_h = E_K + \Delta V_{h(K=low)} - 2 \frac{\Delta V_{h(K=low)} - \Delta V_{h(K=high)}}{1 + e^{(K_{co}/K_o)}} \tag{4}$$

[§]Though large chloride conductances may exist, there are no important primary active chloride-pumps and therefore in all cases $E_{Cl} = V_{rest}$.

[&]As the "potassium conductance" g_K, according to experimental observations, depends linearly on the square root of K_o, and in the first approximation $g_K = c_{membrane} P_K$ the relationship between P_K and K_o is modeled by the reciprocal of the square root of K_o. Therefore g_K decreases by the reduction of K_o and P_K increases due to this element in eq. (3).

[£]The term $4P_{Na}$ represents a residual P_K so that V_m=-50mV when IKR is closed.

All the adaptations of these non-linear elements together in response to variation in K_O produce a new homeostatic set of intensive values such as V_m, K_i, Na_i, Cl_i and extensive values such as J_{pump}, I (electrical current across the membrane: $I = \sum_X I_X$ (with X = Na$^+$, K$^+$ or Cl$^-$) and form the new "homeostatic adjustment" of the cell to new outside conditions. Insertion of P_K in eq. (1) will not lead to a simple analytical solution and it was decided to simulate by numerical integration the behavior of V_m as a function of time with different K_O. For simulation the Macintosh application STELLA (and later STELLA II) was used; a short introduction was given during the meeting Biothermokinetics in 1990 [6]. Integration of the fluxes of the ionic species gives the concentrations K_i, Na_i and Cl_i. Integration (multiplied with the properly scaled charge contribution to give the current contribution I_X) in the membrane capacitor (C_m) the fluxes produces $C_m V_m = \int I_X dt$.

Essentially, the response of V_m on variations of K_O is correctly reproduced by the model. Though the singularity did not occur at $K_{sing} = 0.7$ mM and not at 1 mM, it has not been tried to adjust parameters to change this for several reasons mentioned later. The occurrence of the singularity can be understood also in semi-quantitative terms. The power of the Na/K-pump is reduced by reduction of K_O. The permeability P_K and driving force for K$^+$ increase, but also the driving force for Na$^+$ ($E_{Na} \approx +50$ mV) due to the hyperpolarization. To balance the Na$^+$ currents ($I_{Na} = g_{Na}(V_m - E_{Na})$) and the reduction of g_K in the K-currents ($I_K = g_K(V_m - E_K)$) the value of V_m will differ more and more from E_K (and therefore from V_h) till this difference becomes so large that the increase of the denominator of eq. (3) will exceed that of the nominator and P_K starts to decrease. With eq. (1) one can easily realize that this depolarization of V_m will close the IKR regenerative: the difference between V_m and V_h (eq. 4) amplifies its own action. Sensitivity analysis using other functions for $V_h(K_O)$ (unpublished data) than eq. 4 demonstrated that the occurence of this switch is not dependent on the exact formulation of V_h as long as ($V_m - V_h$)/v becomes large enough to effect this transition.

This model was used to calculate the control coefficients of those parameters that were considered the most influential in determining V_m near the singularity. According to their definition[7] control coefficients were calculated as:

$$(\Delta V_m / V_m)/(\Delta e/e) . \tag{5}$$

Identified as elements comparable with "enzymes" in metabolic control theory, the sodium permeability (P_{Na}), potassium permeability (P_K) and maximal pump activity (J_{max}) were used in the simulation as the entity e. Additionally, as some authors mention that "the Na/K-pump can become more powerful" the control by outside potassium K_{mK} is included. Variations ($\pm\Delta e$) of 10 % were introduced after attainment of the steady state at a specific value of K_O: smaller values appeared inconvenient. For instance, a control coefficient of 10% determined in this way equals a variation in V_m of 0.3 to 0.8 mV, values in the range of the long-term stability of the measuring system. Additionally, a computational problem was that near the singularity (switching point from hyperpolarized to depolarized V_m values) even after long-term calculation (simulating over one hour experimental time) the model still drifts slowly. To overcome this problem the sensitivity analysis of STELLA II was used to simulate V_m with three different values: $0.9e$, e and $1.1e$. The values $10^{(1-(V_m \pm \Delta V_m)/V_m)}$ were calculated as function of time and curve-fitted by a single exponential curve and the value for $t=\infty$, and $+\Delta e$ was taken as the control coefficient. In the calculations, the values with $+\Delta e$ and $-\Delta e$ did not differ more then a few percent.

3. Results

The results of the calculations are plotted in Figure 1 on a logarithmic scale for K_O. In eq.

5 and the calculations the relative entity $\Delta V_m/V_m$ is not transformed to $\Delta \ln|V_m|$ for two reasons. According to eq. 1, V_m already behaves as a complex logarithmic function of concentrations and permeabilities. As V_m is negative ($-20\ mV > V_m > -120\ mV$), it was preferred to keep this fact still observable in the control coefficients. When a positive variation in the input led to a *depolarization* ($\Delta V_m > 0$) this is registered as a *negative control coefficient*. The value $K_0 = 0.7\ mM$ is presented in Fig. 1 as a line; it is difficult to obtain an exact value for the potassium concentration where the singularity occurs: $K_0 = K_{sing}$. Experimentally this is impossible for reasons of measuring inaccuracy and variability of the preparation.

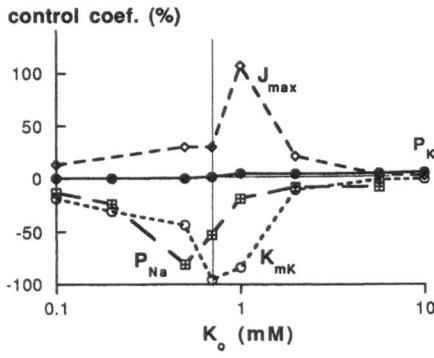

Figure 1. The control coefficients of some elements of the model on the membrane potential (V_m) as a function of the outside potassium concentration (K_0): J_{max} is the control by the maximal activity of the Na/K-pump, P_K by the potassium permeability, K_{mK} by the affinity constant of the Na/K-pump for K_0 and P_{Na} for the sodium permeability. All control coefficients are calculated according to the procedure described in the text and expressed in percents. The vertical line is the value $K_0 = 0.7\ mM$.

Computationally, it turned out that at K_0 values near K_{sing}, simulations keep on drifting:characteristic time constants of 2000 to 3000 s were found. Even after a long simulation time, one cannot conclude what the exact K_{sing} is. This aspect is also observed experimentally: after a long apparent stable hyperpolarization the cell might still switch to depolarization (see also Fig. 2 in Ref. 2). This incertainty, whether $K_{sing} = 0.7\ mM$, precludes the conclusion that the largest control coefficients presented are really the maximal values. Presented here on a logarithmic scale for K_0, they nevertheless demonstrate several remarkable aspects.

K_{mK} ($\approx 1\ mM$ [8]) exerts a negative control because lowering K_{mK} increases the affinity of the Na/K-pump. Apparently the maximal influence of K_{mK} is observed at slightly lower K_0 then J_{max}, most likely due to a different dependence, according to eq. 2. The influence of K_{mNa} has not been calculated, because it does not depend directly on K_0 and will produce similar results as K_{mK} according to eq. 2. The control coefficients for P_{Na} are negative as should be expected; the largest value was found below K_{sing}.

Though it has not been investigated in detail, these results suggest that one might "play around" with these three parameters to "adjust" K_{sing} of the model to experimental data, were it not that it is extremely difficult to specify this value. The control exerted by P_K is small, possibly because P_K itself depends rather steeply on V_m (cf. eq. 3).

4. Conclusions and Discussion

From the results it can be concluded that the sensitivity of V_m for variations in parameters of the model can be calculated. They suggest that one can find differences in the sensitivity of the system above or below $K_o = K_{sing}$. It has to be determined whether this difference in sensitivity reflects the behavior of the muscle fibres themselves. A question not answered, however, is whether one might give these data the name control coefficient, and consider them as metabolic control coefficient in the sense that the summation theorems can be applied. Anyhow, they can be positive and negative in these calculations, for good reasons. This shows that here summation theorems do not guarantee completeness of the description. The regenerative character of the depolarization of V_m also leads to the question whether in such processes description in terms of control coefficients is allowed.

The question put forward in the previous section, whether the influence of affinity constants can be described by the theorems is of major physiological importance. Many hormones and other modulating substances exert their action on cells by changing enzyme affinities (or membrane channel gating properties)[9]. The possibilities to use control theory in (electro)physiology would increase considerably if a well-tailored formalism existed for this field. It should include a description of how the influence of a voltage sensor in a membrane channel can be modeled.

References

1. O.P. Hamill, A. Marty, E. Neher, B. Sakmann and F.J. Sigworth, Improved patch-clamp techniques for high-resolution current recording from cells and cell-free membrane patches, *Pflügers Arch.* **391**:85 (1981).
2. J. Siegenbeek van Heukelom, Role of the anomalous rectifier in determining membrane potentials of mouse muscle fibres ar low extracellular K^+, *J. Physiol. (London)* **434**:549 (1991).
3. B. Hille. "Ionic Channels of Excitable Membranes," (2nd Edition), Sinauer Associates Inc., Sunderland (Mass.) (1992).
4. C. Miller, Annus mirabilis of potassium channels, *Science* **252**:1092 (1991).
5. P. Läuger. "Electrogenic Ion Pumps," Sinauer Associates Inc., Sunderland (Mass). (1992).
6. J. Siegenbeek van Heukelom, Modelling the membrane potential of muscle fibres, *in*: "Proceedings of the 4th BioThermoKinetics Meeting," H.V. Westerhoff, ed., Intercept, Andover, in press.
7. H.V. Westerhoff and K. van Dam. "Thermodynamics and control of biological free-energy transduction," Elsevier, Amsterdam (1987).
8. O.M. Sejersted, Maintenance of Na,K-homeostasis by Na,K-pumps in striated muscle, *in*: "The Na^+,K^+-pump, Part B: Cellular aspects," J.C. Skou, J.G. Nørby, A.B. Maunsbach and M. Esmann, eds, Alan R. Liss, New York (1988).
9. H. van Mil and J. Siegenbeek van Heukelom, Control in electrophysiology and signal transduction, this volume.

IV. CONTROL AND REGULATION
OF OXIDATIVE PHOSPHORYLATION

EXPERIMENTAL DISCRIMINATION BETWEEN PROTON LEAK ACROSS THE MITOCHONDRIAL MEMBRANE AND SLIP IN THE PROTON PUMPS OF THE ELECTRON TRANSPORT CHAIN

Martin D. Brand[1] and Philippe Diolez[2]

[1]Department of Biochemistry
 University of Cambridge
 Tennis Court Road
 Cambridge CB2 1QW, England
[2]Biochimie Fonctionelle des Membranes Végétales
 CNRS
 91198 Gif-sur-Yvette, France

1. Introduction

It has been known since 1974 that when mitochondrial State 4 respiration on succinate is titrated with a respiratory inhibitor such as malonate, the relationship between protonmotive force (Δp) and respiration rate (J_o) is non-linear. This has been interpreted in a number of different ways, but there are two current models to explain the non-linearity[1,2]. There is either

(a) a non-ohmic proton leak (an increase in the proton conductance of the inner membrane at high Δp), so that the rate of respiration is disproportionately high at high potential, but the number of protons pumped by the chain per oxygen consumed (the H^+/O ratio) remains constant, or

(b) a slip reaction in the electron transport chain, so that electron flow is not always accompanied by proton pumping and the H^+/O ratio is decreased at higher Δp, leading to a faster respiration rate at high potential to drive a proton leak that depends linearly on potential.

A situation in which there is a contribution by both (a) and (b) is also possible. We present here an experiment that distinguishes between these interpretations[3].

2. Theory

The proton conductance catalyzed by CCCP (C^{CCCP}) is known to be independent of Δp. In other words, the proton flux catalyzed by the uncoupler depends linearly on potential; it is ohmic. If we measure the relationship between Δp and J_O in isolated mitochondria in State 4, and then add a low concentration of CCCP and repeat the measurement, then (as long as the endogenous proton leak pathway is unaffected by CCCP addition) the difference between the two curves represents the relationship between Δp and J_O catalyzed by the uncoupler. The extra proton influx will be given by C^{CCCP} multiplied by the value of Δp, and will be equal to the extra proton efflux in the steady state. This in turn will be given by the extra respiration rate (ΔJ_O) multiplied by the number of protons pumped per oxygen (H+/O). Therefore

$$\Delta J_o = \Delta p \cdot C^{CCCP} \Big/ H^+/O \;. \tag{1}$$

Since C^{CCCP} is a constant for a given concentration of CCCP, a graph of ΔJ_O against Δp will pass through the origin and will have a slope inversely proportional to H+/O. If H+/O remains constant as Δp is varied (in other words, if the original non-linear titrations of Δp and J_O are explained entirely by changes in proton conductance) then this derived plot will give a straight line through the origin. If, however, slip occurs to any significant extent, H+/O will vary with Δp and the derived plot will give a curve, with slope rising as Δp rises.

We can also test whether the proton pumps slip as a function of turnover rate, in other words whether the H+/O ratio decreases as the rate of electron transport through the respiratory complexes increases at a given value of Δp. If we increase the electron transport rate by increasing the concentration of CCCP, then there will be a disproportionate increase in the respiration rate at a given value of Δp if H+/O decreases with rate, but a proportionate increase in respiration rate if H+/O is independent of rate. The easiest way to display this is to normalize the plot with respect to the concentration of added CCCP by dividing both sides of eq. (1) by [CCCP]. Thus

$$\frac{\Delta J_o}{[CCCP]} = \frac{\Delta p \cdot C^{CCCP}}{H^+/O \cdot [CCCP]} \tag{2}$$

i.e.

$$\frac{\Delta J_o}{[CCCP]} = \frac{\Delta p \cdot \text{constant}}{H^+/O} \;. \tag{3}$$

A graph of $\Delta J_o/[CCCP]$ against Δp will have slope inversely proportional to H+/O. Once again, if the endogenous proton conductance is independent of CCCP addition and the H+/O ratio remains constant as Δp is varied (in other words, if the original non-linear titrations of Δp and J_O are explained entirely by Δp-dependent changes in proton conductance), then this derived plot will give a straight line through the origin. If, however, slip occurs to any significant extent, H+/O will vary with Δp and the derived plot will give a curve, with slope rising as Δp rises. Secondly, if slip depends on electron transport rate then the lines representing higher concentrations of CCCP will be steeper than those representing lower uncoupler concentrations; if rate does not influence the H+/O ratio then these lines will superimpose. Thirdly, if the endogenous leak is affected by CCCP addition (e.g. if the endogenous leak is dependent on the rate of the redox proton pumps) then the lines obtained with different CCCP concentrations will not superimpose.

3. Results

The Figures show the results of the experiment. Fig. 1 shows a titration of respiration and Δp with malonate in the absence of CCCP, and two titrations with sub-maximal concentrations of added CCCP. In the absence of CCCP there is the expected non-linear relationship between Δp and J_O; in the presence of CCCP the non-linearity is less pronounced.

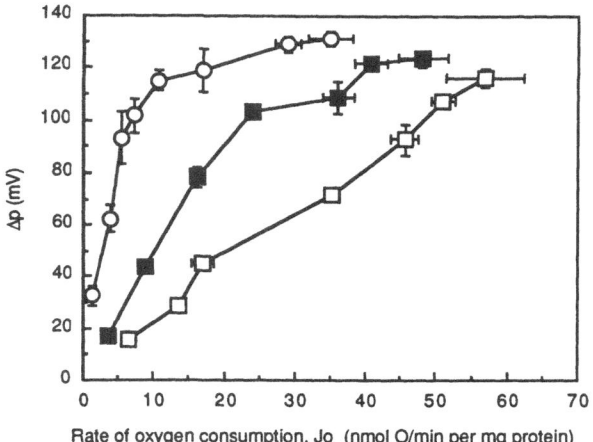

Figure 1. Malonate titrations of State 4 respiration of rat liver mitochondria on 4 mM succinate in the presence of zero (circles), 40 nM (full squares) and 80 nM (open squares) CCCP.
The medium contained 100 mM NaCl, 3 mM Hepes, 5 mM P_i, 1 mM EGTA, 5 μM rotenone, 200 nM valinomycin and 2 μg/ml oligomycin, 37 °C, pH 7.0. Points are mean ± SEM from 4 repeats on 2 rat liver mitochondrial preparations. To avoid complications from possible variations in the binding of lipid-soluble organic probes such as methyltriphenylphosphonium (TPMP$^+$), membrane potential was measured in a sodium-based medium using ^{86}Rb in the presence of valinomycin. ΔpH was measured with [^{14}C]-methylamine. In this medium, even in the presence of 5 mM phosphate, ΔpH was reversed, with a mean value of 21.5 mV. Both ΔpH and mitochondrial volume were constant during the titrations. Δp is quite low under these conditions, but the State 4 titration is clearly markedly non-linear. The respiratory control ratio (rate with excess CCCP/rate with oligomycin) was greater than 9.

Fig. 2 shows the derived secondary plot of ΔJ_O/[CCCP] against Δp. The plot is linear, as predicted if all of the non-linearity of the State 4 titration is caused by the endogenous non-ohmic proton leak. As a comparison, the curve expected if all of the non-linearity of the titration in Fig. 1 is due to slip is shown by open circles. It was generated by assuming an ohmic endogenous proton leak with a slope given by the lowest three points of the State 4 titration in Fig. 1, calculating the relative value of H$^+$/O that would be needed at each higher potential to produce the other points on this titration line, then using these H$^+$/O values and the lowest points of the other curves in Fig. 1 to predict the increment in respiration rate at each potential expected following addition of the ohmic uncoupler. It is clear that the experimental points show no tendency to follow the line predicted for slip; the data is not consistent with the hypothesis of slip in the proton pumps that increases at higher values of membrane potential.

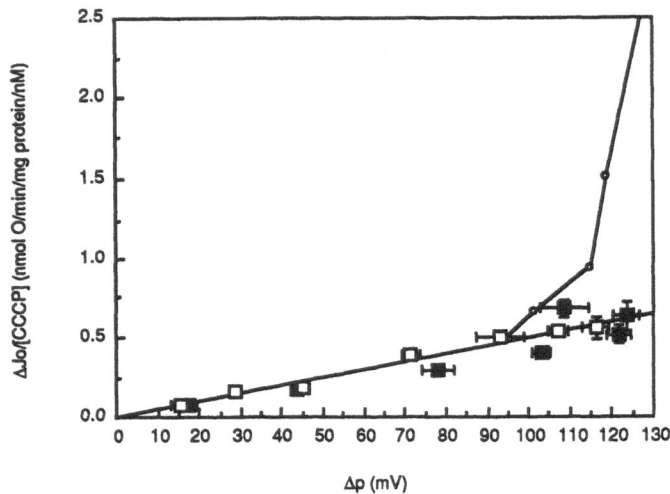

Figure 2. $\Delta J_O/[CCCP]$ versus Δp. ΔJ_O is the difference between the measured value of J_O with CCCP present and the interpolated value at the same Δp with CCCP absent, from Fig. 1. Concentration of CCCP was 40 nM (full squares), 80 nM (open squares). The rising curve (small open circles) shows the result expected if the non-linearity of the State 4 titration is caused by slip (variable H^+/O) superimposed on a small ohmic proton leak.

The points for the two titrations with different concentrations of CCCP superimpose completely, showing that there is no change in H^+/O ratio dependent on the rate of electron transport or the concentration of CCCP. There is also no increase in the endogenous proton conductance as the turnover rate of the proton pumps increases, as this too would cause the line representing the higher uncoupler concentration to be steeper.

4. Conclusions

We conclude that the State 4 titration in Fig. 1 is completely explained by the non-ohmic nature of the endogenous leak of protons across the mitochondrial inner membrane, and that slip reactions are insignificant. Neither the H^+/O ratio nor the endogeneous proton conductance of the inner membrane is affected by the turnover rate of the redox proton pumps.

References

1. M.P. Murphy, Slip and leak in mitochondrial oxidative phosphorylation, *Biochim. Biophys. Acta* **977**:123-141 (1989).
2. G.C. Brown, The leaks and slips of bioenergetic membranes, *FASEB J.* **6**:2961-2965 (1992).
3. M.D. Brand and P. Diolez, Experimental discrimination between proton leak across the mitochondrial membrane and slip in the proton pumps of the electron transport chain, *EBEC Reports* **7**:64 (1992).

DOES REDOX SLIP
CONTRIBUTE SIGNIFICANTLY
TO MITOCHONDRIAL RESPIRATION?

Keith D. Garlid, Craig Semrad and Valerie Zinchenko

Oregon Graduate Institute
19600 N.W. van Neumann Drive
Beaverton, Oregon 97006-1999, USA

1. Introduction

Redox slip is an intrinsic leak of protons that occurs within the electron transport mechanisms, thereby causing oxygen consumption without proton pumping. Controversy continues to surround the mechanism of energy dissipation during resting respiration: does it result primarily from ion leak or from slip in the proton pump machinery?

The existence of slip is based on the following experimental evidence[1]:

(i) The rate of state 4 respiration is 2-3 times faster than can be accounted for by leak at the same $\Delta\Psi$.
(ii) A similar quantitative discrepancy is observed with certain uncouplers, called slip-inducers.
(iii) Respiratory inhibitors have a much greater effect on respiration than on $\Delta\Psi$, such that the flux-voltage plots for respiration are more highly non-linear at high voltages than are those for leak.
(iv) Flux-voltage plots are found to exhibit distinct characteristics for individual elements of the electron transport chain.

Our own experiments confirm these findings, which have been taken to support, if not require, the concept that slip is a major contributor to resting respiration. Indeed, slip has heretofore been the only coherent mechanistic explanation for these phenomena. Whereas Brown and Brand[2] have argued that proton leak is sufficient to account for respiration rate, they have offered no alternative hypothesis to explain the same data in terms of proton leak.

2. The Theory of Proton Leaks in Mitochondria

The large electrical potential difference, $\Delta\Psi$, generated by the redox chain, will drive influx of cations and H^+ across the permeability barrier of the inner membrane. As we have shown[3], the theory of ionic diffusion across thin bilayer membranes is well-understood and

Modern Trends in Biothermokinetics, Edited by
S. Schuster *et al.*, Plenum Press, New York, 1993

readily applied to experiments in mitochondria. Following is a brief review.

Consider an ion located in an aqueous energy well adjacent to the hydrophobic barrier of the membrane. This ion can reach the center of the membrane if it has sufficient energy to overcome the unfavorable Gibbs energy of transfer into the hydrophobic medium. Because penetration to the center of the membrane is the limiting step in ion transport, net flux will be proportional to the differential probability of getting to the center from one side *versus* the other.

The probability of an ion in water having sufficient energy to move to the center of the membrane is given by a Boltzmann function, $\exp(\tilde{\mu}_p / RT)$, where $\Delta\tilde{\mu}_p \equiv \tilde{\mu}_p - \tilde{\mu}_{aq}$ is the Gibbs energy of the ion at the peak (p) relative to its value in the aqueous well (aq). The resulting flux equation, derived previously[3], is

$$J = P\,[C_1 \exp(u/2 - C_2 \exp(-u/2)]\,, \qquad (1)$$

where $u \equiv -zF\Delta\Psi / RT$, z is the valence of the ion, $\Delta\Psi$ is the electrical membrane potential, and C_1 and C_2 are concentrations in the surface energy wells. Concentrations are used rather than activities, because the effect of the membrane on the potential energy of the ion is included in $\Delta\tilde{\mu}_p^o$, the standard Gibbs energy of transfer of the ion to the hydrophobic center (the activation energy). $\Delta\tilde{\mu}_p^o/RT$ represents the height of the barrier that must be overcome in order for the ion to traverse the membrane at equilibrium. P is the permeability, given by

$$P \equiv k\exp(-\Delta\tilde{\mu}_p^o / RT)\,. \qquad (2)$$

For a nonelectrolyte ($z=0$) or an ion in the absence of an electric field ($\Delta\Psi=0$), eq. (1) reduces to Fick's law:

$$J = P\,(C_1 - C_2)\,. \qquad (3)$$

When $\Delta\Psi \leq 25$ mV, eq. (1) reduces to an ohmic relationship:

$$J = P\overline{C}\{(C_1 - C_2)/\overline{C} + u\}\,. \qquad (4)$$

Eq. (4), in turn, is a generalization of the more narrowly restricted phenomenological approximation, which is *only* valid near equilibrium:

$$J = L\left(\Delta\tilde{\mu}_{H^+} + u\right)\,. \qquad (5)$$

This analysis shows that eqs (4) and (5) are approximations and that linearity should not be viewed as the expected behavior in mitochondria, where $\Delta\Psi$ greatly exceeds values at which these equations apply. It must be emphasized that ohmic equations for ion leaks *are not valid* in mitochondria. The same comment applies to the Fick flux equation (3).

3. Practical Flux Equations

At high values of $\Delta\Psi$, contributions from back-flux are negligible, and eq. (1) reduces to a simple function of $\Delta\Psi$:

$$J = P\,C_1 \exp(u/2)\,. \qquad (6)$$

At $\Delta\Psi$=180 mV and $C_1=C_2$, this approximation is accurate to within 0.1%. The factor 1/2 in eqs (1) and (6) reflects the expectation that the maximum energy barrier will be sharp and at the center of the membrane. We removed this assumption (3) to accommodate other geometries of energy barriers within ion transport pathways, including those in channels, by introducing the general parameter, β:

$$J = P \, C_1 \exp(\beta u) \,. \tag{7}$$

Experimental data are conveniently analyzed and plotted in the semi-logarithmic version of eq. (7):

$$\ln J = \ln (PC_1) + \beta u \,. \tag{8}$$

Eq. (8) provides a practical means for estimating fundamental parameters, ß and $J_o \equiv PC_1$, from experimental flux-voltage plots. J_o is the exchange flux at $\Delta\Psi = 0$.

4. Experiments on Ion Flux Across Bilayers

Our experimental results confirm that ion fluxes are log-linear with $\Delta\Psi$ and agree entirely with the predictions of non-ohmic theory[3]. In three cases where ion leak is the mechanism, β=1/2, in agreement with the single-barrier model of eq. (1). These cases are: H^+ leak in non-respiring mitochondria[4], tetraethylammonium (TEA^+) leak in respiring mitochondria[5], and SCN^- leak in liposomes[6]. In three additional cases, where ions cross the membranes through channels, β=1/4, indicating an energy well near the center of the membrane. These cases are: Cl^- flux through the uncoupling protein[6], polyamine flux through the spermine channel[7], and K^+ flux through the K_{ATP} channel[8].

5. Inferences from Studies on Ion Leaks in Mitochondria

A number of workers have assumed that proton leak is a function of protonmotive force, $\Delta\tilde{\mu}_{H^+}$, even at high $\Delta\Psi$. Eq. (6) shows that this is incorrect. The pH gradient can have no effect on proton leak at high values of $\Delta\Psi$, where back-flux makes a negligible contribution.

It might be thought that the exponential flux-voltage relationship only applies to ions with very low permeability. To the contrary, it applies equally to diffusing ions with high permeability, as demonstrated by our results with SCN^-.

The take-home lesson is that ion diffusion across the bilayer can never be ohmic at high $\Delta\Psi$. Non-ohmicity arises because biomembranes are extremely thin, so that edge effects, dominated by the exponential Boltzmann function, predominate. As shown by both theory and experiment, eq. (7) accurately reflects the physicochemical system and furthermore does not appreciably complicate extraction of useful parameters from the data.

6. The Logical Weakness in the Slip Argument

The principal basis for the conclusion that slip is a major contributor to energy dissipation is that state 4 respiration is 2 to 3 times faster than can be accounted for by leak at the same $\Delta\Psi$. Respiration, in H^+ equivalents, equals slip plus leak. Measured H^+ leak only accounts for 30-50% of respiration. Therefore, the remainder is due largely to slip. This syllogism contains a fallacy.

We measure the rate of electron transport in H^+ equivalents, J_{RESP}, and the rate of proton leak, J_{INH}, in inhibited mitochondria and compare these two values at the same $\Delta\Psi$; then

$$J_{RESP} = J^+{}_{leak} + J^+{}_{slip} ,$$

$$J_{INH} = J^-{}_{leak} ,$$

$$J_{RESP} - J_{INH} = J^+{}_{slip} + \Delta J_{leak} ,$$

where superscripts +/- refer to presence and absence of respiration, respectively. The logical fallacy resides in the unstated assumption that $\Delta J_{leak}=0$; that is, that extrinsic leak is

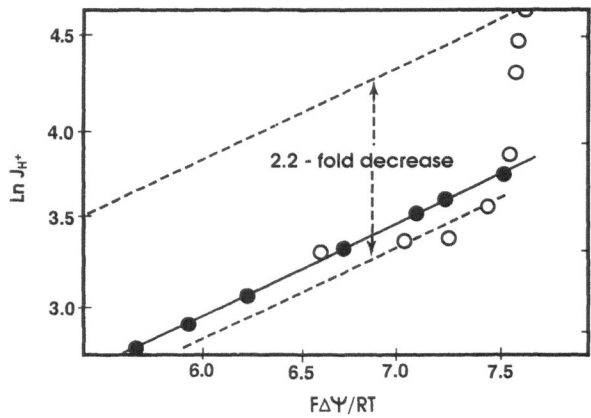

Figure 1. Flux-voltage curves for J_{RESP} and $J^-{}_{leak}$ in rat liver mitochondria. $\ln(J_{H^+})$ is plotted versus the reduced voltage, $F\Delta\Psi/RT$. J_{H^+} was measured as the rate of proton ejection during inhibition by malonate (open circles) or as the rate of proton leak into fully inhibited mitochondria as $\Delta\Psi$ was varied by varying $[K^+]_o$ in the presence of valinomycin (closed circles). The dashed and solid lines have a slope 1/2. Respiration was measured in triethylamine salts of 10 mM succinate, 50 mM acetate, 5 mM phosphate, 5 mM Cl, 0.5 mM EGTA, 10 mM TES buffer, pH 7.4. Medium also contained rotenone and 0.5 mM $MgCl_2$ and was adjusted to 400 mOsm with sucrose so that the osmolality of impermeant solutes was about 270 mOsm. An H^+/O stoichiometry of 7 was assumed.
$\Delta\Psi$ and J_{H^+} were measured in identical medium with a glass K^+ electrode in the presence of valinomycin. Mitochondria were present at 2 mg/ml in both assays.

independent of respiratory chain activity. This assumption has never been addressed and has led to the unverified conclusion that the difference, $J_{RESP}-J_{INH}$, is due entirely to slip.

Fig. 1 contains semi-logarithmic flux-voltage plots from experiments in which J_{H^+} was measured as respiration (open circles) and as leak in inhibited mitochondria (closed circles). It can be seen that $J^-{}_{leak}$ follows theory with slope 1/2 and that J_{RESP} approaches the same curve after respiration has been inhibited by about 50%. Initially, however, J_{RESP} deviates sharply from J^-_{leak}. The only way to explain the deviation in terms of ion leak, given that $\beta = 1/2$, is to imagine that the respiration curve is traversing a series of parallel log-linear curves with different intercepts, $\ln(PC_1)$. This cannot be tested with proton leak, because it cannot be measured directly during respiration. On the other hand, any respiration-dependent changes in permeability, being extrinsic to the proton pump machinery, should affect transport of all ions, and fluxes of cations and ionophores *can* be measured during uncoupled and inhibited respiration. This provides an experimental means to test whether the quantitative discrepancies previously observed can be attributed to changes in extrinsic ion leak.

7. Cation Permeability Depends on the Rate of Respiration

Fig. 2 contains semi-logarithmic flux-voltage plots from experiments in which tetraethylammonium flux (J_{TEA^+}) was measured as respiration was varied with uncoupler

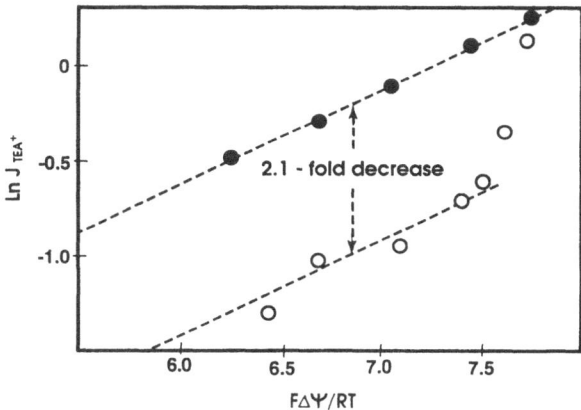

Figure 2. Flux-voltage curves for J_{TEA^+} during uncoupled and inhibited respiration. Influx of TEA$^+$, J_{TEA^+}, was determined by quantitative light scattering as $\Delta\Psi$ was varied by CCCP (closed circles) or malonate (open circles). The two lines have a slope 1/2.
Assay medium was 110 mosm and contained succinate, phosphate, EGTA, TES, MgCl$_2$, as above, plus 40 mM TEA acetate. $\Delta\Psi$ was determined in parallel experiments using DiS-C$_3$(5) fluorescence. Mitochondria were present at 0.1 mg/ml in both assays.

(open circles) and inhibitor (closed circles). As with respiration itself (Fig. 1), J_{TEA^+} deviated from log-linearity at high $\Delta\Psi$. On the other hand, J_{TEA^+} followed theory with slope 1/2 when $\Delta\Psi$ was reduced with uncoupler and approached a parallel curve when $\Delta\Psi$ was reduced with inhibitor. At the same value of $\Delta\Psi$, TEA$^+$ flux differed by a factor of 2.0 depending on how $\Delta\Psi$ was varied.

This result, which has been reproduced in 4 separate experiments, cannot be attributed to slip. Instead, it shows that extrinsic cation permeability is modified by activity of the respiratory enzymes. Indeed, it follows that as respiration is inhibited, J_{TEA^+} is traversing a series of parallel log-linear curves with different intercepts, $\ln(PC_1)$.

8. Ionophore Permeability Depends on the Rate of Respiration

Each uncoupling modality we have studied, including the combination valinomycin plus nigericin, was found to stimulate respiration to a greater degree than can be accounted for by its stimulation of passive leak measured in the absence of respiration. J_{RESP} was linearly related to J_{INH} with slope of 2 (CCCP) to 4 (bupivacaine). The slopes for individual uncouplers were constant, even down to low $\Delta\Psi$, indicating that ΔJ_{leak} is constant as respiration is increased from state 4 value. Pietrobon and coworkers[9] concluded that FCCP interacts with the proton pumps to induce slip. Arguing against this conclusion:

(1) Slip should decrease as $\Delta\Psi$ is lowered by uncouplers;
(2) A "slip-inducer" must interact specifically with elements of the electron-proton transport apparatus; whereas we show that the effect is independent of the chemical nature of the uncouplers used. From a pharmacological point of view, if *all* uncouplers are slip-inducers, then none are.

9. Conclusions

These results show that ion permeability is independent of $\Delta\Psi$, but exhibits a transition that depends on the fraction of active respiratory elements. This means that the assumption that extrinsic leak is independent of respiratory chain activity is incorrect, and ΔJ_{leak} is non-zero. The results account for three of the four evidences for the existence of slip summarized by Luvisetto and coworkers[1]. Given that ΔJ_{leak} (the active component of leak) depends on activity of electron transport proteins, it follows that its value should differ when different segments of the chain are studied. This hypothesis, relating to the fourth evidence for slip, will be examined in future experiments.

How does the permeability transition arise? J_{RESP} and J_{INH} are measured in two entirely different states of the membrane. There is no *a priori* reason to assume that H^+ leak will be unaffected by catalytic activity within membrane proteins. Ion leak occurs primarily at protein-lipid interfaces, where the energy barrier is perturbed and resistance is much lower than in the pure bilayer. Activity of the respiratory enzymes may increase this perturbation and increase the activation entropy for diffusion. Thus, the conclusion that permeability to ions and ionophores increases when respiration is turned on is not as surprising as it may seem at first glance.

The persistence of the slip *versus* leak controversy despite 12 years of study may be due in part to the use of holistic models containing excessive numbers of adjustable parameters[10]. Such models have failed to resolve the problem because they have not correctly described the system. Integrated models, such as Hill diagrams[10], control theory[11] or nonequilibrium thermodynamics[12], are generally incapable of resolving mechanistic questions such as slip versus leak. If one is asking questions about mechanism, reductionism is the more profitable approach and is very likely essential for correct answers. Integrated models may eventually prove useful when they contain a proper formulation of the leak component.

References

1. S. Luvisetto, E. Conti, M. Buso and G. F. Azzone, Flux ratios and pump stoichiometries at sites II and III in liver mitochondria, *J. Biol. Chem.* **266**:1034 (1991).
2. G. C. Brown and M. D. Brand, On the nature of the mitochondrial proton leak, *Biochim. Biophys. Acta* **1059**:55 (1991).

3. K. D. Garlid, A. D. Beavis and S. K. Ratkje, On the nature of ion leaks in energy-transducing membranes, *Biochim. Biophys. Acta* **976**:109 (1989).

4. X. Sun, and K. D. Garlid, On the mechanism by which bupivacaine conducts protons across the membranes of mitochondria and liposomes, *J. Biol. Chem.* **267**:19147 (1992).

5. A. Toninello, G. Miotto, D. Siliprandi, N. Siliprandi and K. D. Garlid, On the mechanism of spermine transport in liver mitochondria, *J. Biol. Chem.* **263**:19407 (1988).

6. K. D. Garlid, New insights into mechanisms of anion uniport through the uncoupling protein of brown adipose tissue mitochondria, *Biochim. Biophys. Acta* **1018**:151 (1990).

7. A. Toninello, L. Dalla Via, D. Siliprandi and K. D. Garlid, Evidence that spermine, spermidine and putrescine are transported electrophoretically in mitochondria by a specific polyamine uniporter, *J. Biol. Chem.* **267**:18393, 1992.

8. P. Paucek, G. Mironova, F. Mahdi, A. D. Beavis, G. Woldegiorgis and K. D. Garlid, Reconstitution and partial purification of the glibenclamide-sensitive, ATP-dependent K^+ channel from rat liver and beef heart mitochondria, *J. Biol. Chem.* **268**, in the press.

9. D. Pietrobon, S. Luvisetto and G. F. Azzone, Uncoupling of oxidative phosphorylation, *Biochemistry* **26**:7339 (1987).

10. D. Pietrobon and S. R. Caplan, Flow-force relationships for a six-state proton pump model, *Biochemistry* **24**:5764 (1985).

11. M. D. Brand, R. P. Hafner and G. C. Brown, Control of respiration in non-phosphorylating mitochondria is shared between the proton leak and the respiratory chain, *Biochem. J.* **255**:535 (1988).

12. D. Pietrobon, M. Zoratti, G. F. Azzone, J. W. Stucki and D. Walz, Non-equilibrium thermodynamic assessment of redox-driven H^+ pumps in mitochondria, *Eur. J. Biochem.* **127**:483 (1982).

MECHANISTIC STOICHIOMETRY OF YEAST MITOCHONDRIAL OXIDATIVE PHOSPHORYLATION: A BEHAVIOR OF WORKING ENGINE

Valérie Fitton, Michel Rigoulet and Bernard Guérin

Institut de Biochimie Cellulaire
C.N.R.S.
Université de Bordeaux II
1 rue Camille Saint-Saëns, 33077 Bordeaux Cédex, France

1. Introduction

From many studies conducted to determine the stoichiometry of mitochondrial oxidative phosphorylation employing a remarkable variety of techniques and discarding several common systematic errors, there has been an increasing consensus in recent years. In mammalian mitochondria, the measured values of ATP/2e⁻ were close to 1, 0.5 and 1 at the three "coupling sites", respectively[1,2]. However, such determinations have been made under conditions where the electron flux through each respiratory unit is maximum (State 3) even if the number of functional units is changed by using inhibitor titration.

Mitochondria isolated from *Saccharomyces cerevisiae* in the exponential phase of growth lack a "phosphorylation site" corresponding to "coupling site" 1 of animal mitochondria. In previous work, we have shown that the ATP/O ratio is about 1.5 with ethanol or succinate and 1 with substrates oxidized at site 3 (TMPD + ascorbate or lactate) when we used classical methods[3]. However, we have also shown that the value of ATP/O increases when respiratory flux slows down in response to a limitation of substrate supply. For this study, we used an external NADH regenerating system and the ability of yeast mitochondria to oxidize exogenous NADH by a NADH dehydrogenase located towards the outer face of the inner membrane. However, under these particular conditions, the ATP/O ratio measured at maximal respiratory rate is rather low (approximately 1).

The aim of the present study is to reevaluate the mechanistic stoichiometry of yeast mitochondrial oxidative phosphorylation when the electron supply to the respiratory chain is changed as a function of the different matricial dehydrogenase activities.

2. Materials and Methods

Cells of diploid wild-strain *Saccharomyces cerevisiae* (yeast foam) were grown aerobically at 28 °C in a complete medium (pH 4.5) with lactate as carbon source. The cells were harvested in logarithmic growth phase and mitochondria were isolated from protoplasts as described in Ref. 4. ATP synthesis and oxygen consumption rates, ΔpH and $\Delta\Psi$ measurement were performed as described in Ref. 3.

3. Results and Discussion

Respiratory rate in State 4 supported by various substrates is very different, from 6 ± 1 natom $O \cdot min^{-1} \cdot mg^{-1}$ protein for ß-hydroxybutyrate to 160 ± 14 for ethanol whereas the Δp measured, under these conditions, increases from 116 ± 15 mV to 197 ± 4 mV (Table 1). Such a dependence between respiratory rate and Δp has been previously described and explained as a consequence of both processes: a non-ohmic proton conductance of the inner membrane and a saturation of the redox proton pump slipping when respiratory rate increases[5].

On isolated mitochondria, it is generally shown that the transition State 4 - State 3 corresponds to a decrease in both the protonmotive force and the level of reduced coenzyme (i.e. NADH). Such a phenomenon is observed on yeast mitochondria when ethanol is the respiratory substrate (not shown). With other substrates leading to NADH formation, the transition State 4 - State 3 is linked to an increase in NADH level. Moreover the stimulation of respiratory rate by ADP addition is insensible to oligomycin with malate, pyruvate or 2-oxoglutarate (compare Tables 1 and 2, and see also Ref. 6).

With ß-hydroxybutyrate as substrate, the transition State 4 - State 3 corresponds to an increase in protonmotive force (not shown). All these facts indicate that the kinetic control of respiratory rate is mainly due to the activities of the different matricial dehydrogenases (except for alcool dehydrogenase).

Table 1. Protonmotive force in State 4 of respiration with different substrates[*].

Substrates	V_{O_2} (State 4) (n·atom/min/mg prot.)	ΔpH (mV)	$\Delta\Psi$ (mV)	Δp (mV)
ß-hydroxy-butyrate	6 ± 01	13 ± 03	103 ± 12	116 ± 15
pyruvate	11 ± 01	53 ± 12	113 ± 09	166 ± 21
malate	19 ± 01	24 ± 05	123 ± 08	147 ± 13
pyruvate + ADP + oligo	19 ± 01	57 ± 01	124 ± 06	181 ± 07
malate + ADP + oligo	24 ± 02	47 ± 05	127 ± 02	174 ± 07
pyruvate-malate	60 ± 03	36 ± 02	145 ± 03	181 ± 05
2-oxoglutarate	62 ± 01	41 ± 03	155 ± 02	196 ± 05
isocitrate	78 ± 03	37 ± 02	150 ± 03	187 ± 05
succinate	139 ± 01	38 ± 04	156 ± 03	194 ± 07
ethanol	160 ± 01	33 ± 02	164 ± 02	197 ± 04

[*]Mitochondria (3 mg/ml) were incubated in basal medium: 0.65 M mannitol, 0.36 mM EGTA, tris-maleate 10 mM, tris-Pi 5 mM supplemented with valinomycin (0.2 µg/ml), oligomycin (25 µg/ml), rubidium (5 µM) in the presence of a respiratory substrate (5 mM) except for ethanol (109 mM).
$\Delta\Psi$ and ΔpH were determined by distribution of (^{86}Rb) and (^{3}H) - acetate respectively, and matrix space was determined by using (^{3}H) - water and inner-membrane impermeable (^{14}C) - mannitol.
The values presented are from, at least, five different experiments carried out with three different mitochondria preparations.

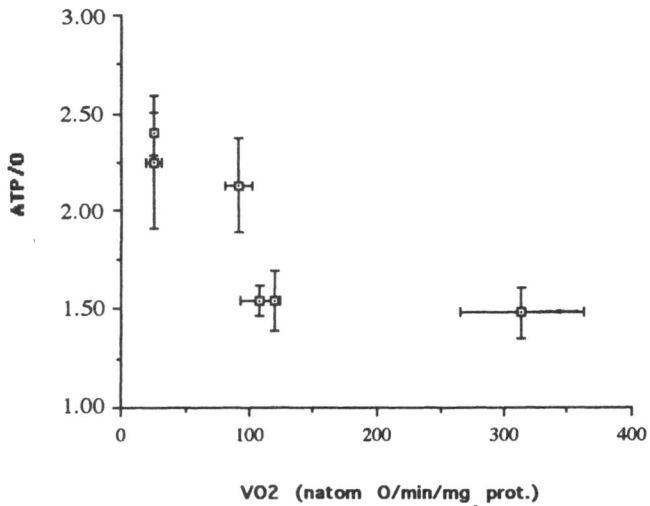

Figure 1. Dependence of ATP/O on respiratory rate. Mitochondria (0.5 mg/ml) except for malate and ß-hydroxy-butyrate (1.25 mg/ml) were incubated in basal medium supplemented with hexokinase (70 units/ml), glucose (10 mM), $MgCl_2$ (1 mM) and a respiratory substrate (5 mM) except for ethanol (109 mM). Phosphorylation was induced by adding ADP (1 mM). The points shown are from six different experiments carried out with three different mitochondria preparations.

Table 2. Protonmotive force in State 3 of respiration with different substrates[*].

Substrates	V_{O_2} (State 3) (n.atom/min/mg prot.)	ΔpH (mV)	$\Delta \Psi$ (mV)	Δp (mV)
ß-hydroxy-butyrate	25 ± 02	26 ± 02	111 ± 05	137 ± 07
malate	25 ± 01	27 ± 02	101 ± 06	128 ± 08
isocitrate	91 ± 06	21 ± 01	112 ± 09	133 ± 10
2-oxoglutarate	108 ± 16	32 ± 04	115 ± 03	147 ± 07
succinate	120 ± 05	28 ± 06	105 ± 11	133 ± 17
ethanol	314 ± 15	20 ± 04	110 ± 03	130 ± 07

[*]Mitochondria (3mg/ml) were incubated in basal medium supplemented with valinomycin (0.2 µg/ml), rubidium (5 µM), hexokinase (70 units/ml), glucose (10 mM), $MgCl_2$ (1 mM) and a respiratory substrate (5 mM) except for ethanol (109 mM).
Phosphorylation was induced by adding ADP (1 mM). $\Delta \Psi$ and ΔpH were determined by distribution of (^{86}Rb) and (^{3}H)-acetate respectively, and matrix space was determined by using (^{3}H)-water and inner membrane impermeable (^{14}C)-mannitol.
The values presented are from, at least, five different experiments carried out with three different mitochondria preparations.

Fig. 1A shows that the ATP/O ratio is 1.5 when respiratory rate is greater than 100 natom O·min^{-1}·mg^{-1} protein. Below this value, ATP/O increases and reaches 2.4 for ß-hydroxy-butyrate as substrate. In contrast, Δp is nearly constant regardless of the value of respiratory rate (Table 2).

Unless phosphorylation rate is misestimated (see Ref. 3 for the precautions in the determination of the oligomycin sensitive phosphorylation rate), two possible systematic errors must be considered. The first one concerns the diffusion of O_2 into the cell which can be a source of error particularly for a low respiratory rate. The facts that

(i) the use of a high amount of protein (until 5 mg.ml^{-1}) does not change the ATP/O ratio measured with malate, and that

(ii) at low respiratory rate obtained with ethanol plus myxothiazol (comparable to that observed with malate) the ATP/O ratio is similar to that measured with ethanol alone (1.5),

avoid the possibility of overestimating of the ATP/O ratio by underestimating oxygen consumption.

The second source of error is the absence of correction for proton leak in State 3. Such a correction has been done by some authors (see for instance Ref. 2) assuming that the rate of proton leak and slip is only controlled by the electrochemical proton gradient. However, if other parameters participate in slip control, i.e. the value of the flux through each respiratory chain unit[3,5], such a correction is not valid.

Nevertheless, we can consider three proton fluxes during oxidative phosphorylation,

(i) proton efflux catalyzed by the respiratory chain:

$$J_{H^+_{out}} = n \, J_O \, ,$$

where n is the stoichiometry H^+/O.

(ii) proton influx through ATP synthase:

$$J_{H^+_p} = n' \, J_{ATP} \, ,$$

where n' is the stoichiometry H^+/ATP linked to ATP synthesis *per se* and the transport processes.

(iii) proton influx trough the proton leakage :

$$J_{H^+_l} = L_{H^+} \, \Delta p \, ,$$

where L_{H^+} is the conductance of the membrane for protons.

At steady state:

$$J_{H^+_{out}} = J_{H^+_p} + J_{H^+_l} \, ,$$

$$n \, J_O = n' \, J_{ATP} + L_{H^+} \, \Delta p \, ,$$

and

$$\frac{ATP}{O} = \frac{J_{ATP}}{J_O} = \frac{J_{H^+_p}}{J_{H^+_{out}}} \cdot \frac{n}{n'} = \frac{n}{n'} \cdot \frac{J_{H^+_p}}{J_{H^+_p} + J_{H^+_l}} \, .$$

Two hypotheses may be considered :

(1) First, if $J_{H^+_l} \ll J_{H^+_p}$,

$$\frac{ATP}{O} \approx \frac{n}{n'}$$

Two cases occur : $\dfrac{ATP}{O}$ is constant, then $\dfrac{n}{n'}$ is constant;

$\dfrac{ATP}{O}$ varies, then $\dfrac{n}{n'}$ varies.

(2) If $J_{H^+_1}$ is not negligible relative to $J_{H^+_p}$, the lower the difference between $J_{H^+_p}$ and $J_{H^+_1}$, the lower will be the $\dfrac{ATP}{O}$ ratio.

Thus, the increase in ATP/O ratio when respiratory rate decreases can be explained only by, at least, a change in the mechanistic stoichiometry of one of the proton pump involved leading to an increase in efficiency while the flux slows down. Such an increase in the efficiency of the overall oxidative phosphorylation seems also linked to a total kinetic control above the respiratory chain, i.e. dehydrogenase activities.

This flux-yield dependence is in favor of an indirect coupling at the proton pump level and is a general behavior of a working engine.

References

1. C.D. Stoner, Determination of the P/2e⁻ stoichiometries at the individual coupling sites in mitochondrial oxidative phosphorylation, *J. Biol. Chem.* **262**:10445-10453 (1987).
2. P.C. Hinkle, M.A. Kuman, A. Resetar and D.L. Harris, Mechanistic stoichiometry of mitochondrial oxidative phosphorylation, *Biochemistry* **30**:3576-3582 (1991).
3. R. Ouhabi, M. Rigoulet and B. Guérin, Flux-yield dependence of oxidative phosphorylation at constant $\Delta\bar{\mu}_{H^+}$, *FEBS Lett.* **254**:199-202 (1989).
4. B. Guérin, P. Labbe and M. Somlo, Preparation of yeast mitochondria (*Saccharomyces cerevisiae*) with good P/O and respiratory control ratios, *Methods Enzymol.* **55**:149-159 (1979).
5. R. Ouhabi, M. Rigoulet, J.L. Lavie and B. Guérin, Respiration in non-phosphorylating yeast mitochondria. Roles of non-ohmic proton conductance and intrinsic uncoupling, *Biochim. Biophys. Acta* **1060**: 293-298 (1991).
6. M. Rigoulet, J. Velours and B. Guérin, Substrate-level phosphorylation in isolated yeast mitochondria *Eur. J. Biochem.* **153**:601-607 (1985).

MITOCHONDRIAL CREATINE KINASE: BIOTHERMOKINETICS OF THE REACTION AND ITS ROLE IN THE REGULATION OF ENERGY FLUXES IN MUSCLE CELLS *IN VIVO*

V.A. Saks[1], G.V. Elizarova[1], Yu. O. Belikova[1],
E.V. Vasil'eva[1], A.V. Kuznetsov[1],
L. Petrova[1], N.A. Perov[1] and F.N. Gellerich[2]

[1]Laboratory of Bioenergetics
 Cardiology Research Center
 3rd Cherepkovskaya 15A, Moscow 121552, Russia
[2]Institut für Biochemie
 Medizinische Akademie Magdeburg
 Leipziger Str. 44, 3090 Magdeburg, Germany

1. Introduction

Both the kinetics and thermodynamics of the creatine kinase reaction has been studied in great detail[1-3]. In these studies the soluble or solubilized isoenzymes of creatine kinase have been used. However, in the cells *in vivo* the creatine kinase isoenzymes function mostly in the coupled state both in mitochondria and myofibrils as well as at the cellular membranes forming the intracellular phosphocreatine pathway, or "phosphocreatine circuit" for the energy channeling[4-9]. In this work we investigated the thermodynamic and kinetic aspects of coupling of the creatine kinase to the oxidative phosphorylation in mitochondria.

The second purpose of this study was an investigation of possible reasons of the retarded intracellular diffusion of ADP in cardiac cells[9-11]. Two possible mechanisms were considered: intermediate binding to myosine filaments and limitation of ADP movement via porin channels in the outer membrane of mitochondria. The permeabilized cell technique was used in combination with selective extraction of myosin or hypo-osmotic swelling of mitochondria to disrupt its outer membrane. The results show that there may be multiple sites of retardation of the ADP diffusion, but one of the major sites of this type is the mitochondrial outer membrane.

The dependences of all kinetic constants of the forward mitochondrial creatine kinase reaction in the presence and absence of oxidative phosphorylation are determined. The role of the coupled system adenine nucleotide translocase—creatine kinase in the regulation of cellular energetics is investigated.

Modern Trends in Biothermokinetics, Edited by
S. Schuster *et al.*, Plenum Press, New York, 1993

2. Materials and Methods

Isolation of rat heart mitochondria was carried out as described earlier[12]. Liver mitochondria were isolated as described by Schneider[13].

Skinned myocardial and skeletal muscle fibers were prepared by treatment of bundles of fibers, 0.3-0.4 mm in diameter and 5-7 mm in length, isolated from the endocardial surface of left ventricle with 50 μg/ml of saponin[14].

"Ghost fibers", deprived of the myosin were produced according to the protocol described by Kakol et al.[15] by treatment of saponin-skinned fibers with 800 mM KCl.

Osmotic shock of skinned fibers was carried out with the purpose of disruption of the outer mitochondrial membrane according to the description by Stoner and Sirak[16].

Cytochrome c test for estimation of the extent of the mitochondrial outer membrane disruption was carried out in the oxygraph medium in the presence of 125 mM KCl. After addition of 1 mM ADP the maximal respiration was recorded, then 8 μM cytochrome c was added and the respiration rate was recorded again.

Respiration rates of isolated mitochondria and skinned fibers were determined by using a Yellow Spring Instruments (USA) oxygraph and a Clark electrode as described previously[14].

The kinetics of the mitochondrial creatine kinase in rat heart mitochondria was analyzed in detail in the presence and absence of oxidative phosphorylation as described earlier by Jacobus and Saks[17]. The values of all kinetic constants were determined at different temperatures in the range of 19 - 33 $^\circ$C with the aim of calculating the thermodynamic parameters of the reaction in the coupled and uncoupled states.

For electron microscopy, the skinned fibers were fixed in 2.5% glutaraldehyde solution in 0.1 M phosphate buffer and processed as described earlier[14]. Protein concentration was determined by the biuret or the Coomassie method. Statistic analyses of data in double reciprocal plots were made by using methods of linear regression. K_m mean values and their standard deviations are given.

3. Results

3.1. Free energy profiles of the coupled mitochondrial creatine kinase reaction: Biothermokinetic evidence for direct substrate channeling

The mechanism of the forward creatine kinase (CK) reaction

$$MgATP + Cr \Longleftrightarrow PCr + MgADP + H^+$$

is of quasi-equilibrium random binding BiBi type[1,12]. This mechanism is characterized by four substrate-enzyme complex dissociation constants, and by a rate constant k for conversion of the ternary enzyme-substrate complex into the enzyme-product complex (see the scheme in the upper part of Fig. 1), which is furthermore characterized by four dissociation constants (for details see Ref. 18). The dissociation constants are practically true equilibrium constants, and therefore, they obey the rules of classical thermodynamics:

$$\Delta G^o = - 2.303 \, RT \log K_d$$

and, since:

$$\Delta G^o = \Delta H^o - T \, \Delta S^o ,$$

$$\log K_d = - \frac{\Delta H^o}{2.303RT} + \frac{\Delta S^o}{2.303R} .$$

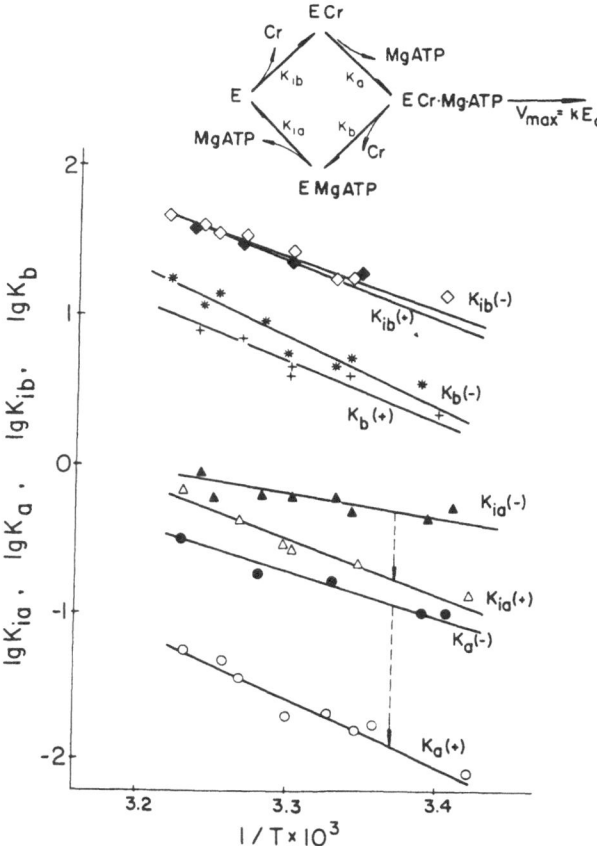

Figure 1. Temperature dependences of the dissociation constants of the enzyme-substrate complexes (shown by the upper scheme) for mitochondrial creatine kinase in the absence (-) and presence (+) of oxidative phosphorylation. For details see Ref. 17.

The slope of the linear dependence of $\log K_d$ on $1/T$ gives us the value of ΔH^o, and the ordinate intercept gives the value of $\Delta S^o/R$ and enables us thus to calculate ΔS^o.

The temperature dependence of $V_{max} = k \cdot E_0$ is described empirically by the Arrhenius equation:

$$k = k_0 \cdot \exp(-E_a / RT) \, ,$$

or, since $V_{max} = k \cdot E_0$:

$$V_{max} = V_0 \cdot \exp(-E_a / RT) \, ,$$

where V_0 is a constant. The empirical activation energy E_a is connected to the activation enthalpy and free energy by the absolute reaction rate theory, but for the sake of simplicity it is not considered here. In the Figures we have used the value of E_a.

Thus, the analysis of the temperature dependences of the dissociation constants and V_{max} allows to describe in detail the energy profile. Besides, since the mitochondrial creature kinase is functionally coupled to the adenine nucleotide translocase[5-8,12,17], such an analysis of kinetic data obtained under the conditions of oxidative phosphorylation may show which step of the creatine kinase reaction is mostly influenced by oxidative phosphorylation and in this way may give us some important information about the mechanism of such a coupling.

The results of a complete kinetic analysis of the mitochondrial creatine kinase in the membrane-bound state in the absence and in the presence of oxidative phosphorylation are shown in Figs 1-3. It becomes clear from Fig. 1, which shows the results for the forward reaction, that oxidative phosphorylation does not influence the binding and dissociation of creatine: its dissociation constants from the binary complex E.Cr, K_{ib}, and from the ternary complex E.Cr.MgATP, K_b, are the same in the presence and absence of oxidative phosphorylation at any temperature. However, the situation is different as for the second substrate, MgATP. There are minor changes in the value of K_{ia}, the dissociation constant for MgATP from the binary complex E.MgATP, induced by oxidative phosphorylation. However, the dissociation constant from the ternary complex E.MgATP.Cr, K_a, is decreased by one order of magnitude at any temperature studied. Thus, oxidative phosphorylation induces very specific changes in the kinetics of the forward creatine kinase reaction. From this and similar analyses of the reverse reaction we constructed the free energy profile (Fig. 2). This profile shows that the free energy changes occur at the steps of the substrate binding but not of phosphoryl transfer, in accordance with some earlier data[3].

The free energy change at the elemental step of formation of the active ternary complex E.Cr.MgATP, is strongly influenced by oxidative phosphorylation (Fig. 3): it is lowered by $\Delta\Delta_2 = 5.33$ kJ per mole. Note that at the step of formation of the binary complex E.MgATP, the free energy profile is changed to a much smaller extent, by the value $\Delta\Delta_1 = 2.44$ kJ per mole (Fig. 3). The facts that K_{ia} and K_a are not equally changed and that $\Delta\Delta_1 \neq \Delta\Delta_2$ point to a dependence of the effect of oxidative phosphorylation on the state of creatine kinase that may most probably show the direct channeling of ATP from translocase to that enzyme (see Discussion). E_a was not changed by oxidative phosphorylation.

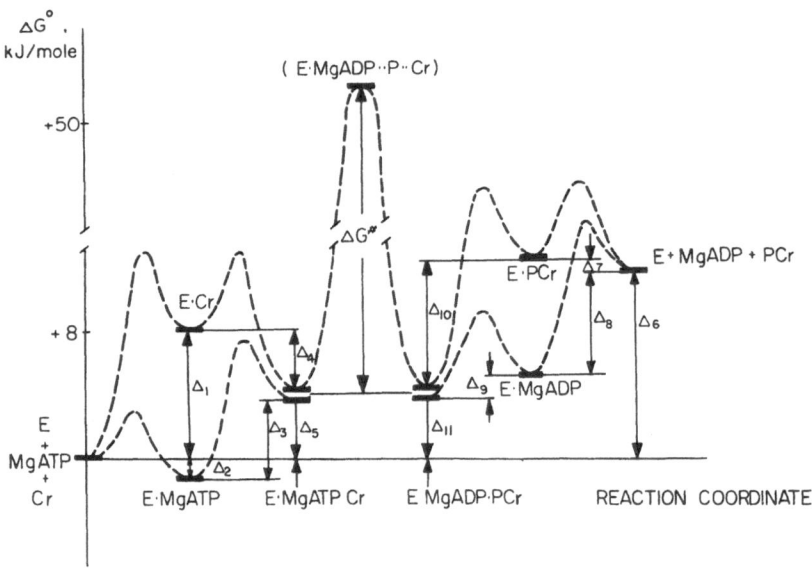

Figure 2. Free energy profile for the reaction catalyzed by the mitochondrial isoenzyme of creatine kinase in the absence of oxidative phosphorylation. The free energy levels of enzyme-substrate complexes were calculated from the data of Fig. 1. The activation energy barrier was determined, from V_{max}, only for the phosphoryl transfer. For the substrate binding step this barrier is unknown and shown arbitrarily.

304

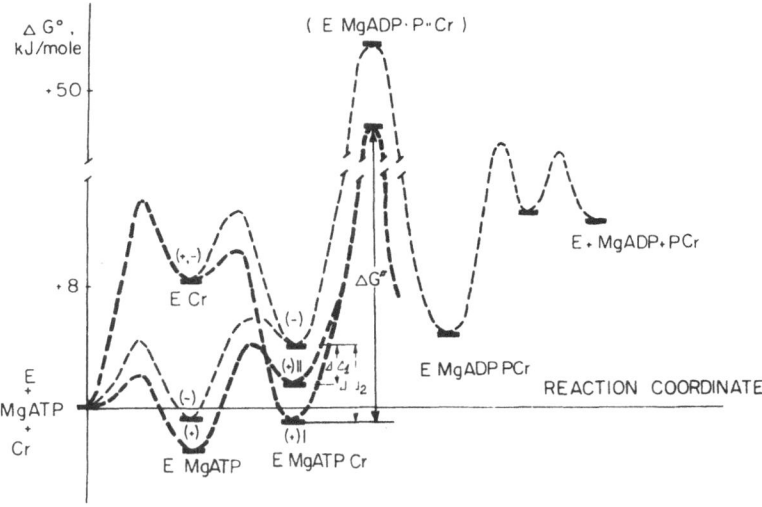

Figure 3. Free energy profiles for the mitochondrial creatine kinase reaction in the absence (-, thin dashed line) and presence (+, thick dashed lines) of oxidative phosphorylation.

3.2. The dependence of cellular respiration on the extracellular ADP concentration: Role of the mitochondrial outer membrane in restricting ADP diffusion

Fig. 4 shows the results with saponin-skinned cardiac fibers when respiration rate was measured in the physiological salt solution at different ADP concentrations. Curve 1 shows that the maximal rate is achieved only at millimolar external ADP levels, but the respiration rate at any ADP concentration is significantly accelerated in the presence of creatine due to activation of mitochondrial creatine kinase. This is in complete accordance with our earlier results[9-11]. This is due to a very significant decrease of the apparent K_m value for ADP at constant values of V_{max} (Fig. 4B). However, when these experiments were repeated in a 125 mM KCl medium, which is known to release mitochondrial creatine kinase from the inner membrane (Fig. 5), the stimulating effect of creatine on the cellular respiration was completely absent. Detachment of mitochondrial creatine kinase from the surface of the

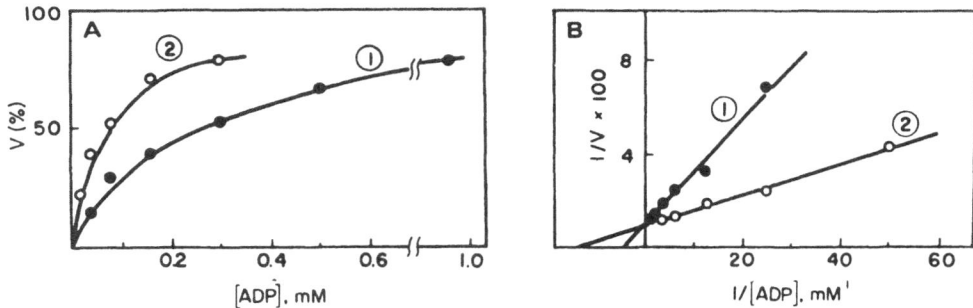

Figure 4. (A) Dependence of the rate of respiration of the skinned cardiac fibers in medium "B" on the external ADP concentration in the absence (curve 1) and presence (curve 2) of 25 mM creatine.
(B) Linearization of the dependences shown in (A) in a double reciprocal plot. Note the change in the value of apparent K_m for ADP by creatine at constant V_{max}.

inner mitochondrial membrane by KCl treatment destroys the functional coupling between creatine kinase and adenine nucleotide translocase[19]. Thus, the stimulating effect of creatine on the respiration of saponin-skinned cardiac fibers is directly related to the tight functional coupling between these two proteins .

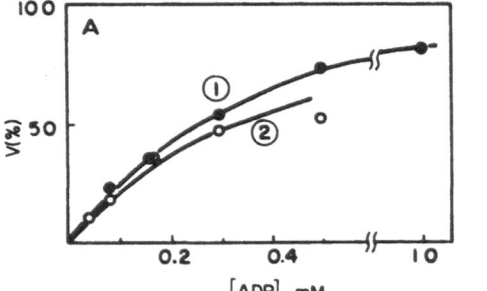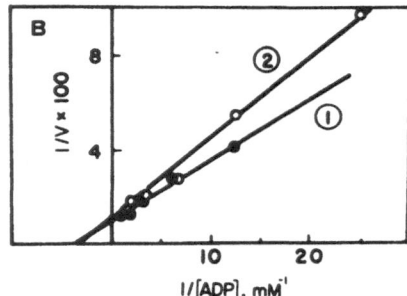

Figure 5. (A) Dependence of the rate of respiration of the skinned cardiac fibers in medium "B" containing additionally 125 mM KCl on the external ADP concentration in the absence (curve 1) and presence (curve 2) of 25 mM creatine.
(B) Linearization of the data from (A) in a double reciprocal plot.

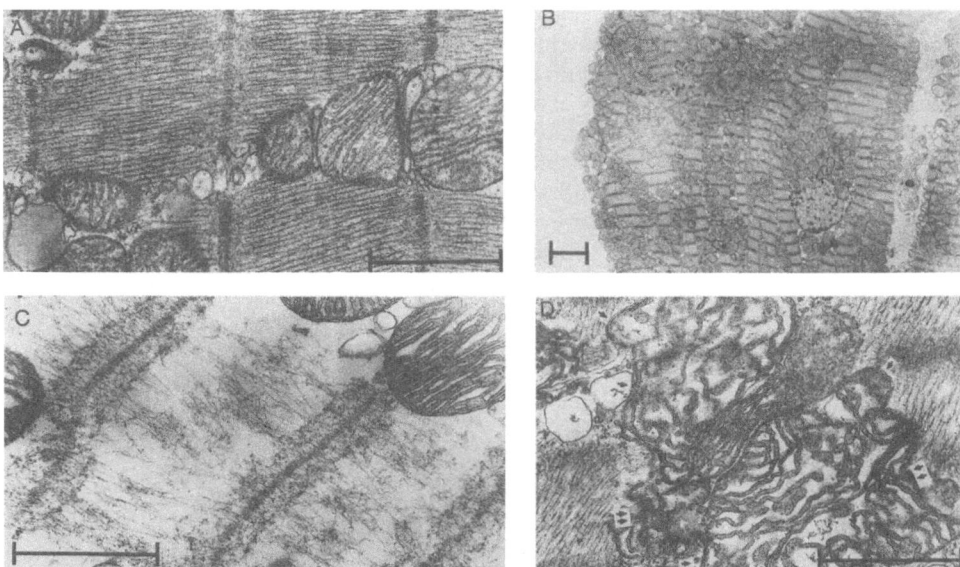

Figure 6. Electron micrographs of cardiac fibers. The magnification is indicated by scale bars, which correspond to 1 μm (A, C, D) and 2 μm (B). (A) Control. (B) After saponin treatment. The phospholipid bilayer of the sarcolemma is destroyed and vesicularized, but intracellular structures - mitochondria and myofibrils - are normal and have an appearance characteristic for these structures in the hyperosmotic physiological salt solution. (C) The saponin-skinned fibers after treatment with 800 mM KCl solution. Note the completely intact structure of the mitochondria. (D) The effect of hypo-osmotic treatment on the mitochondrial structures in the saponin-skinned cardiac fibers. Fibers after 30 min in a 40 mosM solution. For control before osmotic shock see (A) and (B). Mitochondrial outer membrane rupture is obvious (double arrows); however, in some places well preserved outer membranes (single arrows) are seen.

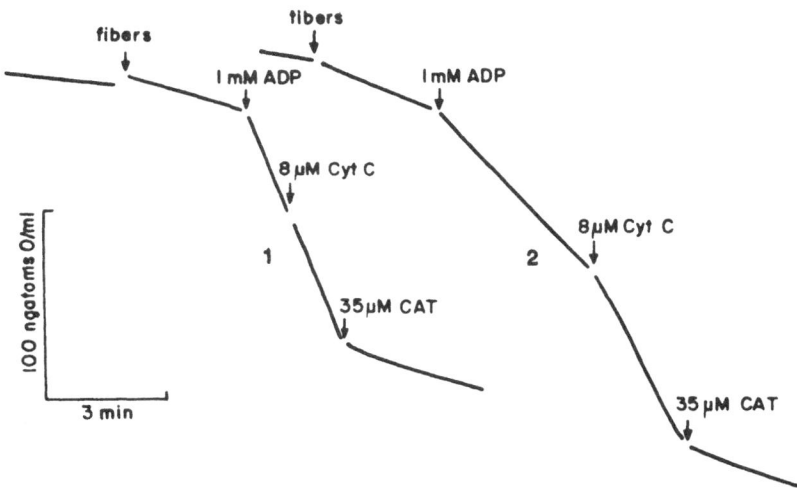

Figure 7. Oxygraph traces of recording of the respiration of skinned cardiac fibers before (A) and after (B) osmotic shock in a 40 mosM solution. Respiration rates were recorded in a medium containing 125 mM KCl to detach the cytochrome c from the membrane and in this way to test the intactness of the outer mitochondrial membrane.

To find the cellular structures responsible for the retarded diffusion of ADP resulting in the very high apparent K_m for this substrate, we treated the saponin-skinned fibers with a 0.8 M KCl solution to dissolve thick myosin filaments and to obtain so-called "ghost fibers"[15].

Using the well-controlled conditions for heart mitochondrial swelling in a hypo-osmotic medium worked out by Stoner and Sirak[16] we also performed the experiments with disruption of the outer mitochondrial membrane in the saponin-skinned cardiac fibers with careful morphological and biochemical analysis of the extent of disruption.

Fig. 6 shows electron micrographs of the cardiac fibers after treatment with saponin and then with KCl or after osmotic shock. After treatment with saponin the sarcolemma bilayer is dissolved and forms vesicles under the glycocalyx layer which is significantly removed from sarcomers and is in some places disrupted (Fig. 6B). However, both sarcomers and mitochondria are normal (Fig. 6B). After 0.8 M KCl treatment the thick filaments of the sarcomer are completely dissolved but the structures of the Z-line and thin filaments are more or less preserved (Fig. 6C). This very hyperosmotic treatment, however, does not change the structure of mitochondria which preserve their normal appearance and perfectly intact outer membrane (Fig. 6C).

On the other hand, Fig. 6D shows that incubation of skinned cardiac fibers in a 40 mosM solution in fact results in a severe disruption of the outer membrane of mitochondria inside the cells. On the micrographs the disrupted areas are clearly seen in many places (double arrows in Fig. 6D). However, significant areas of preservation of the outer membrane could also be observed (single arrows in Fig. 6D). Obviously, after hypo-osmotic treatment there are two populations of mitochondria in the skinned fibers - those with intact and those with disrupted membranes. For a rough estimation of the ratio of these populations we used the cytochrome c test described in Section 2. Fig. 7 shows that before hypoosmotic treatment (trace 1) addition of exogenous cytochrome c has no effect on the respiration of skinned fibers at maximal ADP levels: due to the intact outer membrane, endogenous cytochrome c is not released in spite of the presence of KCl and also the

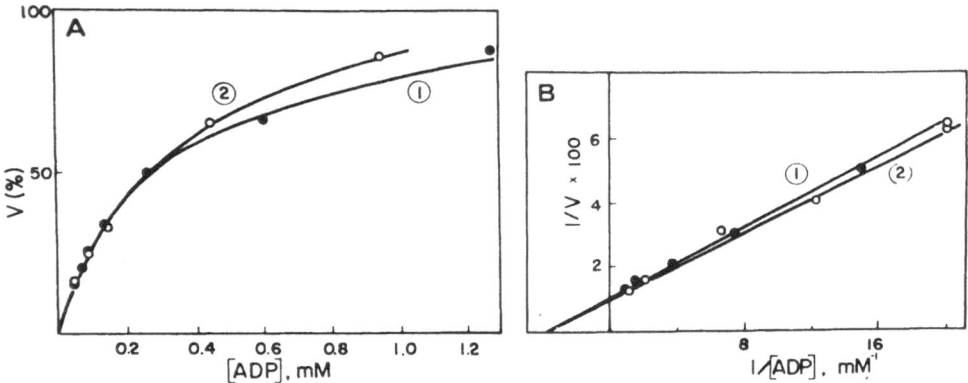

Figure 8. (A) Dependence of the respiration rate of the skinned and "ghost" fibers on the external ADP concentration. 1, control; 2, ghost fibers, obtained after treatment of skinned fibers with 800 mM KCl solution. (B) Linearization of the dependences shown in Fig. 7A in double reciprocal plot.

exogenous one cannot cross the membrane and is not necessary. On the contrary, after osmotic shock the maximal rate of respiration is significantly decreased (trace 2) but can be restored by adding exogenous cytochrome c to the final concentration of 8μM. The effect of exogenous cytochrome c on respiration is a stimulation by about 50 per cent. This rough estimation shows that probably about half of the mitochondrial population has disrupted outer membranes.

Fig. 8 shows a kinetic analysis of the ADP dependence of the respiration of "ghost fibers" after removal of myosin. There is no difference between the experimental dependences before (curve and line 1) and after (curve and line 2) myosin thick filament dissolution. In both cases the value of the apparent K_m for ADP is high and equal to 381± 67 μM. Thus, the binding of ADP to the myosin ATPase active centers is not the reason for the retarded diffusion of ADP in these cells. As shown in Fig. 9, the osmotic shock destroying the mitochondrial outer membrane significantly changes the dependence of respiration rate on the external ADP concentration: at low concentrations the respiration is higher after osmotic shock, and an analysis of these dependences in double-reciprocal plots clearly shows the existence of two types of processes of respiration regulation in these

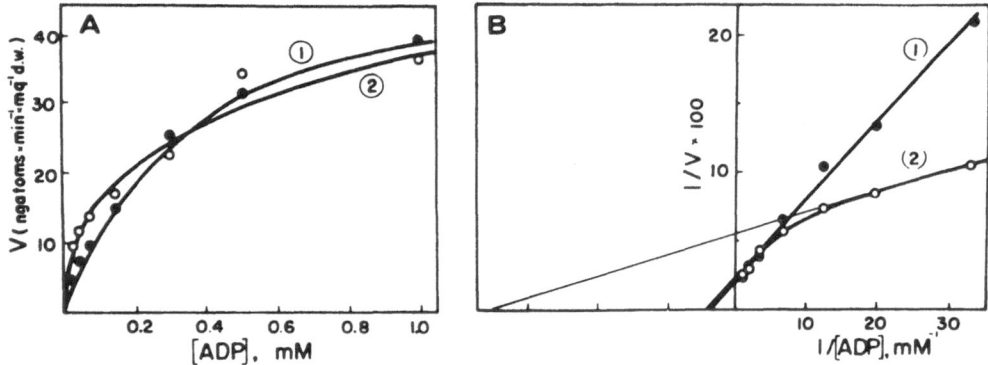

Figure 9. (A) Dependence of respiration rates of the skinned cardiac fibers before (curve 1) and after (curve 2) osmotic shock in a 40 mosM solution on the external ADP concentration.
(B) Linearization of the dependences shown in Fig. 8A in in double reciprocal plot. 1, control; 2, after osmotic shock in the 40 mosM solution. In the latter case two different kinetics are seen corresponding to two popul-ations of mitochondria with disrupted and intact outer membranes (cf. Fig. 6).

fibers: one is characterized by low apparent affinity to external ADP equal to that of fibers before osmotic shock, and there is a second process characterized by an apparent K_m for ADP equal to 35 μM that is rather close to the parameter for isolated mitochondria, or for mitoplasts deprived of the outer membrane (Table 1). These results demonstrated in Fig. 9 led us to the conclusion that the outer membrane of mitochondria is a major intracellular structure which retards ADP diffusion into the intermembrane space. In isolated mitochondria with morphologically intact outer membrane, however, there is no diffusion barrier for ADP (see Table 1).

4. Discussion

The results of this study provide more detailed information on the cellular mechanism of compartmentation of adenine nucleotides and of the important role of the creatine kinase system in the integration of muscle energy metabolism by interconnecting multiple cellular adenine nucleotide compartments. A key role is played by mitochondrial creatine kinase which significantly amplifies the minor ADP signal which may pass through outer membrane channels and in this way controls mitochondrial oxidative phosphorylation.

The high permeability of the outer membrane of isolated mitochondria *in vitro* for substances of low molecular weight has been known for a long time. First it was assumed that this is just a "leaky" membrane. Detailed studies during the last decade have shown that this high permeability is due to the existence of protein pores, or voltage-dependent anion selective channels, VDAC, with pore sizes in the range of 2 μm.[20-25] The 35 kDa protein forming these pores has been isolated from different sources and characterized including molecular cloning and is called "mitochondrial porin"[23-25]. In a series of publications from Brdiczka's and Wallimann's laboratories[7,8,24,25] it has been found that mitochondrial porin channels may be involved in the dynamic contacts between the inner and outer mitochondrial membranes, in particular with participation of the mitochondrial creatine kinase octamers. Such a supramolecular complex of the porin-creatine kinase-adenine nucleotide translocator assumes direct channeling of substrates - ATP and ADP - and a very efficient phosphocreatine production from mitochondrial ATP and cytoplasmic creatine[7].

The present results demonstrate - to our knowledge for the first time - the role of the outer membrane of mitochondria in the regulation of ADP diffusion from the cytoplasm into the intermembrane space in muscle cells *in vivo*. In contrast to isolated mitochondria with a morphologically well-preserved membrane but with a high permeability to ADP[10,11], in the muscle cells *in vivo* the permeability of this membrane is low. Clearly, there is at least one important regulatory factor decreasing the VDAC permeability to ADP inside the cells. Most probably, there exists a still unknown intracellular factor associated with the outer membrane *in vivo* which controls its permeability.

Besides the outer membrane, still some importance for retarded ADP diffusion may be ascribed to the unstirred layers in the cells and probably to some binding to the actin filaments since the K_m for ADP in the presence of creatine is higher in the cells with larger diameters (skeletal muscle vs. cardiac muscle cells, see Table 1). All these possible processes retarding ADP diffusion are shown schematically in Fig. 10.

In living cells the oncotic pressure due to the high cytoplasmic protein concentration might be an additional factor decreasing the VDAC permeability for ADP. This means that at physiological ADP concentrations the outer membrane may be almost impermeable to ADP. The striking differences between the mitochondrial functional behavior with respect to ADP *in vitro* and *in vivo* (see Table 1) observed in this work take us back into the year 1978 when Sjostrand[26] published his observations of the mitochondrial membrane system *in vivo* and *in vitro* using carefully controlled conditions of tissue fixation. He concluded that the "isolated mitochondrion is structurally different from the mitochondrion in its native

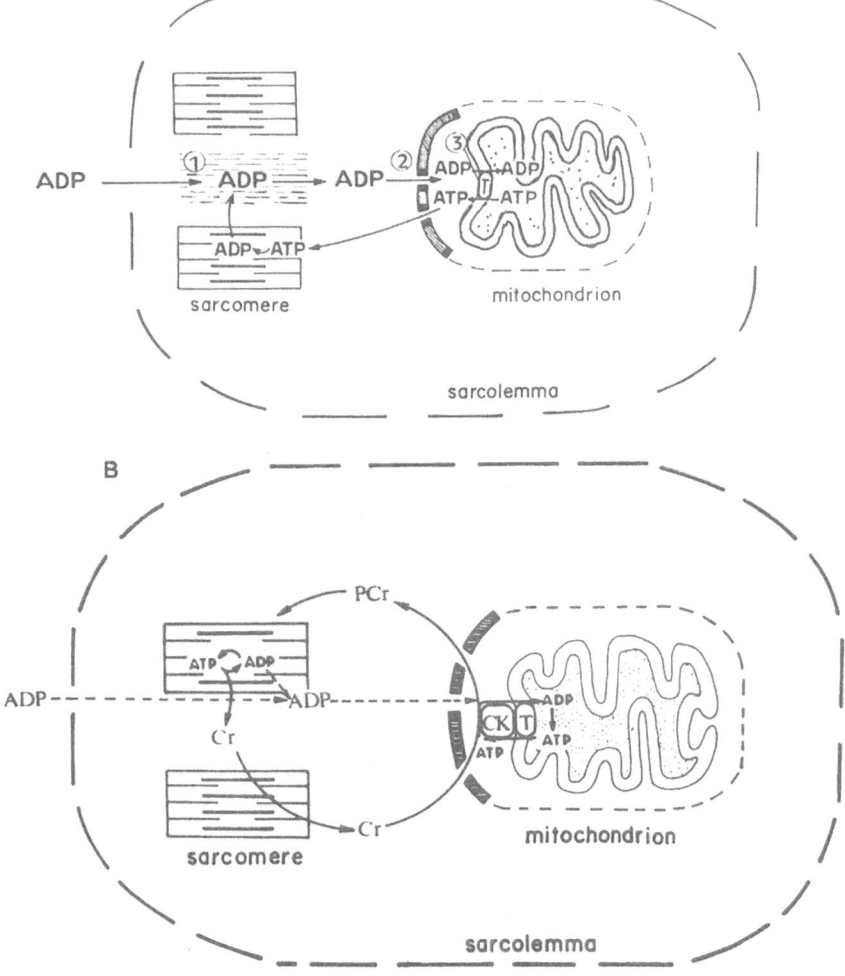

Figure 10. (A) Presentation of the intracellular mechanisms of retarded ADP diffusion. Three processes are controlling the ADP movement in the cells: (1) diffusion through unstirred layers and possible binding to actin filaments, (2) ADP movement through the porin channels in the outer membranes which is strictly controlled by some intracellular factors, (3) translocation across the mitochondrial inner membrane by adenine nucleotide translocase into the matrix for rephosphorylation.

(B) The coupled creatine kinase systems in mitochondria and in myofibrils control the ADP movement and overcome the retarded diffusion of ADP via the outer membrane. In myofibrils ADP is rephosphorylated at the expense of phosphocreatine, and creatine diffuses into the mitochondria along the large concentration gradient. In mitochondria the creatine kinase coupled to oxidative phosphorylation via translocase highly amplifies the weak ADP signal which may come in a controlled manner through the outer membrane, and creatine itself exerts an acceptor control on oxidative phosphorylation. This seems to be a central regulatory mechanism of respiration in muscle cells.

state *in situ*. An outer (intermembrane) compartment is formed by the two surface membranes becoming separated as a consequence of isolation"[26]. Consequently, there may be a continuous "contact system" between the outer and inner mitochondrial membranes *in vivo*, and the supramolecular structure translocase-creatine kinase-porin seems to be inevitable. Further Sjostrand supposed that his continuous membrane phase "with high viscosity is obviously an area of increased diffusion restriction" for adenine nucleotides and the effect of high viscosity may be eliminated by components of multienzyme systems being aggregated to form structural complexes or assemblies[26].

Thus only a very limited amount of cytoplasmic ADP may cross the mitochondrial outer membrane in a controlled manner to activate oxidative phosphorylation as coupled to creatine kinase. This reaction is a powerful amplification mechanism regulating and controlling the rate of oxidative phosphorylation (Fig. 10B). For this mechanism to be efficient, the creatine kinase should stay at the membrane to ensure a direct and precise substrate channeling. In the present work such a channeling has been verified by a kinetic and thermodynamic analysis of the energy profiles of the reaction. If we assume that only the outer membrane is important for "dynamic compartmentation", we can predict that the increased ATP concentration in the intermembrane compartment equally decreases the value of K_{ia} and K_a (as compared to the isolated enzyme, the value of these constants should differ by the ratio of ATP concentration in the intermembrane space and outside). Figs 1-3 show that this is not the case; the explanation is that only if ATP is directly channeled to the enzyme after binding of creatine it is completely and rapidly converted into the reaction products. Thus, the direct channeling of ATP by translocase to creatine kinase seems to be the most possible mechanism of preferable use of mitochondrial ATP for aerobic phosphocreatine synthesis[5-8,12]. In a very recent work[27] further thermodynamic evidence for such a close connection between phosphocreatine production and oxidative phosphorylation was presented. In very good accordance with several earlier observations[19] the detachment of creatine kinase from the membrane by a phosphate solution (20 mM) eliminated the thermodynamic coupling in Soboll's work[27]. In the present study, a KCl solution eliminated the accelerating effect of creatine on respiration (Fig. 2) due to the same effect of enzyme detachment. Thus, for an effective regulation of respiration the intact structure of the membrane-bound complex of creatine kinase with translocase is necessary, with possible involvement of the outer membrane VDAC (porin) protein.

Table 1. K_m values for ADP in the regulation of respiration in different preparations. For explanations see text.

Preparation	K_m^{app} for ADP, µM	
	- creatine	+ creatine, 25 mM
1. Skinned cardiac fibers	263.7±57.4	79±8
2. Skinned cardiac fibers in KCl, 125 mM	225±25	354±7.5
3. Skinned skeletal muscle fibers	334±54	105±15
4. Ghost cardiac fibers (without myosin)	381±67	-
5. Skinned cardiac fibers with swollen mitochondria	I II 315±23; 32.3±5	-
6. Isolated mitochondria heart	17.6±1.0	13.6±4.4
liver	18.4	

These very newly discovered mechanisms of cellular regulation of mitochondrial oxidative phosphorylation should be taken seriously into account in further biothermokinetic and mathematical analyses of the control of mitochondrial function, in cells *in vivo*. A first attempt of this kind of mathematical description has been undertaken by Aliev and Saks[18].

References

1. J.F. Morrison and E. James, The mechanism of the reaction catalyzed by adenosine triphosphate-creatine phosphotransferase, *Biochem J.* **97**:37 (1965).
2. W.E. Teague, Jr. and G.P. Dobson, Effect of temperature on the creatine kinase equilibrium, *J. Biol. Chem.* **267**:14084 (1992).
3. B.D. Nageswara Rao and M. Cohn, ^{31}P NMR of enzyme-bound substrates of rabbit muscle creatine kinase, *J. Biol. Chem.* **256**:1716 (1981).
4. S.P. Bessman and P. Geiger, Transport of energy in muscle: the phosphorylcreatine shuttle, *Science* **211**:448 (1981).
5. W.E. Jacobus, Respiratory control and the integration of heart high-energy phosphate metabolism by mitochondrial creatine kinase, *Ann. Rev. Physiol.* **47**:707 (1985).
6. V.A. Saks, L.V. Rosenshtraukh, V.N. Smirnov and E.I. Chazov, Role of creatine phosphokinase in cellular function and metabolism, *Can. J. Physiol. Pharmacol.* **56**:691 (1978).
7. T. Walliman, M. Wyss, D. Brdiczka, K. Nicolay and H.M. Eppenberger, Intracellular compartmentation, structure and function of creatine kinase isoenzymes in tissues with high and fluctuating energy demands: the "phosphocreatine circuit" for cellular energy homeostasis, *Biochem. J.* **281**:21 (1992).
8. M. Wyss, J. Smeitink, R.A. Wevers and T. Walliman, Mitochondrial creatine kinase: a key enzyme of aerobic energy metabolism, *Biochim. Biophys. Acta* **1102**:119 (1992).
9. V.A. Saks, Yu.O. Belikova, A.V. Kuznetsov, Z.A. Khuchua, T.H. Branishte, M.L. Semenovsky and V.G. Naumov, Phosphocreatine pathway for energy transport: ADP diffusion and cardiomyopathy, *Am. J. Physiol. Suppl.* **261**:30 (1991).
10. V.A. Saks, E. Vasil'eva, Yu.O. Belikova, A.V. Kuznetsov, S. Lyapina, L. Petrova and N.A. Perov, Retarded diffusion of ADP in cardiomyocytes: possible role of mitochondrial outer membrane and creatine kinase in cellular regulation of oxidative phosphorylation, *Biochim. Biophys. Acta*, in press.
11. V.A. Saks, Yu.O. Belikova and A.V. Kuznetsov, *In vivo* regulation of mitochondrial respiration in cardiomyocytes, *Biochim. Biophys. Acta* **1074**:302 (1991).
12. V.A. Saks, G.B. Chernousova, D.E. Gukovsky, V.N. Smirnov and E.I. Chazov, Studies of energy transport in heart cells. Mitochondrial isoenzyme of creatine phosphokinase: kinetic properties and regulatory action of Mg^{2+} ions, *Eur. J. Biochem.* **57**:273 (1975).
13. W.C. Schneider, Isolation of liver mitochondria, *J. Biol. Chem.* **176**:259 (1948).
14. V.I. Veksler, A.V. Kuznetsov, V.G. Sharov, V.I. Kapelko and V.A. Saks, Mitochondrial respiratory parameters in cardiac tissue: a novel method of assessment by using saponin-skinned fibers, *Biochim. Biophys. Acta* **892**:191 (1987).
15. I. Kakol, Y.S. Borovikov, D. Szczesna, V.P. Kirillina and D.I. Levitsky, Conformational changes of F-actin in myosin-free ghost single fibers induced by either phosphorylated or dephosphorylated heavy meromyosin, *Biochim. Biophys. Acta* **913**:1 (1987).
16. C.D. Stoner and H.D. Sirak, Osmotically-induced alterations in volume and ultrastructure of mitochondria isolated from rat liver and bovine heart, *J. Cell Biol.* **43**:521 (1969).
17. W.E. Jacobus and V.A. Saks, Creatine kinase of heart mitochondria: changes in its kinetic properties induced by coupling to oxidative phosphorylation, *Arch. Biochem. Biophys.* **219**:167 (1982).
18. M. K. Aliev and V. A. Saks, Mathematical model of the creatine kinase reaction coupled to adenine nucleotide translocation and oxidative phosphorylation, this volume.
19. A.V. Kuznetsov, Z.A. Khuchua, E.V. Vassil'eva, N.V. Medvedeva and V.A. Saks, Heart mitochondrial creatine kinase revisited: the outer mitochondrial membrane is not important for coupling of phosphocreatine production to oxidative phosphorylation, *Arch. Biochem. Biophys.* **268**:176 (1989).
20. C.A. Manella and H. Tedeschi, Importance of the mitochondrial outer membrane channel as a model biological channel, *J. Bioenerget. Biomembr.* **19**:305 (1987).
21. H. Tedeschi and K.W. Kinnaly, Channels in the mitochondrial outer membrane: evidence from patch clamp studies, *J. Bioenerget. Biomembr.* **19**:321 (1987).
22. J. Zimmerberg and V.A. Parsegian, Water movement during channel opening and closing, *J. Bioenerget. Biomembr.* **19**:351 (1987).
23. V. De Pinto, O. Ludwig, J. Krause, R. Benz and F. Palmieri, Porin pores of mitochondrial outer membranes from high and low eukaryotic cells, *Biochim. Biophys. Acta* **894**:109 (1987).
24. D. Brdiczka, Contact sites between mitochondrial envelope membranes. Structure and function in energy- and protein-transfer, *Biochim. Biophys. Acta* **1071**:291 (1991).
25. M. Kottke, V. Adam, I. Riesinger, G. Bremm, W. Bosch, D. Brdiczka, G. Sandri and E. Panfili, Mitochondrial boundary membrane contact sites in brain: points of hexokinase and creatine kinase location, and control of Ca^{2+} transport, *Biochim. Biophys. Acta* **935**:87 (1988).
26. F.S. Sjostrand, The structure of mitochondrial membranes, *J. Ultrastruct. Res.* **64**:217 (1978).
27. S. Soboll, A. Conrad, M. Keller and S. Hebisch, The role of the mitochondrial creatine kinase system for myocardial function during ischemia and reperfusion, *Biochim. Biophys. Acta* **1100**:27 (1992).

EFFECT OF MACROMOLECULES ON ADP-CHANNELING INTO MITOCHONDRIA

Frank N. Gellerich[1], Michael Wagner[1],
Matthias Kapischke[1] and Dieter Brdiczka[2]

[1]Institut für Biochemie
  Medizinische Akademie Magdeburg,
  Leipziger Str. 44, 3090 Magdeburg, Germany
[2]Fakultät für Biologie,
  Universität Konstanz, 7750 Konstanz, Germany

1. Metabolite Channeling Into Mitochondrial Intermembrane Space

In the resting muscle and even in the working one, the cytosolic ADP is in the micromolar range while the concentrations of ATP, creatine phosphate and creatine are between 5 and 20 mM.[1] The low cytosolic ADP concentration is advantageous to the thermodynamic efficiency of the ATP splitting reactions but does not allow an optimal stimulation of oxidative phosphorylation. For mitochondria from heart[2], liver[3] and brain[4] it was shown that the ADP supply to oxidative phosphorylation was privileged by mitochondrial creatine kinase or adenylate kinase compared to the extramitochondrial ADP supply by added enzymes as yeast hexokinase. That points to a dynamic AdN compartmentation in the mitochondrial intermembrane space[5,6]. From these results it was concluded that the transport of the extramitochondrially formed ADP into mitochondria is a crucial problem in cellular bioenergetics. This ADP transport can be realized by metabolite channeling into the mitochondrial intermembrane space.

The general mechanism of such a channeling basing on proposals of Wittenberg[7] and Meyer and coworkers[8] is shown in the upper part of Fig.1. Metabolite A is reversibly transformed into metabolite B. Both A and B only diffuse into the intermembrane space increasing the transport rate of A into the compartment. A precondition for the function of the metabolite shuttle is that the reaction A ⟷ B is directed differently in both compartments. In contrast to Wittenberg[7] and our previous publications[2,3,6] we now avoid the term facilitated diffusion, since no protein in theses shuttles is involved in the direct transport like the carrier proteins in the glucose transport through cell membranes or myoglobin in the intracellular oxygen transport. In contrast to other well established metabolite shuttles such as the malate/aspartate shuttle (transporting metabolites through the mitochondrial inner membrane into the matrix space) metabolite shuttles into the mitochondrial intermembrane space are

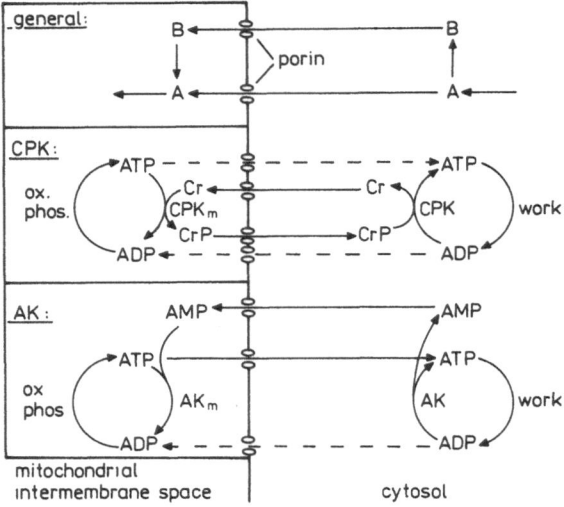

Figure 1. Scheme representing the mechanism of ADP transport into the mitochondrial intermembrane space by means of a metabolite shuttle.

General: General mechanism according to Meyer et al.[8] A,B, metabolites. CPK: CrP-shuttle; CrP creatine phosphate; Cr, creatine; Ox. Phos., oxidative phosphorylation; CPK_m, CPK, mitochondrial and cytosolic creatine kinase. AK: adenylate kinase shuttle; AK_m, AK mitochondrial and cytosolic adenylate kinase. Dashed arrows: transport rates of minor importance at low substrate concentrations.

considered here. Because the metabolites of these shuttles can diffuse through the porin pores, they do not require translocator proteins. In the lower part of Fig.1, two shuttles are shown which can be regarded as special forms of the general mechanism. One is the widely accepted creatine phosphate shuttle [9]. As a second ADP-shuttle we propose an adenylate kinase shuttle acting in a way similar to that of the creatine phosphate shuttle because mitochondrial and cytosolic adenylate kinase isoenzymes are compartmentalized in a way similar to that of the creatine kinase isoenzymes. It may operate in tissues with sufficiently high adenylate kinase activities such as liver, some types of spermatocoa or muscles. In this shuttle, AMP carries ADP equivalents (like creatine) into mitochondria: AMP formed in the cytosol from ADP via cytosolic adenylate kinase diffuses into the intermembrane space forming ADP there via mitochondrial adenylate kinase.

2. AMP-Mediated Transport of ADP-Equivalents into the Intermembrane Space

It was previously shown that ADP formed by mitochondrial creatine kinase or adenylate kinase stimulates mitochondrial respiration of rat heart mitochondria even in the presence of exceeding activities of pyruvate kinase[3,6]. To compare the extent of dynamic ADP compartmentation caused by the action of adenylate kinase with that caused by creatine kinase we performed experiments, the results of which are shown in Fig.2. Here the active rate of heart mitochondrial respiration was adjusted by addition of either yeast hexokinase (HK-system), 25 mM creatine (CPK-system) or 2 mM AMP (AK-system). Then pyruvate kinase was added step by step.

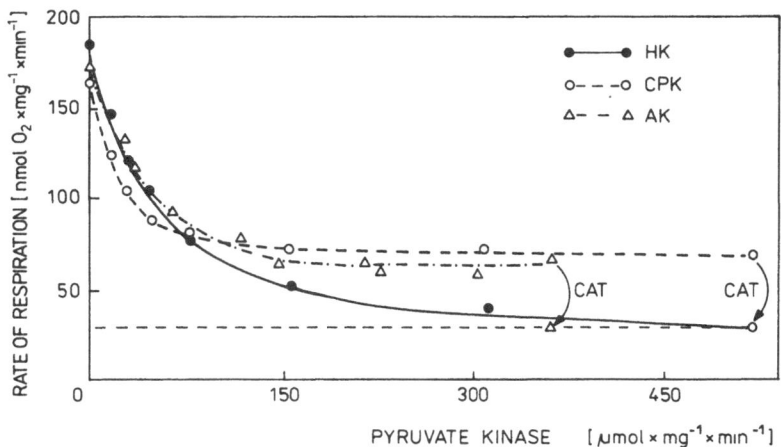

Figure 2. ADP compartmentation in the intermembrane space of rat heart mitochondria.
Rat heart mitochondria were incubated in a standard medium[3] additional containing 4 mM glutamate, 2 mM malate, 8 mM Mg-acetate, 5mM PEP and 2mM ATP. The active rate of respiration was adjusted either by addition of 1.8 U yeast hexokinase/mg (HK), 25 mM creatine (CPK) or 2 mM AMP (AK). Then the rate of respiration was diminished step by step by addition of pyruvate kinase. The stationary rates of respiration were measured by means of a ratemeter.

Only in the HK-system the resting rate of respiration was adjustable while in both the AK- and the CPK-system, the adjustable minimum rate of respiration was about twice the resting state, which indicates that AMP is as able as creatine to transport ADP equivalents into the mitochondrial intermembrane space.

3. Effect of Macromolecules on Dynamic ADP Compartmentation

Due to the dynamic concentration gradients of adenine nucleotides between the intermembrane space and the bulk phase[2,6] mitochondrial and extramitochondrial isoenzymes of the metabolite shuttles can work in different directions. However, we assume that these gradients are still more pronounced *in vivo* due to the action of cytosolic proteins.

We simulated the action of these cellular cytosolic proteins by addition of 10% (g/V) dextran to the reconstituted systems, which consist of functional intact mitochondria and different soluble enzymes[10].

As shown previously, the presence of dextran increased the frequency of contact sites up to 156 % (active state) and 253 % (resting state) in comparison to the controls, while the volume of the intermembrane space strongly decreased[11]. We investigated whether or not these structural changes affect the exchange of extra- and intramitochondrial adenine nucleotides through the porin pores. The addition of 10 % dextran did affect neither basic mitochondrial parameters nor kinetic properties of soluble kinases[10]. In contrast to that, we observed with all tested reactions which enclose a metabolite diffusion step through the porin pores that dextran caused a significant increase of the Michaelis constant (see Ref. 10, Table 1). The K_{ADP} of mitochondrial compartmentalized adenylate kinase increased from 118 to 192 μM ADP and the K_{ATP} of creatine kinase increased from 341 to 553 μM ATP in the presence of dextran. Similar effects were observed for the apparent K_{ADP} of oxidative phosphorylation of heart and liver mitochondria (Table 1). If it is so that increased Michaelis constants are caused by diffusion limitations, these effects ought to be dependent on the rate of enzymatic

Table 1. Effect of 10 % dextran on kinetic constants of homogenous and compartmentaiized reactions.[1]

Reaction	Control		10% Dextran	
	K_{AdN}	V_{max}	K_{AdN}	V_{max}
	[μM]ADP	[U/mg]	[μM]ADP	[U/mg]
AK RLM	118±1	0.85±0.02	192±20	0.76±0.03
AK Soluble	168±9	0.90±0.02	172±9	0.90±0.02
	[μM]ATP	[U/mg]	[μM]ATP	[U/mg]
CPK RHM	341±50	3.1±0.1	553±65	3.1±0.1
CPK MP	248±30	1.9±0.2	237±35	1.8±0.1
CPK Soluble	181±19	1.8±0.0	180±16	1.8±0.03
	[μM]ADP	[nmolO$_2$/mg/min]	[μM]ADP	[nmolO$_2$/mg/min]
OX. PH. RLM	46± 5	67±3	60± 7	66±4
OX. PH. RHM	40± 5	132±10	85±11	140±10

[1]The effect of 10 % dextran 70 was measured at rat liver mitochon-dria (RLM), rat heart mitochondria (RHM), rat heart mitoplasts(MP) and released enzymes[10]. The activities of adenylate kinase (AK) and creatine kinase (CPK) were measured by means of the optical test[10] The rate of respiration (OX.PH.) was measured by means of a custom built ratemeter in the presence of exceeding yeast hexokinase[3].

Table 2. Temperature dependence of the effect of dextran on K_{ATP} of mitochondrial compartmentalized creatine kinase.*

Temperature [°C]	K_{ATP} [μM] ATP		
	Control	Dextran	Difference
20	240 ± 74	365 ± 112	125
25	284 ± 51	435 ± 123	151
30	341 ± 50	553 ± 65	212
36	443 ± 81	689 ± 70	246

*Kinetic constants were determined and calculated as described in Table 1 at different temperatures as indicated.

reactions. As the rate of mitochondrial creatine kinase may easily be changed by temperature, we measured the dependence of the apparent K_{ATP} on temperatures between 20 and 36 °C (See Table 2). As expected the difference between the K_{ATP} with and without dextran increased from 125 μM at 20 °C to 246 μM at 36 °C.

Since mitochondria in the intact cell are embedded in the cytosol which is a 30 % protein solution we did additional experiments to investigate the influence of higher dextran concentrations on the dynamic AdN compartmentation. We measured the kinetic constants of compartmentalized creatine kinase in the presence of 10 % and of 30 % dextran and compared it to the control. K_{ATP} increased from 332 over 525 to 641 μM ATP and V_{max} decreased with increasing dextran concentrations from 1.98 over 1.71 to 1.08 U/mg. The decreased V_{max} which is probably caused by an increased creatine phosphate inhibition of creatine kinase raised to the original rate by increased creatine concentrations (Gellerich, unpublished results). These data show that in the intact cell remarkable concentration gradients should exist between the intermembrane space and the cytosol.

Assuming that the mitochondrial outer membrane is essential to the dextran effects on dynamic AdN compartmentation, we investigated the kinetic properties of mitochondrial creatine kinase in heart mitochondria with disrupted outer membranes in the absence and in the presence of dextran. As shown in Table 1, after disruption of the mitochondrial outer membrane the dextran effect on K_{ATP} disappeared completely indicating that the net transport through the homogeneous phase was not influenced by dextran. These data support our recent findings which show that the removing of the mitochondrial outer membrane diminishes the dynamic ADP compartmentation induced by mitochondrial creatine kinase[12].

To obtain informations about the mechanism of the dextran effects we varied the oncotic pressure using dextrans differing in their molecular weight (20, 40, 70, 500 kDa) at comparable concentrations (mM) or contents (mg/ml) and measured the kinetic constants of mitochondrial compartmentalized creatine kinase, the dynamic ADP compartmentation[10], the conductivity of porin pores inserted into black membranes (see Table 3) and the oncotic pressure[10].

Table 3. Effect of different dextrans on the conductivity of porin pores inserted into "black membranes" as well as on kinetic constants of mitochondrial compartmentalized creatine kinase.[§]

Addition			K_{ATP} [μM]	High conductivity porin [%]
Control			255 ± 50	33
Dextran	20	10 %	480 ± 60	2
		0.5 mM	294 ± 89	37
Dextran	70	10 %	593 ± 80	5
		0.5 mM	316 ± 120	32
Dextran	500	10 %	503 ± 50	5
		0.5 mM	680 ± 69	5

[§]Kinetic constants were determined and calculated as described in Table 1. The methods used for the "black" lipid bilayer experiments have been described previously[13]. The current at a membrane potential of 10 mV increased stepwise because of the insertions of pores into the membrane. 40 - 100 single current steps at 10 mV were recorded for each condition. The percentage of pore insertions of a conductance of about 5 nS is given in the table.

The oncotic pressure correlated like observed with the effects on the K_{ATP} of mitochondrial creatine kinase and the conductivity of porin pores with the content of dextran regardless of differences in the molecular weight. We found a similar behavior of dextran effects on the dynamic ADP compartmentation in rat liver mitochondria[10]. Results show that mitochondrial porin pores obviously occur in the intact cell at a low permeability state for ADP due to the oncotic pressure exerted by cytosolic proteins. That is why it is an important problem in cellular energy transport to shuttle ADP across the porin pores into the intermembrane space. It can be done either by direct binding of ADP-generating enzymes such as hexokinase or glycerol kinase to porin[11,14] or by metabolite shuttles into the intermembrane space such as the creatine kinase shuttle or the newly proposed adenylate kinase shuttle.

Acknowledgements

This work was supported by grants of the Deutsche Forschungsgemeinschaft given to F.N. Gellerich (Ge 663/1-1) and D.Brdiczka (Br 773/3-2).

References

1. J. A. Hoerter, C. Lauer, G. Vassort and M. Gueron, Sustained function of normoxic hearts depleted in ATP and phosphocreatine: a ^{31}P-NMR study, *Am. J. Physiol.* **255** (Cell Physiol. 24):C192 (1988).

2. F. N. Gellerich, M. Schlame, R. Bohnensack and W. Kunz, Dynamic compartmentation of adenine nucleotides in the mitochondrial intermembrane space of rat heart mitochondria, *Biochim. Biophys.Acta* **722**:381 (1987).

3. F. N. Gellerich, The role of adenylate kinase in dynamic compartmentation of adenine nucleotides in the mitochondrial intermembrane space, *FEBS Lett.* **297**:55 (1992).

4. M. Kottke, V. Adams, T. Wallimann, V. K. Nalam and D. Brdiczka, Location and regulation of octameric mitochondrial creatine kinase in the contact sites, *Biochim.Biophys. Acta* **1061**:215 (1991).

5. F. N. Gellerich and V. A. Saks, Control of heart mitochondrial oxygen consumption by creatine kinase: The importance of enzyme localization, *Biochem. Biophys. Res. Commun.* **105**:1473 (1982).

6. F. N. Gellerich, R. Bohnensack and W. Kunz, Role of the outer membrane in dynamic compartmentation of adenine nucleotides, in: "Anion Carriers of Mitochondrial Membranes," A. Azzi, K. A. Nalecz, M. J. Nalecz and L. Woijtczak, eds, Springer, Berlin (1989).

7. J. B. Wittenberg, Myoglobin - facilitated oxygen diffusion: Role of myoglobin in oxygen entry in the muscle, *Physiol. Rev.* **50**: 559 (1970).

8. R. A. Meyer, H. L. Sweeny and M. J. Kushmerick, A simple analysis of the "phosphocreatine shuttle", *Am. J. Physiol.* **246**:C365 (1984).

9. T. Wallimann, M. Wyss, D. Brdiczka, K. Nicolay and H. M. Eppenberger, Significance of intracellular compartmentation, structure and function of creatine kinase isoenzymes for cellular energy homeostasis: The phospho-creatine circuit, *Biochem. J.* **281**: 215 (1992).

10. F. N. Gellerich, M. Wagner, M. Kapischke, U. Wicker and D. Brdiczka, Effect of macromolecules on the regulation of the mitochondrial outer membrane pore and the activity of adenylate kinase in the intermembrane space, *Biochim. Biophys. Acta*, in the press.

11. U. Wicker, K. Bücheler, F. N. Gellerich, M. Wagner, M. Kapischke and D. Brdiczka, Effect of macromolecules on the structure of the mitochondria intermembrane space and the regulation of hexokinase, *Biochim. Biophys. Acta*, in the press.

12. F. N. Gellerich, Z. A. Khuchua and A. V. Kuznetsov, Influence of the mitochondrial outer membrane and the binding of creatine kinase to the mitochondrial inner membrane on the compartmentation of adenine nucleotides in the intermembrane space of rat heart mitochondria, *Biochim. Biophys. Acta*, in the press.

13. R. Benz, M.Kottke and D. Brdiczka, The cationically selective state of the mitochondrial outer membrane pore: a study with intact mitochondria and reconstituted mitochondrial porin, *Biochim. Biophys. Acta*, **1022**:311 (1990).

14. V. Adams, L. Griffin, J. Towbin, B, Gelb, K. Worley and E. R. B. McCabe, Porin interactions with hexokinase and glycerol kinase: metabolic microcompartmentation at the outer mitochondrial membrane, *Biochem. Med. Metab. Biol.* **45**:271 (1991).

CONTROL OF ATP FLUX
IN RAT MUSCLE MITOCHONDRIA

Laurence Jouaville, François Ichas,
Thierry Letellier, Monique Malgat and Jean-Pierre Mazat

Université de BORDEAUX II
G.E.S.B.I
146 rue Léo Saignat, 33076 Bordeaux Cedex, France

1. Introduction

Most of the studies performed on the control of oxidative phosphorylation deal with the control on the oxygen consumption flux. Nevertheless, a relevant flux for the cell is that of ATP production, since it is important for the homeostasis of the phosphate potential of the cell.

The control of ATP flux resulting from oxidative phosphorylation has been investigated partially in yeast[1]. In that study, the rate of ATP synhesis was measured by radioactive incorporation of P_i, thus picturing the ATP changes in the mitochondria.

We here present a method based on the enzymatic monitoring of the ATP synthesized by oxidative phosphorylation using firefly luciferase. It allows us to monitor *continuously* ATP formation by mitochondria and, because of its high sensitivity, with a very low amount of mitochondria.

Using this method we determined control coefficients of the various steps involved in our network, on the ATP flux extruded, via the translocator, out of mitochondria as the result of oxidative phosphorylation. We found that the most important part of the control is exerted by complex III. This result is in agreement with those found in our laboratory on the oxygen consumption flux[2].

2. Methods

2.1. Isolation of mitochondria

Rat muscle mitochondria were isolated by differential centrifugation as described by Morgan-Hughes and coworkers[3]. Muscle of the two hindlegs are collected in the isolation medium I (mannitol 210 mM, sucrose 70 mM, Tris/HCl 50 mM pH 7.4, EDTA 10 mM), digested by trypsin (0.5 mg/g of muscle) during 30 minutes. The reaction is stopped by addition of trypsin inhibitor (soja bean 3:1 inhibitor to trypsin). The homogenate was centrifuged at 1000 g for 5 minutes. The supernatant was strained on gauze and recentrifuged at 7000 g for 10 minutes. The resulting pellet was resuspended in the ice-cold isolation medium II (mannitol 225 mM, sucrose 75 mM, Tris/HCl 10 mM pH 7.4, EDTA 0.1 mM) and a new series of centrifugations (1000 g and 7000 g) were performed. The last pellet of mitochondria was resuspended into a minimum volume of isolation medium II in order to obtain a mitochondria concentration lying between 20 and 40 mg/ml. Protein concentration was estimated by the biuret method using bovin serum albumin as standard.

2.2. ATP measurement

This assay is based upon the quantitative measurement of a level of light produced as a result of the following enzyme reaction :

$$ATP + LUCIFERIN \xrightarrow[\text{Mg}^{2+}]{\textit{Luciferase}} OXYLUCIFERIN + AMP + PPi + CO_2 + h\upsilon$$

The amount of the light produced per unit of time (RLU/min) was continuously monitored, and the rate of light production was calculated by linear regression.

The ATP-monitoring reagent used (AMR) was commercially supplied as a lyophilized mixture containing: fyrefly luciferase, D-luciferine, bovine serum albumine 50 mg, magnesium acetate 500 µmoles, inorganic pyrophosphate 0.1 µmoles, (Bio Orbit).

The AMR was dissolved in 10 ml of a solution (pH 7.5) containing: Saccharose (180 mM), KH$_2$PO$_4$ (35mM), EDTA (1mM).

Mitochondrial ATP synthesis was monitored at 30 °C in the 200 µl wells of the mechanically stirred luminometer microplate (Luminoskan, Labsystems).

The mitochondria (2,5-10 µg/ml) were incubated in the same buffer containing 10 mM pyruvate + 10 mM malate as substrate.

For each inhibition curve, the inhibitor was preincubated with mitochondria before the ADP stimulation.

2.3. Calculation of control coefficients

The control coefficients of various steps involved in oxidative phosphorylation were determined with specific inhibitors of these steps, according to the definition :

$$C_i = \left. \frac{\dfrac{\partial \ln J}{\partial I}(I = 0)}{\dfrac{\partial \ln v_i}{\partial I}(I = 0)} \right|_{\text{steady state}}$$

i.e., the ratio of the initial slope of the systemic inhibition curve $J(I)$ and the initial slope of the inhibition curve of the isolated step $v_i(I)$ in the same conditions as in the network.

In actual practice we used the model of Gellerich and coworkers[4] to determine the control coefficients. This method is based on non-linear regression to fit the experimental data to a model assuming non-competitive inhibition. The equation of the model used to fit the experimental data involves the dissociation constant of the inhibitor, the concentration of the titrated enzyme and the control coefficient. The non-linear fitting was performed using the program "SIMFIT"[5]. In each case, the parameters determined with the inhibition curve of the whole flux were also used to draw the theoretical inhibition curve of the isolated step.

3. Results and Discussion

3.1. Bioluminescent assay of mitochondrial ATP synthesis

This method has been used by other authors[6,7], with an instability of the light emission by the firefly luciferase reagent. Therefore we checked the signal stability of the new firefly luciferase reagent used in this study by monitoring the light emission produced by the activity of the luciferase during ten minuts. The light decrease, in the presence of a constant ATP concentration, was found to be less then 1% per minut. Moreover we have checked the linearity of light production with respect to the amount of ATP. We found that the signal is linear up to 10 µM of ATP.

A significant activity of adenylate kinase was observed both within the firefly reagent and the mitochondrial suspension[8]. The background noise produced by this enzyme has to be subtracted from the total ATP formation rates when calculating the ATP synthesis depending on oxidative phosphorylation. As the adenylate kinase ATP synthesis is independent of the presence of the respiratory substrate we measured the ATP synthesis in the absence as well as in the presence of 10 mM pyruvate + malate (Fig. 1). The difference between the two curves gives the value of the ATP formation depending on oxidative phosphorylation, which reaches a maximum activity for an ADP concentration ranging betwen 30 and 50 µM ADP. Therefore, the ADP concentration of 30 µM was chosen to ensure a maximal ATP synthesis by oxidative phosphorylation with a minimal ATP synthesis independent of oxidative phosphorylation. The latter representing 10 % of the registered total rate ATP synthesis is responsible for the residual not inhibited ATP synthesis which can be encountered in most of the flux titration curves (Fig. 3).

In order to know wether the activity of the luciferase modifies the stationary state we studied the rate of ATP synthesis as a function of the amount of AMR present in the medium. Adding AMR to the system had no influence on the rate of ATP synthesis. Therefore, the weak ATP consumption by luciferase does not affect the stationary state.

The rate of ATP synthesis we measured with this method was 750±70 nmoles/mn/mg protein. This value is close to those found in the literature[8].

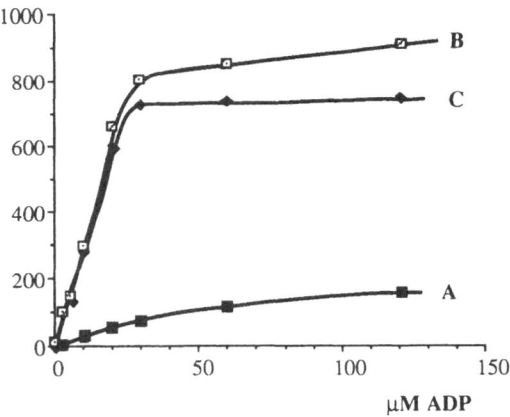

Figure 1. ATP synthesis rate as a function of ADP concentration.
Mitochondria (2,5 to 5 µg/ml) were incubated in 200 µl of buffer in the abscence (A) or in the presence of 10 mM pyruvate + malate. The state of ATP synthesis was started by the addition of the indicated ADP concentrations and the measure by the addition of AMR. All the experiments were monitored at 30 °C and the microplate was mechanically stirred. The difference between curves A and B is the synthesis owing to oxidative phosphorylation (C).

3.2. Determining control coefficients

The control coefficients of the various steps were determined using specific inhibitors. Because of the enzymatic character of the luciferase reaction, we checked the possible effect of all the inhibitors on the luciferase reaction. We found that they had no effect on the linearity of the signal, nor on its proportionality to ATP concentration.

In addition, owing to the fact that some inhibitors had to be prepared in ethanol, we examined the inhibition of the ATP flux by this alcohol (Fig. 2). The inhibition appears from 1 % of ethanol on. Therefore we used dilutions of these inhibitors in order not to exceed 0.5% of ethanol in the final volume.

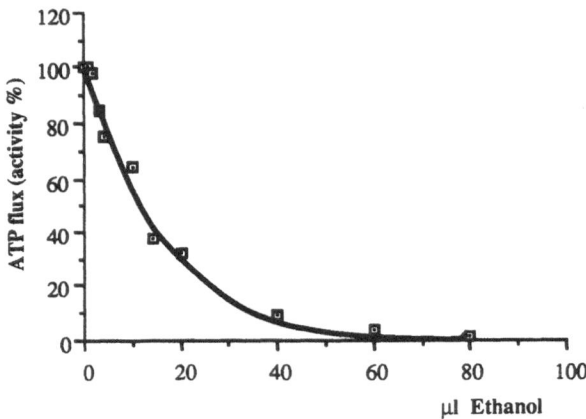

Figure 2. ATP flux inhibition by ethanol.
Mitochondria (2,5 to 5 µg/ml) were incubated in 200 µl of buffer containing 10 mM pyruvate + malate and the indicated volume of ethanol. The state of maximal ATP synthesis was started by the addition of 30 µM ADP as in Fig. 1.

Finally, since we used for the first time α-cyano-4-hydroxycinnamate (C4H), wich is a non-competitive inhibitor of the pyruvate carrier[9,10], we examined its possible effects on the other complexes involved in oxidative phosphorylation by titrating the ATP synthesis, in the presence of succinate 10 mM as substrate (data not shown). Non-specific inhibition appears above 1 mM wich is at least 200 times more than the concentration needed for maximal inhibition of the ATP flux on pyruvate plus malate as substrate.

The titration curves of the ATP synthesis with specific inhibitors of several enzymes of oxidative phosphorylation are shown in Fig. 3. The control coefficients and the amounts of inhibitors required to obtain maximal flux inhibition (I_{max}) are given in Table 1.

Except in the case of the pyruvate carrier, all control coefficients were calculated according to the model of Gellerich and coworkers[4] using the non-linear fitting procedure of Holzhütter and Colosimo[5]. In the case of the pyruvate carrier, although the inhibitor used, α-cyano-4-hydroxycinnamate (C4H), was described as a non-competitive inhibitor[9,10] it was impossible to fit the flux titration curve with "SIMFIT"[5]. Therefore, the control strength of the pyruvate carrier was calculated graphically assuming the inhibitor was irreversible.

The highest control coefficient is the one of complex III, while the other respiratory complexes (I and IV)) as well as the phosphate carrier and the pyruvate carrier, exert a low control on the ATP flux. On the other hand, a significant control is exerted by the ATP synthase and adenine nucleotide translocator (Table 1).

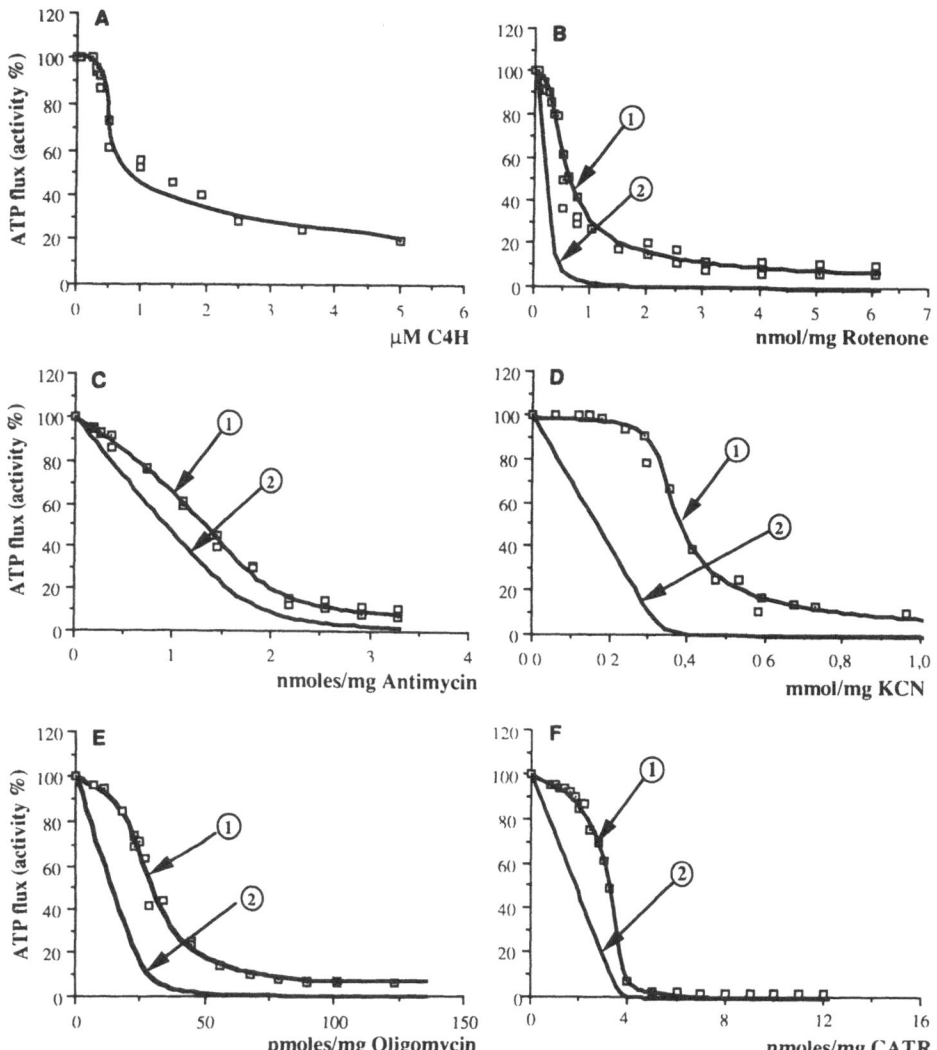

Figure 3. Inhibition of the ATP flux by several specific inhibitors in muscle mitochondria. Mitochondria (2,5 to 5 µg/ml) were incubated in 200 µl of buffer containing 10 mM pyruvate + malate and the indicated concentrations of inhibiteurs for 5 mn. The state of maximal ATP synthesis was started by the addition of 30 µM ADP and the measurement by the addition of AMR. All the experiments were monitored at 30 °C and the microplate was mechanically stirred. Measured ATP flux inhibition (▢), flux inhibition curve drawn from the model of Gellerich et al.[4], isolated step inhibition curve drawn from the model.

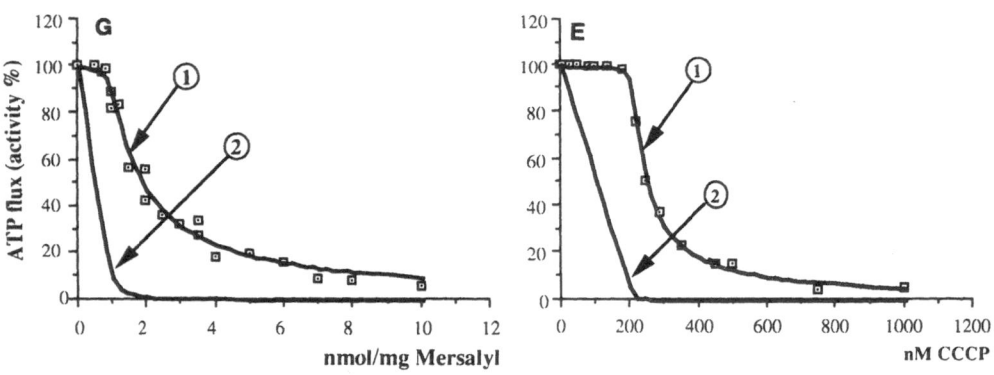

Figure 3. (continued).

Table 1. Flux control coefficients of various steps in oxidative phosphorylation from rat muscle mitochondria.[§]

Process	Control coefficient	Inhibitor
Pyruvate carrier	0.016±0.02	α-cyano-4-hydroxycinnamate
Complex I	0.07±0.02	Rotenone
Complex III	0.38±0.02	Antimycine
Complex IV	0.02±0.02	KCN
ATP synthase	0.18±0.04	Oligomycine
ATP/ADP carrier	0.12±0.01	CATR
Phosphate carrier	0.02±0.01	Mersalyl
Leak	0	CCCP (Activator)

$$\Sigma = 0.81$$

[§]Mitochondria were incubated as in Fig. 3. Except for the pyruvate carrier, flux control coefficients were calculated according to the model of Gellerich et al.[4] using the non-linear fitting procedure of Holzhütter and Colosimo[5]. For the pyruvate carrier, control coefficients were calculated according to the formulation proposed by Groen et al.[11] owing to the fact that C4H is a non-competitive inhibitor.

The control coefficient of the passive permeability of the mitochondrial membrane to protons, regarded as a divergent step of our network, was investigated using the CCCP as uncoupler. In accordance with studies on oxygen consumption flux by Groen and coworkers[11] and on ATP synthesis flux by Mazat and coworkers[1], we found that this control was very low.

All these results were compared with those obtained in our laboratory on the oxygen cunsumption flux[2]. This investigation was carried out on the same type of mitochondria with 10 mM of pyruvate plus malate as substrate and 1.6 mM of ADP. As well as in the present study it was found that the most important part of control is exerted by complex III. However, the control on the oxygen consumption flux was found to be broadly distributed among all the steps of the pathway. In the case of the ATP synthesis flux, there are three important control coefficients, those of the complex III, ATP synthase and adenine nucleotide translocator. In contrast, the remaining steps were found to exert a very low control.

It is worth noting that the sum of control coefficients experimentally determined is less than one in the case of the ATP synthesis flux, as well as in the case of the oxygen cunsumption flux[2]. This can be explained by the fact that some important steps may not have been taken into account in the studied network.

4. Implications for Mitochondrial Cytopathies

Another aim was to develop a technique to enable studies of mitochondrial activities with a very low amount of mitochondria. Because the muscular biopsy used for the diagnosis of a mitochondrial disease usually weighs a few hundred milligrams the amount of isolated mitochondria is too low to allow a wide polarographic investigation. In contrast, the method presented here allowed us to measure ATP synthesis on samples of mitochondrial protein concentration as low as 2.5 µg/ml instead of the usual 0.5 to 1mg/ml needed for polarography studies.

References

1. J.P. Mazat, E. Jean-Bart, M. Rigoulet and B. Guérin, Control of oxydative phosphorylations in yeast mitochondria. Role of the phosphate carrier, *Biochim. Biophys. Acta* **849**:7-15 (1985).
2. T. Letellier, M. Malgat and J.P. Mazat, Control of oxidative phosphorylation in rat muscle mitochondria. Applications to mitochondrial myopathies, *Biochim. Biophys. Acta* **1141**: 58-64 (1993).
3. J.A. Morgan-Hughes, P. Darveniza, S.N. Kahn, D.N. Landon, R. M. Sherratt, J.M. Land and J.B. Clark, A mitochondrial myopathy characterized by a deficiency in reducible cytochrome b, *Brain* **100**:617-640 (1977).
4. F.N. Gellerich, W.S. Kunz and R. Bohnensack, Estimation of flux control coefficients from inhibitor titrations by non-linear regression, *FEBS Lett.* **274**:167-170 (1990).
5. H.G. Holzhütter and A. Colosimo, Simfit: A microcomputer software toolkit for modelistic studies in biochemistry, *Comp. Appl. Biosci.* **6**:23-38 (1990).
6. J.J. Lemasters and C.R. Hackenbrock, Continuous measurement in mitochondrial suspensions by firefly luciferase luminescence, *Biochem. Biophys. Res. Commun.* **55**:1262-1270 (1973).
7. J.J. Lemasters: Dynamic measurement of ATP with firefly luciferase bioluminescence, *in*: "Bioluminescence and Chemiluminescence: Basic Chemistry and Analytical Applications," M.A. De Luca and W.D. McElroy, eds, Academic Press, New York, pp. 197-202 (1981).
8. R. Wibom, A Lundin and E. Hultman, A sensitive method for measuring ATP-formation in rat muscle mitochondria, *Scand. J. Clin. Lab. Invest.* **50**:143-152 (1990).
9. A.P. Halestrap, The mitochondrial pyruvate carrier: Kinetics and specificity for substrates and inhibitors, *Biochem. J.* **148**:85-96 (1975).
10. A.P. Halestrap and R.M. Denton, The specificity and metabolic implications of inhibition of pyruvate transport in isolated mitochondria and intact tissue preparations by α-cyano-4-hydroxycinnamate and related compounds, *Biochem. J.* **148**:97-107 (1975).
11. A.K. Groen, R.J.A. Wanders, H.V. Westerhoff, R. van der Meer and J.M. Tager, Quantification of the contribution of various steps to the control of mitochondrial respiration, *J. Biol. Chem.* **257**:2754-2757 (1982).

CONTROL OF RESPIRATION OF MITOCHONDRIA FROM *EUGLENA GRACILIS* DETERMINED BY MEANS OF METABOLIC CONTROL ANALYSIS: COMPARISON OF OXIDATION OF MATRIX- AND CYTOSOLIC NADH

Adolf Kesseler, Stefan Kisteneich and Klaus Brinkmann

Botanisches Institut
Universität Bonn
Kirschallee 1, 5300 Bonn 1, Germany

1. Introduction and Goals

The mitochondria of Euglena have an outstanding capacity to oxidize cytosolic NADH as efficiently coupled to oxidative phosphorylation as matrix NADH: both paths reveal the same sensitivity to rotenone and identical P/O values[1]. However, they strongly differ in the rotenone sensitivity of ferricyanide reduction, indicating two separate NADH-oxidases at the inside and the outside of the inner mitochondrial membrane (see Fig. 1).

The strange similarity in phosphorylation efficiency of both oxidations raises the question as to whether there is a difference between the control of respiration on matrix- and cytosolic NADH. We investigated the distribution of control at each step by applying the "bottom-up" approach of metabolic control theory.

2. Isolation and Characterization of Mitochondria

Mitochondria were isolated by means of tryptic digestion from the bleached mutant K_2 of *Euglena gracilis* strain 1224-5/9 grown under a 12:12 dark-light cycle on a glutamate/ malate medium, harvested at the late exponential phase of growth[1]. There is evidence that isolated mitochondria are refusioned vesicles of a mitochondrial net[2], which highly vary in volume and density, but the mitochondria are intact with respect to the following features: spontaneous State 3 - State 4 transitions, trapping of matrix enzymes and cytochrome c and trapping of adenylate kinase in the intermembrane space. Fig. 2 shows an experiment with simultaneous recording of O_2-consumption and the nucleotides ATP, ADP and AMP after initiation of the first State 4 - State 3 transition by adding AMP during malate respiration.

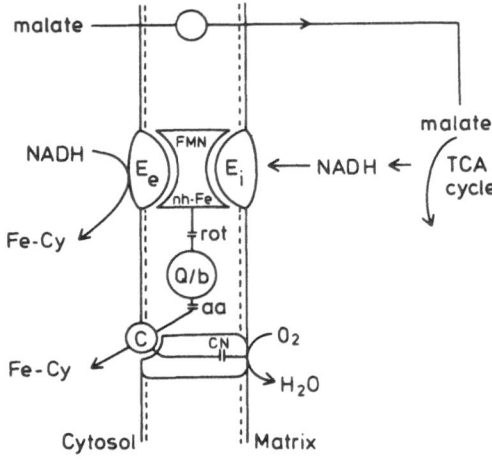

Figure 1. Scheme of the electron chain in mitochondria of *Euglena gracilis* (adapted from Ref. 1).

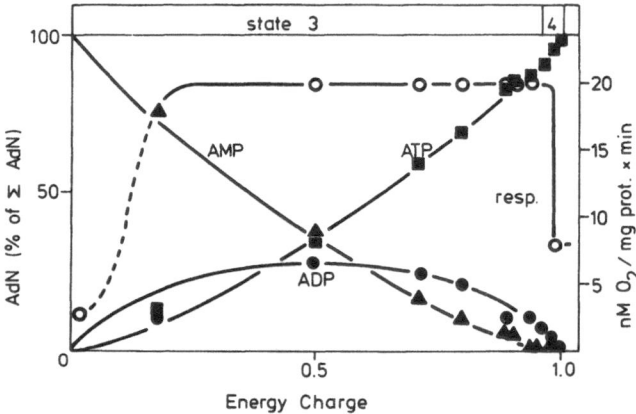

Figure 2. Adenine nucleotides and respiration after stimulation of malate respiration by AMP. Time proceeds from left to right, but data are plotted versus energy charge.

A constant State-3 respiration is reached within about 30 seconds after addition of AMP; it is maintained for about 5 min. until exhaustion of ADP. Between the energy charges 0.2 and 0.96 the K_{app} of the adenylate kinase was fitted to the value of 1.66.

The adenylate kinase is easily released from the intermembrane space by hypoosmotic disruption.

3. Control Analysis

We investigated the control at the single steps at State 3 (maintained by 1mM ADP) and State 4 respiration of extramitochondrial NADH (1mM) and matrix NADH (supplied with 6.75 mM malate). Respiration rates were measured using a "Clark"-type O_2-electrode. The control coefficients dJ/dI over dv_i/dI have been calculated according to Refs 3,4, with J being the overall flux, v_i representing the flux through the single step and I being the concentration of an inhibitor, which has to be specific for this particular step.

4. Results

The control coefficients are shown in Table 1. As an example, Fig. 3 shows inverted KCN titration curves of complex IV at State 4. v_i was measured by the oxidation of ascorbate/TMPD. Since KCN binds competitively, determination of the initial slope by simple linear regression in the original titration curves is in general inaccurate, because the curves show inverse "Michaelis-Menten" kinetics, due to competitive inhibition. The initial slope can be expressed more accurately by the first derivative of these curves at $I=0$, the initial point of each titration, which is maximum inhibition over K_i.

The control coefficient is therefore calculated as $(inhibition_{max}/K_i)_J$ over $(inhibition_{max}/K_i)_{v_i}$. Those parameters can be determined either from the original titration curves, or from double reciprocal Lineweaver-Burk plots (see Fig.3).

In the case of using irreversible inhibitors, measurements of the isolated step are not required and the control coefficients are given by dJ/dI over dI/I_{max}. The control of complex III and ATPase was determined by titrating respiration with the irreversible

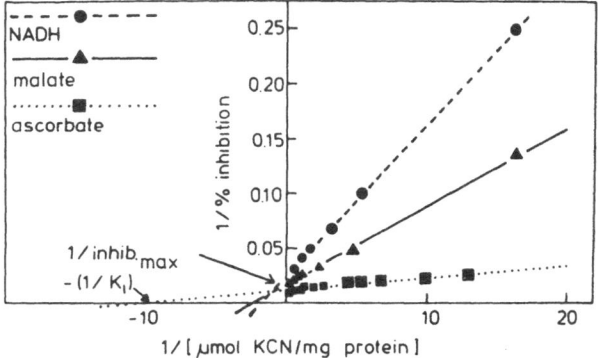

Figure 3. Inverted KCN titration curves (Lineweaver-Burk plots) of NADH and malate State 4 respiration (representing the overall flux through the system) and ascorbate State-4 respiration (representing the flux through cytochrome-c oxidase).

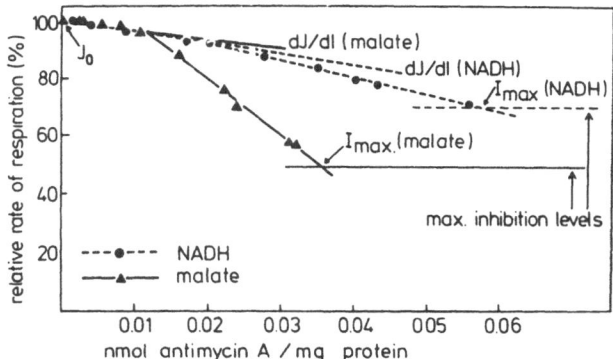

Figure 4. Titration curves of State 3 respiration using antimycin A. All parameters required for the calculation of the control coefficient of complex III can directly be derived from the titration of the overall flux (respiration rate). The initial slope dJ/dI and I_{max} (the minimal amount of inhibitor giving maximal inhibition) were determined by linear regression.

inhibitors antimycin A and oligomycin. Fig. 4 exemplarily shows the State-3 titration curves with antimycin A, revealing the control of complex III under this condition.

The control exerted by the "proton leak" has been calculated by CCCP titrations without (Fig. 5) and in presence of saturating oligomycin (Fig. 6). According to Groen and coworkers[4] the control coefficient is given by dJ/J over $dCCCP/CCCP^*$. It is now known that the extrapolation of $CCCP^*$ to represent the amount of uncoupler virtually necessary to induce unaffected State-4 respiration (Fig. 6) is not valid[5]. Since an alternative method[6] hitherto failed with *Euglena* (because an online record of membrane potential ist not yet possible), we use this CCCP-based index at least to represent the "capacity" of control to be exerted by all proton fluxes. There was no effect of butylmalonate under any condition up to concentrations of 5 μmol/mg protein. At higher concentrations the effect is no longer specific. It is therefore concluded that the dicarboxylate carrier exerts no control on malate respiration. Also the adenine nucleotide carrier does not control respiration, which was clearly demonstrated by the lack of inhibition at low concentrations of carboxyatractyloside (0-0.8 μmol/mg protein at State 4; 0-0.3 μmol/mg protein at State 3), although respiration is finally inhibited to 60 % at concentrations of 4 μmol/mg protein.

Finally we found a minor initial inhibition by mersalyl, indicating a control by the phosphate carrier. Since a suitable method was not available to measure the individual reaction rate for this step, control could not be quantified.

Figure 5. CCCP titration curves of NADH and malate respiration. The values of $dJ/dCCCP$ were determined by linear regression.

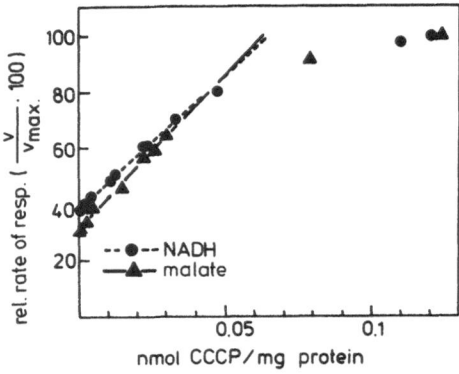

Figure 6. Titration of State 4 respiration with CCCP. Oxidative phosphorylation was completely inhibited by saturating concentrations of oligomycin. The figure shows the linear extrapolation, that was used in order to calculate the value of $CCCP^*$, which is required to calculate the control coefficient.

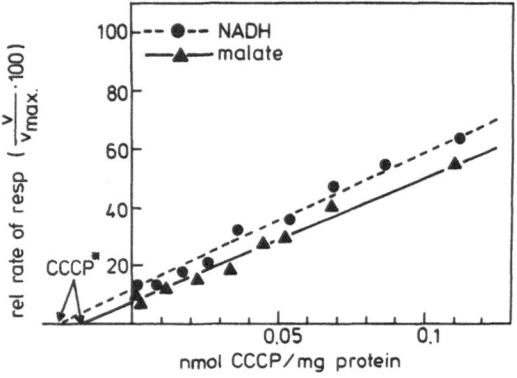

5. Comparison of Mitochondria from Euglena, Higher Plants and Animals

Concerning the pattern of control, the data of Table 1 characterizes the mitochondria from *Euglena gracilis* roughly halfway in between higher plant and animal mitochondria.

Compared with higher plant mitochondria[7,8], control exerted by the steps of phosphorylation, especially by ATPase is higher and the control of the respiratory chain is lower in Euglena; no control by the adenine nucleotide- and the dicarboxylate carrier is typical for organelles from both sources.

In animal mitochondria the control exerted by the electron chain is comparably weak, while the control of the oxidative phosphorylation over respiration may rise up to 60 %.[4,9] Those control features characterize mitochondria predominantely serving the ATP supply of the cytosol, which is typical for strictly heterotrophic organisms.

Table 1. Comparison of flux control coefficients over respiration in mitochondria from *Euglena gracilis* in States 3 and 4.[¶]

Step	Inhibitor	Flux control coefficient			
		State 3		State 4	
		NADH	Malate	NADH	Malate
Proton leak	CCCP	0	0	0.63±0.21	0.61±0.19
cyt.-bc complex	antimycin A	0.22±0.11	0.14±0.06	0.03±0.03	0.02±0.02
cyt.-c oxidase	KCN	0.04±0.03	0.19±0.09	0.07±0.03	0.13±0.04
H$^+$-ATPase	oligomycin	0.20±0.06	0.22±0.06	0.14±0.04	0.11±0.05
DC-carrier	butylmalonate	0	0	0	0
AdN-carrier	CATR	0	0	0	0
Summation		0.46	0.55	0.87	0.87

[¶]Distributions of flux control coefficients (mean value ± standard deviation) of several steps in mitochondria of *Euglena gracilis* at State 3 and State 4. The control that was not detected is supposed to be shared between complex I and the phosphate carrier. Mean uninhibited respiration rates were 20.5 ± 5.2 (State 4), 71.0 ± 18.1 (State 3) for NADH and 10.0 ± 3.9 (State 4) and 33.0 ± 9.1 (State 3) for malate (rates in nmol O_2/min.∗mg protein).

In plants the tricarboxylate cycle also (or even predominantly) serves anaplerotic reactions supplying intermediates to a high net synthesis on the expense of energy mainly from photosynthesis. Under these conditions, the tricarboxylic acid cycle (or parts of it) have to be kept in an active state, although the demand for ATP is low, and NADH has to be recycled with low and flexible coupling to phosphorylation. Euglena represents a state in between animals and plants, as it grows either photoautotrophically or heterotrophically, depending on the light and growth medium. In our case, mitochondria have been isolated from a heterotrophic mutant. Thus, if a modification of the control type of mitochondria exists depending on interactions with chloroplasts, the data represents the uppermost 'animal like' state of organization that is possible for *Euglena*.

6. Critical Points

We want to point out that the "bottom-up" approach of metabolic control analysis, used here and by others[4,7] contains a pitfall, in so far as it is assumed that the inhibitors used to titrate a single step are specific to this particular enzyme. The presence of another binding site, either affecting or not affecting the respiration rate, finally results in miscalculations of the control coefficients. Due to the lack of specific inhibition, we refrained from quantifying the control at the phosphate carrier by means of titrations with mersalyl (a SH group reagent), but also the specificity of the other inhibitors may be doubtful. For example, if an inhibitor also affects the different NADH dehydrogenases, this would be a possible explanation for the different maximum inhibition levels of cytosolic- and matrix NADH respiration (Figs 3 and 4).

7. Summarizing Remarks

Comparing the oxidation of external and internal NADH, Table 1 shows no significant differences in the control coefficients.

Thus, in *Euglena* mitochondria not only the phosphorylation efficiency of oxidizing cytosolic- and matrix NADH is the same: the control of respiration is very similar as well. The bulk of missing control in State 3 must be due to complex I, since high initial inhibition rates occur with rotenone. These inhibition rates, however, do not differ either using internal and external NADH.

References

1. U. W. Kümmel and K. Brinkmann, The oxidation of exogenous NADH by mitochondria of *Euglena gracilis*, *Planta* **176**:261-268 (1988).
2. Y. Hayashi and K. Ueda, The shape of mitochondria and the number of mitochondrial nucleotides during the cell cycle of *Euglena gracilis*, *J. Cell Sci.* **43**:137-166 (1989).
3. R. Heinrich and T. A. Rapoport, A linear steady-state treatment of enzymatic chains; general properties, control and effector strength, *Eur. J. Biochem.* **42**:89-95 (1974).
4. A. K. Groen, R. J. A. Wanders, H. V. Westerhoff, R. van der Meer and J. M. Tager. Quantification of the con-tribution of various steps to the control of mitochondrial respiration, *J. Biol. Chem.* **257**:2754-2757 (1982).
5. M. D. Brand, R. P. Hafner and G. C. Brown, Control of the respiration in non-phosphorylating mitochondria is shared between the proton leak and the respiratory chain, *Biochem. J.* **255**:535-539 (1988).
6. G. C. Brown, R. P. Hafner and M. D. Brand, A 'top-down' approach to the determination of control coefficients in metabolic control theory, *Eur. J. Biochem.* **188**:321-325 (1990).
7. A. C. Padovan, I. B. Dry and J. T. Wiskich, An analysis of the control of phosphorylation coupled respiration in isolated plant mitochondria, *Plant Physiol.* **90**:928-933 (1989).
8. A. Kesseler, P. Diolez, K. Brinkmann and M. D. Brand, Characterization of the control of respiration in potato tuber mitochondria using the top-down approach of metabolic control analysis, *Eur. J. Biochem.* **210**:775-784 (1992).
9. R. P. Hafner, G. C. Brown and M. D. Brand, Analysis of the control of respiration rate, phosphorylation rate, proton leak rate and protonmotive force in isolated mitochondria using the top-down approach of metabolic control theory, *Eur. J. Biochem.* **188**:313-319 (1990).

TOP DOWN METABOLIC CONTROL ANALYSIS
OF OXIDATIVE PHOSPHORYLATION
AT DIFFERENT RATES
IN POTATO TUBER MITOCHONDRIA

Adolf Kesseler[1], Philippe Diolez[2],
Klaus Brinkmann[1], and Martin D. Brand[3]

[1]Botanisches Institut der Universität Bonn
 Abteilung für Experimentelle Oekologie
 5300 Bonn, Germany
[2]Biochimie Fonctionnelle des Membranes Végétales, CNRS
 91198 Gif-sur-Yvette, France
[3]Dept. of Biochemistry
 University of Cambridge
 Cambridge CB2 1QW, England

Mitochondria isolated from plant tissues have specific properties (such as exogenous NADH oxidation and the alternative oxidase) that may result in a distribution of control over oxidative phosphorylation that is different from the control in mitochondria from other sources. However relevant data are scarce[1,2] and no complete description of the control of phosphorylation in plant mitochondria is available. This study is an attempt to establish the basis for the study of the regulation of mitochondrial function in the plant cell. As a model system we used mitochondria purified from potato tuber since they appear straightforward with respect to bioenergetic processes[3,4].

The top-down approach of metabolic control analysis[5] has been used as it allows an overview of the complete system. Attention has been focused on the intermediate states of phosphorylation (ADP-limited respiration) which are likely to be the conditions *in vivo*. The analysis has been performed around the proton-motive force considered as the obligatory intermediate.

Fig. 1 presents the system under study, with the specific features of the plant mitochondrial respiratory chain. The use of exogenous NADH has a great advantage since the oxidation of this substrate occurs on the external face of the inner membrane and therefore does not involve substrate transport which would otherwise be included in the system. The alternative oxidase, a non-electrogenic branched pathway with these substrates, must be excluded from the system as it diverts electrons from the cytochrome pathway (Δp producers) and consumes O_2. However, it must be borne in mind that further studies must include the

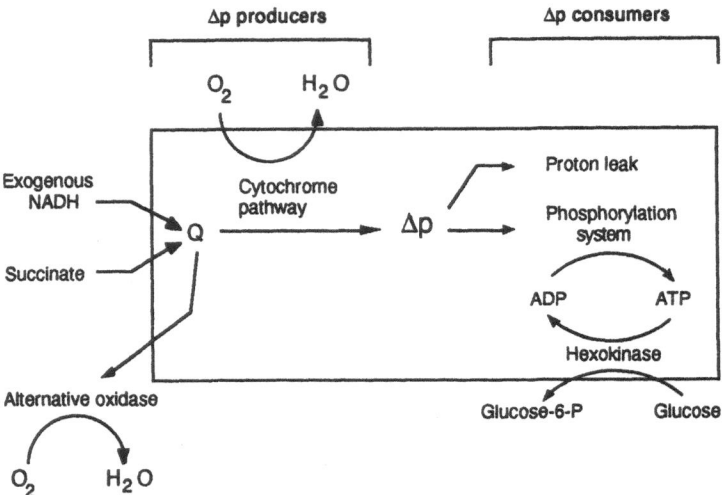

Figure 1. Scheme of oxidative phosphorylation in plant mitochondria. The system analyzed here consists of all the reactions that are contained in the box.

alternative pathway as a regulatory mechanism in physiological conditions. Purified potato mitochondria have been reported to contain an unusually low cyanide-resistant oxidase activity[6]. However, these data were deduced from the comparison of the maximal activities of the two repiratory pathways and a careful examination showed that a non-negligible activity may be demonstrated under State-4 conditions (Fig. 2).

Two different approaches to measure alternative oxidase activity were used: insensitivity to cytochrome pathway inhibitors (Fig. 2a) and sensitivity to salicylhydroxamic acid (SHAM), an inhibitor of the alternative oxidase (Figs 2b and 2c).

In Fig. 2a the curves represent oxygen consumption to drive proton leak plus any extra oxygen consumption due to other processes such as the alternative oxidase. With the inhibitors acting on the oxidizing site of quinones (Q) (KCN, myxothiazol, antimycin A) the curves for succinate were displaced by about 30 nmol oxygen/min per mg protein compared to the curves with NADH as substrate. This oxygen-consuming activity was not inhibited by saturating concentrations of these inhibitors, suggesting that the alternative oxidase was present to this extent with succinate but not with NADH (Fig. 2a). By acting on the reducing site of Q, malonate induced a progressive shift to the line for NADH as Q becomes more oxidized.

The second approach allowed estimation of the alternative oxidase activity between the resting (State 4) and the maximal oxidation rate (Figs 2b and 2c). An unexpected uncoupling effect of SHAM is evident from these results. With NADH as substrate, SHAM induced an increase in the oxidation rate (full triangles compared to open ones in Fig. 2c), but no change in the relationship between the oxidation rate and Δp. Effectively, a pure uncoupling effect would not interfere with titrations of the respiratory chain since they represent the kinetic response of the respiration rate to Δp regardless of the mechanism used to vary Δp. On the other hand, an inhibition of the oxygen consumption would shift the relationship towards lower values of respiration rate, as observed with succinate where both effects occured (Fig. 2b). The same value of about 30 nmol/min per mg protein is observed for this shift.

It may be concluded that succinate oxidation takes place with a nearly constant alternative oxidase activity while no activity is observed with NADH. These conclusions agree with those of others[2, 6-8]. For the analysis of the pattern of control with succinate as substrate, this

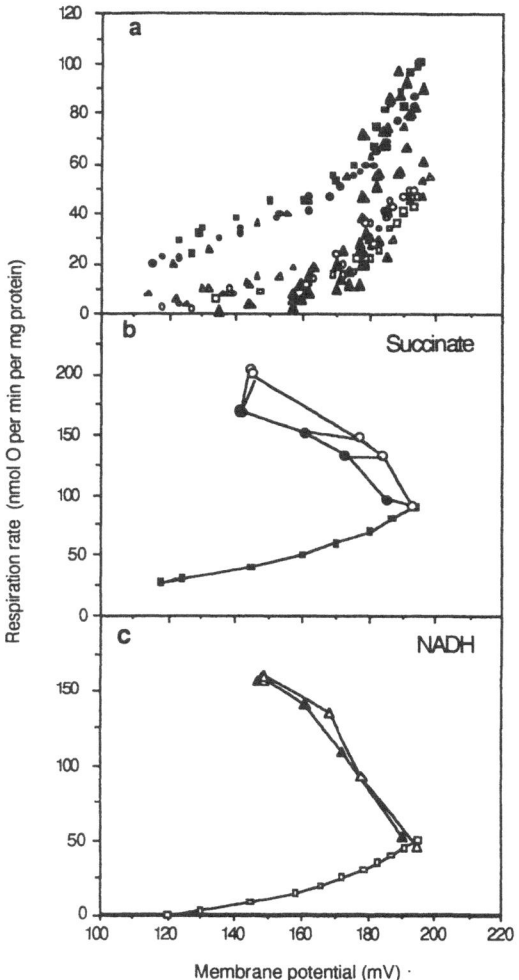

Figure 2. Estimation of the alternative oxidase activity. (a) State 4 titrations of mitochondria oxidizing NADH (open symbols) or succinate (closed symbols) using different inhibitors. State 4 was titrated using cyanide (represented by triangles), myxothiazol (circles) or antimycin A (squares). Succinate respiration was also titrated with malonate (full triangles). (b) Effect of SHAM on hexokinase titrations of State 4 with succinate as substrate. The average curve of succinate State 4 titrations with KCN, antimycin A and myxothiazol (indicated by full squares) is from Fig. 2a. The hexokinase titration of State 4 was carried out in the absence (open circles) and in the presence (full circles) of 1.25 mM SHAM. (c) Effect of SHAM on hexokinase titrations of State 4 with NADH as substrate. The average curve of NADH State 4 titrations with KCN, antimycin A and myxothiazol (open squares) is from Fig. 2a. The hexokinase titration of State 4 was carried out in the absence (open triangles) and in the presence (full triangles) of 1.25 mM SHAM.

activity was subtracted from the oxygen consumption. A possible reason for the absence of alternative oxidase with NADH is that Q is always too oxidized under these conditions to trigger the alternative pathway.

Control over respiration rate, phosphorylation rate, proton leak rate and over Δp exerted by the respiratory chain, oxidative phosphorylation and the proton leak was measured over a range of phosphorylation rates from resting (State 4) to maximal (State 3) with NADH or succinate as substrates. These rates were obtained using the hexokinase-glucose system or oligomycin inhibition of ATP synthase. Small differences may be observed depending on the substrate used, and the presence of hexokinase slightly increased the control exerted by the oxidative phosphorylation at intermediate rates[9]. The results presented here were obtained with NADH (Fig. 3) or succinate (Fig. 4) as substrates, using the hexokinase-glucose system to vary the rate of phosphorylation.

The results indicate that whatever is the substrate used, the control in potato mitochondria is predominantly exerted by the respiratory chain under all conditions, except close to State 4.

Respiration in State 3 (Figs 3a and 4a) is controled almost entirely by the respiratory chain (indicated by full circles), with a flux control coefficient aproaching 1.0. As we move to intermediate rates of respiration, this flux control drops but always remains as high as 0.5. The most striking change is the increase in the flux control coefficient of the leaks which rises to 0.6-0.8 (open circles) in State 4. Figs 3b and 4b show that control over the rate of phosphorylation is shared between the respiratory chain (full circles, predominant in State 3) and the phosphorylation system (open squares, predominant in State 4). The proton leak has almost no control over the flux through the phosphorylation branch. In contrast, the phosphorylation system has high negative control over the leaks (Figs 3c and 4c) so that an increase in the activity of the phosphorylation system results in a decrease in flux through the proton leak pathway.

All control coefficients over the 'concentration' of the common intermediate Δp are small (Figs 3d and 4d), reflecting the way the system is set up to keep Δp fairly constant over a wide range of rates of ATP synthesis.

These results can be compared with the results for succinate respiration by rat liver mitochondria also using hexokinase for adjustment of intermediate respiration rates[10]. The main difference is the higher control by the respiratory chain over all the fluxes in most

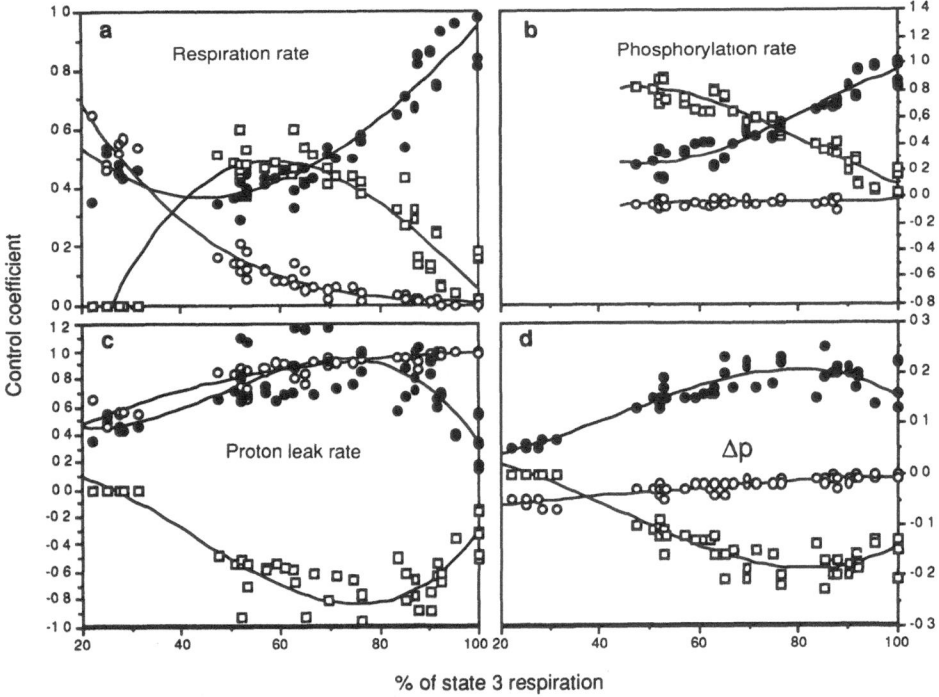

Figure 3. Control coefficients using NADH and hexokinase. Control coefficients over (a) mitochondrial respiration rate, (b) phosphorylation rate, (c) proton leak rate and (d) Δp, exerted by the respiratory chain (represented by full circles), the phosphorylation system (open squares), and the proton leak (open circles) with NADH as substrate were calculated as described previously[9].

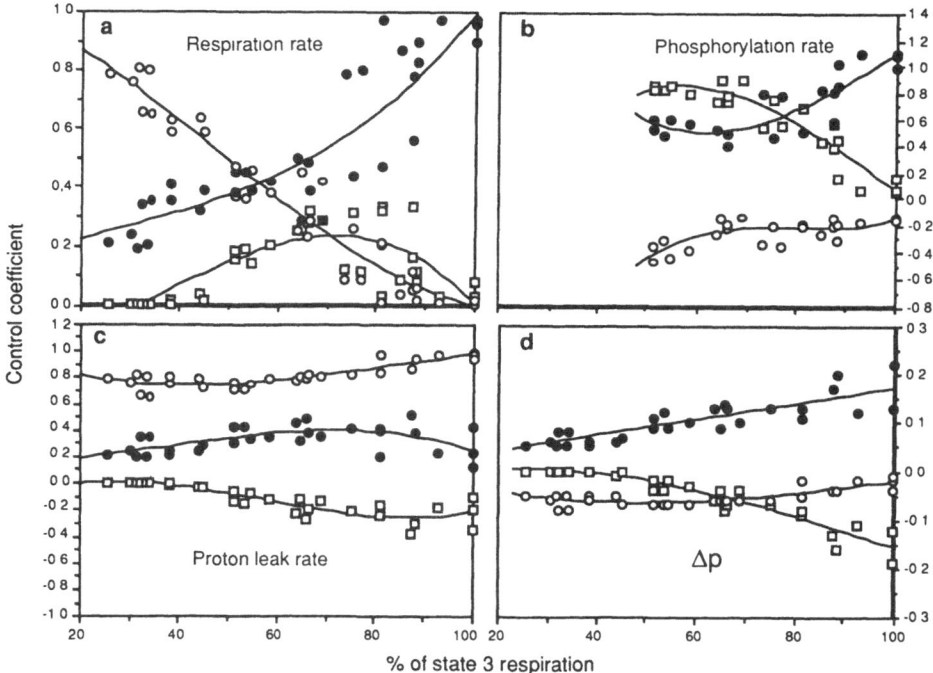

Figure 4. Control coefficients using succinate and hexokinase. Symbols as in Fig. 3 with succinate as substrate.

conditions in potato mitochondria. However, potato tuber is a metabolically resting organ and a different pattern may be expected for mitochondria from an active tissue. Because of this high control at the level of the respiratory chain it will be interesting to carry out further work on the distribution of control in the presence of several substrates as well as within the respiratory chain. The non coupled-branch represented by the alternative oxidase pathway must also be considered as a physiological means to modify the control pattern in plant mitochondria.

References

1. R.C. Padovan, I.B. Dry and J.T. Wiskich, An analysis of the control of phosphorylation coupled respiration in isolated plant mitochondria, *Plant Physiol.* **90**:928-933 (1989).
2. A.L. Moore, I.B. Dry and J.T. Wiskich, Regulation of electron transport in plant mitochondria under state 4 conditions, *Plant Physiol.* **95**:34-40 (1991).
3. P. Diolez and F. Moreau, Correlation between ATP synthesis, membrane potential and oxidation rate in plant mitochondria, *Biochim. Biophys. Acta* **806**:56-63 (1985).
4. P. Diolez and M.D. Brand, Proton fluxes during oxidative phosphorylation in potato mitochondria, *in:* "Molecular Biochemical and Physiological Aspects of Plant Respiration," H. Lambers and L.H.W. van der Plas, eds, SPB Academic Publishing, The Hague, in press.
5. G.C. Brown, R.P. Hafner and M.D. Brand, A 'top-down' approach to the determination of control coefficients in metabolic control theory, *Eur. J. Biochem.* **188**:321-325 (1990).
6. A.L. Moore and J.N. Siedow, The regulation and nature of the cyanide resistant alternative oxidase of plant mitochondria, *Biochim. Biophys. Acta* **1059**:121-140 (1991).
7. I.B. Dry, A.L. Moore, D.A. Day and J.T. Wiskich, Regulation of alternative pathway activity in plant mitochondria: non-linear relationship between electron flux and the redox poise of the quinone pool, *Arch. Biochem. Biophys.* **273**:148-157 (1989).

8. D.A. Day, I.B. Dry, K.L. Soole, J.T. Wiskich and A.L. Moore, Regulation of alternative pathway activity in plant mitochondria, *Plant Physiol.* **95**:948-953 (1991).

9. A. Kesseler, P. Diolez, K. Brinkmann and M.D. Brand, Characterisation of the control of respiration in potato tuber mitochondria using the 'top-down' approach of metabolic control analysis, *Eur. J. Biochem.* **210**:775-784 (1992).

10. R.P. Hafner, G.C. Brown and M.D. Brand, Analysis of the control of respiration rate, phosphorylation rate, proton leak rate and protonmotive force in isolated mitochondria using the 'top-down' approach of meta-bolic control theory, *Eur. J. Biochem.* **188**:313-319 (1990).

THE INFLUENCE OF PHOSPHATE ON
THE DISTRIBUTION OF FLUX CONTROL
AMONG THE ENZYMES
OF OXIDATIVE PHOSPHORYLATION
IN RAT SKELETAL MUSCLE MITOCHONDRIA

Wolfram S. Kunz, Elke Wisniewski and Frank Norbert Gellerich

Institut für Biochemie
Medizinische Akademie Magdeburg
Leipziger Str. 44, 3090 Magdeburg, Germany

1. Introduction

The regulation of oxidative phosphorylation in muscle is still a matter of dispute. To clarify the discrepancies in the literature, a detailed reinvestigation of the distribution of flux control in the process of oxidative phosphorylation in muscle mitochondria seems to be of importance. Moreover, large variations in the phosphate concentration are known to occur in muscle[1]. Possible effects of those changes have to be taken into consideration due to observations made with rat liver mitochondria that the distribution of flux control seems to depend on phosphate concentration[2,3].

The flux control coefficients (C_i) of the adenine nucleotide translocase, the phosphate transporter and the H^+-ATPase were determined in rat skeletal muscle mitochondria using glutamate plus malate as substrates and soluble F_1-ATPase as load enzyme. The determination of C_i-values was performed by nonlinear regression analysis of the complete inhibitor titration curves[4].

2. Results

In Fig. 1, two titrations with carboxyatractyloside, an irreversible inhibitor of the adenine nucleotide translocase, are shown, one at 10 mM phosphate (filled circles) and one at 1 mM phosphate (open circles), respectively.

At an identical rate of intermediate state respiration (120 nmol O_2/min/mg), adjusted with F_1-ATPase, the same amount of carboxyatractyloside had a more pronounced inhibitory effect at 10 mM phosphate. This is an indication that the flux control coefficient of the adenine nucleotide translocase is larger in the presence of 10 mM phosphate.

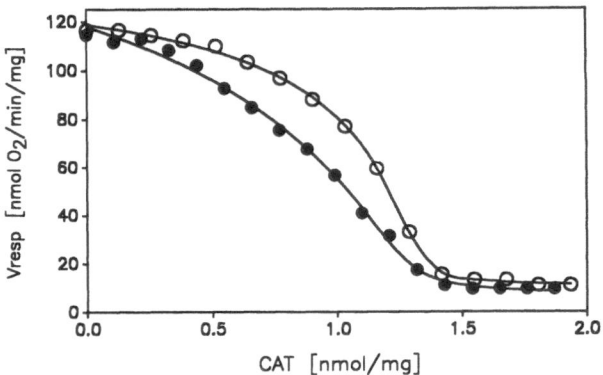

Figure 1. Titration of respiration of rat skeletal muscle mitochondria with carboxyatractyloside. Rat skeletal muscle mitochondria (0.9 mg/ml) were incubated in a standard oxygraph medium[3] in the presence of 1 mM phosphate (open circles) or 10 mM phosphate (filled circles).The stationary rate of respiration (118 nmol O_2/min/mg) was adjusted by addition of 0.48 U/mg (open circles) or 0.21 U/mg (filled circles) F_1-ATPase, respectively. The experimental points were fitted to the titration curves (according to Ref. 4) with the follow-ing parameters (± asymptotic standard deviation), flux control coefficient, 0.16 ± 0.01, maximal amount of inhibitor, 1.32 ± 0.01 nmol/mg, K_d = 0.001 ± 0.0004 nmol/mg (open circles); flux control coefficient, 0.36 ± 0.04 , maximal amount of inhibitor, 1.28 ± 0.04 nmol/mg, K_d = 0.003 ± 0.002 nmol/mg (filled circles).

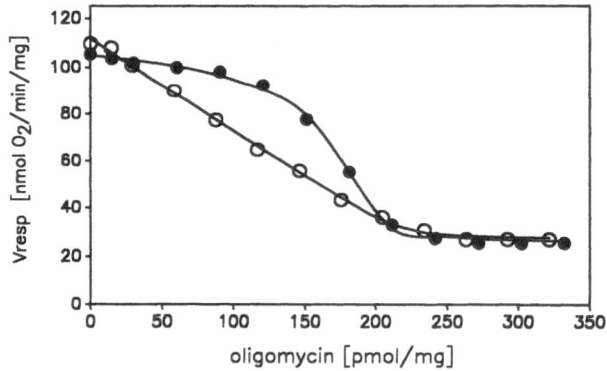

Figure 2. Titration of respiration of rat skeletal muscle mitochondria with oligomycin. Rat skeletal muscle mitochondria (0.8 mg/ml) were incubated in a standard medium[3] at 1 mM phosphate (open circles) and 10 mM phosphate (filled circles). The stationary rates of respiration were adjusted by the addition of 0.48 U/mg (open circles) or 0.21 U/mg (filled circles) F_1-ATPase, respectively. The experimental points were fitted to the titration curves (according to Ref. 4) with the following parameters (± asymptotic standard deviation), flux control coef-ficient, 0.76 ± 0.07, maximal amount of inhibitor, 0.216 ± 0.009 nmol/mg, K_d = 0.85 ± 0.51 pmol/mg (open circles); flux control coefficient - 0.11 ± 0.01 , maximal amount of inhibitor - 0.226 ± 0.003 nmol/mg, K_d = 0.26 ± 0.07 pmol/mg (filled circles).

To get a correct value for the flux control coefficient, the experimental points were fitted by means of nonlinear regression using a mathematical model, describing the action of a non-competitive and mostly irreversible inhibitor[4]. This method, accounting for the dissociation equilibrium of the inhibitor, allows an exact description of the complete inhibitor titration curve. An overestimation of the value of the flux control coefficient, as usually done with the graphical method (cf. Ref. 4) is therefore excluded. For the titration curves presented in Fig. 1, flux control coefficients of 0.36 and 0.16 were obtained at 10 mM phosphate and 1 mM phosphate, respectively.

A similar experiment was performed with the inhibitor of the H^+-ATPase oligomycin. The result of a typical experiment (of three) is shown in Fig. 2. At 10 mM phosphate in the incubation medium a strong sigmoid titration curve was observed yielding, as a result of the curve fitting procedure, a flux control coefficient of 0.13 (filled circles). At a phosphate concentration of 1 mM the shape of the titration curve was completely different (open circles). Applying the curve fitting program to this experiment, a flux control coefficient of 0.76 for the H^+-ATPase was determined.

The flux control coefficient of the load enzyme F_1-ATPase at comparable rates of respiration did depend on the phosphate concentration too, whereas flux control exerted by the phosphate transporter determined by titrations with mersalyl was not affected.

At a flux rate of approximately 120 nmol O_2/min/mg the sum of flux control coefficients of adenine nucleotide translocase, H^+-ATPase, phosphate transporter and the load enzyme F_1-ATPase is very close to unity (see Table 1). This is an indication that under the conditions of intermediate rates of respiration all other reactions have a negligible controlling influence on oxidative phosphorylation in skeletal muscle mitochondria.

Table 1. Flux control coefficients of some enzymes of oxidative phosphorylation in rat skeletal muscle mitochondria determined at two different phosphate concentrations in the incubation medium.¶

enzyme	1 mM phosphate	10 mM phosphate
AdN translocase	0.16±0.01	0.36±0.04
phosphate transporter	0.13±0.02	0.16±0.03
H^+-ATPase	0.76±0.07	0.11±0.01
F_1-ATPase	0.10±0.01	0.40±0.02
sum of flux control	1.15±0.11	1.03±0.10

¶The experimental error indicated is the standard deviation of the flux control coefficients determined in three independent experiments. All data correspond to a rate of respiration of approximately 120 nmol O_2/min/mg and to the experimental conditions of the legend to Fig. 1.

Acknowledgements

This work was supported by the grant Ge 664/3-1 (Deutsche Forschungsgemeinschaft).

References

1. A.H.L. From, S.D. Zimmer, S.P. Michurski, P. Mohanakrishnan, V.K. Ulstad, W.J. Thoma and K. Ugurbil, Regulation of the oxidative phosphorylation rate in the intact cell, *Biochemistry* **29**:3731-3743 (1990).

2. R. Moreno-Sanchez, Contribution of the translocator of adenine nucleotides and the ATP synthase to the control of oxidative phosphorylation and arsenylation in liver mitochondria, *J. Biol. Chem.* **260**:12554-12560 (1985).
3. W. Kunz, F.N. Gellerich, L. Schild and P.Schönfeld, Kinetic limitations in the overall reaction of mito-chondrial oxidative phosphorylation accounting for flux-dependent changes in apparent $\Delta G^{ex}_p/\Delta\mu H^+$ ratio, *FEBS Lett.* **233**:17-21 (1988).
4. F.N. Gellerich, W.S. Kunz and R. Bohnensack, Estimation of flux control coefficients from inhibitor titrations by nonlinear regression, *FEBS Lett.* **274**:167-170 (1990).

THE MOLECULAR BASIS
OF THRESHOLD EFFECTS
OBSERVED IN MITOCHONDRIAL DISEASES

Jean-Pierre Mazat, Thierry Letellier and Monique Malgat

Université Bordeaux II
G.E.S.B.I.
146 rue Léo-Saignat, F-33076 Bordeaux Cedex, France

1. Introduction

Whether or not a metabolic defect expresses itself as a recognizable clinical disease will depend upon the extent to which it affects the metabolic pathway in question. This could lead to a threshold expression of the disease state.

Threshold effects have been recently described in mitochondrial diseases, particularly by Wallace and coworkers[1,2] and were shown by these authors to be related to the balance between normal and mutant mitochondrial DNA. We think that a crucial stage in the expression of a threshold in clinical diseases, can lie, at the molecular level, in the action of a localized defect in a given step on the global flux of a metabolic network. Thus it was observed by Bindoff[3] both in a patient with cytochrome c oxidase deficiency and in an animal model (copper-deficient rat) that a lowering of complex IV activity by more than 50 % did not affect respiratory flux.

We have considered this problem from an experimental point of view by progressively inhibiting the activity of cytochrome c oxidase of isolated mitochondria from rat muscle with increasing concentrations of potassium cyanide (KCN), thus mimicking the effect of a defect in the activity of this enzyme.

2. Material and Methods

Muscle mitochondria were isolated by differential centrifugations as described by Morgan-Hughes and coworkers[4]. Protein concentration was estimated by the biuret method using bovin serum albumin as a standard. Mitochondrial oxygen consumption was monitored at 30 °C in a 1 ml thermostated closed vessel with rapid stirring and equipped with a Clark oxygen electrode, in the following buffer: Mannitol 75 mM, Sucrose 25 mM, KCl 100 mM, Tris Pi 10 mM, Tris/HCl 10 mM pH 7.4, EDTA 50 μM, with pyruvate 10 mM and malate 10 mM as respiratory substrate to record the respiration rate of the whole chain, or in the presence of ascorbate 0.50 mM and TMPD (Tetramethyl-p-phenylenediamine) 0.25 mM and antimycin 10 µg/ml to record the activity of cytochrome c oxidase alone. The mitochondrial concentration used for this study was 1 mg/ml, and State 3 was obtained by addition of ADP 1.2 mM.

3. Results

It is possible to record, in parallel experiments, the remaining cytochrome c oxidase activity and the ensuing value of respiration rate. Both curves are shown in Fig. 1A. It is of interest to note the shape of these curves. The inhibition curve of cytochrome c oxidase activity is quasi-linear with an inhibition nearly proportional to the KCN concentration added. On the contrary, the variation of respiratory rate as a function of KCN concentration is sigmoïdal. The difference in these shapes is the basis of a threshold effect. For instance, we can see that a 50 % inhibition of cytochrome c oxidase only gives 10 % inhibition of the whole respiratory flux. As much as 75 % inhibition of the enzyme only leads to 20 % inhibition of the flux. However, when the inhibition of cytochrome c oxidase activity increases beyond this value, the respiratory flux decreases sharply, giving rise to a threshold effect. This is well illustrated in Fig. 1B, which represents respiratory rate as a function of cytochrome c oxidase activity.

These results are in accordance with the observations of Bindoff[3] on a patient with cytochrome c oxidase deficiency and in the animal model of copper-deficient rats. We found similar results with the modulation of complex I activity by rotenone and of complex III activity by antimycin (unpublished results).

4. Discussion and Conclusion

These results are entirely consistent and in accordance with Metabolic Control Theory[5,6]. In the framework of this theory, a control coefficient measures the effect of a small perturbation of a given step on a global flux in a metabolic network under steady-state conditions. This parameter can be evaluated in the present experiment by the ratio of the initial slope of the V_{O_2} inhibition curve (the observed perturbation of the flux) towards the initial slope of the cytochrome c oxidase activity (the step perturbation). In our case, we find a control coefficient of cytochrome c oxidase on mitochondrial respiration (V_{O_2}) of 0.12. In Control Theory, there are strong constraints on the control coefficients (e.g. summation theorem) so that the values of control coefficients of individual steps are usually low; and this is the case in oxidative phosphorylation[7-9]. This means that the initial slope to the pathway flux inhibition curve is much weaker than the corresponding slope to the step inhibition curve; this is the case in Fig. 1A. However, both curves must converge at high inhibitor concentration, when the step activity becomes very low. Thus the flux inhibition curve must reach the low activity region more steeply than the step activity does, hence its sigmoïdal pattern. In this sense, a threshold effect is necessarily associated with a low control coefficient. This is well illustrated in Fig. 1.

How can this direct threshold effect be related to the heteroplasmy of mtDNA, which may be observed in mitochondrial diseases? As a first approximation, it can be thought that the activity decrease of a given step (e.g. cytochrome c oxidase) is proportional to the percentage of mutant mt DNA, in which case the profile of Fig. 1B would remain unchanged when plotting the percentage of respiration rate as a function of the percentage of heteroplasmy.

A threshold effect of DNA heteroplasmy has been observed earlier by Flint and coworkers[10-11]. They studied the rate of arginine production in heterokaryons of *Neurospora*

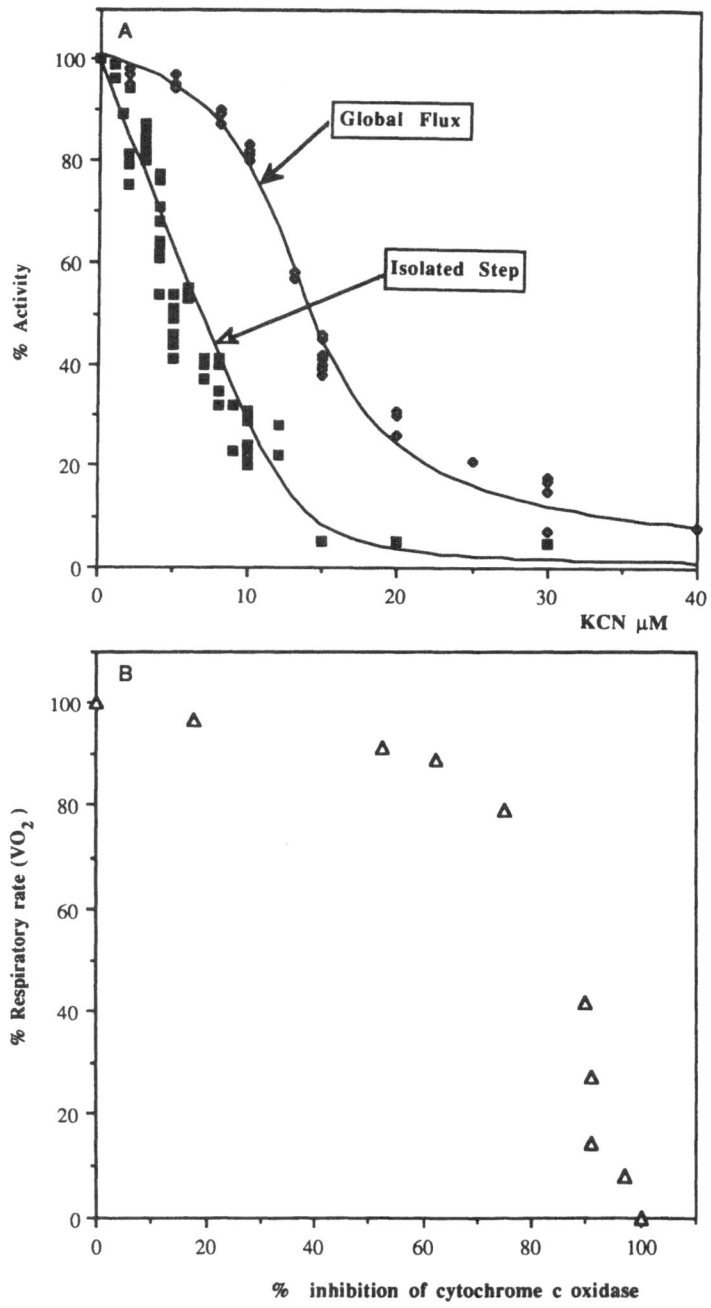

Figure 1. KCN inhibition of cytochrome c oxidase activity (□) and of respiratory rate (○) (A). Respiratory rate as a function of cytochrome c oxidase inhibition (B). The points are the means of the data of Fig. 1A corresponding to the same KCN concentrations.

crassa with different proportions of a mutant gene of the arginine pathway. This phenomenon was extensively discussed by Kacser and Burns[12].

A more precise quantification of the heteroplasmy effect at the level of the activity of a given step would require measurement of its influence on the amount of transcribed normal mtDNA on the one hand and of translated normal transcripts on the other. At these particular stages, a threshold effect could occur as well[13].

Another question arises as to what are the effects of deficient mitochondria on cellular metabolism as a whole. If mitochondrial metabolism is considered as if it were a single step in the whole cellular metabolism it may be hypothesized that a further threshold effect occurs which is analogous to that observed at the level of the whole mitochondrial metabolism.

Thus, we are in fact faced with at least three levels at which threshold effects may occur. The first one is the expression of the heteroplasmy of DNA at the level of a given enzymatic step, the second is the threshold effect observed in mitochondrial metabolism as a result of a decrease in a given mitochondrial activity; this is the subject of the present paper. The third one may occur in the expression of deficient mitochondria in the whole cellular metabolism. At each level the threshold effect will reinforce the other. Thus when we consider the global effect of mtDNA heteroplasmy on total cellular metabolism and its clinical expression, a very sharp threshold would be anticipated.

Acknowledgments

The authors are grateful to Dr. J. Clark and to Brenda Groen for critically reading the manuscript. This work was supported by the Association Française contre les Myopathies (A.F.M), the Université Bordeaux II, the CNRS and the Région Aquitaine.

References

1. D. C. Wallace, Mitotic segregation of mitochondrial DNAs in human cell hybrids and expression of chloramphenicol resistance, *Som. Cell Molec. Genet.* **12**:41 (1986).
2. J. M. Shoffner, M.T. Lott, A.M.S. Lezza, P. Seibel, S.W. Ballinger and D.C. Wallace, Myoclonic Epilepsy and ragged-red fiber disease (MERRF) is associated with a mitochondrial DNA tRNA^{Lys} mutation, **61**:931-937 (1990).
3. L.A. Bindoff, PhD Thesis, Newcastle upon Tyne University (1990).
4. J.A. Morgan-Hughes, D.J. Hayes, J.B. Clark, D.N. Landon, M. Swash, R.J. Strak, P. Rudge, Mitochondrial encephalomyopathies. Biochemical studies in two cases revealing defects in the respiratory chain, *Brain* **105**:553-582 (1982).
5. H. Kacser and J. A. Burns, The control of flux, *Symp. Soc. Exp. Biol.* **27**:65-104 (1973).
6. R. Heinrich and T. A. Rapoport, A linear steady-state treatment of enzymatic chains. General properties, control and effector strength, *Eur. J. Biochem.* **42**:89-95 (1974).
7. A.K. Groen, R.J.A. Wanders, H. V. Westerhoff, R. van der Meer and J.M. Tager, Quantification of the contribution of various steps to the control of mitochondrial respiration, *J. Biol. Chem.* **257**:2754-2757 (1982).
8. J.M. Tager, R.J.A. Wanders, A.K. Groen, W. Kunz, R. Bohnensack, U. Küster, G. Letko, G. Boehme, J. Duszynski and L. Wojtczak, *FEBS Lett.* **151**:1-9 (1983).
9. J.-P. Mazat, E. Jean-Bart, M. Rigoulet and B. Guérin, Control of oxidative phosphorylation in yeast mitochondria. Role of the phosphate carrier, *Biochim. Biophys. Acta* **849**:7-15 (1986).
10. H.J. Flint, D.J. Porteous and H. Kacser, Control of the flux in the arginine pathway of Neurospora crassa, *Biochem. J.* **190**:1-15 (1980).
11. H. J. Flint, R.W. Tateson, I.B. Barthelmess, D.J. Porteous, W.D. Donachie and H. Kacser, Control of the flux in the arginine pathway of Neurospora crassa, *Biochem.J.* **200**:231-246 (1981).
12. H. Kacser and J. A. Burns, The molecular basis of dominance, *Genetics* **97**:630-666 (1981).
13. J.-I. Hayashi, S. Ohta, A. Kikuchi, M. Takemitsu, Y-I. Goto and I. Nonaka, *Proc. Nat. Acad. Sci. U.S.A.* **88**:10614-10618 (1991).

CONTROL OF RESPIRATION IN HEART MITOCHONDRIA: COMPARATIVE STUDY OF OXIDATION OF SUCCINATE AND NAD-DEPENDENT SUBSTRATES

V. Mildažienė[1], V. Borutaitė[1], Z. Katiliūtė[1], R. Petroliūnaitė[1],
L. Ivanovienė[1], A. Toleikis[1] and B. Kholodenko[2]

[1]Biomedical Research Institute
 Kaunas Medical Academy
 Eiveniu st. 4, 3007 Kaunas, Lithuania
[2]A. N. Belozersky Institute of Physico-Chemical Biology
 Moscow State University
 Moscow 119899, Russia

1. Introduction

The experimental application of the quantitative theory of metabolic regulation[1,2] allowed to demonstrate that the control of mitochondrial respiration is not concentrated at one step and various processes prevail in control at different metabolic states[3-5]. The problem was investigated mainly on mitochondria oxidizing succinate. This metabolite is generated within mitochondria during oxidation of other substrates, so it can be considered as substrate for mitochondria only with some reservations[6]. At the same time the question arises as to the distribution of control in the case of oxidation of pyruvate and fatty acids, which are the main fuel for mitochondrial respiration in the heart.

The aim of our study was to compare the control of oxidation of succinate and main physiological substrates (pyruvate plus malate, and palmitoylcarnitine) in the whole range of respiratory rates between State 4 and State 3.

2. Methods

Mitochondria from rabbit heart were isolated with trypsin[7]. The rate of respiration was measured polarographically[7] at 37 °C. The creatine kinase[7] and glucose-hexokinase (in some experiments) systems were used for the regeneration of ADP. In the case of glucose-hexokinase system maximal rate of mitochondrial respiration was achieved by adding 100-

Modern Trends in Biothermokinetics, Edited by
S. Schuster *et al.*, Plenum Press, New York, 1993

200 µM ADP and excess hexokinase to the incubation medium containing 10 mM Tris-HCl, 150 mM KCl, 5 mM KH$_2$PO$_4$, 1.25 mM MgCl$_2$, 10 mM glucose, pH 7.2. The rate of mitochondrial respiration in State 3 was registered in the same medium except that 1 mM ADP was added instead of glucose and hexokinase.

The following concentrations of oxidizable substrates were used: 20 mM succinate (+1 µM rotenone), 1 mM pyruvate plus 1 mM malate, 20 µM palmitoyl-L-carnitine. The concentration of mitochondrial protein in the incubation medium was 0.5 mg/ml in the case of succinate and pyruvate plus malate, and 1 mg/ml in the case of palmitoylcarnitine oxidation. The control coefficients of mitochondrial enzymes over respiration rate were calculated from titration curves of mitochondrial respiration with specific inhibitors[3] - carboxyatractyloside (for the ANT[&]), thenoyltrifluoroacetone or carboxine (for SDH), oligomycine (for H$^+$-ATPase), and azide (for cytochrome oxidase). The control coefficient of proton leak was determined by titration with uncoupler[3].

3. Results and Discussion

The control pattern of mitochondrial respiration depends on the contribution of the ADP-regenerating system, which is normally used for the maintenance of steady state[4,5]. It was shown[6,7] that quantitative estimation of the actual regulatory roles of mitochondrial enzymes *per se* is possible only if the concentrations of extramitochondrial adenylates are stabilized. The creatine kinase ADP-regenerating system with excess creatine kinase was proposed for this purpose[7]. The same system is used in the present work.

Fig. 1 shows the dependence of the control coefficient of ANT on the relative rate of respiration with succinate (A), puruvate plus malate (B) and palmitoylcarnitine (C).

As can be seen from Fig. 1B, it was impossible in the creatine kinase system to achieve maximal rates of pyruvate plus malate oxidation corresponding to State 3. To overcome this problem the glucose-hexokinase ADP-regenerating system was used. In this case excess hexokinase was added to the medium, so the ADP-regenerating system did not exert any control on respiration, such as the creatine kinase system at lower rates of oxygen uptake.

For comparative purposes, the titration with carboxyatractyloside of succinate and palmitoylcarnitine oxidation rate near State 3 was also performed in the glucose-hexokinase system. The results confirm that there is no difference between the control coefficient values for the ANT measured in two distinct ADP-regenerating systems under the experimental conditions used (Figs 1B and 1C).

Table 1. Dependence of the control coefficient of inner membrane proton leak (C_L) on the relative rate of mitochondrial respiration with different substrates[¶].

v [%]	C_L	
	succinate	pyruvate + malate
15-25	0.76	0.89
26-50	-	0.17±0.1
51-75	0.59±0.07	0.08±0.05
76-90	0.22±0.03	-
91-100	0.08	-

[¶]The average results of 1-4 parallel experiments are presented. Standard error is not shown when only one value of C_L was obtained in the indicated interval of relative rate of respiration.

[&]Abbreviations: ANT, adenine nucleotide translocator; SDH, succinate dehydrogenase.

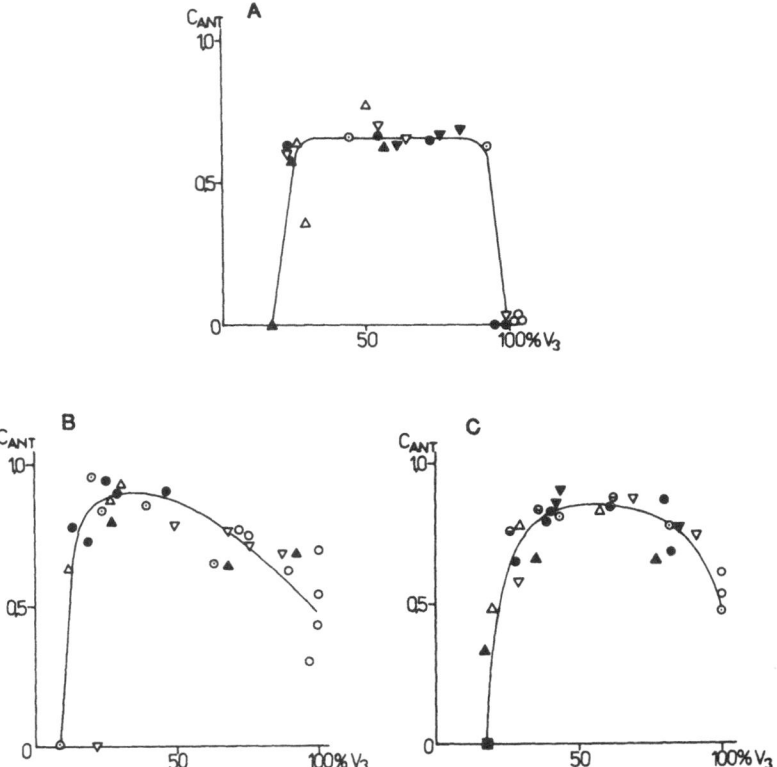

Figure 1. Dependence of the control coefficient of ANT over the relative rate of mitochondrial respiration with succinate (A), pyruvate + malate (B) and palmitoyl-L-carnitine. Along abscissa: relative rate of mitochondrial respiration in percent of the rate in State 3. Each open circle represents a separate experiment in the hexokinase system, each set of other symbols a separate experiment in the creatine kinase system. The mean mitochondrial respiration rate in State 3 was 389 ± 29 with succinate as substrate (five experiments), 500 ± 79 with pyruvate + malate as substrate (five experiments) and 209 ± 18 with palmitoylcarnitine as substrate (eight experiments) (all values in natom O/min per mg protein).

We estimated the control coefficient values of proton leak at various relative rates of respiration of heart mitochondria, oxidizing succinate and pyruvate plus malate (see Table 1). In the case of succinate our results are in agreement with data of Groen and coworkers[3]; the contribution to the control of proton leak decreases from the value 0.76 near State 4 practically to zero near State 3 as the relative rate of repiration increases. The dependence of the control coefficient of proton leak on the rate of respiration with pyruvate plus malate is similar to such a dependence in the case of succinate, but the value of this coefficient at the same relative rate of respiration is significantly lower in the first case.

As follows from Table 1 and Fig. 1, the sum of control coefficients of proton leak and of the ANT over respiration is close to unity in the range of respiratory rates from State 4 to approximately 70% of the rate in State 3. The results suggest that the contribution of these processes to the control of respiration is greatest in comparison with all other steps of oxidative phosphorylation. The values of control coefficients of H⁺-ATPase, SDH and cytochrome oxidase over succinate oxidation presented in Table 2 confirm this conclusion.

Table 2. Control coefficients (C_i) of mitochondrial enzymes over rate of succinate oxidation in different metabolic states[*].

Metabolic state of mitochondria	C_i		
	SDH	H$^+$-ATPase	cytochrome oxidase
State 3	0.7±0.06 (5)	0.1±0.095 (4)	0.07±0.04 (4)
Intermediate between State3 and State 4	0.13±0.1 (4)	0 (3)	0.1±0.01 (4)

[*]Control coefficients were determined in the creatine kinase system, using the following ratios of creatine/creatine phosphate in the medium: 45mM/5mM (State 3) and 10mM/40mM (intermediate state). The number of experiments is indicated in parentheses. The results of thenoyltrifluoroacetone titration experiments are presented, close values of C_{SDH} were obtained by titration with carboxine.

Our data show that the respiration of heart mitochondria is almost completely controlled by the ANT at intermediate values (corresponding to the physiological range) of rates of oxidation both of FAD- and NAD-dependent substrates, while the regulative role of the ANT is higher and that of proton leak is significantly lower in the case of pyruvate and malate and palmitoylcarnitine oxidation in comparison with succinate.

At maximal rates of succinate oxidation the ANT completely loses control over respiration. This is associated with an increase of the regulative role of SDH (see Table 2). In the case of NAD-dependent substrates, the contribution of the ANT to the regulation decreases near State 3 as well, but not to zero. The degree of control of proton leak becomes negligibly small in this region (control coefficient of proton leak is close to zero both in the case of succinate and pyruvate plus malate).

The data obtained suggest that although general features of the control pattern of heart mitochondrial respiration are principally the same for all substrates studied, certain differences between the regulation of oxidation of succinate and physiological substrates exists (especially large in State 3).

References

1. H. Kacser and J.A. Burns, The control of flux, *Symp. Soc. Exp. Biol.* **27**:65-104 (1973).
2. R. Heinrich and T.A. Rapoport, A linear steady-state treatment of enzymatic chains. General properties, control and effector strength, *Eur. J. Biochem.* **42**:89-95 (1974).
3. A. K. Groen, R.J.A. Wanders, H.V. Westerhoff, R. van der Meer and J.M. Tager, Quantification of the contribution of various steps to the control of mitochondrial respiration, *J. Biol. Chem.* **257**:2754-2757 (1982).
4. F.N. Gellerich, R. Bohnensack and W. Kunz, Control of mitochondrial respiration; the contribution of the adenine nucleotide translocator depends on the ATP- and ADP-consuming enzymes, *Biochim. Biophys. Acta* **722**:381-391 (1983).
5. R. J.A. Wanders, A. K. Groen, C.W.T. van Roermund and J.M. Tager, Factors determining the relative contribution of the adenine nucleotide translocator and the ADP-regenerating system to the control of oxidative phosphorylation in the isolated rat liver mitochondria, *Eur. J. Biochem.* **142**:417-424 (1984).
6. J.E. Davis and W.J. Davis-van Thienen, Rate control of phosphorylation-coupled respiration by rat liver mitochondria, *Arch. Biochem. Biophys.* **233**:573-581 (1984).
7. B. Kholodenko, V. Zilinskiene, V. Borutaite, L. Ivanoviene, A. Toleikis and A. Praskevicius, The role of adenine nucleotide translocator in the regulation of oxidative phosphorylation in heart mitochondria, *FEBS Lett.* **233**:247-250 (1987).

MECHANISMS OF THE EFFECTS OF HYPOTHYROIDISM AND HYPERTHYROIDISM IN RATS ON RESPIRATION RATE IN ISOLATED HEPATOCYTES

Mary-Ellen Harper and Martin D. Brand

Department of Biochemistry
University of Cambridge
Tennis Court Road
Cambridge CB2 1QW, England

1. Introduction

Thyroid hormone-induced increases in oxygen consumption and heat production are consistent properties of many mammalian tissues including the liver. Hepatocytes from hypothyroid rats respire more slowly than those from control animals; cells from hyperthyroid rats respire faster. In liver mitochondria there are two major effects of thyroid hormones on oxygen consumption - one occurs in State 3 respiration where ATP turnover is increased[1] and the other occurs in State 4 respiration where there is an increase in the leak of protons across the inner mitochondrial membrane[2].

In intact hepatocytes comparatively little is known about the mechanisms which mediate the effects of thyroid hormones on oxygen consumption despite many studies indicating effects of thyroid hormones on various component enzymes and pathways involved in cellular energy metabolism (see Refs 3-5 for reviews). However, recent results show that a decrease in the proton leak accounts for 30% of the difference in resting respiration rate between hepatocytes from hypothyroid rats and euthyroid control rats[6].

In the present work we have addressed two questions:

1) What are the significant sites of action of thyroid hormones which affect oxygen consumption rate?
2) Which groups of energy dissipating reactions change and what is their relative importance in terms of the changed resting oxygen consumption rate?

Modern Trends in Biothermokinetics, Edited by
S. Schuster *et al.*, Plenum Press, New York, 1993

2. Methods

Hypothyroidism was induced in 11 male rats by treatment with 6-n-propyl-thiouracil (0.05% w/v in the drinking water). Hyperthyroidism was induced in 12 male rats by 10 daily injections of triiodothyronine (15 μg T_3/100g body weight). There were 12 littermate paired controls for each of the hypothyroid and hyperthyroid groups.

Hepatocytes were isolated from fed rats as described by Berry[7] and incubated as described by Nobes[6] except that our incubation medium included additions of lactate (10 mM) and pyruvate (1 mM). Oxygen consumption by hepatocytes was measured with Clark-type oxygen electrodes and $\Delta\Psi_m$ was measured using the distribution of TPMP+ and correction factors as described by Nobes and coworkers[6].

Top-down elasticity analysis, an extension of top-down control analysis developed in our laboratory[8], was used as our experimental approach. It involves dividing a system conceptually into blocks of reactions that produce and consume a common intermediate (such as mitochondrial membrane potential, $\Delta\Psi_m$, in Fig. 1). To answer question 1, we used top-down elasticity analysis to identify those blocks of reactions whose overall kinetic response (*i.e.* elasticity) to $\Delta\Psi_m$ has been significantly altered by treatment with an effector (*i.e.* absence, or excess, of thyroid hormones). Analyses involved measuring the changes in the overall kinetic responses to $\Delta\Psi_m$ of the block of reactions that produce $\Delta\Psi_m$ ('substrate oxidation' pathway in Fig. 1) and those that consume it ('proton leak' and 'phosphorylating system' in Fig. 1).

To answer question 2, we used the overall elasticity titration curves (Fig. 2) to calculate the oxygen consumption used in the resting state for energy dissipating reactions: non-mitochondrial, proton leak-dependent and ATP turnover-dependent. From these results we determined which reactions changed (and by how much) in hypothyroid and hyperthyroid cells as compared to cells from littermate controls.

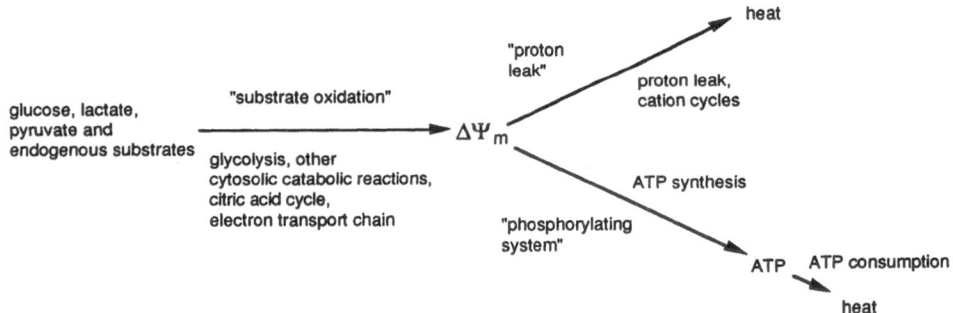

Figure 1. The branched system of $\Delta\Psi_m$ production and consumption. $\Delta\Psi_m$ is produced by 'substrate oxidation' which consists of all of the steps from added glucose, lactate, pyruvate and endogenous substrates to $\Delta\Psi_m$. Then $\Delta\Psi_m$ is consumed by the 'phosphorylating system' which includes ATP production using $\Delta\Psi_m$, and all cellular ATP-consuming reactions, or consumed by the 'proton leak' which includes the leak of protons and any cation cycles across the mitochondrial inner membrane.

3. Results and Discussion

3.1. Resting oxygen consumption differences

Mean oxygen consumption rates were 5.35±0.25 (SEM, n=11) and 6.15±0.22 (SEM, n=12) nmol O_2/min/mg dry weight of hepatocytes from hypothyroid rats and paired euthyroid control rats, respectively. With cells from hyperthyroid and paired euthyroid controls, rates were 9.88±0.48 (SEM, n=12) and 5.98±0.27 (SEM, n=12) nmol O_2/min/mg dry weight cells. Rates were significantly lower (p=0.03) with cells from hypothyroid rats and significantly higher (p < 0.001) with cells from hyperthyroid rats than with the corresponding euthyroid cells.

3.2. Top-down elasticity analysis of the kinetics of 'substrate oxidation', 'proton leak' and the 'phosphorylating system' to $\Delta \Psi_m$

In hepatocytes from hypothyroid rats the only block of reactions to have a different kinetic response to $\Delta \Psi_m$ was the mitochondrial proton leak (compare Figs 2A and B). In hepatocytes from hyperthyroid rats there were significant differences in the kinetic response of the 'phosphorylating system' and of the 'proton leak' to $\Delta \Psi_m$ (Figs 2B and C). We did not determine the kinetics of the 'substrate oxidation' reactions because of secondary effects of oligomycin (via cytosolic ATP depletion) on 'substrate oxidation' reactions.

3.3. Top-down control analysis

Top-down elasticity analysis provides all of the data necessary to complete a top-down control analysis of the system being studied[9]. In cells from hypothyroid rats elasticities to $\Delta \Psi_m$ and flux control coefficients were similar to those from euthyroid controls. The results from both hypothyroid and euthyroid cells in the resting state indicate that the phosphorylating system (0.4 - 0.5) and 'substrate oxidation' (0.3 - 0.4) exert most of the control over mitochondrial oxygen consumption while the remainder of the control is through the 'proton leak' (0.2 - 0.3). These results are similar to those obtained with hepatocytes from euthyroid rats[10]. A more comprehensive description of the application of top-down elasticity analysis to the study of the effects of thyroid hormone status on hepatocyte respiration is provided elsewhere[11].

3.4. Quantitative analysis of the effects of thyroid hormones on resting oxygen consumption

Fig. 2 shows the titrations of respiration in cells from hypothyroid and paired control rats and from hyperthyroid rats (results from hyperthyroid-paired controls are not shown but were similar to those from hypothyroid-paired controls). These titration curves were used to determine the elasticities of the three pathways as described in Section 2. From the titration curves it is also possible to calculate the oxygen consumption used in the resting state (furthest point to the right of each titration curve) for energy dissipating reactions (non-mitochondrial oxygen consumption, proton leak and ATP turnover) as described by Brand[8]. Non-mitochondrial oxygen consumption was identified as that which was insensitive to saturating amounts of myxothiazol and oligomycin. Oxygen consumption used to drive the proton leak cycle across the mitochondrial inner membrane in the resting state was measured as the rate of the proton leak (oxygen consumption in the presence of saturating amounts of oligomycin, titrated with myxothiazol) at the same value of $\Delta \Psi_m$ as in resting cells, as previously described[6,10]. The remaining oxygen consumption was used to drive ATP synthesis.

The resting oxygen consumption rate attributable to the activities of each of the three pathways is given in the bar graph in Fig. 2. The results in this graph show that of the difference in resting oxygen consumption between hepatocytes from hypothyroid and euthyroid rats (*i.e.* 0.80 nmol O_2/min/mg), 52% was attributed to differences in non-mitochondrial oxygen consumption processes; the remainder of the difference, 48%, was due

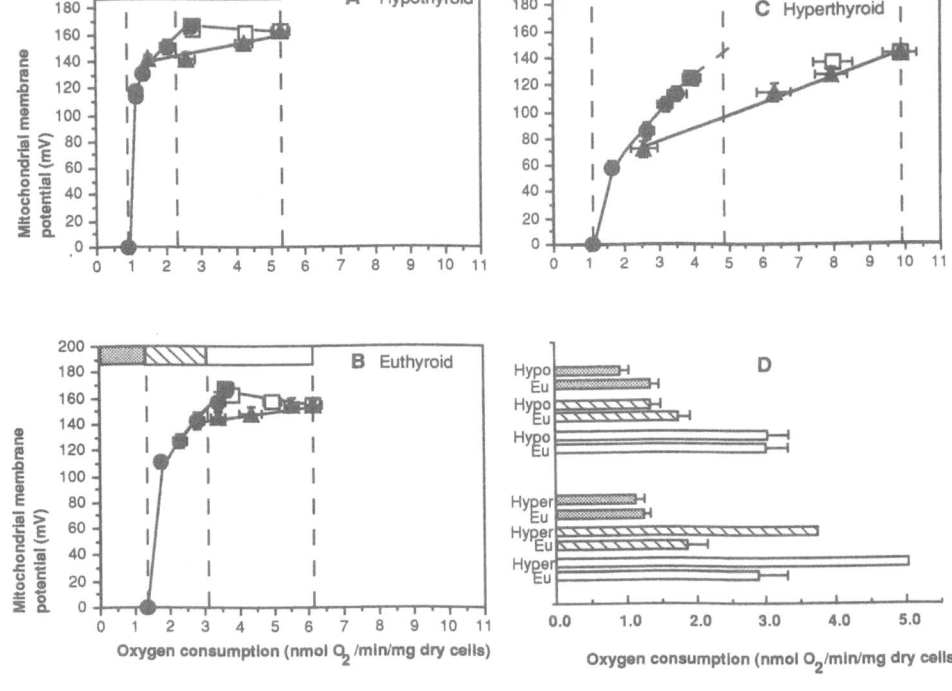

Figure 2. Relationship between $\Delta\Psi_m$ and respiration rate in hepatocytes from hypothyroid, euthyroid and hyperthyroid rats and histogram showing the differences in oxygen consumption by non-mitochondrial, proton leak and phosphorylation pathways. (A-C) Titrations were carried out from the resting state (furthest point on the right). The response of the $\Delta\Psi_m$-producers (substrate oxidation) to $\Delta\Psi_m$ (open squares) was measured by titrating the $\Delta\Psi_m$-consumers with oligomycin (0.1-1.0 µg/ml). The elasticity of the proton leak (full circles) to $\Delta\Psi_m$ was measured by titrating with myxothiazol (0.05-0.2 µM) in the presence of saturating oligomycin (1.0 µg/ml). The combined elasticity of the $\Delta\Psi_m$-consumers (ATP turnover, plus proton leak) (full triangles) to $\Delta\Psi_m$ was measured from titrations of myxothiazol alone (0.05-0.2 µM). To determine the non-mitochondrial rate of oxygen consumption cells were incubated in the presence of oligomycin (1.0 µg/ml), myxothiazol (1.0 µM), valinomycin (0.1 µM) and FCCP (20 µM). The titration graph shown for euthyroids is that for the hypothyroid-paired control group; the graph for the hyperthyroid-paired control group (not shown) was similar. The histogram (D) shows the proportions of resting oxygen consumption calculated to be due to non-mitochondrial (shaded bars), leak-dependent (hatched bars) and ATP turnover-dependent (clear bars) reactions. All values are mean ± SEM for 11, 12, 12 and 12 determinations for hypothyroid, hypothyroid-paired controls, hyperthyroid and hyperthyroid-paired controls. It was not possible to determine SEM for the means of the leak-dependent and ATP turnover-dependent proportions of resting oxygen consumption (D) as it was necessary to extrapolate the curve representing the kinetics of the proton leak.

to a difference in mitochondrial oxygen consumption. All of the difference in mitochondrial oxygen consumption was accounted for by a decrease in proton leak-dependent oxygen consumption. The mechanisms responsible for the changes in non-mitochondrial oxygen consumption are not known but they may involve altered microsomal and/or peroxisomal oxidative pathways. Altered microsomal oxygen consumption could include altered activities of the P450-dependent reactions. It was recently shown that hypothyroidism in rats led to a depletion of approximately 80% of hepatic microsomal P450 reductase activity and protein[12]. It is thus possible that at least some of the decrease in non-mitochondrial oxygen consumption in hypothyroid hepatocytes occurs as a result of decreased P450 reductase activity. Of the increase in resting oxygen consumption in hepatocytes from hyperthyroid rats (*i.e.* 3.90 nmol O_2/min/mg), 48% could be accounted for by differences in the rate of proton leak and 55%

could be accounted for by differences in the rate of ATP turnover. The difference in the rate of ATP turnover could be due to increased activities of the adenine nucleotide carrier[13], phosphate carrier[14] and/or of ATP-consuming reactions such as Na,K-ATPase[15]. There was no significant difference (3% decrease) in non-mitochondrial oxygen consumption.

The mechanisms through which thyroid hormones alter the proton leak across the inner mitochondrial membrane in isolated liver mitochondria and hepatocytes are unknown. However, they are probably indirect and act through gene expression as the *in vivo* effects of thyroid hormones on the proton leak in liver mitochondria require at least 24 hours[2]. Moreover much of the observed changes in proton leak that are induced by hypothyroidism and hyperthyroidism are likely explained by thyroid hormone-induced changes in inner membrane surface area[16] and the intrinsic permeability of vesicles formed from mitochondrial phospholipids[17].

4. Conclusions

Differences in oxygen consumption rate between hepatocytes from hypothyroid and euthyroid rats are entirely due to changes in the properties of the mitochondrial proton leak and in non-mitochondrial oxygen consumption. The control over oxygen consumption is similar in hepatocytes from hypothyroid and euthyroid rats. Differences in oxygen consumption rate between hepatocytes from hyperthyroid and euthyroid rats are caused by changes in the kinetics of the 'proton leak' and 'phosphorylating system'. The effects on the kinetics of the 'substrate oxidation' pathway could not be determined for technical reasons. Thus in both hypothyroidism and hyperthyroidism alterations in the rate of mitochondrial proton leak are responsible for approximately 50% of the differences in resting cellular oxygen consumption compared to controls. The remaining 50% is due to decreased non-mitochondrial oxygen consumption (hypothyroidism) or to increased ATP turnover (hyperthyroidism).

References

1. R.P. Hafner, G.C. Brown and M.D. Brand, Thyroid hormone control of state 3 respiration in isolated rat liver mitochondria, *Biochem. J.* **265**:731-734 (1990).
2. R.P. Hafner, C.D. Nobes, A.D. McGown and M.D. Brand, Altered relationship between protonmotive force and respiration rate in non-phosphorylating liver mitochondria isolated from rats of different thyroid hormone status, *Eur. J. Biochem.* **178**:511-518 (1988).
3. J.H. Oppenheimer, H.L. Schwartz, C.N. Mariash, W.B. Kinlaw, N.C.W. Wong, and H.C. Freake, Advances in our understanding of thyroid hormone action at the cellular level, *Endocrinol. Rev.* **8**:288-308 (1987).
4. M.D. Brand and M.P. Murphy, Control of electron flux through the respiratory chain in mitochondria and cells, *Biol. Rev.* **62**:141-193 (1987).
5. M.J. Dauncey, Thyroid hormones and thermogenesis, *Proc. Nutr. Soc.* **49**:203-215 (1991).
6. C.D. Nobes, G.C. Brown, P.N. Olive and M.D. Brand, Non-ohmic proton conductance of the mitochondrial inner membrane in hepatocytes, *J. Biol. Chem.* **265**:12903-12909 (1990).
7. M.N. Berry, High-yield preparation of morphologically intact isolated parenchymal cells from rat liver, *Methods Enzymol.* **32**:625-632 (1974).
8. M.D. Brand, The proton leak across the mitochondrial inner membrane, *Biochim. Biophys. Acta.* **1018**: 128-133 (1990).
9. G.C. Brown, R.P. Hafner and M.D. Brand, A 'top-down' approach to the determination of control coefficients in metabolic control theory, *Eur. J. Biochem.* **188**:321-325 (1990).
10. G.C. Brown, P.L. Lakin-Thomas and M.D. Brand, Control of respiration and oxidative phosphorylation in isolated rat liver cells, *Eur. J. Biochem.* **192**:355-362 (1990).

11. M.-E. Harper and M.D. Brand, The quantitative contributions of mitochondrial proton leak and ATP turnover reactions to the changed respiration rates of hepatocytes from rats of different thyroid hormone status, *J. Biol. Chem.*, submitted.

12. P.A. Ram and D.J. Waxman, Thyroid hormone stimulation of NADPH P450 reductase expression in liver and extra hepatic tissues, *J. Biol. Chem.* **267**:3294-3301 (1992).

13. B.M. Babior, S. Creagen, S.H. Ingbar and R.S. Kipnes, Stimulation of mitochondrial adenosine diphosphate uptake by thyroid hormones, *Proc. Natl. Acad. Sci.* **70**:98-102 (1973).

14. J.-P. Clot and M. Baudry, Effect of thyroidectomy on oxidative phosphorylation mechanisms in rat liver mitochondria, *Mol. Cell. Endocrinol.* **28**:455-469 (1982).

15. C.D. Nobes, P.L. Lakin-Thomas and M.D. Brand, The contributions of ATP turnover by the Na^+/K^+-ATPase to the rate of respiration of hepatocytes. Effect of thyroid status and fatty acids, *Biochim. Biophys. Acta* **976**:241-245 (1989).

16. S. Jacovcic, H.H. Swift, N.J. Gross and M. Rabinowitz, Biochemical and stereological analysis of rat liver mitochondria in different thyroid states, *J. Cell Biol.* **77**: 887-901 (1978).

17. M.D. Brand, D. Steverding, B. Kadenbach, P.M. Stevenson and R.P. Hafner, The mechanism of the increase in mitochondrial proton permeability induced by thyroid hormones, *Eur. J. Biochem.* **206**: 775-781 (1992).

SUBSTRATE AND CALCIUM DEPENDENCE OF THE PYRIDINE NUCLEOTIDE AND FLAVINE REDOX STATE IN RAT HEART MITOCHONDRIA RESPIRING AT DIFFERENT METABOLIC STATES

Russell C. Scaduto, Jr.[*] and Alexander V. Panov

Department of Cellular and Molecular Physiology
The Milton S. Hershey Medical Center
The Pennsylvania State University
Hershey, PA 17033, U.S.A.

1. Introduction

The role of the mitochondrial pyridine nucleotide redox state in the regulation of oxidative phosphorylation in the perfused heart has recently received renewed attention. Measurements of oxygen consumption, high-energy phosphate metabolites and redox state of the pyridine nucleotides during alterations in heart work have challenged the classical models of regulation of oxidative phosphorylation in which regulation was considered solely dependent on changes in ADP concentration or the ATP/ADP ratio[1-3]. An alternative hypothesis has placed substrate dehydrogenation and the supply of redox potential as important control parameters in the regulation of cardiac oxygen consumption[4-9]. Increased calcium within the mitochondrial matrix upon an increase in cardiac work is thought to activate dehydrogenase activity and generate a higher matrix NADH/NAD$^+$ ratio to a level sufficient to blunt perturbations in the ATP/ADP ratio[10].

This study was conducted in order to investigate the effect of mitochondrial calcium on the steady-state generation of matrix reducing potential generated by various substrates. These measurements were made at respiration rates ranging from State 2 to State 3 by the control of ADP availability. The data obtained illustrate a marked substrate dependence on the maintenance of mitochondrial redox potential at a given state of respiration. Furthermore, both the respiratory activity and intramitochondrial NAD(P)H display very high sensitivity to changes in the external ADP concentration. However, at a given respiratory rate, calcium greatly enhances the reduction of mitochondrial NAD(P)H. In effect, this functionally lowers the K_m of ADP for oxygen consumption.

---

[*]To whom correspondence should be addressed.

Modern Trends in Biothermokinetics, Edited by
S. Schuster *et al.*, Plenum Press, New York, 1993

357

2. Methods

2.1. Isolation and incubation procedure

Rat heart mitochondria were isolated by differential centrifugation and were further purified by centrifugation through a percoll gradient. Mitochondria were incubated in a basic medium containing (in mM) 130 KCl, 5 NaCl, 20 MOPS, 1 $MgCl_2$, 2 KH_2PO_4, 1 EGTA, pH 7.0 at 28 °C. To some incubations 0.7 $CaCl_2$ was also added. We previously determined that this Ca^{2+}-EGTA buffer system generates a mitochondrial matrix free calcium concentration of approximately 1 μM, whereas the matrix free calcium is near or below the limit of detection (<0.05 μM) in mitochondria incubated in the basic medium with EGTA alone[11]. Substrates were added at the concentrations indicated. All substrates were used in the presence of 2 mM of malate. In some experiments, ADP was added in the presence of 7 U of hexokinase and 20 mM glucose.

2.2. Determination of mitochondrial pyridine and flavine fluorescence

Fluorescence measurements were made using a custom fluorimeter designed in this laboratory. NAD(P)H was estimated using an excitation of 340 ± 10 nm. Emission was detected using an interference filter having a bandpass from 480 to 520 nm and a Hamamatsu R2560 photomultiplier tube operating in a photon counting mode. Data was collected at 0.032 second intervals and was later compressed using a 10 to 1 rolling average algorithm. Full oxidation of mitochondrial pyridine nucleotides (PN) was achieved by incubation in the absence of substrates and in the presence of 0.5 μM CCCP. The fully reduced state for a given substrate was obtained by addition of 1 μg/mg rotenone for pyridine nucleotides and 0.5 mM KCN for flavine reduction after 4 min. incubation in the presence of a substrate.

2.3. HPLC determination of NAD^+ and $NADP^+$

For HPLC determinations, rat heart mitochondria were incubated at a concentration of 5 mg/ml. After 4 min. of incubation, the suspension was transferred into vials containing cold perchloric acid. Separation of the nucleotides was achieved using a 25 cm Whatman Particil 10 SAX analytical column as described elsewhere[12]. The reduction of intramitochondrial NAD^+ and $NADP^+$ was calculated from the difference between the concentration obtained in mitochondria incubated under fully oxidized conditions and the concentration obtained in incubations with a substrate as described previously[11].

3. Results and Discussion

3.1. The redox state of mitochondrial pyridine nucleotides under State 2 conditions

Since fluorescence measurements can not distinguish between NADH and NADPH, the redox state of the NAD- and NADP-couples was determined using HPLC techniques. Fig. 1 illustrates the reduction pattern of the intramitochondrial NAD- and NADP-couples as a percentage of the total nucleotide concentration in rat heart mitochondria oxidizing various substrates in the absence of oxidative phosphorylation. These data indicate that $NADP^+$ was highly reduced (>95%) under all substrate conditions and was independent of the presence of calcium. However, the redox state of NAD^+ was highly dependent on the particular substrate employed and on the presence of calcium ions. The highest level of reduction was obtained with succinate alone or with succinate plus glutamate in both the presence and absence of rotenone. In general, the steady-state reduction of NAD$^+$ was lower in media without added calcium. The largest effect of calcium upon the redox state of the NAD$^+$-couple was observed with 2-oxoglutarate or glutamate as substrate. However, even the highest levels of NAD$^+$ reduction obtained with 10 mM 2-oxoglutarate in the presence of calcium comprised only 60-70 % of the total amount of NAD$^+$.

In Fig. 2, it is shown that there was a close correlation between NAD$^+$ measured by HPLC and NAD(P)H measured fluorimetrically, illustrating the validity of the latter measurement in assessment of the mitochondrial redox state of the NAD$^+$ couple. Thus the fluorescence contributed by NADPH is a constant under energized conditions.

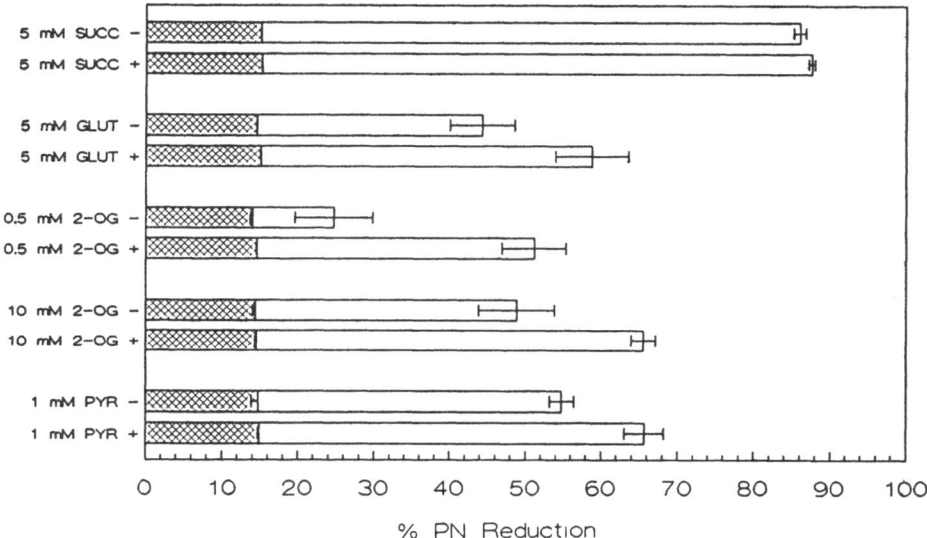

Figure 1. Influence of various substrates and calcium on the steady-state reduction of NAD$^+$ and NADP$^+$. Mitochondria were incubated under State 2 conditions for 4 min. as described in Methods. Substrates were used at the indicated concentration. All conditions contained 2 mM malate. The data are presented as a percent of reduction of NAD$^+$ or NADP$^+$ normalized to the total content of pyridine nucleotides. The reduction observed with 5 mM succinate was not increased further by the addition of 5 mM glutamate or rotenone. The + and — refer to incubations conducted in the presence or absence of calcium, respectively. Abbreviations: SUCC, succinate; GLUT, glutamate; 2-OG, 2-oxoglutarate; PYR, pyruvate. N=4-8 incubations.

The typical fluorescence pattern of NAD(P)H in rat heart mitochondria oxidizing various substrates in State 2 is shown in Fig. 3. There was a considerable quenching of fluorescence with 10 mM of 2-oxoglutarate, the magnitude of which can be deduced from the difference between the fluorescence obtained in mitochondria oxidizing 10 mM 2-oxoglutarate and 0.5 mM 2-oxoglutarate plus rotenone. The highest levels of NAD(P)H fluorescence were observed with mitochondria oxidizing 5 mM succinate both in the presence and absence of 5 mM glutamate and 2 mM malate, and with the NAD-dependent substrates in the presence of rotenone.

Of all substrates used in this study, the reduction of NAD(P)$^+$ by mitochondria oxidizing glutamate in the presence of 2 mM malate was distinct in that it continuously diminished during the course of incubation after the initial burst of reduction (Fig. 3). The rate of the decline in fluorescence was faster in aged mitochondria and was significantly suppressed in the presence of calcium (not shown). Both the pyridine nucleotide (PN) fluorescence and its

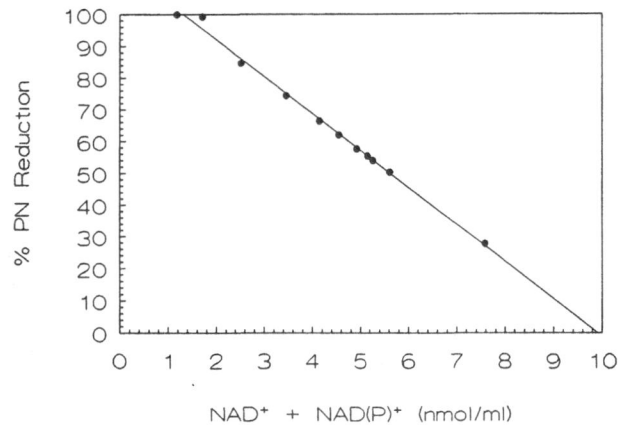

Figure 2. Relationship between pyridine nucleotide reduction determined fluorimetrically and as measured by HPLC. Percent PN reduction was measured fluorimetrically and calculated as described in Methods. These data are expressed as a function of the NAD$^+$ and NADP$^+$ determined by HPLC. The points indicate the redox status attained after a 4 min. incubation with a particular substrate under State 2 conditions. The substrates, in order of increasing PN fluorescence, were: 0.5 mM 2-oxoglutarate - Ca^{2+}; 5 mM glutamate - Ca^{2+}; 10 mM 2-oxoglutarate - Ca^{2+}; 0.5 mM 2-oxoglutarate + Ca^{2+}; 1 mM pyruvate - Ca^{2+}; 5 mM glutamate + Ca^{2+}; 10 mM 2-oxoglutarate + Ca^{2+}; 1 mM acetyl-carnitine + Ca^{2+}; 5 mM succinate + Ca^{2+}; 5 mM succinate - Ca^{2+}; 5 mM glutamate - Ca^{2+}; where - Ca^{2+} and + Ca^{2+} refer to incubations done in the absence or presence of calcium, respectively.

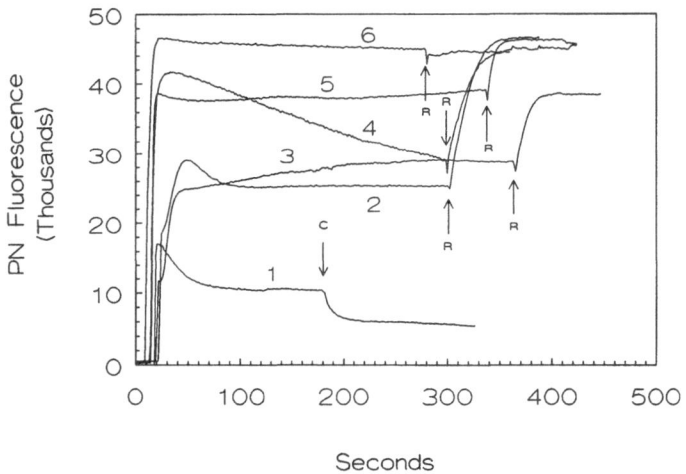

Figure 3. Influence of various substrates on mitochondrial pyridine nucleotide fluorescence. Incubation conditions were as described in Methods. Substrates were as follows: 1) no substrate; 2) 0.5 mM 2-oxoglutarate; 3) 10 mM 2-oxoglutarate; 4) 5 mM glutamate; 5) 1 mM pyruvate; 6) 5 mM succinate. All incubations contained 2 mM malate. Arrows refer to the addition of 1 µg of rotenone (R) or 0.5 µM CCCP (C).

rate of decline in mitochondria oxidizing glutamate were independent of the concentration of glutamate used under State 2 conditions. This decline in the PN fluorescence can be explained by inhibition of aspartate-glutamate transaminase by 2-oxoglutarate which accumulates during the incubation[13]. Prior studies have shown that 2-oxoglutarate competitively inhibits oxaloacetate transamination to aspartate and that calcium, by stimulating α-ketoglutarate dehydrogenase, lowers the rate of 2-oxoglutarate accumulation by mitochondria oxidizing glutamate[11,13].

Thus under steady-state conditions of resting respiration in the absence of oxidative phosphorylation, different substrates generate distinct redox states. This is presumably due to the different thermodynamic efficiency of the enzyme systems involved in transport and oxidation of these substrates. Almost complete reduction of the mitochondrial NADP-couple, irrespective of the substrate used, is understandable in view of the fact that the energy-dependent transhydrogenase is a transmembrane proton pump[14,15] and thus generation of NADP is directly coupled to the membrane potential.

3.2. The redox state of the mitochondrial pyridine nucleotides at intermediary states of respiration

Upon titration of respiration with ADP in the presence of excess hexokinase and glucose, the pyridine and flavine nucleotides became progressively more oxidized. Fig. 4 illustrates a typical titration with ADP of rat heart mitochondria oxidizing acetyl-carnitine 1 mM plus malate 2 mM. In parallel incubations, oxygen consumption was also measured. The pyridine nucleotide reduction state and the effect of calcium at these intermediate respiration rates with either glutamate or 2-oxoglutarate as substrate are shown in Figs 5, 6, and 7.

With glutamate as substrate, the PN reduction state at a given concentration of ADP was markedly dependent on the presence of calcium. This effect was observed with use of 5 or 20 mM glutamate (Figs 5A, 6A). In fact, the PN reduction at any concentration of ADP was largely independent of the glutamate concentration employed. Although the degree of PN reduction was similar with either concentration of glutamate, the rates of oxygen consumption at higher concentrations of ADP were markedly greater in the presence of 20 mM glutamate provided that calcium was present (Figs 5B and 6B). With 20 mM glutamate, State 3 respiration rates were doubled by the presence of calcium. At low concentrations of ADP,

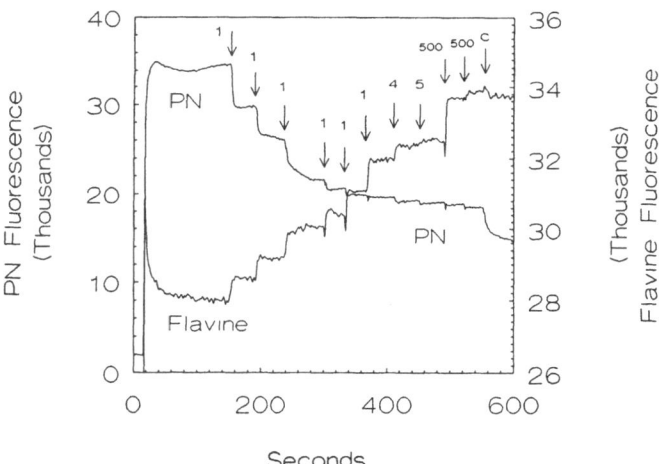

Figure 4. Pyridine nucleotide and flavine fluorescence of rat heart mitochondria during titration with ADP. Incubation conditions as described in Methods, except 20 mM glucose and 7 U/ml hexokinase were also present. Substrates were 1 mM acetyl-carnitine plus 2 mM malate. Arrows illustrate the time of addition of the indicated quantity of ADP (nmoles). "C" refers to the addition of 0.5 μM CCCP. Fluorescence is in arbitrary units.

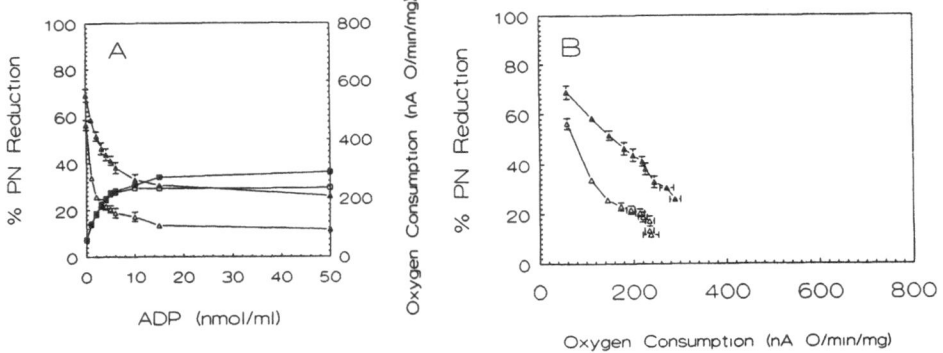

Figure 5. Influence of calcium and ADP on pyridine nucleotide reduction and respiratory activity of rat heart mitochondria oxidizing 5 mM glutamate plus 2 mM malate. Incubation conditions as described in Methods, except 20 mM glucose and 7 U/ml hexokinase were also present. Shown are the percent pyridine nucleotide (PN) reduction (triangles) and the rate of oxygen consumption (squares) during incubation in the absence (open symbols) or presence (filled symbols) of calcium. Standard errors are shown when the magnitude of the standard error was greater than the symbol size. Panel A, data plotted as a function of ADP concentration; Panel B, data plotted as a function of respiratory rate.

however, respiration rates with 20 mM glutamate were only slightly greater than with 5 mM. This effect indicates that calcium increased also the mitochondrial respiratory capacity. In the absence of calcium, only very low concentrations of ADP (1-2 μM) were required to elicit a marked PN oxidation (see Figs 5B, 6B). However, this enhanced respiratory capacity was only evident in the presence of high glutamate concentrations. Presumably, with lower concentrations of glutamate, there are kinetic restrictions on glutamate metabolism that are not relieved by the presence of calcium. One such restriction could be the transport of glutamate on the glutamate/aspartate exchange carrier since the K_m for glutamate in cardiac mitochondria is high - about 4 mM.[16]

Among the NAD-dependent substrates used in this study, the most pronounced effects of calcium upon reduction of NAD(P)+ were obtained with glutamate and 2-oxoglutarate as substrate. As shown in Fig. 7, calcium addition to mitochondria oxidizing 0.5 mM 2-oxoglutarate caused similar effects to those observed with glutamate. At all rates of oxygen consumption, calcium increased the PN reduction state. Similarly, respiration rates at equivalent concentrations of ADP were greater in the presence of calcium. This effect of calcium was also significant with 10 mM 2-oxoglutarate (not shown). With pyruvate as substrate, calcium caused a statisticially higher respiratory rate at all concentrations of ADP, but there was a relatively small influence of calcium upon reduction of NAD(P)+ (not shown). Acetyl-carnitine (1 mM) did not show significant effects of calcium (not shown). These observations support previous conclusions illustrating α-ketoglutarate dehydrogenase as the primary site of calcium activation of cardiac mitochondrial metabolism[11].

Taken together, these data illustrate clearly that at a given concentration of ADP, respiration rates, and hence the rate of ATP synthesis, are stimulated significantly by calcium in a substrate-dependent fashion. The data suggest that at a given concentration of ADP, the lower rates of respiration in the absence of calcium are due to a restriction in the capacity for PN reduction. This limitation can be overcome, at least in part, by an increase in substrate concentration.

Respiration and oxidation of NAD(P)H both display very high sensitivities to changes in the concentration of extramitochondrial ADP. Koretsky and Balaban[8] have studied the reduction pattern of liver mitochondrial NAD(P)H in the presence of 1 mM ADP and have also observed different reduction states during oxidation of various substrates. Our data indicate that although various substrates produce different levels of NAD(P)H at full State 3 respiration, higher rates of respiration are achieved at lower concentrations of external ADP in the presence of calcium. This implies that small changes in concentrations of cytosolic ADP might cause larger changes in respiratory rate and the steady-state reduction of pyridine nucleotides if they are accompanied by an increase in matrix calcium. The fact that a significant change in the adenine nucleotides pattern of the perfused heart is not observed upon an increase in work[5-7] could be explained by a shift in the sensitivity for ADP due to the increased presence of calcium.

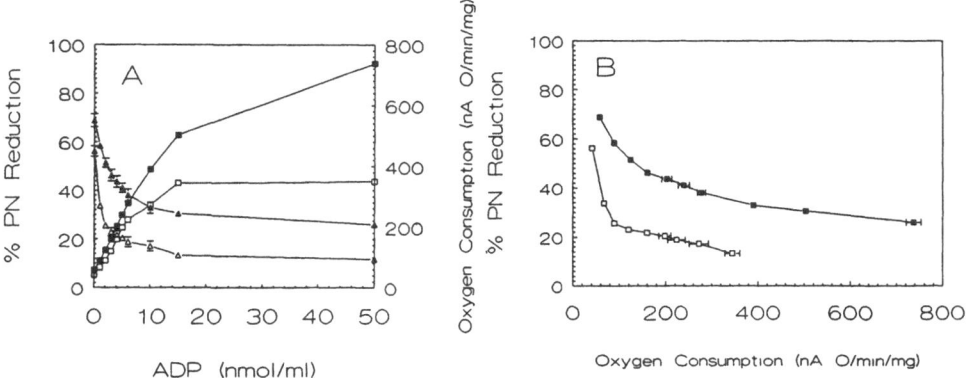

Figure 6. Influence of calcium and ADP on pyridine nucleotide reduction and respiratory activity of rat heart mitochondria oxidizing 20 mM glutamate plus 2 mM malate. Incubation conditions and symbols were as described in the legend to Fig. 5.

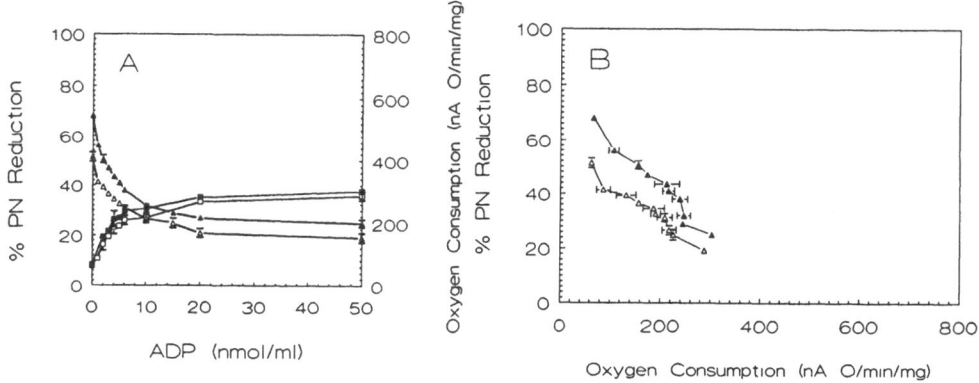

Figure 7. Influence of calcium and ADP on pyridine nucleotide reduction and respiratory activity of rat heart mitochondria oxidizing 0.5 mM 2-oxoglutarate plus 2 mM malate. Incubation conditions and symbols were as described in the legend to Fig. 5.

It is of importance that the greatest effect of calcium on the reduction of NAD(P)+ was obtained with 2-oxoglutarate or glutamate as substrate. The pathway for glutamate oxidation by rat heart mitochondria represents the intramitochondrial half of the malate-aspartate shuttle. This shuttle is the only significant pathway for the mitochondrial oxidation of cytosolic NADH[17]. When glucose is used as the sole exogenous substrate by the perfused heart, stimulation of respiration must be accompanied by a parallel increase in the shuttle flux and pyruvate oxidation. This increase in cardiac work causes an increase in the NAD(P)H surface fluorescence of the heart[4]. Thus the stimulation of glutamate oxidation to aspartate and malate-aspartate shuttle flux may be in part responsible for the enhanced reduction observed under these conditions.

3.3. Influence of calcium and different substrates on the flavine redox state

An advantage of these fluorescence measurements is that it allows the simultaneous determination of the mitochondrial pyridine and flavine redox state, as shown in Fig. 4. Addition of cyanide caused a maximal reduction of both redox systems, whereas addition of rotenone caused reduction of the PN fluorescence and oxidation of flavine fluorescence (not shown). The extent of flavine oxidation with rotenone was similar to that caused by CCCP, illustrating that the majority of the flavine fluorescence was due to the flavines involved in the inner membrane electron transport. Using cyanide and CCCP to establish the reduction range, the percent flavine reduction was plotted as a function of PN reduction. Fig. 8 illustrates that there was a close correlation between the flavine and pyridine nucleotide redox states generated by different substrates under State 2 conditions. Under all substrate conditions, the pyridine nucleotides were more reduced than the flavines. It is of interest to note that in State 2, almost complete flavine oxidation occurred with a poorly oxidized substrate (0.5 mM 2-oxoglutarate) while the pyridine nucleotides remained 50 % reduced. Under conditions of maximal PN reduction (mitochondria oxidizing succinate plus malate) the flavines are reduced only by about 65 %. This observation, in part, may explain the inability of excess ADP to cause complete PN oxidation. Complete flavine oxidation occurs with excess ADP, but it is not associated with complete PN oxidation. In titration experiments with ADP, as exemplified in Fig. 4, the oxidation of flavine caused by ADP addition paralleled the PN oxidation.

In summary, we have shown that in the absence of oxidative phosphorylation (state 2), different substrates maintain different steady-state levels of PN reduction. The PN redox state displays a high sensitivity to the presence of extramitochondrial ADP. In the presence of calcium, however, the PN are more reduced at any concentration of ADP. This effect of calcium is most prominent with glutamate or 2-oxoglutarate (plus malate) as substrate and is

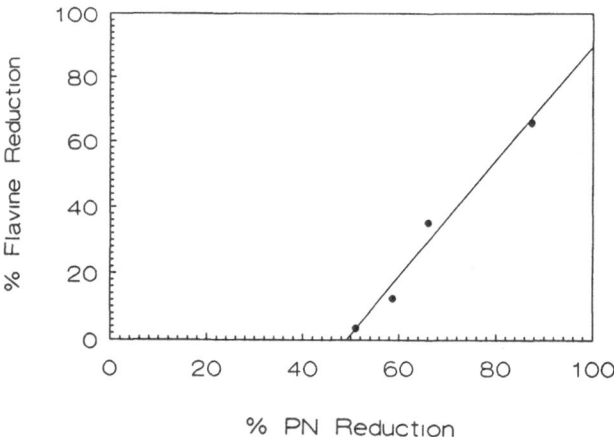

Figure 8. Relationship between reduction of flavines and pyridine nucleotides in rat heart mitochondria oxidizing various substrates. The percentages of flavines and pyridine nucleotides reduction were calculated as described in Methods. Symbols, in order of increased reduction, indicate incubations conducted with either 0.5 mM 2-oxoglutarate, 5 mM glutamate, 1 mM pyruvate, or 0.5 mM succinate.

much smaller with pyruvate and is absent during oxidation of acetyl-carnitine. This is in keeping with the activation by calcium of α-ketoglutarate dehydrogenase. With 20 mM glutamate as substrate, calcium was essential for maximal rates of State 3 respiration. Significant PN oxidation occurred with relatively low concentrations of ADP in the absence of calcium. Simultaneous measurements of flavine fluorescence revealed that ADP-induced oxidation of the flavines occurred in parallel to PN oxidation. This indicates that the primary site of calcium in the stimulation of cardiac respiration is at the substrate level and is proximal to any control exerted by the electron transport chain.

Acknowledgements

This work was supported by Public Health Service grant HL43215. The authors acknowledge the expert technical assistance of Lee W. Grotyohann and Pamela Young.

References

1. B. Chance and G.R. Williams, The respiratory chain and oxidative phosphorylation, *Adv. Enzymol.* **17**:65-134 (1956).
2. W.E. Jacobus, R.W. Mereadith and K.M. Vandegaar, Mitochondrial respiratory control. Evidence against the regulation of respiration by extramitochondrial phosphorylation potentials or by [ATP]/[ADP] ratios, *J. Biol. Chem.* **257**:2397-2402 (1982).
3. E.J. Davis and W.I.A. Davis-Van Thienen, Control of mitochondrial metabolism by the ATP/ADP ratio, *Biochem. Biophys. Res. Comm.* **83**:1260-1266 (1978).
4. L.A. Katz, A.P. Koretsky and R.S. Balaban, Respiratory control in the glucose perfused heart. A ^{31}P NMR and NADH fluorescence study, *FEBS Lett.* **221**:270-276 (1987).
5. L.A. Katz, A.P. Koretsky and R.S. Balaban, Activation of dehydrogenase activity and cardiac respiration: A ^{31}P-NMR study, *Am J. Physiol.* **255**:H185-H188 (1988).
6. R.S. Balaban, Regulation of oxidative phosphorylation in the mammalian cell, *Am J. Physiol.* **258**:C377-C389 (1990).
7. L.A. Katz, J.A. Swain, M.A. Portman and R.S. Balaban, Relation between phosphate metabolites and oxygen consumption in heart *in vivo*, *Am J. Physiol.* **256**:H265-H274 (1989).
8. A.P. Koretsky and R.S. Balaban, Changes in pyridine nucleotide levels alter oxygen consumption and extra-mitochondrial phosphates in isolated mitochondria: a ^{31}P-NMR and NAD(P)H fluorescence study, *Biochim. Biophys. Acta* **893**:398-408 (1987).
9. R. Moreno-Sanchez, B.A. Hogue and R.G. Hansford, Influence of NAD-linked dehydrogenase activity on flux through oxidative phosphorylation, *Biochem. J.* **268**:421-428 (1990).
10. R.G. Hansford, Relation between mitochondrial calcium transport and control of energy metabolism, *Rev. Physiol. Biochem. Pharmacol.* **102**:2-72 (1985).
11. B. Wan, K.F. LaNoue, J.Y. Cheung and R.C. Scaduto, Jr., Regulation of citric acid cycle by calcium, *J. Biol. Chem.* **264**:13430-13439 (1989).
12. M.S. Lui, R.C. Jackson and G. Weber, Enzyme pattern-directed chemotherapy. Effects of antipyrimidine combinations on the ribonucleotide content of hepatomas, *Biochem. Pharmacol.* **28**:1189-1195 (1979).
13. T. Strzelecki, D. Strzelecka, C.D. Koch and K.F. LaNoue, Sites of action of glucagon and other Ca^{2+} mobilizing hormones on the malate aspartate cycle, *Arch. Biochem. Biophys.* **264**:310-320 (1988).
14. A.E. Dontsov, L.L. Grinius, A.A. Jasaitis, I.I. Severina and V.P. Skulachev, A study on the mechanism of energy coupling in the redox chain. I. Transhydrogenase: The fourth site of the redox chain energy coupling, *J. Bioenerget.* **3**:277-303 (1972).
15. B. Hoeberg and J. Rydstrom, Purification and molecular properties of reconstitutively active nicotinamide nucleotide transhydrogenase from beef heart mitochondria, *Biochem. Biophys. Res. Commun.* **78**:1183-1190 (1977).
16. K. LaNoue and J. Duszynski, The influence of protons on the glutamate-aspartate carrier in the mitochondrial membrane, *in*: "The Proton and Calcium Pumps," G.F. Azzone, M. Avron, J.C. Metcalfe, E. Quagliariello and N. Siliprandi, eds, Elsevier/North-Holland Biomed. Press, Amsterdam, pp. 297-307 (1978).

17. Y. Jong, E.J. Davis, Reconstitution of steady-state in cell-free systems. Interactions between glycolysis and mitochondrial metabolism: Regulation of the redox and phosphorylation states, *Arch. Biochem. Biophys.* **222**:179-191 (1983).

V. INVESTIGATION OF CELL PROCESSES

MODIFICATION OF MITOCHONDRIAL METABOLISM BY AMINO ACID-INDUCED CELL SWELLING

Pascal Espié, Michel Rigoulet and Bernard Guérin

Institut de Biochimie Cellulaire du C.N.R.S.
Université de Bordeaux II
1 rue Camille Saint-Saëns, 33077 Bordeaux Cédex, France

1. Introduction

It is now well known that changes in cell volume *per se*, both in perfused liver and isolated hepatocytes, affect fluxes through metabolic pathways, acting like an anabolic signal, stimulating and inhibiting respectively the biosynthesis and degradation pathways[1]. In the present work cell volume increase is a consequence of the Na^+-cotransported aminoacids uptake by the hepatocytes[2].

While intermediary metabolism is greatly studied and some mechanisms linking cell volume changes and the alterations of metabolic functions have been proposed[3], little is known in this field about the mitochondria *in situ* and the bioenergetics parameters.

However we know from previous work that on isolated mitochondria, matrix volume influences their metabolism. For example hormone induced swelling[4] increases mitochondrial ATP/ADP ratio at the same time as stimulating pyruvate carboxylation, succinate oxidation and the respiratory chain activity[5].

In the liver, the mitochondrial respiratory chain is stimulated by Ca^{2+} mobilizing hormones too, acting via changes in mitochondrial volume. The mechanisms involved in such a stimulation, are: a rise in the mitochondrial $[Ca^{2+}]$, an inhibition of matrix pyrophosphatases, and an increase in pyrophosphates which stimulates K^+ uptake inside the mitochondria. A rise in the mitochondrial volume follows which activates the respiratory chain[5,6]. Hormonal stimulation of respiration also involves an increased rate of mitochondrial NADH production as the result of an activation of the Ca^{2+}-sensitive mitochondrial dehydrogenases[7]. As cell volume changes in hepatocytes are the consequence of the modification in the intra- or extracellular osmolarity, this may greatly affect the mitochondrial compartment. Our aim is to study the energy state of the mitochondria when the cell swells i.e. when increasing the intracellular osmotic pressure.

2. Materials and Methods

2.1. Preparation and incubation of hepatocytes

Hepatocytes from 20-24h starved male wistar rats, were isolated by the method of Berry and Friend[8]; as modified by Groen and coworkers[9].

Hepatocytes (8-10mg dry weight/ml) were incubated in 20ml stopped vials in a shaking waterbath. The basic incubation medium was a Krebs-Henseleit-bicarbonate buffer at pH 7.4 (see Ref. 10) containing 2.5mM Ca^{2+}, 2% defatted bovine serum albumin, 20mM glucose, 0.1µM $TPMP^+$, 0.5mM mannitol, and 0.2mg/ml inuline. Amino acids (proline, alanine, glutamine) when added were used at 10mM, exept for leucine 2.5mM. All media were in equilibrium with a gas phase containing O_2:CO_2 (95:5). Temperature was 37 °C.

2.2. Measurement of respiration rate

After 15 min. incubation, the cell suspension was transfered to an oxygen electrode for determination of respiration rate.

2.3. Measurement of cell and mitochondria volume *in situ*

Hepatocytes were incubated in the presence of (i) 2.5µCi/ml ^3H_2O and 0.1 µCi/ml [^{14}C]carboxymethylinulin (added 1 min before sampling) for measurement of cell volume (ii) 2.5 µCi/ml ^3H_2O and 0.1 µCi/ml [^{14}C]mannitol for measurement of mitochondrial volume[11]. After 25 min. incubation, 1ml of cell suspension was transferred to an Eppendorf tube and centrifuged for 5 seconds at 7600 g. Within 10 seconds, 100 µl of the supernatant is removed into a scintillation vial containing 5ml of a liquid scintillation cocktail, the remaining is aspired. The pellet is resuspended with 300 µl $HClO_4$ 10%, centrifuged, and 200 µl of the supernatant treated as before. The apparent volume of each isotope in the pellet is $(dpm_c \cdot 3/2)/(dpm_s/100)$ in µl.

2.4. Measurement of the mitochondrial membrane potential *in situ*

Hepatocytes were incubated in the presence of (i) 1 µCi/ml [^3H]$TPMP^+$ for measurement of $TPMP^+$ accumulation (ii) 0.1µCi/ml $^{36}Cl^-$ for measurement of Cl^- distribution[12]. Cell-samples are treated as above. The [^3H]$TPMP^+$ accumulation ratio and the $^{36}Cl^-$ distribution are calculated as in Ref. 13. Thus, the mitochondrial membrane potential is

$$\Delta\Psi_{m(i-e)} = -60 \log \frac{V_c a_m}{V_m a_c} \left(\frac{[TPMP^+]_i}{[TPMP^+]_e} \frac{[Cl^-]_i}{[Cl^-]_e} \frac{(V_m + V_c)a_c}{V_c a_e} - \right. \tag{1}$$

(see also Refs 13-16). V_c, V_m, a_e, a_c, and a_m refer, respectively, to cellular and mitochondrial volume, and apparent activity coefficients for $TPMP^+$ in the extracellular medium, the cytoplasm and the mitochondrial matrix. We have redetermined a_e (because of our different extracellular medium), as well as a_c and a_m (see Refs 13,17 for experimental procedures), and we will take a_e=0.85, a_c=0.43, a_m=0.38.

2.5. Measurement of the mitochondrial ΔpH

The mitochondrial $\Delta pH_{(i-e)}$ was calculated from the [Pi] gradient accross the mitochondrial inner membrane, assuming an electrochemical equilibrium of H_3PO_4 (the phosphate carrier is thought to operate close to equilibrium in mammals[18,19]). If the anion is translocated in an electroneutral manner associated with an equivalent amount of protons, and $[Pi_{tot}]_m = [Pi^{2-}]_m + [Pi^-]_m$:

$$\log \frac{[Pi^-]_m}{[Pi^-]_c} = pH_m - pH_c , \tag{2}$$

$$pH_m = pKa_m + \log\frac{[Pi^{2-}]_m}{[Pi^-]_m} . \tag{3}$$

Eq. (3) is rearranged:

$$[Pi^-]_m = \frac{[Pi_{tot}]_m \cdot 10^{pKa_m}}{10^{pH_m} + 10^{pKa_m}} . \tag{4}$$

Substitution into eq. (2) gives:

$$10^{-13.7} \cdot (10^{pH_m})^2 + 10^{-7} \cdot 10^{pH_m} - \frac{[Pi_{tot}]_m}{[Pi^-]_c} = 0. \tag{5}$$

We have assumed that pH_c remained constant at 7, $pKa_m = 6.7$ and $pKa_c = 6.9$. Thus, $[Pi^-]_c = 0.44 \, [Pi_{tot}]_c$. Subscripts m, c, tot, refer to: mitochondrial, cytosolic and total, respectively.

2.6. Measurement of the intra- and extramitochondrial phosphate potentials

The mitochondrial and cytosolic distribution of ATP, ADP and Pi was studied by using the digitonin fractionation method as described by Zuurendonk and Tager[20] (see also Ref. 21, with a slight modification concerning the volumes used).

2.7. Measurement of the intramitochondrial redox potential

The intramitochondrial NADH/NAD$^+$ ratio was determined by the metabolite indicator method[21] assuming the ß-hydroxybutyrate dehydrogenase reaction to be in near-equilibrium ($K_{app} = [AcAc] \cdot [NADH]/[ßHOBu] \cdot [NAD^+] = 4.93 \cdot 10^{-2}$).

2.8. Assays

ß-hydroxybutyrate, acetoacetate, ATP, and ADP were measured spectrophotometrically (i) in neutralized HClO$_4$-extracts for ß-hydroxybutyrate, acetoacetate and the intra-mitochondrial adenine nucleotides[22,23], (ii) in phenol/chloroform/isoamyl alcohol extracts[24,25] for the cytosolic adenine nucleotides.

Pi was measured according to the method of Berenblum and Chain[26].

Table 1. Effects of amino acids on the cellular and mitochondrial volumes, oxygen consumption, mitochondrial and cytosolic ATP/ADP ratios and Pi concentrations[*].

Amino acid added	V_c µl.mg PT^{-1}	V_m µl.mg PM^{-1}	V_m/V_c %	J_{O_2} n at./min/ mg dw	ATP/ADP$_m$	ATP/ADP$_{cyt}$	Pi$_m$ mM	Pi$_{cyt}$ mM
Control	2.75	0.98	13.04	39.0	2.19	6.14	8.20	5.50
Proline	3.27	0.43	4.81	54.5	2.26	10.51	20.7	3.56
L-alanine	3.36	0.50	4.45	51.5	1.81	10.03	12.5	3.75
Glutamine	3.11	0.43	5.06	49.5	2.22	6.16	12.1	2.78
Glutamine+Leucine	3.47	0.45	4.75	55.5	1.75	6.82	14.7	2.22

[*]Subscripts m, c, cy refer to mitochondrial, cellular and cytosolic respectively. PM=mitochondrial protein, PT=total protein.
Hepatocytes (8-10 mg dw) were incubated with 20mM glucose and different amino acids (10mM, exept for leucine 2.5mM) as indicated. For the determination of the volumes, J_{O_2}, ATP/ADP ratios and Pi concentrations see Materials and Methods. Values are means for at least 2 or 3 cell preparations.

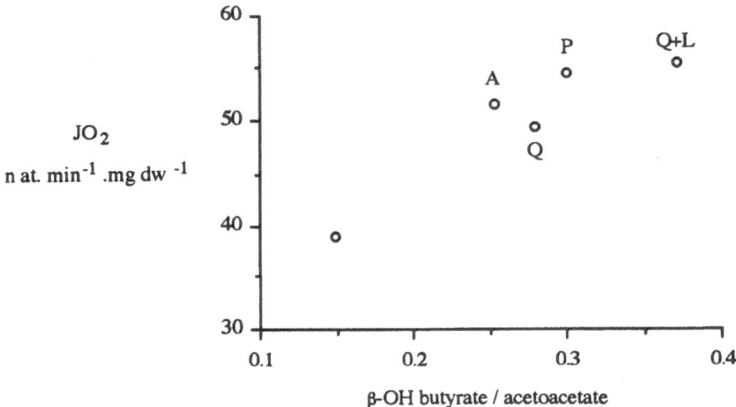

Figure 1. Relation between the flux of oxygen consumption and the ß-OH butyrate/acetoacetate ratio. Hepatocytes were incubated with glucose (20mM) and different amino acids (10mM, exept for leucine 2.5mM) as indicated. The flux of oxygen consumption and the ketone bodies were determined as described in Materials and Methods. Abbreviations: P, proline; A, L-alanine; Q, glutamine; L, leucine.

3. Results and Discussion

3.1. The mitochondrial volume

As we see in Table 1, hepatocytes swell when incubated in the presence of Na$^+$ cotransported amino acids (except leucine). The mitochondrial compartment answers differently: it shrinks. This can be explained by an increase in the cytosolic osmolarity caused by the amino acid uptake.

3.2. Cell swelling and oxygen consumption

When hepatocytes swell and mitochondria shrink, respiration is enhanced. This result seems to be inconsistent with those of Halestrap[27]. Indeed, Ca^{2+}-mobilizing hormones (phenylephrin, vasopressin, glucagon) all three raise the mitochondrial volume and the respiration rate by increasing the mitochondrial Ca^{2+} uptake. In fact, in our experiments, we do not change the quantity of Ca^{2+} inside the matrix space, but decreasing the mitochondrial volume raises the concentration of Ca^{2+} and enhances the respiration through an activation of the matrix dehydrogenases. So it seems that the increase in oxygen consumption is probably due to a decrease in the matrix volume rather than an increase in the cell volume.

The activation of the dehydrogenases is confirmed (Fig. 1): when mitochondria shrink the ß-OH butyrate/acetoacetate raises and is proportional to the respiration rate.

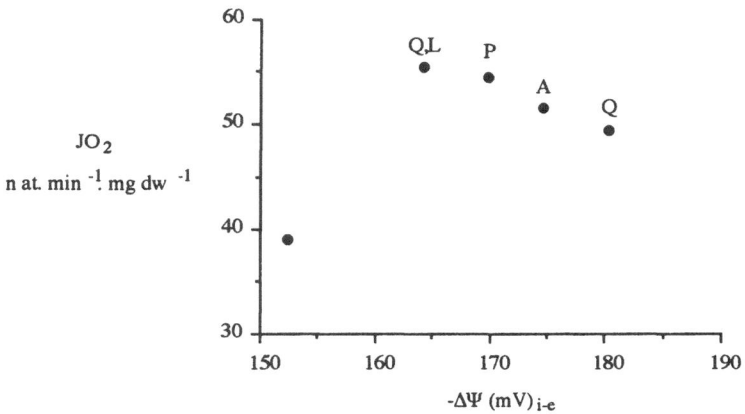

Figure 2. Relation between the flux of oxygen consumption and the mitochondrial transmembrane potential. Hepatocytes were incubated with glucose (20mM) and different amino acids (10mM, exept for leucine 2.5mM) as indicated. The flux of oxygen consumption and the membrane potential were determined as described in Materials and Methods. P=proline, A=L-alanine, Q=glutamine, L=leucine.

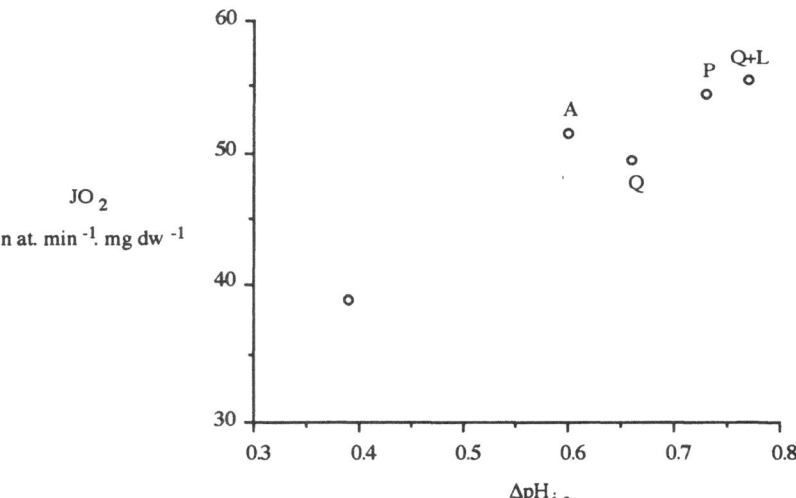

Figure 3. Relation between the flux of oxygen consumption and the mitochondrial transmembrane pH difference. Hepatocytes were incubated with glucose (20mM) and different amino acids (10mM, exept for leucine 2.5 mM) as indicated. The flux of oxygen consumption and the mitochondrial ΔpH were determined as de-scribed in Materials and Methods. P=proline, A=L-alanine, Q=glutamine, L=leucine.

3.3. The proton-motive force

In the presence of amino acids the mitochondrial membrane potential and the ΔpH are both increasing (Fig. 2 and 3), but they do not vary in the same way when the flux of oxygen consumption raises: $\Delta\Psi$ decreases and ΔpH increases. The increase in the ΔpH is the consequence of an increase in the NADH supply to the respiratory chain (via the activation of the matrix dehydrogenases).

The relationships between J_{O_2} and $\Delta\Psi$ are generally controlled by the ratio between the fluxes of ATP synthesis and the fluxes of substrate supply to the respiratory chain. In our situation, at a given redox potential, the kinetic control of the energy-spendthrift systems (which dissipate the $\Delta\Psi$) vary from one amino acid to another, which explains that $\Delta\Psi$ decreases when J_{O_2} increases.

3.4. The phosphate potential

In the presence of amino acids, when the respiration rate is enhanced the cytosolic phosphate potential increases in a proportional way (Fig. 4). This result can be explained if we consider the redox potential as the most important controlling force of the system.

Concerning the cytosolic ATP/ADP ratio, for two situations ("glutamine" and "glutamine+leucine", Table 1), it does not change when the respiration increases. In those two cases one can think that the ATP-consuming systems are more important compared to the two other situations ("proline" and "alanine").

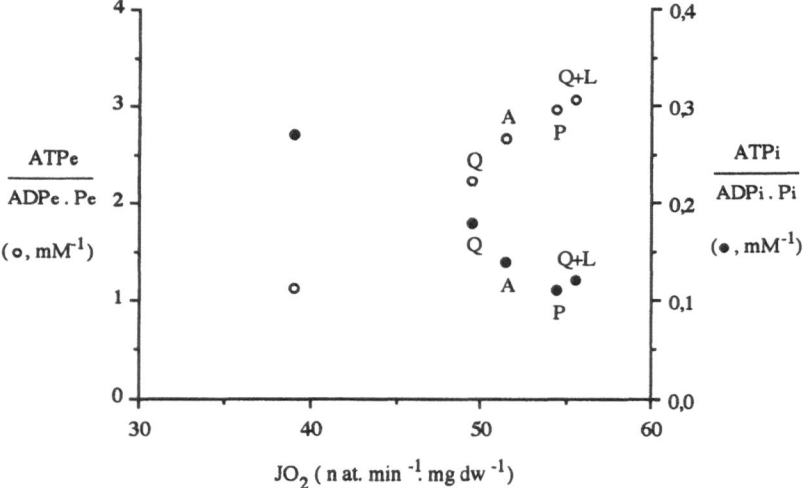

Figure 4. Relation between the flux of oxygen consumption and the mitochondrial and cytosolic phosphate potentials.
Hepatocytes were incubated with glucose (20mM) and different amino acids (10mM, exept for leucine 2.5mM) as indicated. The flux of oxygen consumption, the mitochondrial and cytosolic phosphate potentials were determined as described in Materials and Methods. P=proline, A=L-alanine, Q=glutamine, L=leucine.

Table 1 show us the Pi concentrations too, inside the mitochondria and in the cytosolic compartment. The increase in the mitochondrial Pi concentration (linked to ΔpH increase) can be related both to the increase in the respiration rate and the decrease of the mitochondrial volume. On the other hand, as the cytosolic compartment increases and respiration raises, the cytosolic Pi concentration lowers.

4. Conclusions

The results described in this paper show how the mitochondrial volume shrinkage can impose a determinate energetic metabolism through the activation of the matrix dehydrogenases and the increase of the mitochondrial Ca^{2+} concentration. Indeed, an increase in the matrix NADH/NAD ratio is well correlated to an increase in the respiration rate, proton-motive force and cytosolic phosphate potential.

Acknowledgements

P.E. thanks the Conseil Regional d'Aquitaine for financial support.

References

1. D. Häussinger, and F. Lang, Cell volume in the regulation of hepatic function: a mechanism for metabolic control, *Biochim. Biophys. Acta*, **1071**:331-350 (1991).
2. A. Baquet, L. Hue, A.J. Meijer, G.M. van Woerkom, and P.J.A.M.Plomb, Swelling of rat hepatocytes stimulates glycogen synthesis, *J. Biol. Chem.*, **265**:955-959 (1990).
3. A. J. Meijer, A. Bacquet, L. Gustafson, G. M. van Woerkom, and L. Hue, Mechanism of activation of liver glycogen synthase by swelling, *J. Biol. Chem.*, **267**:5823-5828 (1992).
4. A.P. Halestrap, A.M. Davidson, and W.D. Potter, Mechanisms involved in the hormonal regulation of mitochondrial function through changes in the matrix volume, *Biochim. Biophys. Acta*, **1018**:278-281 (1990).
5. A.P. Halestrap, and J. L. Dunlop, Intramitochondrial regulation of fatty acid ß-oxidation occurs between flavoprotein and ubiquinone. A role for changes in matrix volume, *Biochem. J.*, **239**:559-565 (1986).
6. A.P. Halestrap, The nature of the stimulation of the respiratory chain of rat liver mitochondria by glucagon pretreatment of animals, *Biochem. J.*, **204**:37-47 (1982).
7. R. M. Denton and J. G. McCormack, Ca^{2+} as a second messenger within mitochondria of the heart and other tissues, *Annu. Rev. Physiol.*, **52**:451-466 (1990).
8. N.M. Berry and D.S. Friend, High-yield preparation of isolated rat liver parenchymal cells, a biochemical and fine structural study, *J. Cell. Biol.*, **43**:506-520 (1969).
9. A.K. Groen, H.J. Sips, R.C. Vervoorn and J.M. Tager, Intracellular compartmentation and control of alanine metabolism in rat liver parenchymal cells, *Eur. J. Biochem.*, **122**:87-93 (1982).
10. H.A. Krebs and K. Henseleit,Urea formation in the animal body, *Hoppe-Seyler's Z. Physiol. Chem.*, **210**: 33-66 (1932).
11. P.T. Quinlan, A.P. Thomas, A.E. Armston and A.P. Halestrap, Measurement of the intramitochondrial volume in hepatocytes without cell disruption and its elevation by hormones and valinomycin, *Biochem. J.*, **214**:395-404 (1983).
12. C.D. Nobes and M.D. Brand, A quantitative assessment of the use of $^{36}Cl^-$ distribution to measure plasma membrane potential in isolated hepatocytes, *Biochim. Biophys. Acta*, **987**:115-123 (1989).
13. C.D. Nobes, G.C. Brown, P.N. Olive and M.D. Brand, Non-Ohmic proton conductance of the mitochondrial inner membrane in hepatocytes, *J. Biol. Chem.*, **265**:12903-12909 (1990).
14. I. D. Scott and D.G. Nicholls, Energy transduction in intact synaptosomes, *Biochem. J.*, **186**:21-33 (1980).
15. M.N. Berry, R.B. Gregory, A.R. Grivell, D.C. Henly, C.D. Nobes, J.W. Phillips and P.C. Wallace, Intracellular mitochondrial membrane potential as an indicator of hepatocyte energy metabolism: further evidence for thermodynamique control of metabolism, *Biochim. Biophys. Acta*, **936**:294-306 (1988)
16. J.B. Hoek, D.G. Nicholls and J.R. Williamson, Determination of the mitochondrial protonmotive force in isolated hepatocytes, *J. Biol. Chem.*, **255**:1458-1464 (1980).
17. M.P. Murphy and M.D. Brand, The control of electron flux through cytochrome oxydase, *Biochem. J.*, **243**: 499-505 (1987).
18. K. van Dam, H.V. Westerhoff, K. Krab, R. Van der Meer and J.C. Arents, Relationship between chemiosmotic flows and thermodynamic forces in oxidative phosphorylation, *Biochim. Biophys. Acta.*, **591**:240-250 (1980).
19. E. Ligeti, G. Brandolin, Y. Dupont and P. Vignais, Kinetics of Pi-Pi exchange in rat liver mitochondria. Rapid filtration experiments in the millisecond time range, *Biochemistry*, **24**:4423-4428 (1985).
20. P.F. Zuurendonck and J.M. Tager, Rapid separation of particulate components and soluble cytoplasm of isolated rat liver cells, *Biochim. Biophys. Acta*, **333**:393-399 (1974).
21. T.P.M. Akerboom, R. Van der Meer and J.M. Tager, Techniques for the investigation of intracellular com-

partmentation, *Techniques in Metabolic Research*, **B205**:1-33 (1979).

22. H.U. Bergmeyer. "Methods of Enzymatic Analysis," 2nd edition, Verlag Chemie, Academic Press, London (1970).

23. H.U. Bergmeyer. "Methoden der Enzymatischen Analyse," Verlag Chemie, Weinheim (1962).

24. E.C. Slater, J. Rosing and A. Mol, The phosphorylation potential generated by respiring mitochondria, *Biochim. Biophys. Acta.*, **292**:534-553 (1973).

25. R.J.A. Wanders, G.B. Van den Berg and J.M. Tager, A re-evaluation of conditions required for an accurate estimation of the extramitochondrial ATP/ADP ratio in isolated rat liver mitochondria, *Biochim. Biophys. Acta.*, **767**:113-119 (1984).

26. I. Berenblum and E. Chain, The colorimetric determination of phosphate, *Biochem. J.*, **32**:286-294 (1938).

27. A.P. Halestrap, The regulation of the matrix volume of mammalian mitochondria *in vivo* and *in vitro* and its role in the control of mitochondrial metabolism, *Biochim. Biophys. Acta*, **973**:355-382 (1989).

ESTIMATION OF THE MECHANISM
OF PEPTIDE BINDING
TO HLA CLASS II MOLECULES

E. Furuichi[1], N. Kamikawaji[2], T. Sasazuki[2] and S. Kuhara[3]

[1]Fukuoka Women's Junior College
4-16-1 Gojo, Dazaifu, Fukuoka 818-01, Japan
[2]Department of Genetics, Medical Institute of Bioregulation, Kyushu University
3-1-1 Maidashi, Higashi-ku, Fukuoka 812, Japan
[3]Graduate School of Genetic Resources Technology, Kyushu University
6-10-1 Hakozaki, Higashi-ku, Fukuoka 812, Japan

1. Introduction

HLA (human leukocyte antigen) class II molecules interact with antigenic peptides and form complexes that are recognized by T cells. The molecules have a large groove structure which seems to serve as the antigenic peptide binding site. To elucidate the mechanism of peptide binding to HLA class II molecules, we evaluate a complex model of three class II molecules and nine peptides used in binding experiments.

There are at least four steps before antigenic peptides are recognized by CD4[+] T cells, namely uptake by antigen presenting cells, degradation, binding to HLA class II molecules and transport to the cell surface membrane, as shown in Fig. 1.

The immune response to foreign antigens is regulated in humans by HLA (human leukocyte antigen) molecules. HLA class I molecules bind to peptides derived from newly synthesized intracellular antigens to activate CD8[+] T cells, while class II molecules associate with extracellular-antigen-derived peptides to activate CD4[+] T cells.

The three-dimensional structures of two HLA class I molecules, A2 and Aw68, have been determined by X-ray crystallographic analysis[1,2]. These molecules have in common a large groove structure which seems to form the antigenic peptide binding site. An analysis[3] of a complex of HLA-B27 with a nonameric peptide revealed that the groove forms an antigen binding site and holds the foreign antigen in the cleft between α_1 and α_2 domains.

There has been no documentation on structural data of class II molecules because of the difficulty in crystallization. These class II molecules are also expected to have a groove structure similar to that of class I molecules because there is a similarity in amino acid sequences between the putative binding regions of class I and II molecules[4].

As a first step at elucidation, we built a structural model of complex of HLA class II molecule and each of nine peptides, as shown in Fig. 2.[5]

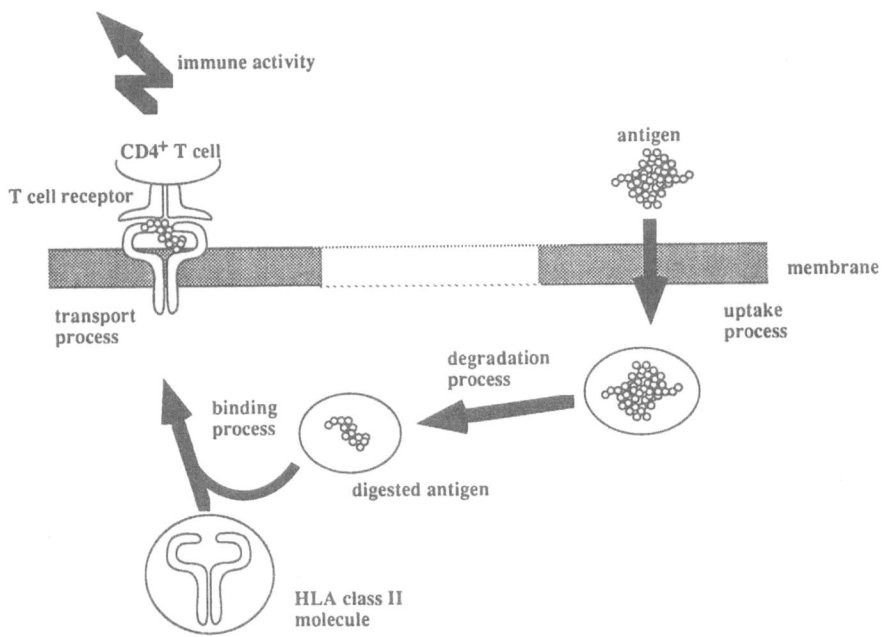

Figure 1. Schematic representation of the pathways of antigen processing.

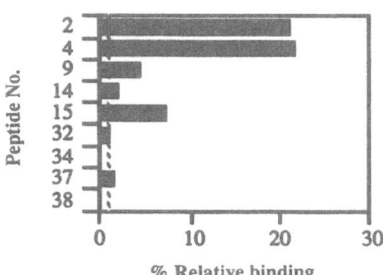

Figure 2. Binding of HLA-DR14 to peptides derived from M protein.

2. Methods

In HLA class I molecules, α_1 and α_2 domains formed the peptide-binding cleft, as seen on X-ray crystallography[3]. Conversely, putative binding clefts of class II molecules are considered to consist of α_1 and β_1 domains. The three-dimensional structure of this binding groove (α_1 and β_1 domains) of class II molecules (DR2, DR14, DQw6) was constructed, based on the coordinates of α_1 and α_2 domains of HLA-A2 (PDB entry 3HLA)[6], class I, by energy minimizing in AMBER[7].

The three-dimensional structure of peptides was constructed while assuming the structure of all nine peptides (peptide 2, 4, 9, 14, 15, 32, 34, 37, 38) to be an α-helix, because almost all similar sequences obtained in PDB were α-helical.

Each complex of HLA class II molecule and one of nine peptides was again evaluated by energy minimizing in AMBER, and binding potentials were compared with those obtained in binding experiments.

Table 1. Partially charged amino acid residues with possible electrostatic interaction between peptide and HLA-DR 14.[¶]

HLA-DR 14	binding peptide					non-binding peptide			
	2	4	9	4	15	32	34	37	38
α-Gln9			Glu16						
α-Glu11				Glu16					
α-Met36		Tyr5							Gln9
α-Glu47		Tyr5 Trp9							Gln9
α-Gln57		Ser2		Gln2	Glu4				
α-Glu71	Glu13	Gln13	Glu13	Thr14 Asn17	Glu11	Thr13	Glu12	Glu13	Glu13
α-Thr74			Gln20		Gln18			Gln20	Glu13
α-Ser77								Gln20	
α-Asn78	Asn22	Asp20 Gln23	Glu24	Glu23	Asp21				
α-Tyr79	Tyr19								
α-Thr83		Gln23							
β-Ser13	Tyr15								
β-Asp28	Tyr15			Glu16		Glu15		Glu15	Glu15
β-Tyr30	Tyr19	Arg19	Glu16	Glu16					
β-Glu59					Tyr24				
β-His60	Asn22	Glu22		Glu23					
β-Trp61	Asn22 Glu23	Glu22		Glu23	Asp21				
β-Gln64	Asp21	Asn18	Asn18	Asn17 Asp21		Glu21	Gln21	Glu18	
β-Asp66	Gln18 Tyr19		Asp18		His17		Ser19		Glu15
β-Glu69	Gln18	Asn18	Glu14 Asp18				Gln14	Glu18	
β-Arg72			Glu14		Asp9	Asn11	Glu11 Gln14		
β-Glu74	Tyr15								
β-Asp76	Arg11						Glu11		
β-Thr77		Gln12	Glu11		Asp9	Asn11			
β-Arg80						Asn11	Asp7	Glu7	Glu7
β-His81		Tyr5							
β-Tyr83						Gln4	Asn1		Glu4

[¶] Peptides binding to HLA-DR14 are indicated by bold-face symbols. Partially charged residues of HLA-DR14 in relation to binding peptide are italicized. For further explanations see text.

3. Results and Discussion

Nine peptides were divided into two groups, based on results of the binding experiments shown in Fig. 2, one is a binding peptide and the other is non-binding. The group of binding peptides was classified according to calculation coinciding with the results of binding experiments. Table 1 lists the partially charged residues of HLA-DR14 within 5 Å from the peptides after energy minimizing of the complex.

The difference between the interaction of binding peptides to HLA-DR14 and that of non-binding peptides is shown by bold-face symbols in Table 1. Those partially charged residues of peptides with binding activity (numbers 2, 4, 9, 14, 15) possibly interact with the residues (α-Gln[9], α-Glu[11], α-Asn[78], α-Tyr[79], α-Thr[83], α-Ser[13], β-Tyr[30], β-Glu[59], β-His[60], β-Trp[61]) of HLA-DR14. These residues of HLA-DR14 are mainly located at the end of the groove and form pockets (near α-Asn[78], β-Tyr[30], β-His[60]), as shown in Fig. 3.The non-binding peptides (numbers 32, 34, 37, 38) showed no significant interaction with these pockets at the end of the groove. Similar results were obtained in cases of HLA-DR2 and HLA-DQw6.

We tentatively conclude that pockets at the end of the groove may play an important role in binding to peptides.

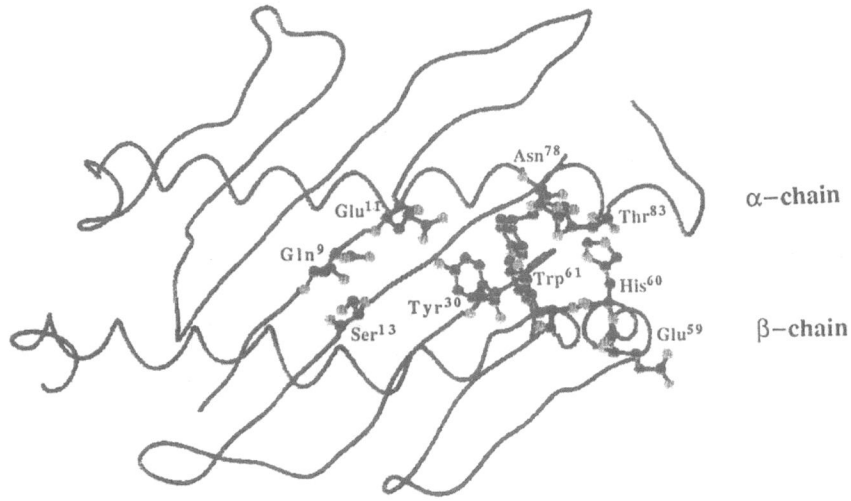

Figure 3. Model structure of HLA-DR14. Residues interacting with the binding peptides are represented as ball and stick.

Acknowledgements

We thank M. Ohara for valuable advice on the presentation. This work was supported by a Grant-in Aid for Scientific Research on Priority Areas, "Genome Informatics", from the Ministry of Education, Science and Culture of Japan.

References

1. P. J. Bjorkman, M. A. Saper, B. Samraoui, W. S. Bennett, J. L. Strominger and D. C. Wiley, Structure of the human class I histocompatibility antigen, HLA-A2, *Nature* **329**:506 (1987).
2. T. P. J. Garrett, M. A. Saper, P. J. Bjorkman, J. L. Strominger and D. C. Wiley, Specificity pockets for the side chains of peptide antigens in HLA-Aw68, *Nature* **342**:692 (1989).
3. D. R. Madden, J. C. Gorga, J. L. Strominger and D. C. Wiley, The structure of HLA-B27 reveals nonamer self-peptides bound in an extended conformation, *Nature* **353**:321 (1991).
4. J. H. Brown, T. Jardetzky, M. A. Saper, B. Samraoui, P. J. Bjorkman and D. C. Wiley, A hypothetical model of the foreign antigen binding site of Class II histocompatibility molecules, *Nature* **332**:845 (1988).
5. Manuscript under preparation.

6. F. C. Bernstein, T. F. Koetzle, G. J. B. Williams, E. F. Meyer, M. D. Brice, J. R. Rodgers, O. Kennard, T. Shimanouchi and M. Tasumi, The Protein Data Bank: a computer-based archival file for macromolecular structures, *J. Mol. Biol.* **112**:535 (1977).
7. S. J. Weiner, P. A. Kollman, D. A. Case, U. C. Singh, C. Ghio, G. Alagona, S. Profeta and P. Weiner, A New Force Field for Molecular Mechanical Simulation of Nucleic Acids and Proteins, *J. Am. Chem. Soc.* **106**:765 (1984).

SHIFT OF A PEPTIDE BOND EQUILIBRIUM TOWARDS SYNTHESIS CATALYZED BY α-CHYMOTRYPSIN

Brigitte Deschrevel and Jean-Claude Vincent

Laboratoire "Polymères, Biopolymères, Membranes"
U.R.A. 500, C.N.R.S.
Université de Rouen
B.P. 118, 76134 Mont Saint Aignan Cedex, France

1. Introduction

Under physiological conditions, reactions catalyzed by proteases or other hydrolases, usually shift far over towards hydrolysis. Suitable conditions, such as reduction of the concentration of water (more exactly its activity) in the reaction medium, are thus required to shift the equilibrium towards synthesis. Reduction of the water concentration may be ensured by addition of organic solvents in the reaction medium[1-3]. This possibility has been examined to synthesize highly valuable compounds such as biologically active peptides[3-5]. It may also be used to build a new artificial model of active transport and such models have been worked out in our laboratory[6-8]. They were based on reversible enzyme reactions occurring in gel slabs in which the asymmetrical distribution of enzyme activities was ensured by maintaining a difference in pH between the two faces of the barrier. The same kind of functional asymmetry may be induced by a difference in water concentration between the two faces if water participates in the reversible reaction.

As a model reaction, we selected the synthesis-hydrolysis reaction (1) of the peptide bond involved in N-Cbz-L-tryptophanyl-glycineamide, catalyzed by α-chymotrypsin.

$$
\begin{array}{c}
\text{N-Cbz-L-tryptophanyl-glycineamide} \\
\text{+ water}
\end{array}
\underset{}{\overset{\alpha\text{-chymotrypsin}}{\rightleftharpoons}}
\begin{array}{c}
\text{N-Cbz-L-tryptophan} \\
\text{+ glycineamide}
\end{array}
\qquad (1)
$$

Homandberg and coworkers[9] have tested reaction (1) in media composed of water and water-miscible organic solvents. From their data, we selected a water/1,4-butanediol reaction medium to study this reaction in more detail. The presence of 1,4-butanediol in the reaction medium encourages the synthesis reaction by decreasing the water concentration and by reducing the ionization of the substrates[9,10]. Under the experimental conditions used, there is no significant ester formation between N-Cbz-L-tryptophan and 1,4-butanediol[9]. Here we deal with the effect of the solvent composition of the reaction medium on the kinetic and thermodynamic aspects of reaction (1).

2. Materials and Method

As N-Cbz-L-tryptophanyl-glycineamide (the dipeptide) was not commercially available, we prepared[10] it by enzymatic synthesis and purified it by high performance liquid chromatography (HPLC). All the other reagents were purchased: glycineamide (Sigma, G7378), N-Cbz-L-tryptophan (Sigma, C5377), α–chymotrypsin (EC 3.4.21.1.) from bovine pancreas (Sigma, C4129), 1,4-butanediol (99%) (Fluka, 18960), methanol (99.9%) (Carlo Erba, 414 818) and Tris Base (Tris-(hydroxymethyl)-aminomethane) (Prolabo, 28 812 232).

All the reaction assays were performed in stirred solutions. The temperature was close to 23 °C. The reaction media were buffered with Tris Base (200 mM) and their apparent pH was 8.1. For each reaction studied, the initial concentration of α–chymotrypsin was 0.1 mg ml^{-1}. The reaction kinetics was studied by sampling aliquots of the reaction medium at increasing intervals of time, and determining the concentrations of N-Cbz-L-tryptophan and of the dipeptide. These measurements were performed by HPLC with a spectrophotometric detection. Home-made software was used for data treatment.

3. Results and Discussion

3.1. Kinetic aspect

The initial rates of reaction, V_{hi} (for hydrolysis of the peptide bond), and, V_{si} (for its synthesis), were computed from the kinetic curves and plotted against the water content of the reaction medium (Fig. 1). Using constant and non-saturating concentrations of substrates, both V_{hi} and V_{si} exponentially increased when the water content was increased.

In order to better understand the effect of the solvent composition of the reaction medium on the α–chymotrypsin activity, substrate-dependences of hydrolysis were performed in media containing various water contents. The hydrolysis reaction catalyzed by α–chymotrypsin being of the Michaelis-Menten type[11,12], the results of the substrate-dependences were treated by using the double reciprocal plot of Lineweaver and Burk. The apparent maximum rate, $V_{mh\ app}$, and the apparent Michaelis constant for the dipeptide, $K_{mhp\ app}$, were determined for each curve. Fig. 2 shows that in a purely aqueous medium $V_{mh\ app}$ was nearly four times higher than in a medium containing 20 % (v/v) water. On the other hand, $K_{mhp\ app}$ exponentially decreased on increasing the water content of the reaction medium (Fig. 3): $K_{mhp\ app}$ values ranged from 59.8 to 0.58 mM when the water content ranged from 20 to 100 % (v/v).

The initial rate of the hydrolysis reaction is expressed by:

$$V_{hi} = \frac{V_{mh}}{1 + \dfrac{K_{mhp}}{C_p} + \dfrac{K_{mh\ H_2O}}{C_{H_2O}}} , \qquad (2)$$

where V_{mh} is the maximum rate, K_{mhp} and $K_{mh\ H_2O}$ are the Michaelis constants with regard to the dipeptide and water, respectively. C_p and C_{H_2O} are the initial concentrations of dipeptide and water, respectively.

The expressions of $V_{mh\ app}$ and of $K_{mhp\ app}$ deduced from eq. (2) are given by:

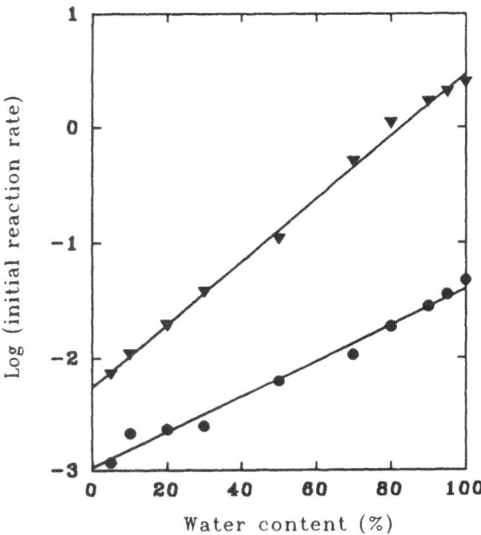

Figure 1. Logarithm of V_{hi} (full triangles) (initial rate of hydrolysis) and logarithm of V_{si} (full circles) (initial rate of synthesis), plotted against the water content of the reaction medium. V_{hi} and V_{si} are expressed in $mmol \cdot h^{-1} \cdot mg^{-1}$. Initial concentrations: dipeptide 1 mM and glycineamide 19 mM, for hydrolysis; N-Cbz-L-tryptophan 1 mM and glycineamide 20 mM, for synthesis.

$$V_{mh\,app} = \frac{V_{mh}}{1 + \dfrac{K_{mh\,H_2O}}{C_{H_2O}}}, \tag{3}$$

$$K_{mh\,app} = \frac{K_{mhp}}{1 + \dfrac{K_{mh\,H_2O}}{C_{H_2O}}}. \tag{4}$$

Considering eq. (3), it appears that the observed increase in $V_{mh\,app}$ was due, at least partly, to the increase in the water concentration in the reaction medium. On the contrary, eq. (4) did not allow any explanation of the exponential decrease in $K_{mhp\,app}$ on increasing the water content; according to eq. (4), increasing the water content of the reaction medium should lead to an increase in $K_{mhp\,app}$. Furthermore, the ratio $K_{mhp\,app}$ over $V_{mh\,app}$, which is equal to the ratio K_{mhp} over V_{mh}, should be independent of the water content. Since, using our experimental data, this ratio decreased with an increase in the water content of the reaction medium (Fig. 4), it implies that some or all of the intrinsic kinetic parameters are affected by the solvent composition of the reaction medium.

Variations of V_{mh} may be due to conformational changes of the enzyme, induced by the water-miscible organic cosolvent[1,13] itself and/or by the decrease in the water concentration in the vicinity of the enzyme molecules; this decrease, which depends on the nature of the organic cosolvent, may be responsible for an insufficient hydration of enzyme molecules[2,13]. Both these effects tend to decrease the V_{mh} values, via progressive enzyme inactivation. In our study, α–chymotrypsin kept some activity in a reaction medium containing as little as 2% (v/v) water, whereas, in pure 1,4-butanediol, the enzyme was totally inactive. So, it seems reasonable to assume that V_{mh} was poorly affected by the solvent composition of the reaction medium, except for very low water contents.

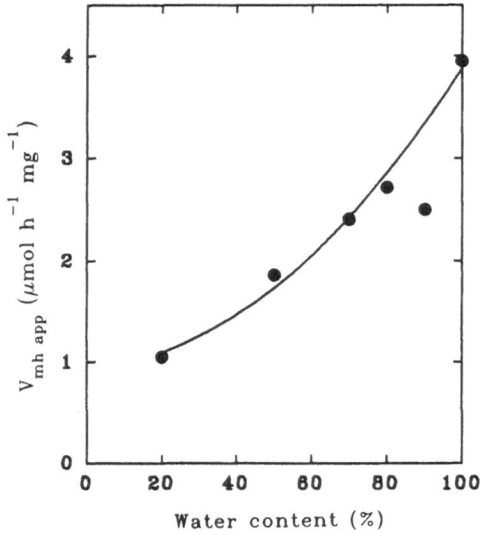

Figure 2. Apparent maximum rate, $V_{\text{mh app}}$, plotted against the water content of the reaction medium.

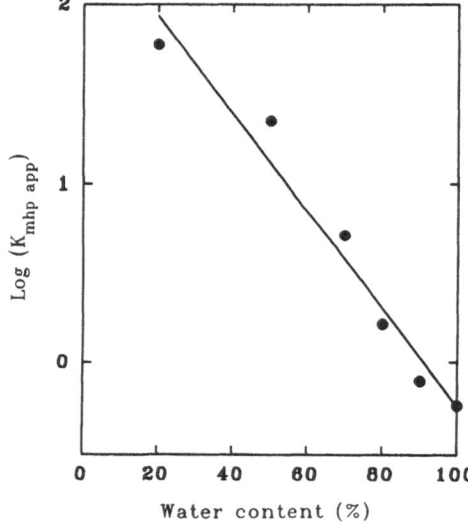

Figure 3. Logarithm of the apparent Michaelis constant for the dipeptide, $K_{\text{mhp app}}$, plotted against the water content of the reaction medium.

The K_m depends on the interactions of the substrate with the active site of the enzyme: hence it may be affected by the partitioning of the substrate between the active sites and the solvent. For α-chymotrypsin, the interactions of the substrate with the active site are mainly hydrophobic[11,12]. The dipeptide being a rather hydrophobic compound, it is more soluble in 1,4-butanediol than in water. This means that increasing the water content of the reaction

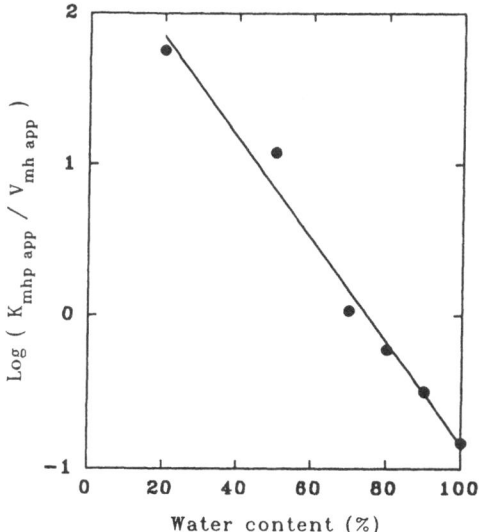

Figure 4. Logarithm of $K_{mhp\ app}/V_{mh\ app}$ as a function of the water content of the reaction medium.

medium tends to weaken the interactions of the dipeptide with the solvent and thus, to favor those of the dipeptide with the active site of the enzyme. As a consequence, K_{mhp} is decreased. Comparable effects of the presence of a water-miscible organic cosolvent on the K_m values have been reported[14-16].

Moreover, Maurel[17] showed that the free energy for the transfer of a substrate from an aqueous solution to a water/organic solvent mixture, as defined by eq. (5), is a linear function of the water content.

$$\Delta G_{tr} = -RT \ln (S_{(mix)}/S_{(water)}) . \tag{5}$$

$S_{(mix)}$ is the solubility limit of the substrate in the water/organic solvent mixture, and $S_{(water)}$ is the particular value in pure water. Since

   i) we have observed that the solubility limit of the dipeptide is an exponential function of the water content, and

   ii) the solubility limit is related to K_{mhp} and consequently to $K_{mhp\ app}$ (see eq. 4),

it is not surprising that the $K_{mhp\ app}$ variations are exponential. As a consequence, the exponential increases of V_{hi} and V_{si} were mainly due to the exponential variation of the K_m's.

3.2. Thermodynamic aspect

In a reaction medium with a given solvent composition, and using appropriate initial concentrations of substrates, whatever the direction of the reaction (hydrolysis or synthesis), the relative concentrations of the reagents at equilibrium were the same. The dependence of the equilibrium concentration of dipeptide (C_p^*) on the water content of the reaction medium is presented in Fig. 5. As expected, the less concentrated in water the reaction medium is, the higher will be the equilibrium concentration of dipeptide.

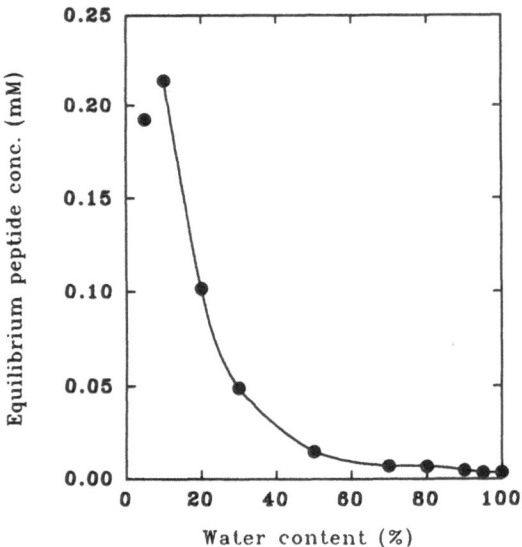

Figure 5. Equilibrium concentration of dipeptide plotted against the water content of the reaction medium. (Initial concentrations of substrates were identical to those indicated in Fig. 1).

Taking into account the neutralization/ionization reactions of N-Cbz-L-tryptophan and glycineamide, the overall process may be described by eq. (6).

$$
\begin{array}{c}
\overset{(K_{syn})}{R_1COOH + R_2NH_2 \rightleftharpoons R_1CONHR_2 + H_2O} \\
\\
\uparrow(K_1) \qquad \uparrow(K_2) \\
\qquad \qquad \searrow H^+ \\
\\
\searrow H^+ \\
R_1COO^- \qquad R_2NH_3^+
\end{array}
\tag{6}
$$

By assuming the activities of the reagents to be equal to their concentrations, the equilibrium constant for the synthesis of the dipeptide may be defined by

$$
K_{syn} = \frac{[R_1CONHR_2][H_2O]}{[R_1COOH][R_2NH_2]} .
\tag{7}
$$

[R₁COOH] and [R₂NH₂] can be expressed by using the total equilibrium concentrations of N-Cbz-L-tryptophan and of glycineamide, C_t^* and C_g^*, and their ionization constants K_t and K_g. Thus, K_{syn} may be written:

$$
K_{syn} = \frac{C_p^*[H_2O]}{C_t^* C_g^* f(H^+)}
\tag{8}
$$

$$\text{with} \quad f(H^+) = \frac{H^+ K_g}{(H^+ + K_t)(H^+ + K_g)} . \tag{9}$$

A first approximation, K_{syn1}, of K_{syn} was calculated using the values of K_t and K_g in pure water (Fig. 6). However, a more realistic approximation, K_{syn2}, was obtained by considering the variations of the ionization constants as a function of the water content in a water/1,4-butanediol mixture. These variations of K_t and K_g were estimated according to Ref. 9. Finally, a value of K_{syn}, K_{syn3}, independent of the solvent composition of the reaction medium was evaluated by considering the activity of water (a_{H_2O}) rather than its concentration (the activity coefficients of water being assumed to be identical to those in water/propanol mixtures[18]).

Using the expressions of the mass conservation law for N-Cbz-L-tryptophan and for glycineamide, K_{syn3} is expressed by:

$$K_{syn3} = \frac{C_p^* a_{H_2O}}{\left(C_t^o - C_p^*\right)\left(C_g^o - C_p^*\right) f(H^+)} , \tag{10}$$

where the superscript ^{o} refers to the initial conditions. Eq. (10) may thus be used for predicting (eq. 11) the dipeptide synthesis yield (R) as a function of the initial substrate concentrations, the pH and the solvent composition of the reaction medium (we consider here, as an example, the cases where C_t is lower than or equal to C_g):

$$R = \frac{C_p}{C_t^o} =$$

$$= \frac{C_t^o + C_g^o + a_{H_2O}/K_{syn3}f(H^+) - \left[\left(C_t^o + C_g^o + a_{H_2O}/K_{syn3}f(H^+)\right)^2 - 4C_t^o C_g^o\right]^{1/2}}{2C_t^o} . \tag{11}$$

Until now, eq. (11) has been proved valid through our experiments concerning the preparative scale dipeptide synthesis[10] for which a yield of 90.9 % was obtained.

4. Conclusion

Due to the partitioning of the substrate between the active site of the enzyme and the solvent, the decrease in the water content of the reaction medium seems to be kinetically rather unfavorable for the hydrolysis and synthesis reactions. However, decreasing the water content allows the use of higher substrate concentrations, which is kinetically favorable. In addition, it seems that 1,4-butanediol does not greatly affect the α–chymotrypsin conformation, except when the water content is very low. Water-miscible organic cosolvents may thus be useful in developing such reaction systems.

From the thermodynamic point of view, the addition of an organic cosolvent to the reaction medium appears a suitable method for shifting equilibrium towards synthesis since synthesis yields as high as 90% may be obtained. Furthermore, the definition and the evaluation of the synthesis equilibrium constant allows quantitative predictions for the synthesis yield.

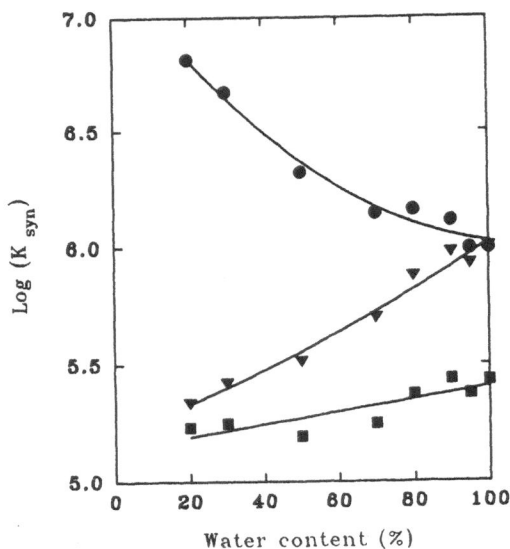

Figure 6. Logarithm of the synthesis equilibrium constant, K_{syn}, plotted against the water content of the reaction medium. K_{syn1} (full circles), K_{syn2} (full triangles), K_{syn3} (full squares).

References

1. L.G. Butler, Enzymes in non-aqueous solvents, *Enz. Microb. Technol.* **1**:253-259 (1979).
2. A.M. Klibanov, Enzymatic catalysis in anhydrous organic solvents, *Trends Biochem. Sci.* **14**:141-144 (1989).
3. H.D. Jakubke, P. Kuhl and A. Könnecke, Basic principles of protease-catalyzed peptide bond formation, *Angew. Chem. Int. Ed. Engl.* **24**:85-93 (1985).
4. W. Kullmann. "Enzymatic Peptide Synthesis," CRC Press, Florida (1987).
5. J.S. Fruton, Proteinase-catalyzed synthesis of peptide bonds, *Advan. Enzymol.* **53**:239-306 (1981).
6. S. Alexandre, "Les cycles enzymatiques: leur fonctionnement simultané ou alterné dans l'espace ou dans le temps," Thèse de Doctorat de l'Université de Paris VI (1988).
7. J.C. Vincent, S. Alexandre and M. Thellier, How a soluble enzyme can be forced to work as a transport system: description of an experimental design, *Arch. Biochem. Biophys.* **264**:405-408 (1988).
8. J.C. Vincent, S. Alexandre and M. Thellier, How a soluble enzyme can be forced to work as a transport system: theoretical interpretation, *Bioelectrochem. Bio energ.* **2 0**:215-222 (1988).
9. G.A. Homandberg, J.A. Mattis and M. Laskowski, Synthesis of peptide bonds by proteinases. Addition of organic cosolvents shifts peptide bond equilibria toward synthesis, *Biochem.* **17**:5220-5227 (1978).
10. B. Deschrevel, J.Y. Dugast and J.C. Vincent, Gram-scale enzymatic synthesis of a peptide bond, *Compt. Rend. Acad. Sci.* Série III **314**:519-524 (1992).
11. C. Walsh. "Enzymatic Reaction Mechanisms," W.H. Freeman and Company, San Francisco (1979).
12. G.P. Hess, Chymotrypsin: chemical properties and catalysis, *in*: "The enzymes," P.D. Boyer, ed., Academic Press, New York (1971).
13. J.S. Dordick, Principles and applications of nonaqueous enzymology, *in*: "Applied Biocatalysis," Vol. 1, H.W. Blanch and D.S. Clark, ed., Marcel Dekker, New York (1991).
14. K. Tanizawa and M.L. Bender, The application of insolubilized α-chymotrypsin to kinetic studies on the effect of aprotic dipolar organic solvents, *J. Biol. Chem.* **249**:2130-2134 (1974).
15. R.M. Guinn, H.W. Blanch and D.S. Clark, Effect of a water-miscible organic solvent on the kinetic and structural properties of trypsin, *Enz. Microb. Technol.* **13**:320-326 (1991).
16. M.M. Fernandez, D.S. Clark and H.W. Blanch, Papain kinetics in the presence of a water-miscible organic solvent, *Biotech. Bioeng.* **37**:967-972 (1991).
17. P. Maurel, Relevance of the dielectric constant and solvent hydrophobicity to the organic solvent effect in enzymology, *J. Biol. Chem.* **253**:1677-1683 (1978).
18. J.T. Edsall and J. Wyman. "Biophysical Chemistry. Thermodynamics, Electrostatics, and the Biological Significance of the Properties of Matter," Academic Press, New York (1958).

MODULATION OF CELLULAR ENERGY STATE AND DNA SUPERCOILING IN *E. coli*

P. R. Jensen[1], N. Oldenburg[1], B. Petra[1],
O. Michelsen[2] and H. V. Westerhoff[1]

[1]The Netherlands Cancer Institute
Plesmanlaan 121, NL-1066 CX Amsterdam
and E. C. Slater Institute for Biochemical Research
University of Amsterdam
NL-1018 TV Amsterdam, The Netherlands
[2]Department of Microbiology, Technical University of Denmark
DK-2800 Lyngby, Denmark

1. Introduction

Changes in DNA supercoiling have been shown to affect the expression of various genes in procaryotes. One of the factors that may control the level of DNA supercoiling in *E. coli* is the enzyme DNA gyrase, which utilizes ATP to introduce negative supercoils in topologically constrained DNA. Furthermore, the activity of DNA gyrase has been demonstrated to be sensitive to the ratio between [ATP] and [ADP] around the enzyme, *in vitro* [1]. Consequently, cellular free energy metabolism may regulate transcription.

We are analyzing this regulatory system experimentally along the lines of Hierarchical Metabolic Control Analysis[2]. We use several different approaches to alter the intra-cellular [ATP]/[ADP] ratio in intact cells of *Escherichia coli*, such as inhibiting oxidative phosphorylation, through addition of protonophores or respiratory inhibitors. Here we present a method that employs *Escherichia coli* strains in which the expression of the chromosomal *atp* operon, encoding H[+]-ATPase, can be modulated though inducible promoter elements. We used these strains to analyze the control of the H[+]-ATPase on various growth variables of *E. coli* and to modulate the intra-cellular [ATP]/[ADP] ratio and level of DNA supercoiling.

2. Materials and Methods

2.1. Bacterial strains and growth of bacterial cultures

The *E. coli* K-12 strain used as the wild type strain in this study, LM3118, has the genotype, F[+], *asnB32*, *thi-1*, *relA1*, *spoT1*, *lacUV5* and *lacY*[am]. In the strain LM3113, the chromosomal *atp* promoters have been exchanged with the *tacI* promoter, see Fig. 1.

Cells were grown aerobically (30 °C) in minimal MOPS (morpholino-propane sulfonic acid) medium[3], supplemented with 1 mg/ml thiamin and either 0.05 % succinate (in the steady state experiments; Table 1, Fig. 2) or 0.5 % succinate and 100 μg/ml ampicillin (in the induction experiment; Fig. 3).

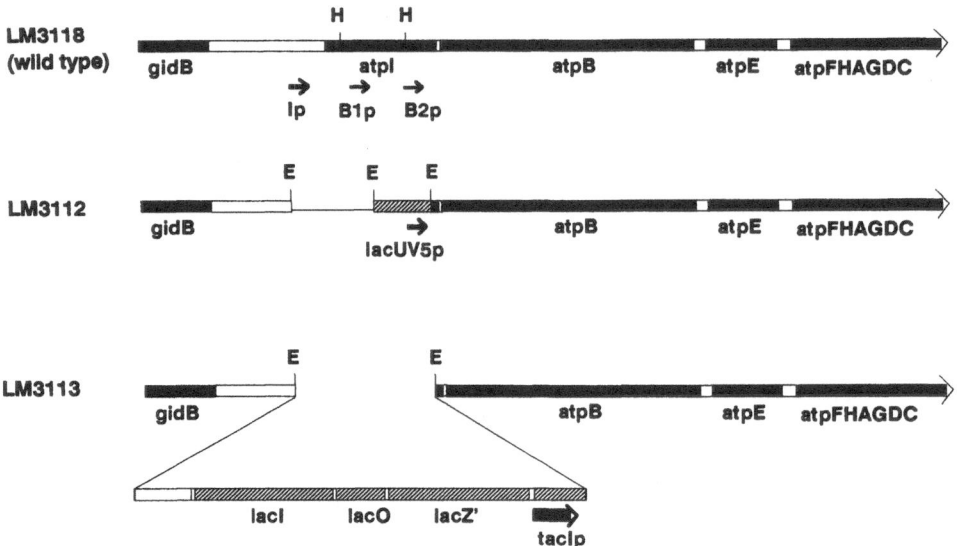

Figure 1. *E. coli* strains used for modulation of the concentration of H[+]-ATPase. Top: LM3118, harboring the wild type chromosomal *atp* operon. Middle: LM3112, the chromosomal *atp* operon transcribed from the *lacUV5* promoter. Bottom: LM3113, the *atp* operon transcribed from the *tacI* promoter (cf. Ref. 9).

2.2. Respiration rates

Respiration rates were determined by transferring 15 ml of exponentially growing culture (OD_{450} = 0.1—0.2) to a reaction tube, and recording the decrease in dissolved oxygen concentration by a Clark type electrode.

2.3. Quantitation of H[+]-ATPase *c* subunit

This was done as described by Von Meyenburg and coworkers[4]. Briefly, the total protein content of the cells was labeled by ^{35}S-methionine and separated by SDS polyacrylamide gel electrophoresis. The content of H[+]-ATPase *c* subunit was quantified relative to lipoprotein (lpp) after optical scanning of the resulting autoradiograms.

2.4. Measurement of [ATP]/[ADP] ratios in cell extracts and isolation of plasmids

1 ml samples were withdrawn from the cultures and immediately mixed with 1 ml phenol

(80 °C, equilibrated with TE buffer (10 mM TRIS, 1mM EDTA, pH 8) and containing 0.1 %
8-hydroxyquinoline) by vortexing for 10 sec. After 5 minutes at room temperature the
samples were again vortexed for 10 sec., cooled on ice, and centrifuged for 5 minutes at
14.000 rpm. The water phase was removed and extracted with chloroform to remove phenol.
At this point, 50 μl was removed for [ATP]/[ADP] measurements (see below) and the rest of
the sample (850 μl) was precipitated by isopropanol and used for analyzing the supercoiling
of the reporter plasmid (pBR322). This was done by gel electrophoresis (1.4 % agarose) in
the presence of 10 μg/ml chloroquine and the topoisomers were visualized after blotting by
hybridization with ^{35}S-dATP labeled pBR322.

The ATP assay was carried out at room temperature, using a luciferin-luciferase ATP
monitoring kit, obtained from and used as recommended by BioOrbit. ADP was determined in
the same sample, after the determination of [ATP], by addition of phospho*enol*pyruvate (3
mM) and pyruvate kinase and recording the increase in [ATP] concentration.

Table 1. Control exerted by the H[+]-ATPase on the physiology of *E. coli*.

c subunit[a]	relative rate, yield (%)				control coefficient[b]
	0.59[a]	**0.88**	**1.59**	**2.4**	
growth rate, μ	97.0	100.2	92.8	84.6	-0.04 (-0.10 to +0.05)[c]
growth yield, Y	87.4	96.8	106.8	109.9	0.21 (0.17 to 0.25)
$J_{succinate}$	111	102	89	84	-0.25 (-0.24 to -0.26)
resp. rate, J_o	111	103.5	87	77	-0.24 (-0.23 to -0.25)

[a]Expression of H[+]-ATPase *c* subunit, relative units compared to the wild type strain, LM3118.
[b]The control coefficients were determined for the wild type level of *c* subunit expression (=1.0), using the data
in column 2-5.
[c]In order to estimate the accuracy of the control coefficients in this experiment, we removed the data points one
by one and re-calculated the coefficient. The range of the coefficients obtained in this way are shown in brackets.

3. Results and Discussion

3.1. Control of the H[+]-ATPase on *E. coli* physiology at the wild type enzyme
level

The H[+]-ATPase plays a central role in *E. coli* free-energy metabolism and hence in *E. coli*
physiology. Does this mean that under physiological conditions the H[+]-ATPase is also in
control of that metabolism?

In order to be able to address this question, we constructed *E. coli* strains in which the
chromosomal *atp* operon has been placed under the control of inducible promoter elements
(see Fig. 1). By changing the concentration of the inducer (IPTG) in the growth medium, we
then modulated the expression of the H[+]-ATPase concentration over a range around its wild
type level and measured the effect on the growth of *E. coli*. Table 1 shows how the growth
rate, growth yield and respiration varied with the concentration of H[+]-ATPase, when the cells
were grown in succinate minimal medium.

We then fitted a second order polynomial curve to these experimental data points, and used the slope of this curve to calculate the control coefficient at the wild type enzyme level, see Table 1, from the equation[5]:

$$C^{\mu}_{[H^+\text{-ATPase}]} = \frac{(d\mu/\mu)}{(d[H^+\text{-ATPase}]/[H^+\text{-ATPase}])} .$$

It appears that the H[+]-ATPase has very little control on the growth rate and a minor positive control on the growth yield. These results imply that the concentration of this enzyme in wild-type *E. coli* is optimal for growth rate, but only at higher concentrations is it optimal with respect to growth yield.

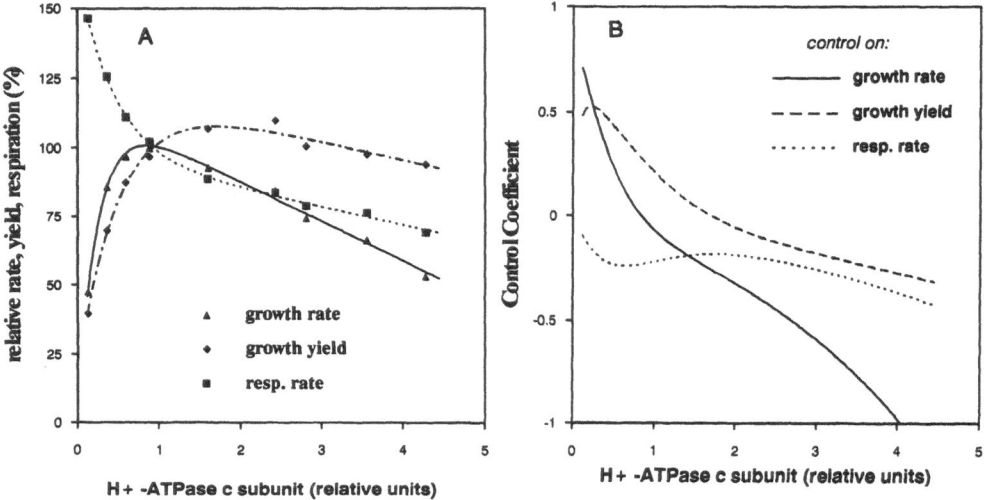

Figure 2. Dependence of *E. coli* physiology on H[+]-ATPase, during steady state growth in succinate minimal medium. (A) The growth rate, growth yield and respiration rate were measured as a function of the expression of H[+]-ATPase *c* subunit. *E. coli* cells were used, in which the expression of the *atp* operon could be controlled by the *tacI* promoter (Fig. 1, strain LM3113), through the addition of different concentrations of IPTG. (B) Control of H[+]-ATPase on *E. coli* physiology. The log-log derivatives of the smooth curves in A) were plotted as a function of the level of H[+]-ATPase (cf. Ref. 10).

The H[+]-ATPase had a small but negative control on the respiration rate. At first this may be a surprising result: a reduced concentration of H[+]-ATPase in the membrane is expected to result in an increased membrane potential (and we have some indications that this does occur[6]) and therefore a reduced respiration rate. However, the negative control exerted on respiration may be explained from the fact that the rate of succinate consumption increased as the concentration of H[+]-ATPase was reduced[10] and this is expected to lead to an increased redox state of the cells, which in turn could activate respiration.

Furthermore, we observed an enhanced expression of cytochromes in a strain which had deleted the *atp* operon[6], and this may also explain the negative control of the H[+]-ATPase on respiration.

3.2. Control of the H⁺-ATPase on *E. coli* physiology at enzyme levels far from the wild type level

The control exerted by an enzyme is expected to increase, if the activity is greatly reduced compared to the wild type level, particularly when the enzyme is essential for growth. This is the situation with *E. coli* when the cells are grown on succinate, since *atp* deletion strains fail to grow on this substrate[7]. In Fig. 2a, we show how the physiology depended on the H⁺-ATPase, for the expression range between 0.15 and 4 times the wild type level.

We then used these data to calculate the control coefficients for the entire range, see Fig. 2b. Indeed, we found a much stronger control at low expression levels. At over-expression of H⁺-ATPase, all three control coefficients became negative, and the control on growth rate was -1 at 4 times the wild type expression level.

3.3. Modulation of the intra-cellular [ATP]/[ADP] ratio and level of DNA supercoiling

When *E. coli* cells are grown in minimal succinate medium, the H⁺-ATPase is expected to be responsible for most of the ATP synthesis. Indeed, we found that when the expression of H⁺-ATPase was reduced to 0.15 times the wild type level, the [ATP]/[ADP] ratio decreased to 3 (compared to 12 in the wild type cell). The cells were still growing, but the growth rate was reduced to 40 % of the wild type rate, see Fig. 3a.

We then added IPTG to induce the expression of the *atp* operon, and observed that both the growth rate and the ratio of [ATP]/[ADP] started to increase within 40 minutes, see Fig. 3. 60 minutes after the induction, this ratio was increased above the ratio observed for the wild type strain. After 150 minutes, the growth rate decreased to approximately the rate that was observed before induction, indicating that the concentration of H⁺-ATPase had become inhibitory for growth (analogous to the steady state experiment described above). The [ATP]/[ADP] ratio remained at a higher value than in the wild type cells.

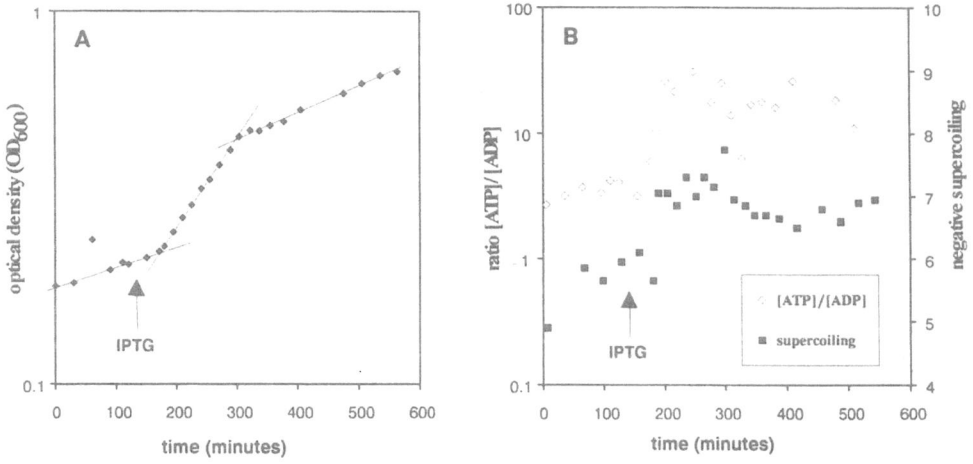

Figure 3. Modulation of A) growth rate and B) [ATP]/[ADP] ratio and DNA supercoiling, by induced bio-synthesis of H⁺-ATPase. Cells of strain LM3113, containing the plasmid pBR322, were grown aerobically (37 °C), in succinate minimal medium without the inducer (IPTG). At t =140 minutes, 1mM IPTG was added to induce maximal expression of the *atp* operon from the *tacI* promoter.

Changes in the ratio of [ATP]/[ADP] were expected to affect the level of negative supercoiling as discussed above. When the expression of H^+-ATPase was increased from 0.15 to 4 times the wild type level, we observed a small but significant decrease in the linking number of a reporter plasmid, see Fig. 3, corresponding to an increased level of negative supercoiling.

In this experiment we have employed a strain with a *tac I* promoter in control of the expression of the *atp* operon. This promoter was leaky, and therefore the lowest concentration of H^+-ATPase that could be attained with this strain was 0.15 times the wild type enzyme level. We are presently employing a strain with a *lacUV5* promoter in control of the *atp* operon, see Fig. 1. This promoter is tighter[8], and preliminary results show that [ATP]/[ADP] ratios below 0.5 can be obtained with this system. Indeed, *in vitro* studies showed that at low ratios, the gyrase dependent supercoiling of plasmid DNA is most sensitive to changes in the [ATP]/[ADP] ratio[1].

Acknowledgements

This study was supported by the Danish Center for Microbiology (CM), the Danish Natural Research Council (SNF) and the Netherlands Organization for Scientific Research (NWO).

References

1. H.V. Westerhoff, M.H. O'Dea, A. Maxwell and M. Gellert, DNA supercoiling by DNA gyrase, *Cell Biophys.* 12:157-181 (1988).
2. H.V. Westerhoff, J. G. Koster, M. van Workum and K.E. Rudd, On the control of gene expression, *in*: "Control of Metabolic Processes," A. Cornish-Bowden and M. L. Cárdenas, eds, Plenum Press, New York, pp. 399-412 (1990).
3. F.C. Neidhardt, P.L. Bloch and D.F. Smith, Culture medium for enterobacteria, *J. Bacteriol.* 119:736-747 (1974).
4. K. Von Meyenburg, B.B. Jorgensen, J. Nielsen and F. G. Hansen, Promoters of the *atp* operon coding for the membrane bound ATP synthase of *Escherichia coli* mapped by Tn*10* insertion mutations, *Mol. Gen. Genet.* 188:240-248 (1982).
5. H. Kacser and J.A. Burns, The control of flux, *Symp. Soc. Exp. Biol.* 27:65-104 (1973).
6. P. R. Jensen and O. Michelsen, Carbon and energy metabolism of *atp* mutants of *E. coli*, *J. Bacteriol.* 174: 7635-7641 (1992).
7. K. Von Meyenburg, B.B. Jorgensen and B. van Deurs, Physiological and morphological effects of over-production of membrane-bound ATP synthase in *Escherichia coli* K-12, *EMBO J.* 3:1791-1797 (1984).
8. H.A. de Boer, L.J. Comstock and M. Vasser, The *tac* promoter: A functional hybrid derived from the *trp* and *lac* promoters, *Proc. Natl. Acad. Sci.* 80:21-25 (1983).
9. P.R. Jensen, H.W. Westerhoff and O.Michelsen, The use of *lac* type promoters in control analysis, *Eur. J. Biochem.* 211:181-191 (1993).
10. P.R. Jensen, H.W. Westerhoff and O.Michelsen, Excess capacity of H^+-ATPase and inverse respiratory control in *E. coli*, *EMBO J.* 12 (No. 4), in the press (1993).

CASCADE CONTROL OF AMMONIA ASSIMILATION

Wally C. van Heeswijk[1,2], Hans V. Westerhoff[1,2] and Daniel Kahn[2,3]

[1]E.C. Slater Institute
University of Amsterdam
Plantage Muidergracht 12, NL-1018 TV Amsterdam, The Netherlands
[2]Division of Molecular Biology
The Netherlands Cancer Institute (H5)
Plesmanlaan 121, NL-1066 CX Amsterdam, The Netherlands
[3]Present address: Laboratoire de Biologie Moléculaire
CNRS-INRA
B.P. 27, 31326 Castanet-Tolosan Cedex, France

In cellular physiology, metabolism is organized at several levels: the level of intermediary metabolism, the level of protein metabolism (synthesis, modification and degradation), and the level of mRNA metabolism (transcription and decay). At the level of protein metabolism we study cascades of enzymes modifying one another covalently. To this purpose metabolic control theory (MCT) has been extended explicitly to include regulatory cascades. The resulting modular control theory[1], which started from the general formalism of Reder[2], will be used to analyze the glutamine synthetase regulatory cascade in *Escherichia coli*. The modular control analysis of this cascade should give insight into how control is distributed over the different hierarchical levels of the cascade, and to quantitate the relative importance of metabolic and genetic regulation of nitrogen assimilation.

In enteric bacteria, assimilation of ammonia occurs primarily via the glutamine synthetase (GS)/glutamate synthase (GOGAT) cycle (for review see Ref. 3). The activity of GS is regulated by a bicyclic nucleotidylation cascade. One cycle consists of a modification of GS, to the less active form GS-AMP, and a demodification of GS-AMP, both reactions being catalyzed by adenylyl-transferase (AT_a and AT_d, respectively). The second cycle involves a uridylylation of protein P_{II} by uridylyl-transferase (UT_u) and a deuridylylation of P_{II}-UMP by the same enzyme (UT_d). The ratio of P_{II}/P_{II}-UMP regulates the activity of adenylyl-transferase, while UT_u and UT_d are regulated allosterically by α-ketoglutarate and glutamine. Thus, uridylyl-transferase serves as a sensor for the nitrogen status of the cell.

This cascade can be divided into three modules (see the boxes in Fig. 1). Note that we ignore, for simplicity, the glutamate dehydrogenase (GDH) route (as we would do with a glutamate dehydrogenase negative mutant). Each module contains a set of reactions wich are connected with each other, but not with reactions from other modules. Therefore separate modules interact solely via effector-type interactions, i.e., substrates from one module may effect reactions in another module without being a substrate or product in this other module.

Modern Trends in Biothermokinetics, Edited by
S. Schuster *et al.*, Plenum Press, New York, 1993

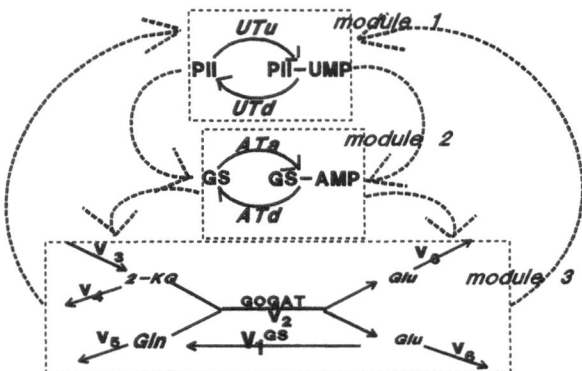

Figure 1. Scheme of the pathways involved in ammonia assimilation in *Escherichia coli*. 2-KG: α-ketoglutarate, Glu: glutamate, Gln: glutamine, GS: glutamine synthetase, GOGAT: glutamine-2-ketoglutarate aminotransferase, ATa: adenylyl-transferase, ATd: deadenylylase, PII: protein II, UTu: uridylyl-transferase, UTd: deuridylylase. Modules are indicated by dashed boxes . Full arrows refer to fluxes, dashed arrows to allosteric effects.

Flux control coefficients are defined as the relative sensitivities of the steady state fluxes (J_i) to changes in rates (v_j):

$$C_j^i = \frac{d \ln J_i}{d \ln v_j}.$$

Following the modular structure of the system, we can construct a matrix of control coefficients **C**, decomposed into 9 blocks, containing all flux control coefficients of the system[1]:

$$\mathbf{C} = \begin{bmatrix} \mathbf{C}_1^1 & \mathbf{C}_2^1 & \mathbf{C}_3^1 \\ \mathbf{C}_1^2 & \mathbf{C}_2^2 & \mathbf{C}_3^2 \\ \mathbf{C}_1^3 & \mathbf{C}_2^3 & \mathbf{C}_3^3 \end{bmatrix}$$

where the submatrix $\mathbf{C}_j^i$ contains the flux control coefficients describing the sensitivities of the fluxes within module *i* to changes in the activities in module *j*. Submatrix $\mathbf{C}_1^3$, for example, describes the sensitivity of the fluxes through GS and GOGAT to changes in the uridylyl-transferase and deuridylylase activity. To measure the flux control coefficients of this submatrix we will change the activity of uridylyl-transferase by modulating expression of *glnD*, the gene coding for uridylyl-transferase, around the wild-type level, and measure the relative change of the fluxes through GS and GOGAT. Similar experiments will be performed with adenylyl-transferase.

To this end, the genes encoding UT (*glnD*) and AT (*glnE*) have been cloned and sequenced[4]. The P_{II} gene *glnB* had been cloned and characterized by others[5]. To modulate their expression, we shall substitute the wild-type promoters of these genes by a *lac*-type promoter in the chromosome, in a similar way as has been done by Jensen and coworkers[6] with the promoter of the *atp* operon.

As a first step, we have characterized the three genes to know if the genes are part of operons and if the genes are transcriptionally regulated. Determination of the transcription starts by RNAse protection assays and primer extension assays revealed that the *glnD* gene is transcribed from its own promoter and that *glnB* and *glnE* are both transcribed as polycistronic messengers[4]. Interestingly, the amino-acid sequence of an open reading frame upstream the *glnB* gene, OrfXB, showed homology with NtrC. NtrC is a protein involved in the transcriptional activation of *glnA*, the gene coding for glutamine synthetase, and of other genes involved in nitrogen assimilation (for review see Ref. 7). The above results indicate that a promoter substitution in the natural chromosomal location of both *glnB* and *glnE* genes with an inducible promoter should be preceded by a transcription terminator. In such a construct, transcription of the upstream gene will stop at the end of the open reading frame without disturbing the properties of the new promoter.

To determine the transcription regulation of these three genes, RNAse protection assays were performed with RNA isolated from *glnD*, *glnB*, *glnE*, *glnA*, and *rpoN* (σ^{54}-factor) mutants grown in either nitrogen-limited or nitrogen-rich medium. This revealed that the ammonia concentration of the medium did not or hardly affect the transcription of *glnD*, *glnB* and *glnE* . Furthermore it showed that UT, P_{II}, AT, GS and σ^{54}-factor did not or hardly affect, directly nor indirectly, expression of the three genes. Thus, by and large, expression of the proteins of the GS adenylylation cascade is constitutive: their activities are regulated solely by covalent modifications or allosteric interactions, and not or hardly genetically. Therefore expression of each of these proteins can be manipulated independently in experiments designed to measure their control coefficients.

Acknowledgements

This work was supported by the Netherlands Organization for Scientific Research (NWO).

References

1. D. Kahn and H.V. Westerhoff, Control theory of regulatory cascades, *J. theor. Biol.* **153**:255-285 (1991).

2. C. Reder, Metabolic control theory: a structural approach, *J. theor. Biol.* **135**:175-201 (1988).

3. S.G. Rhee, W.G. Bang, J.H. Koo, K.H. Min and S.C. Park, 1988, Regulation of the glutamine synthetase activity and its biosynthesis in *Escherichia coli*: mediation by three cycles of covalent modification, *in*: "Enzyme dynamics and regulation," P. Boon-Chock, C.Y. Huang, C. L. Tsou, J.H. Wang, eds, Springer-Verlag, Berlin, pp. 136-145 (1988).

4. W. van Heeswijk, M. Rabenberg, H.V. Westerhoff and D. Kahn, The genes of the glutamine synthetase adenylylation cascade are not regulated by nitrogen in *Escherichia coli*, *Mol. Microbiol.*, in press.

5. H.S. Son and S.G. Rhee, 1987, Cascade control of *Escherichia coli* glutamine synthetase. Purification and properties of P_{II} protein and nucleotide sequence of its structural gene, *J. Biol. Chem.* **262**:8690-8695 (1987).

6. P.R. Jensen, H.V. Westerhoff and O. Michelsen, On the use of *lac*-type promoters in control analysis, *Eur. J. Biochem.*, in press.

7. J. Collado-Vides, B. Magasanik and J.D. Gralla, Control site location and transcriptional regulation in *Escherichia coli*, *Microbiol. Rev.* **55**:371-394 (1991).

PHOTOVOLTAGE MEASUREMENTS AS A PROBE OF LIGHT DISTRIBUTION AND ION MOVEMENTS AROUND THE PHOTOSYNTHETIC APPARATUS

Géza Meszéna[1] and Hans V. Westerhoff[2,3]

[1]Population Biology Group
 Department of Atomic Physics
 Eötvös University
 Puskin u. 5-7, H-1088 Budapest, Hungary
[2]E.C. Slater Institute for Biochemical Research
 University of Amsterdam, The Netherlands
[3]Division of Molecular Biology
 The Netherlands Cancer Institute (H5)
 Plesmanlaan 121, NL-1066 CX Amsterdam, The Netherlands

1. Introduction

In photovoltage experiments a signal is generated because of a light intensity difference between the two sides of the thylakoid vesicula. Fig. 1 shows the most trivial case, when the side facing to the light source gets more light than the other one. In that case, the signal is positive according to the usual sign convention. This sign was detected in the early experiments[1,2]. Wavelength-dependent polarity was reported by flash-lamp excitation[3]. A negative signal was seen using rubin and Nd-YAG laser[4]. The laser generated signal was so fast that it came undoubtedly from the primary charge separation[5,6]. The signal generated by flash lamp was suspected to be a consequence of different mobilities of positive and negative charges[7]. A negative laser signal was also found[8,9].

The goal of the work presented here was:

- to clarify that the wavelength is the factor determining the polarity, and
- to explain this dependence by calculating the light intensity pattern in the thylakoid system.

It turned out that the problem of polarity is strongly connected to the kinetics of the signal.

2. Experimental Results

A tunable dye laser was used as a light source. In the first time, we were able to detect

Modern Trends in Biothermokinetics, Edited by
S. Schuster *et al.*, Plenum Press, New York, 1993

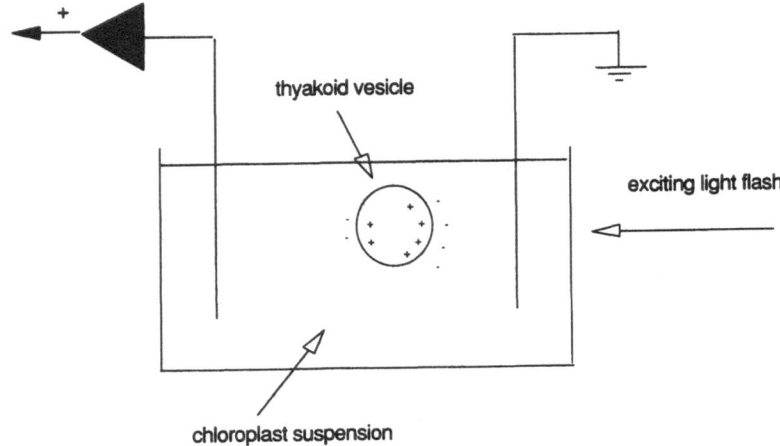

Figure 1. The photovoltage experiment. The polarity corresponds to the simplest case, when the front sheet gets more light than the back one. (This polarity is defined to be the positive one.)

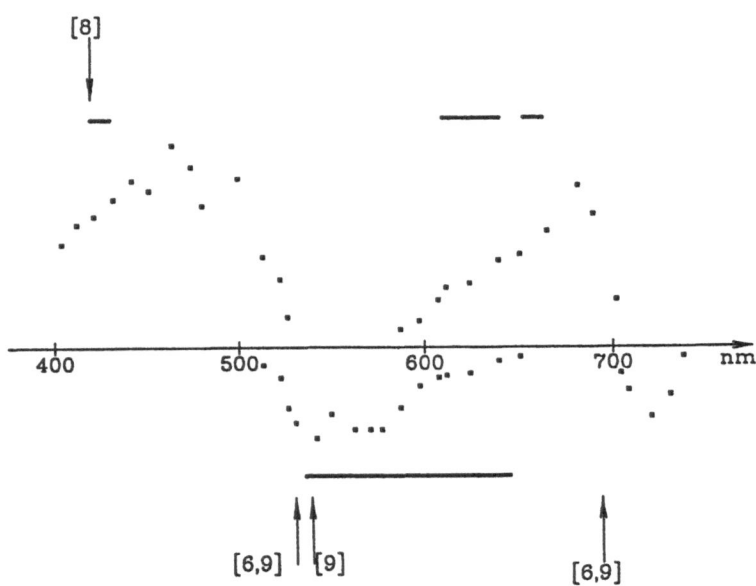

Figure 2. Comparison of the polarities as a function of wavelength. Every item above/below the wavelength scale represents a positive/negative polarity. The discrete points represent an action spectrum measured with flash-lamp[3]. The vertical arrows with serial numbers of references indicate wavelengths where positive or negative laser-generated signals were detected previously. Horizontal bars indicate the polarity of signals observed in the present work.

signals with different polarities in the same set-up using laser excitation. Fig. 2 shows the signals at different wavelengths and a comparison with literature data. It is clear that the laser signal follows the same wavelength dependence as the flash-lamp signal. Roughly speaking, the photovoltage signal is positive within chlorophyll and carotenoid absorption bands and negative outside them. Note the correlation between signal duration and polarity: the time constant of the slowest component is around 50 μsec for the positive signals and 3 μsec for the negative ones. There is a wavelength region where a short negative signal is followed by a long positive one.

3. Theory of Light Distribution

The complex polarizability (e.g. the absorbance and refractive index) of the pigments has a crucial role in determining the interference pattern. In other words, the wavelength-dependent phase of the scattered light (relative to the incident light) determines the interference pattern in the thylakoid system. In classical electrodynamics, interaction of light and matter is described in the following way. The electric field component of the light excites the molecular (classical) oscillators. The oscillating dipoles emit light waves. This emitted light is observed as scattered light from a distance much larger than the size of the scattering piece of material. Inside a continuous medium absorption as well as the refraction are consequences of the interference between the incident and the emitted light. (The refractive index is connected to the real part of the polarizability, while the absorbance is connected to the imaginary part.) Inside an inhomogeneous structure the outcome of this interference is more complicated. This interference (as a 1D pattern) was calculated as a function of the real and the imaginary part of the polarizability for the stroma as well as the grana. The distance between the stroma lamellas is much smaller than the wavelength. According to the calculations, the light intensity difference between the two sheets is dominated by the real component of the polarizability. At the long wavelength side of an absorption maximum it leads to an "inverse" light gradient. (The back sheet gets more light than the front one.) The much thicker grana behave more classically: the absorbance of the thylakoids leads to a "normal" light gradient similar to the prediction of geometrical optics.

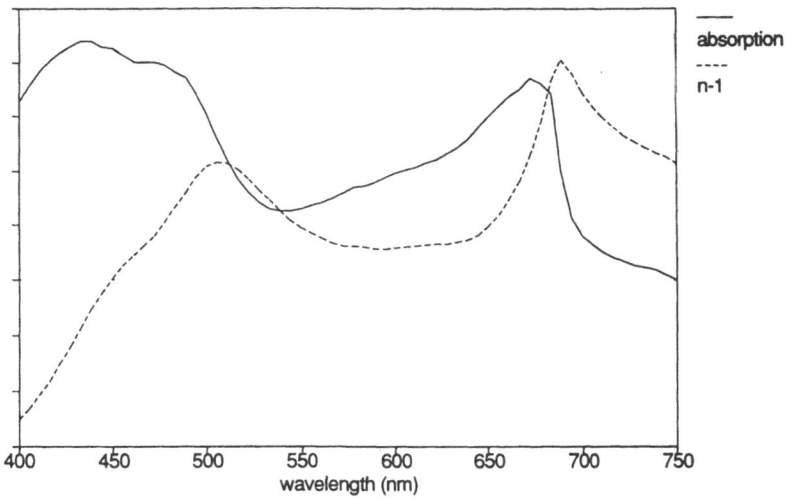

Figure 3. Real and imaginary part of the polarizability as a function of the wavelength (arbitrary units). Imaginary part = empirical absorption curve; real part = refractivity (more exactly: n-1) as calculated from the former one by the Kramers-Kronig relation. (The contribution from the absorption peeks at much higher frequencies was not included.)

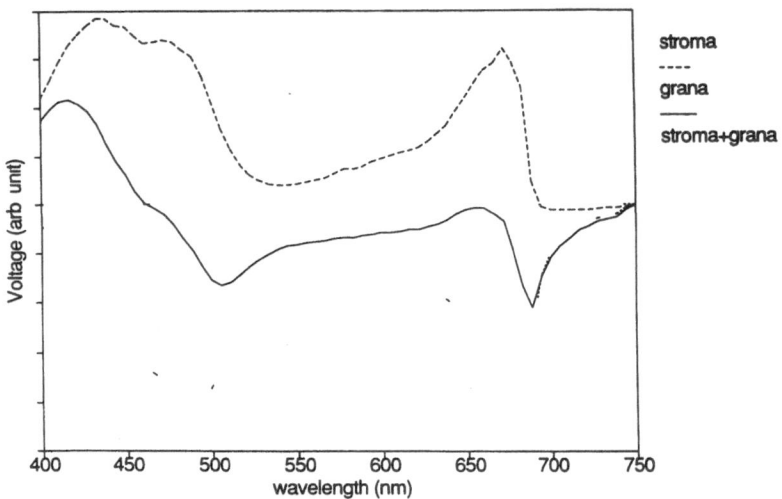

Figure 4. Predicted action spectrum of the photovoltage signal from the stroma as well as the grana based In the stroma plot and grana plot, the distance values on the legends indicate the distance between the sheets and the thickness of the grana stack, respectively

Fig. 3 shows the components of the complex polarizability. The real part (n-1, where n is the refractive index of the thylakoid relative to water) is calculated from the imaginary part (that is from the absorbance) by the Kramers-Kronig relation. Fig. 4 shows the predicted photovoltage from the stroma as well as from the grana based on the polarizability curves.

4. Theory of Kinetics

The time constant of a bulk ionic relaxation in a solution is the following:

$$\tau_{bulk} = \frac{\varepsilon_{sol}}{\sigma} ,$$ (1)

where ε_{sol} is the dielectric constant and σ is the conductivity of the medium.
For a 1 mM KCl solution:

$$\tau_{bulk} \approx 0.16 \ \mu sec .$$ (2)

(It is smaller for larger ionic strength.)
The time constant of the lateral relaxation parallel to a membrane sheet is

$$\tau_{lat} = \frac{L}{d} \cdot \frac{\varepsilon_{membr}}{\varepsilon_{sol}} \cdot \tau_{bulk} ,$$ (3)

where L is the lateral scale, d is the membrane thickness and ε_{membr} is the dielectric constant of the membrane. Using the values ε_{membr}=2, L=700 nm and d=7 nm:

$$\tau_{lat} = 0.4 \ \mu sec$$ (4)

for the same ion concentration.

5. Discussion

Light distribution is strongly different in the stroma and in the grana. The positive signal is expected to come predominantly from the grana, where the absorption causes a 'normal' light gradient. The source of the negative signal is the stroma, where the light back-scattered from the back sheet interferes destructively with the incoming light at the front sheet causing an 'inverse' light gradient. Two experimental facts support this picture:

- In stacked chloroplasts PSII does not contribute to the negative signal at the rubin and the Nd-YAG wavelengths[4], because they are located in the grana.
- There is a kinetic difference between the positive and the negative signal.

The experimental values of the decay time constant are much larger than the calculated ones. The difference is one order of magnitude in the case of the negative signal, and two orders of magnitude in the positive one. This question needs more investigation. Hindered ion relaxation between the appressed membrane sheets or lateral diffusion of membrane proteins are likely candidates for explanation.

6. Summarizing Remarks

The aim of the work presented here is to establish a connection between some aspects of photovoltage measurements and some physiologically relevant properties of the photosynthetic systems.

Photovoltage signals come from a suspension of chloroplasts or photosynthetic bacteria illuminated by a short light flash through a pair of electrodes immersed into the suspension. At the picosecond scale the signal kinetics is well understood. It reflects the exciton trapping as well as the primary charge transfers in the reaction centers.

Measurements in the microsecond scale are presented. Polarity and kinetics of the photovoltage signals depend strongly on the wavelength of the exciting light, despite both of the photocenters transduce a negative charge in the same direction. The dependence of the polarity is explained by wavelength dependence of the interference pattern of light distribution within the chloroplast. Indeed, model calculations of this pattern qualitatively reproduce the observed polarities. We concluded that the positive signal comes from the grana, while the negative one comes from the stroma.

Acknowledgements

Discussions with Győző Garab and Douglas B. Kell were valuable. This work was partly supported by a grant OTKA 2999 and by the Peregrinatio foundation.

References

1. H.T. Witt and A. Zickler, Electrical evidence for the field indicating absorption change in bioenergetic membranes, *FEBS Lett.* **37**:307-310 (1973).
2. H.T. Witt, Energy conversion in the functional membrane of photosynthesis, *Biochim. Biophys. Acta* **505**:355-427 (1979).
3. P. Gräber and H.-W. Trissl, On the rise time and polarity of the photovoltage generated by light gradients in chloroplast suspension, *FEBS Lett.* **123**:95-99 (1981).
4. H.-W. Trissl, U. Kunze and W. Junge, Extremely fast photoelectric signals from suspen-sions of broken chloroplasts and of isolated chromatophores, *Biochim. Biophys. Acta* **682**:364-377 (1982).
5. H.-W. Trissl, J. Breton, J. Deprez and W. Leibl, Primary electrogenic reactions of Photo-system II as probed by the light gradient method, *Biochim. Biophys. Acta* **893**:305-319 (1987).

6. H.-W. Trissl, W. Leibl, J. Deprez, A. Dobek and J. Breton, Trapping and annihilation in the antenna system of Photosystem I, *Biochim. Biophys. Acta* **893**:320-332 (1987).
7. J.F. Becher, N.E. Geacintov and D.E. Swenberg, Photovoltages in suspensions of magnetically oriented chloroplasts, *Biochim. Biophys. Acta* **503**:545-554 (1978).
8. G. Meszéna, E. Papp and G. Fricsovszky, Study of the fast photoelectric signal from a chloroplast suspension, *Stud. biophys.* **126**:77-86 (1988).
9. G. Meszéna and D. DeVault, Investigations of the polarity of the photo-induced electrical signal of chloroplast suspensions, *Photosynth. Res.* **22**:115-122 (1988).

NEURAL NETWORKS APPLIED TO CELL SIGNALING AND CHEMICAL COMMUNICATION IN CELL NETWORKS

J.A. Prideaux[1], J.L. Ware[2],
A.M. Clarke[1] and D.C. Mikulecky[3]

[1]Biomedical Engineering Program
[2]Department of Pathology
[3]Department of Physiology
  Medical College of Virginia Commonwealth University
  Richmond, VA 23298-0551, U.S.A.

1. Introduction

The network approach to complex systems, established and successfully employed in the field of electronics, is becoming increasingly fruitful as an approach to problems of complexity in other fields. The field of *Network Thermodynamics* was created in order to approach a wide variety of systems, including living systems, with the same power as was demonstrated in electronic systems[1-3].

The essentials of the network approach are the use of a systematic combination of conservation laws and topological information about the connectedness of the system, along with constitutive laws, to achieve an analysis of very complicated, heterogeneous *linear* and *nonlinear* systems. Interconnected structures in living systems are ubiquitous. Thus, in a sense, everything can be viewed as a network. A widely recognized manifestation of this is the area of neural networks.

Secondly, and more recently, immune networks have been studied to a significant extent[4,5]. Reaction networks and reaction-diffusion networks have been studied in depth as well[2]. In fact, the extent to which chemical communication among cells, which has an analogous form to networks, exists is far more widespread than these more widely acknowledged examples. These reaction networks are so clearly different from the neural networks which now command considerable attention, that a serious comparison seems almost frivolous. However, in his work on complex systems, Rosen[6-9] points out some similarities which deserve further attention. If we compare metabolic networks with neural networks from Rosen's point of view, a number of striking parallels emerge. Among these is the mathematical form of their description, especially as signals are modulated by various levels of activators and inhibitors. The consequences of this mathematical structure are profound and are what is explored in a preliminary way in this work.

To examine the similarities between metabolic and other chemical networks and neural networks in a systematic manner, network thermodynamics might be used as a tool for the formal description of both. Network thermodynamics is a field based on the dual nature of living systems, namely that as *systems* they are as much a function of their topology as the more readily recognized parameters: rate constants, volumes, pressures, electric potentials, etc.[2,3] These parametric properties are captured by the network elements which are "connected together" in a representation of the system. The topology is more subtle, since it is the pattern of those connections itself. It is, of course, the topology which is most frequently gained using reductionist methods.

One goal of network thermodynamics is to salvage the information gained using reductionist methods and synthesize some holistic composite of the parts. In the simplest interpretation of neural networks, the nodes are analogous to the cell bodies of the neurons and the branches represent the axons and dendrites making the connections between the cells[10-13]. All the axons do is to transmit an all-or-none signal to the synapse and there, in fact, the signal becomes biochemical once more. Although the synaptic event is rich in its detail, artificial neural networks tend to adopt a limited number of stereotypical transfer functions to model these events. What is preserved is a kind of temporal summation which becomes the central pert of the network's function. If the artificial neural network is at all representative of the real thing, then the neural network has a constantly changing set of constitutive relations.

Once we examine the properties of a biochemical network in some more detail, it is clear that the constitutive relations are also variable! These variations are in the form of inhibition and/or activation patterns[6-9]. In fact the example of folate metabolism given above is rich in such possibilities. Rosen focuses on the patterns of inhibition/activation and further modifications of these patterns by agonists/antagonists to establish a very important mathematical feature that both types of system have in common. This feature is the demonstration that in order for these systems to be characterized by the usual notion of states, these patterns of modulation must be reciprocal in their manifestation, a condition which is probably never met. This idea alone is enough to make Rosen sure that the classical Newtonian paradigm does not rigorously hold for these systems. This same idea was recognized by Mason[14] for the electronic systems he models with unistors. Network thermodynamic models of both biochemical and neural networks are consistent with this important attribute.

Biochemical events are becoming more widely recognized as information transmitting processes in their own right, outside their role in synaptic transmission within the nervous system. Modern molecular biology has identified "second messengers" and "signal transduction" mechanisms that extend the traditional role of hormones as information carriers to a more local level. With these ideas as a foundation, it seems feasible to generate models of chemical signaling between cells which are as realistic as necessary to begin to see what role "organization" has in cellular systems. *In fact, as a first generation set of models, it is probably important that biochemical kinetics and other details be suppressed so that the role of topology can be more clearly demonstrated.*

2. An Example: Cell Signaling in a Tumor Cell Population

Tumors are generally phenotypically heterogeneous populations of cells. Tumor cells are able to communicate eith each other by chemical signals. Some tumor cells may secrete and respond to their own growth factors via specific membrane receptors, the *autocrine hypothesis*[15]. The response to these factors and to those secreted by other cells, is often an acceleration of the mitotic rate.

Cells also secrete factors which inhibit growth and these factors also act through membrane receptors. Often, apparently stable tumor populations seem to be destabilized by chemical or surgical deletion of subpopulations of cells. This simple, abstract model will attempt to reproduce many of these observations with a minimum of molecular detail and specificity. In particular, the cells will be genotypically equivalent and no mutations will be considered. This is clearly simplification to the extreme, but deliberately so, in order to explore the effect of simple signals and topology in a cellular network.

2.1. The model

A hypothetical growing population of cells is studied as an information system. The basic pattern of a Hopfield neural network is used with some significant modifications - the most significant being that cells can divide and die. This model was programmed using the object oriented C++ language on an Intel based 486 personal computer. A RAM disk was utilized to store the interconnection weights between the cells because of storage limitations in the 640 K DOS partition. Simulations containing up to 500 cells were possible with only 1 MB of extended memory. The model resembles a Hopfield network in the following ways:

1. Each cell is connected by a weight to every other cell including itself.
2. The output of a cell is a function of the incoming signals.
3. Cells are randomly selected to be updated.

The model differs from a Hopfield network in the following ways:

1. All weights are positive. The outputs can range from -0.5 to 0.5.
2. There are two outputs from each cell. One is stimulatory to growth, the other inhibitory. These can be viewed as two coupled Hopfield nets operating with the same weights and topology.
3. Each cell has a receptor for the stimulatory input, and a receptor for the inhibitory inputs. The receptors are sensitive to the signaling compounds in the vicinity (the input) and modify their affinity for these signal substances accordingly.
4. Random variations in the calculated outputs are constantly introduced. This signal variation is a small percentage of the output, but significant.
5. Cells divide and die as a function of the stimuli received and the receptor values.
6. Runaway growth or extinction of cells can occur if certain properties get out of balance.
7. The weights of the connections between cells are designed to approximate the diffusion distance between cells and its alteration due to tumor growth in three dimensions.

Each cell is only definable as a part of the whole population. As each cell is randomly selected and updated, its new output conditions are determined based on the incoming signals and ist receptor values. The topology of the network dynamically changes with every cell division or cell death. These decisions occur during each cellular update. Rules of cellular survival were chosen based on both known biological considerations and certain simplifying assumptions. The receptors are positively regulated for the concentrations of their respective ligand (input to the cell). The cell will divide, (halving the receptor values) when the receptor values get above a certain threshold. A cell will die when certain variables are "out of bouns". For illustrating the behavior of a chemically connected population of stereotypic cells a limited number of properties are available to group them into categories which can then be studied for their interactions. One choice might be to monito the "state" of cell receptors at each sampling time[16]. In the simulations that follow, the cells are categorized into six subpopulations ("phenotypes") according to their secretion rates - low to high.

2.2. Results

Two types of simulations were conducted. The first of these determined the population behaviors which developed from assignment of different random number seeds. The second evaluated the impact of deletion of one of the six subpopulations and the effect of the timing of that deletion. The random number seed influences the random numbers generated. Randomness is used in the selection of the cellular updates and to break symmetry in the cellular outputs.

The cells are partitioned into six different categories based on the secretion rate of the ligand positively affecting the metabolic rate. Subpopulation one has the lowest secretion rate, subpopulation six has the highest. The six subpopulations were chosen for graphical convenience.

2.2.1. *Growth patterns.* The population is grown from a single cell which immediately divides, adjusting the network topology. After a few divisions, the cells will stabilize and stop dividing. Subsequent behavior of the subpopulations differed with different random number sequences used throughout the simulation. Small amounts of randomness added to the secretion rates greatly affected the distribution of cells among the phenotypes.

Simulations were run using 100 different random number seeds. For each simulation, the program was run for up to 100,000 updates. Three basic growth patterns were observed:

(1) In 15% of the different simulations, the population grew to a certain size and stabilized;
(2) In 30% of the simulations the population remained stable, then suddenly began to grow rapidly until the program's capacity was exceeded.This would be analogous to overgrowth of the host by the tumor; and
(3) In 55% of the simulations the population remained relatively constant, followed by extinction of the whole population.

This would be analogous to the destruction of the tumor. These distributions are identical with those obtained in a similar study where receptor state was the basis for identifying the subpopulations[17].

2.2.2. *Impact of subpopulation deletion.* Both the nature of the subpopulation deletion and the timing of deletion affected tumor population dynamics. Elimination of subpopulation 2 at the indicated time point (late in the program) had only a minor impact on the population as compared with the control. In striking contrast, deletion of subpopulation 2 *earlier* in the simulation dramatically altered the dynamics of the population, sharply reduced subpopulation diversity, and resulted in a different outcome, i.e., "death" of the tumor prior to the end of the program. Thus both quantitative and qualitative changes occurred.

2.2.3. *Effect of connectivity.* Chemical connectivity among the cells was a necessary condition for demonstration of the complex behaviors described. Simulations were run using an altered version of the program in which the interconnections were removed. In each "unconnected" simulation the population parameters remained near an average value. Neither rapid growth nor extinction was observed.

3. Some Ideas Arising from the Model

The model depends on randomness being introduced in the initial conditions as well as at each updating of the cell's state. Neural networks have demonstrated this need as well. Two questions arise once this need is recognized. First, is it realistic? Is the result being supplied artifically by introducing this randomization? Second, if it is realistic, what is its counterpart or its source in an actual biological system? Reaction kinetics are diverse in their nature, but this diversity has certain patterns.

Two very distinct classes of kinetic patterns are those in which the reaction rate is some monotonically increasing function of substrate concentration and those in which it is not. If reaction networks themselves may be the source of a chaotic type of dynamics they can then "supply" a natural form of pseudo-randomness to the system. Thus, to the extent that our results reflect a kind of chaotic dynamics, it appears that chaos may actually feed on chaos in these systems. The same is true for the possibility of the randomness in artificial neural networks really being a substitute for chaotic dynamics in real neural networks.

4. Conclusions

These preliminary studies simply make realizable the idea hypothesized at the onset of these studies, namely that in networks of cell communications, the whole is generally more than, and different from, a simple sum of its parts. It may behoove us to consider diseases such as cancer in this light. Furthermore, the normal behavior of cells in tissues may be stabilized by similar system properties. The study of cellular networks should be a fruitful one.

5. References

1. P. Cruziat and R. Thomas, SPICE - a circuit simulation program for physiologists, *Agronomie* 8:613-623 (1988).
2. D.C. Mikulecky. "Applications of Network Thermodynamics to Problems in Biomedical Engineering," New York University Press, New York (1992).
3. L. Peusner. "Studies in Network Thermodynamics," Elsevier, Amsterdam (1986).
4. N.K. Jerne, Towards a network theory of the immune system, *Ann. Immunol. (Inst. Pasteur)* 125C:373-389.
5. A.S. Perelson, Toward a realistic model of the immune system, *in*: "Theoretical Immunology," A.S. Perelson, ed., Addison-Wesley, Redwood City (CA), pp. 377-401 (1988).
6. R. Rosen, Information and complexity, *Canad. Bull. Fish. Aquat. Sci.* 213:221-233 (1985).
7. R. Rosen, Organisms as casual systems which are not mechanisms: An essay into the nature of complexity, *in*: "Theoretical Biology and Complexity," R. Rosen, ed., Academic Press, New York (1985).
8. R. Rosen. "Anticipatory Systems," Pergamon, London (1985).
9. R. Rosen. "Life Itself," Columbia University Press, New York (1991).
10. W.S. McCulloch and W.H. Pitts, A logical calculus of the ideas immanent in nervous activity, *Bull. Math. Biophys.* 5:115-133 (1943).
11. J.J. Hopfield, Neural networks and physical systems with emergent collective computational abilities, *Proc. Nat. Acad. Sci. U.S.A.* 79:2554-2558.
12. D.E. Rumelhart, J.L. McClelland and the PDP Research Group. "Explorations in the Microstructure of Cognition," MIT Press, Cambridge (MA) (1988).
S. Grossberg (ed.). "Neural Networks and Natural Intelligence," MIT Press, Cambridge (MA) (1988).
14. S.J. Mason and H.J. Zimmermann. "Electronic Circuits, Signals, and Systems," Wiley, New York (1960).
15. M.B. Sporn and A.B. Roberts, Autocrine growth factors and cancer, *Nature* 313:745-747 (1988).
16. J.A. Prideaux, A systems approach to modelling and simulating the interactions between cells in a solid tumor, Master's Thesis, Virginia Commonwealth University, Richmond (VA) (1992).
17. J.A. Prideaux, J.L. Ware, A.M. Clarke and D.C. Mikulecky, From neural networks to cell signalling: Chemical communications among cell populations, *Proc. 14th Ann. Conf. IEEE Eng. Med. Biol. Soc.* (in press).

SYNCHRONIZATION OF
GLYCOLYTIC OSCILLATIONS
IN INTACT YEAST CELLS

Peter Richard[1], Bas Teusink[1],
Hans V. Westerhoff[1,2] and Karel van Dam[1]

[1]E.C. Slater Institute
 University of Amsterdam
 Plantage Muidergracht 12, 1018 TV Amsterdam, The Netherlands
[2]Division of Molecular Biology
 Netherlands Cancer Institute
 Plesmanlaan 121, 1066 CX Amsterdam, The Netherlands

1. Introduction

In the glycolytic pathway, oscillations in metabolite concentrations occur in different organisms under different conditions[1]. In intact yeast cells, oscillations in glycolysis can be monitored by NAD(P)H fluorescence. For the induction of these oscillations, glucose is added to starved cells. After a steady state is reached, blocking respiration leads to oscillations in the NAD(P)H concentration with a frequency of 1.2 min[-1] at a temperature of 25 °C. The damping of the oscillations depends on the state of growth before harvesting. Sustained oscillations can be observed when the cells are harvested at the transition from using glucose to ethanol as a substrate[2]. Furthermore there is evidence that the damping is also influenced by the cell concentration during the NAD(P)H measurement as first recognized by Pye[3]. This was interpreted as a consequence of intercellular communication[4]. In this paper we investigate the influence of the cell concentration on the duration of the oscillations and collect further evidence for intercellular communication.

2. Material and Methods

The yeast *Saccharomyces cerevisiae* (X2180 diploid strain) was grown under semi-aerobic conditions at 30 °C on a rotary shaker in a medium containing 10 g/l glucose, 6.7 g/l yeast nitrogen base (Bacto) and 100 mM potassium phtalate pH 5.0. The cells were harvested around the transition from using glucose to ethanol as a growth substrate. They were harvested by filtration (Schleicher & Schuell 1.2 µm), washed twice with potassium phosphate pH 6.8, resuspended and starved in the same buffer for 3 hours at 30°C on a

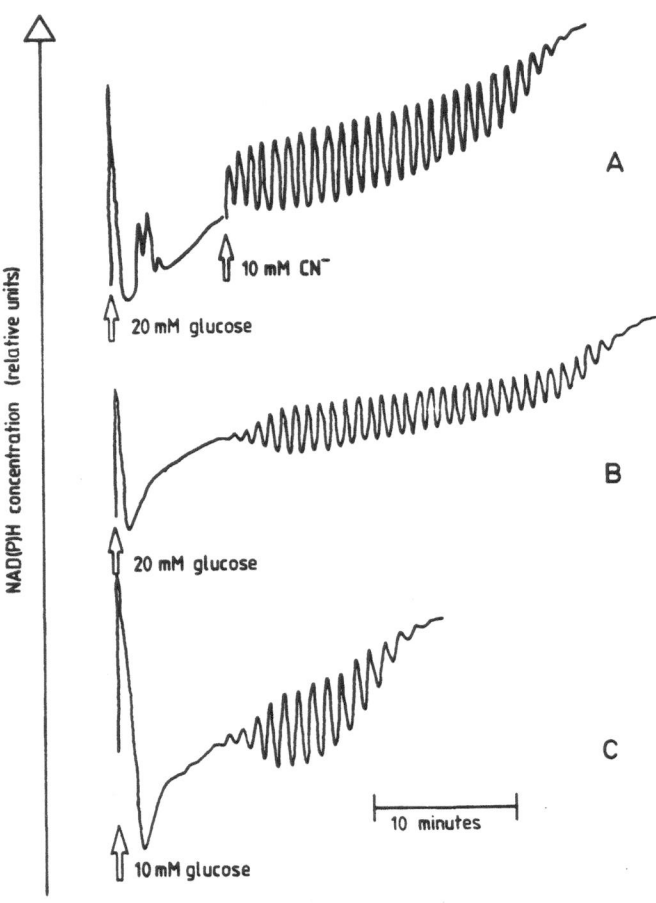

Figure 1. NAD(P)H fluorescence traces of yeast cells. At the points indicated glucose and cyanide were added. Trace A shows a representative experiment with starved cells. Trace B and C show experiments performed on cells in which oscillations have been induced, as in trace A and subsequently exhausted.

rotary shaker. They were washed by filtration and stored on ice until use. Protein was measured according to the method of Lowry and coworkers[5]. Glucose was measured with a commercial glucose kit (Sigma). NAD(P)H fluorescence was measured in an Eppendorf 1011M fluorimeter (excitation 316 nm, emission 400-2000 nm) in a stirred and thermostated (25 °C) cuvette. Oscillations were induced by adding 20 mM glucose to the starved cells and, after 4 minutes, adding 10 mM KCN.

3. Results and Discussion

Transient NAD(P)H oscillations were induced in the yeast cells by first adding 20 mM glucose. After a steady state had been established after about 4 minutes, stopping respiration by adding 10 mM KCN immediately led to oscillations in NAD(P)H concentration. Fig. 1A shows a fluorescence trace of a representative experiment at a cell concentration of 6.25 g protein per liter.

The duration of the oscillations was measured as a function of the cell concentration. To this end the number of cycles was counted after inducing the oscillations by addition of cyanide. Only cycles were counted with an amplitude of at least 10% of the initial amplitude. We varied the cell concentration under the conditions where we measured the NAD(P)H concentration. Cell concentrations between 0.4 and 6.4 g protein per liter (which corresponds to $60 \cdot 10^9$ to $900 \cdot 10^9$ cells per liter, respectively) were used. The result is presented in Fig. 2.

In the range from approximately 2 g/l to 7 g/l the number of oscillations decreased with the cell concentration. The glucose uptake rate was approximately 0.2 U/mg of total protein, so that with a protein concentration of 2 g/l 20 mM glucose should have been exhausted after about 50 minutes or 60 cycles. A doubling of the cell concentration will lead to a doubling in the glucose consumption rate.

Fig. 2 shows the corresponding behavior for cell concentrations above 2 g protein per liter. At a cell concentration of 3.2 g protein per liter 60 cycles, and with a protein concentration of 6.4 g/l, 30 cycles were observed. We have also measured the glucose concentrations during the oscillations and found glucose exhaustion to coincide with the end of the oscillations for cell concentrations above 2 g protein per liter. For lower cell concentrations the situation is different. Here the glucose was not exhausted and an increasing cell concentration led to an increasing number of cycles (Fig. 2). With 0.4 g protein per liter 4 cycles were observed. The number of cycles increased to 70 at 1.6 g protein per liter. Under these conditions glucose was not exhausted at the end of the oscillations.

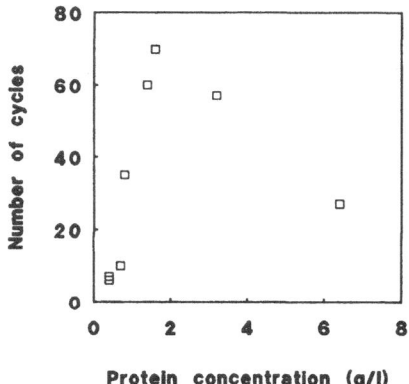

Figure 2. Duration of the oscillations, indicated as the number of cycles, as a function of the cell concentration.

If cells were oscillating as individuals, differences in frequency would lead to a damping of the total signal. Such a process would be independent of the cell concentration. We did observe a protein-concentration dependent damping which points towards a synchronization. This is in accordance with earlier results of Pye[3] and Ghosh and coworkers[6], who reported synchronization phenomena observed with cells which were oscillating out of phase and synchronized after mixing. A synchronization mechanism where the sender secretes the signaling compound and the receiver absorbs it must have a threshold concentration for the signaling compound. Fig. 2 suggests a steep increase in the number of cycles around a cell concentration of 1 g protein per liter. An interpretation of this finding is that the threshold concentration of the signaling compound is reached at a cell concentration of around 1 g protein per liter. Below this protein concentration the macroscopic oscillations may die out because the individual cells run out of phase. Above this cell concentration, due to active synchronization, sustained oscillations are observed.

To check whether the end of oscillations is only due to glucose exhaustion and not to a threshold concentration of products such as ethanol, we added additional glucose to the cells after they had stopped oscillating, as shown in Fig. 1. Trace A in Fig. 1 shows the oscillations induced in starved cells (the arrows show the time of glucose and cyanide additon). Trace B in the Figure shows the response to addition of 20 mM glucose to cells which have stopped oscillating because of glucose exhaustion. Glucose was added right at the end of trace A. During the first 3 to 4 minutes after glucose addition a non-oscillating NAD(P)H signal was observed which changed later into an oscillating signal with quite similar amplitude, frequency and duration as compared to the initial starved cells. Addition of 10 mM rather than 20 mM glucose under the same conditions as in trace B led to the same response exept that the duration was half as long (trace C). The oscillations in trace B and C were not triggered with KCN. They were self-induced and exhibited self-amplification.

Acknowledgements

This work was supported by the Commission of the European Communities and the Netherlands Organization for Scientific Research (NWO).

References

1. B. Hess and A. Boiteux, Oscillatory phenomena in biochemistry, *Ann. Rev. Biochem.* **40**:237-258 (1971).
2. P. Richard, B. Teusink, H.V. Westerhoff and K. van Dam, Around the growth phase transition *S. cerevisiae*'s make-up favours sustained oscillations of intracellular metabolites, *FEBS Lett.* **318**:80-82 (1993).
3. E.K. Pye, Biochemical mechanisms underlying the metabolic oscillations in yeast, *Can. J. Bot.* **47**:271-285 (1969).
4. M.A. Aon, S. Cortassa, H.V. Westerhoff and K. van Dam, Synchrony and mutual stimulation of yeast cells during fast glycolytic oscillations, *J. Gen. Microbiol.* **138**:2219-2227 (1992).
5. O.H. Lowry, N.J. Rosebrough, A.L. Farr and R.J. Randall, Protein measurement with the Folin phenol reagent, *J. Biol. Chem.* **193**:265-275 (1951).
6. A.K. Ghosh, B. Chance and E.K. Pye, Metabolic coupling and synchronization of NADH oscillations in yeast cell populations, *Arch. Biochem. Biophys.* **145**:319-331 (1971).

CONTROL IN ELECTROPHYSIOLOGY
AND INTRACELLULAR SIGNAL TRANSDUCTION

Harald van Mil and Jan Siegenbeek van Heukelom

Dept. of Experimental Zoology
Sect. Cell Biophysics
University of Amsterdam
Kruislaan 320, 1098 SM Amsterdam, The Netherlands

1. Introduction

Any organism must be able to adapt itself to environmental changes and to react to challenges. These adaptations are brought about by changing the activities of the organs that are functions of lower hierarchic organizations on the cellular and metabolic level. For an optimal response of the organism the different organs must be orchestrated. This orchestration of the different changes in cellular activity can be co-ordinated by signal transducers like hormones.

A well-known example is the "Fright, Flight or Fight reaction" (FFF), which is induced during stress and excitement, and starts by neural activation of the adrenal medulla. This gland then secrets a first messenger, epinephrine, into the blood. Binding of epinephrine to the different target tissues results in a variety of responses like increase of muscle power development, cardiac activity and blood pressure. But also the redistribution of blood towards the working muscles, arousal of the central nervous system and mobilization of large energy stores.

All of these responses reflect the induced change on the cellular level. A number of changes is induced by the binding of epinephrine to the β-receptor which will initiate an amplification pathway. This amplification pathway consists of stimulatory G-proteins, adenylate cyclase, which catalyzes the formation of the second messenger cAMP from ATP. cAMP in turn activates protein kinase A (cPKA). In the cell cPKA can phosphorylate all kinds of target proteins that contain the amino acid sequence Arg-Arg-X-Ser (Thr)[1]. Membrane channels and pumps can also be target proteins of the cascade by direct or indirect phosphorylation.† Dephosphorylation of the proteins occurs by a related phosphatase (PHOS). The cPKA/PHOS ratio determines the phosphorylation level of the different target protein pools and changes in this ratio most likely induce the changes in phosphorylation level.

---

†In addition, direct influence of G-proteins on membrane channels have been reported[2].

Modern Trends in Biothermokinetics, Edited by
S. Schuster *et al.*, Plenum Press, New York, 1993

This cascade will change the physiology and activity of the cell. Beside the β-adrenergic receptor, epinephrine stimulates also the pathway of the α-adrenergic receptor although binding occurs with a much lower affinity. These and other effects make the detailed analysis the action of epinephrine on the system very complex.

In many fields of science, sensitivity analysis is applied to get a more quantitative view of the object under study. In the case of biochemical systems, Biochemical systems analysis[3] and Metabolic control theory (MCT)[4] provide a mathematical formalism based on sensitivity analysis, and a number of theorems have been derived to describe the organization of the biochemical kinetic structure of interest.

Our interest is whether similar applications of sensitivity analyses would be effective for electrophysiological systems. This formalism could provide a quantitative topology for comparison of the different steady states of our system.

2. Metabolic Control Theory and Protein Phosphorylation

The conductance changes, as measured as I_i at the same V_m, induced by phosphorylation of membrane channels can occur due to a variety of processes as expressed in the following:

$$I_i = N_t * P_f * P_o * i_{channel} \tag{1}$$

with $i_{channel}$ = total current through single type of channel, N_t = total amount of single type of channel, P_f = part of the N_t that can be opened*, and P_o = probability of an open channel.

There is no *a priori* reason to exclude changes in any of these parameters. In patch clamp experiments phosphorylation of the calcium channels results in an increase of I_i due to increase of P_f and P_o[5]. Phosphorylation of potassium channels by cPKA has been reported in vitro for the delayed potassium rectifier suggesting a change in N_t and the rectification properties[6]. cPKA can also control the active ion translocation across the membrane through phosphorylation of the Na/K-pump, and this has been frequently been identified to be the site where cPKA operates[7].

In the musculus lumbricalis of mice a specific β-adrenergic agonist gives rise to a hyperpolarization of the membrane potential (V_m), whereas in the soleus muscle smaller hyperpolarizations or even depolarizations were observed[7]. These observations point to a different cellular organisation of the membrane constituents, that generate V_m, or related cPKA target proteins in these two types of muscle fibres. Understanding these differences in responses could give a better perception of the action of hormones on electrophysiological variables.

3. Elements of Electrophysiology

These differences in cellular organisation will be reflected in the topology of the matrix that describes the relation of the system in a quantitative manner. One of the first, most frequently used equations for a quantitative description of V_m in electrophysiology is the Goldman-Hodgkin-Katz equation. Although of great importance for this field it can not take the membrane dynamics into account.[8]

$$V_m = \frac{RT}{F} \ln \frac{P_K[K]_o + P_{Na}[Na]_o + P_{Cl}[Cl]_i}{P_K[K]_i + P_{Na}[Na]_i + P_{Cl}[Cl]_o} . \tag{2}$$

---

*Channels can have different closed and open states. For example, in the imaginative simple scheme, $C_1 \longrightarrow C_2 \longrightarrow O$ (where "C" represents a closed and "O" an open conformation) the open conformation "O" can only be reached form C_2.

Though in many cases eq. 2 produces clear and useful results, a number of pitfalls makes it less attractive as soon as one wants to understand the underlying mechanism for the generation of V_m:

1. the permeabilities (P_X) for influx and efflux of an ion (X) are considered to be the same and constant whereas a number of membrane ion channels demonstrate varying conductance that might differ for outward current and for inward current.
2. it does not contain variables to describe the contributions of different active transport-systems in the membrane such as the Na/K-pump or the Na-glucose cotransporter.
3. the solution of the equation is not unique; and the additional condition that the osmolarity in the cell must equal the osmolarity of the medium, is not sufficient to that end.
4. for the steady states of the cell with or without epinephrine this relation cannot give any information on the amount and type of the change that is initiated by the amplification pathway, because several ion channels carrying the same ion can be involved.

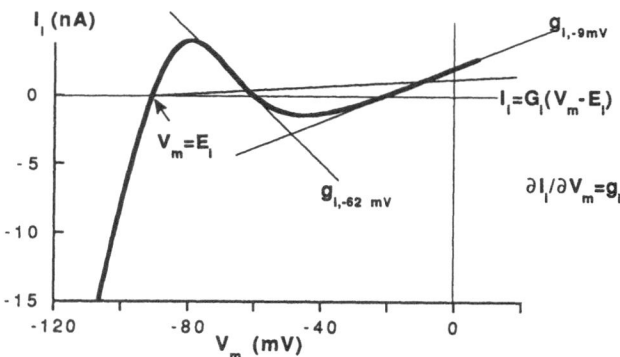

Figure 1. In a highly non-linear, though physiologically realistic I-V characteristics the difference between slope conductance and chord conductance is illustrated. The slope of the line $I_i = G_i (V_m - E_i)$ is the chord conductance (G_i) and is the same for both points: $V_m = -62$ mV and $V_m = -9$ mV of the curve. The slope conductances ($\partial I_i / \partial V_m = g_i$) in both points differ considerably.

Nevertheless, voltage clamp experiments can produce suitable results to be introduced in descriptions common to MCT. Voltage clamp is possible because, by electronic feedback, the potential in a cell is kept constant: the current needed is measured as a function of V_m. In Fig. 1 the result that such a measurement might produce is shown. It is clear from this curve that frequently I_i is a unique function of V_m, whereas V_m is not a unique function of I_i. Contrary to this type of experimentally measured relationships, in the unclamped cell V_m is the resultant of the balancing of all transmembrane currents. Translation from the voltage clamp results into the V_m-$\{\Sigma I_i\}$ relationship of an unclamped cell is as classical as the first voltage clamp results that have been published.[9]

As can be seen from Fig. 1, two types of conductance can be used: *chord conductance* (G_i) and *slope conductance* (g_i). The definition of G_i is written as:

$$I_i = G_i (V_m - E_i) \qquad (3)$$

with E_i denoting the reversal potential of ion species i.

Here G_i is the conductance in a linear, thus Ohmic sense, and is the reciprocal value of the membrane resistance. For the definition of the slope conductance we write:

$$g_i = \frac{\partial I_i}{\partial V_m}, \qquad (4)$$

where g_i describes the slope in the I_i/V_m curve in the point (V_m, I_i). g_i can also be expressed in terms of $V_m - E_i$:

$$g_i = \frac{\partial I_i}{\partial V_m} \frac{\partial (V_m - E_i)}{\partial (V_m - E_i)} = \frac{\partial I_i}{\partial (V_m - E_i)} \frac{\partial (V_m - E_i)}{\partial (V_m)}.$$

By the definition $\frac{\partial V_m}{\partial V_m} = 1$ and $\frac{\partial E_i}{\partial V_m} = 0$ we arrive at:

$$g_i = \frac{\partial I_i}{\partial (V_m - E_i)}. \qquad (5)$$

In two cases, we have $g_i = G_i$: if the relationship between I_i and V_m is Ohmic (linear) or if $V_m = E_i$.

In electrophysiology this linearity is seldom observed because of the rectifying properties of the channels in the membrane through which the I_i/V_m relation becomes non-linear (see Fig. 1).

4. Metabolic Control Theory and Electrophysiology

One of the coefficients used in MCT, the "elasticity coefficient (ε)", is a measure for the responsiveness of an enzyme, or in our case a channel or pump, to an imposed perturbation. The definition of ε, the small relative change in rate induced by a small relative change in a variable or parameter, can be related to electrophysiological terms. For channels and pumps this could be the current through the membrane carried by the ions. Therefore we propose the following expression of ε in electrophysiological terms.

$$\varepsilon_{(V_m - E_i)}^{I_i} = \lim_{(V_m - E_i) \to 0} \frac{\partial I_i}{\partial (V_m - E_i)} \frac{V_m - E_i}{I_i}. \qquad (6)$$

If we combine eqs 3 and 5 with eq. 6 we find:

$$\varepsilon_{(V_m - E_i)}^{I_i} = \frac{g_i}{G_i}. \qquad (7)$$

So $\varepsilon_{(V_m - E_i)}^{I_i}$ is a function of the slope and chord conductances which can directly be obtained by voltage clamp studies. Here the elasticity can have any different values:

- $\varepsilon^{I_i}_{(V_m-E_i)} = 1$ or -1, if there is a linear (Ohmic) relationship ($g_i = G_i$) between I_i and V_m or the channels do not possess any rectifying properties, or the rectifying properties of the different channels are canceling one another.

- $\varepsilon^{I_i}_{(V_m-E_i)} = 0$. For instance, $V_m-E_{Cl} = 0$, because there is no active transport of chloride.

- $\varepsilon^{I_i}_{(V_m-E_i)} \neq 1$ or 0, if conductances are non-linear and ions can be actively transported through the membrane, by the Na/K-pump. The elasticity may be unequal to zero or unity over a wide range of V_m. Here the elasticity can by subdivided into two different outcomes:

- $-1 < \varepsilon^{I_i}_{(V_m-E_i)} < 1$: $g_i < G_i$; the slope of the I_i/V_m curve is smaller then the Ohmic conductance: inward rectification in the range between $V_m > E_i$.

- The absolute value of $\varepsilon^{I_i}_{(V_m-E_i)}$ is greater than unity; $|g_i| > |G_i|$ and the slope of the I_i/V_m curve is negative (see Fig. 1: at $V_m = -62$ mV) or positive ($V_m = -9$ mV).

5. Conclusions

Here we derived a definition for the elasticity coefficient for ion specific conductances. The variables used in these elasticity coefficients can be measured directly using the voltage clamp procedure which is a commonly used technique in electrophysiology[2]. The elasticities can be calculated from the I_i/V_m curves of these experiments like in Fig. 1.

The questions arises as to whether this elasticity is equivalent to the one used in MCT. Whereas the elasticities in MCT are usually derivatives with respect to concentrations, which obey mass balance equations, in our treatment the elasticity coefficient is a derivative with respect to membrane potential, V_m, which is not subject to balance equations. This means that the connectivity theorem will get a different interpretation as in MCT.

We must ask ourselves whether we can still describe electrophysiological systems in terms of, the now classical, MCT. Before we look for further analogies with MCT we must further investigate this fundamental question. Can an electrophysiologically oriented control theory be built on the same framework used in MCT or must we derive new mathematical theorems for its application? Another question is whether the control coefficients will have the same meaning as in MCT[10].

For a suitable linkage, these problems in interpreting electrophysiological data in terms of MCT should be the subject of further investigations.

References

1. R.B. Pearson and B.E. Kemp, Protein kinase phosphorylation site sequences and consensus specificity motifs: Tabulations, *Meth. Enzym.* **201**:62-81 (1991).
2. B. Hille, "Ionic Channels of Excitable Membranes," 2nd ed., Sinauer Associates Inc., Sunderland (Mass.) (1992).
3. M.A. Savageau, E.O. Voit and D.H. Irvine, Biochemical systems theory and metabolic control theory, *Math. Biosci.* **86**:127-169 (1987).
4. H. V. Westerhoff and K. van Dam, "Thermodynamics and Control of Biological Free-energy Transduction," Elsevier, Amsterdam (1987).
5. R. W. Tsien, B. P. Bean, P. Hess, J. B. Lansman, B. Nilius and M. C. Nowycky, Mechanisms of calcium channel modulation by b-adrenergic agents and dihydropyridine calcium agonists, *J. Mol. Cell. Cardiol.* **18**:691-710 (1986).
6. P. B. Bennet, R. Kass and T. Begenisich, Nonstationary fluctuation analysis of the delayed rectifier channel in cardiac Purkinje fibers: action of epinephrine on single channel current, *Biophys. J.* **55**:731-738 (1991).

7. T. Clausen and J. A. Flatman, The effect of catecholamines on Na-K transport and membrane potential in rat soleus muscle, *J. Physiol.* **270**:383-414 (1977).

8. P. Läuger. "Electrogenic Ion Pumps", Sinauer Associates Inc., Sunderland (Mass.) (1992).

9. A. L. Hodgkin and A. F. Huxley, A quantitative description of the membrane current and its application to conduction and excitation in nerves, *J. Physiol.* **117**:500-544 (1952).

10. J. Siegenbeek van Heukelom, Control coefficients in cellular cation homeostasis: a model analysis, this volume.

STRUCTURE-FUNCTION RELATIONSHIP
OF MAGAININS.
STUDIES IN LIPOSOMAL MODEL SYSTEMS

A. Vaz Gomes[1], A. de Waal[1],
J.A. Berden[2] and H.V. Westerhoff[1,2]

[1]The Netherlands Cancer Institute
 H5, Plesmanlaan 121, NL-1066 CX Amsterdam
[2]E.C. Slater Institute for Biochemical Research-UvA,
 University of Amsterdam
 Plantage Muidergracht 12, 1018 TV Amsterdam, The Netherlands

1. Introduction

Magainins are 23 or 21 residues-long peptides, initially detected in the skin of *Xenopus laevis* [2]:

Magainin 2: GIGKFLHSAKKFGKAFVGEIMNS-CONH$_2$,
PGLa: GMASKAGAIAGKIAKVALKAL-CONH$_2$.

They possess broad spectrum antibiotic activity (towards bacteria, fungi, protozoa) but no hemolytic activity[2]. Studies aimed at describing the molecular origin of such cytolytic activity, in model systems carrying a transmembrane electric potential, negative inside, such as rat liver mitochondria, cytochrome oxidase reconstituted into liposomes[3,4] or hamster spermatozoa[5], showed that they stimulate respiration, and also inhibit sperm motility. Impairing of free-energy transduction seems to be an important step in the mechanism of action in such model systems.

An important detail of the effects of magainins in such model systems, is the higher order concentration dependence. Considering the high helical propensity that is expected from the magainins' primary structure and the observation that indeed the helical content of the peptides increases significantly in hydrophobic environments, it is conceivable that the formation of a transmembrane complex of magainin monomers, is at the basis of the effects observed.

We try to elucidate the mechanism of action of the magainin peptides, and understand how it is related to the physico-chemical properties of the target system. Liposomal systems will thus be ideal, to achieve a complete definition of the model system.

2. Experimental Strategy and Results

Small unilamellar liposomes (SUVs) of different lipid composition, were prepared by sonication. Since we want to study magainin-induced membrane permeability, we used an assay where a fluorescent compound was entrapped into the liposomes (calcein) and its quencher was present outside (Co^{2+}), as described by Oku and coworkers[6]. Membrane permeability was then measured by the initial decrease in fluorescence, after addition of the magainin peptides.

We observed that magainin 2 induced membrane permeability to a smaller extent than PGLa, and this to a smaller extent than the mixture of magainin 2 and PGLa (1:1), which showed a synergistic effect, as described in cytochrome oxidase liposomes[7] and in hamster spermatozoa[5]. When the mixture of magainin 2 and PGLa (1:1) is added to liposomes of phosphatidyl choline (PC) and dicetyl phosphate (DCP) at different ratios, not only did the peptides induce stronger membrane permeability in liposomes containing higher amounts of negatively charged lipids, but also cooperativity seemed to increase with the negative charge (cf. Fig. 4 in Ref. 9).

We studied also the possibility of potentiation of the action of magainins, by a transmembrane potential. Such electric potential was present in all the model systems mentioned above, and in the case of target sequences for import of mitochondrial proteins, with which magainins share similarity of size and net charge, it is known that a transmembrane potential, negative inside, can indeed potentiate the insertion of such target sequences[8]. Induction of a transmembrane electric potential, negative inside, by means of the

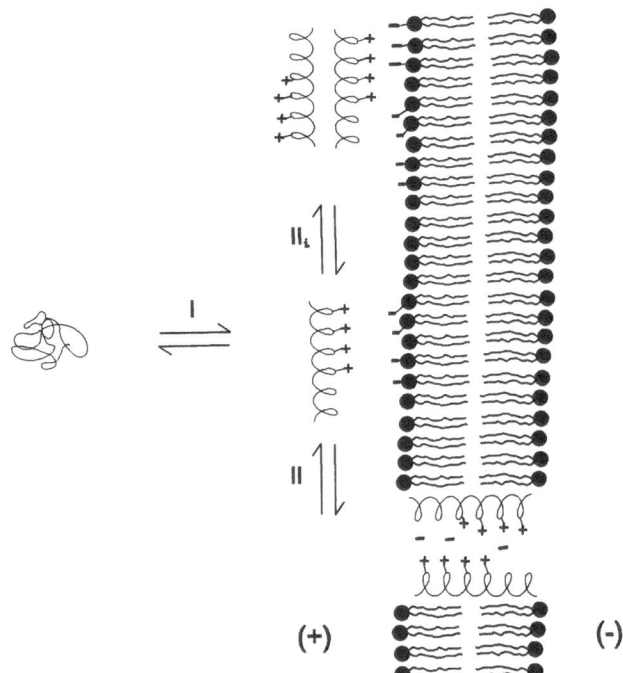

Figure 1. Tentative model of the mechanism of action of magainins. Bound charges are present on the peptide and lipid bilayer, and mobile charges are depicted, across the "pore". While step II (or possibly through several intermediary additional steps) may lead to the formation of a complex, causing functional effects, step II_i may lead to the formation of inactive complexes, or simple insertion of the peptides, parallel to the surface, into the hydrophobic part of the bilayer.

establishment of a K$^+$-diffusion potential, upon addition of valinomycin, on the magainin-induced membrane permeability caused potentiation of the effect of PGLa or the mixture (1:1) of magainin 2 and PGLa, when the diffusion potential was established simultaneously (cf. Fig. 6 in Ref. 9).

3. Discussion

Considering the data obtained until now, we propose the model for the mechanism of action of magainins, depicted in Fig. 1. Magainins do not show defined secondary structure when in solution, but upon in hydrophobic environments, helical conformation is attained. Electrostatic interactions play a determining role in the action of the peptides. "Active" complexes may be attained via step II, possibly through several subsequent steps, each leading to a complex with different intrinsic activity. Through step II$_i$, the formation of inactive complexes is represented, that would not cause functional effects. Experiments are being performed to discriminate between all these possibilities.

Acknowledgements

A. Vaz Gomes is supported by JNICT-PORTUGAL; A. de Waal and H.V. Westerhoff are supported by NWO.

References

1. M. Zasloff, Magainins, a class of antimicrobial peptides from *Xenopus* skin: isolation, characterization of two active forms, and partial cDNA sequence of a precursor, *Proc. Natl. Acad. Sci. USA* **84**:5449-5453 (1987).
2. C.L. Bevins and M. Zasloff, Peptides from frog skin, *Annu. Rev. Biochem.* **59**:395-414 (1990).
3. H.V. Westerhoff, R.W. Hendler, M. Zasloff and D. Juretic, Interactions between a new class of eucaryotic antimicrobial agents and isolated rat liver mitochondria, *Biochim. Biophys. Acta* **975**:361-369 (1989).
4. H.V. Westerhoff, D. Juretic, R.W. Hendler and M. Zasloff, M., 1989b, Magainins and the disruption of membrane-linked free-energy transduction, *Proc. Natl. Acad. Sci. USA* **86**:6597-6601 (1989).
5. A. de Waal, A. Vaz Gomes, A. Mensink, J.A. Grootegoed and H.V. Westerhoff, Magainins affect respiratory control, membrane potential and motility of spermatozoa, *FEBS Lett.* **293**:219-223 (1991).
6. N. Oku, D.A. Kendall and R.C. MacDonald, A simple procedure for the determination of the trapped volume of liposomes, *Biochim. Biophys. Acta* **691**:332-340 (1982).
7. D. Juretic, R.W. Hendler, M. Zasloff and H.V. Westerhoff, Cooperative action of magainins in disrupting membrane-linked free-energy transduction, *Biophys. J.* **55**:572a (1989).
8. D. Roise, S.J. Horvath, J.M. Tomich, J.H. Richards and G. Schatz, A chemically synthesized pre-sequence of an imported mitochondrial protein can form an amphiphilic helix and perturb natural and artificial phospholipid bilayers, *EMBO J.* **5**:1327-1334 (1986).
9. A. Vaz Gomes, A. de Waal, J.A. Berden and H.V. Westerhoff, Electric potentiation, cooperativity and synergism of magainin peptides in protein-free liposomes, *Biochemistry*, in press.

THE COMMUNICATION BETWEEN
THE MITOCHONDRIAL
AND CYTOSOLIC BRANCHES
OF ENERGY METABOLISM

Fanny Laterveer, Ferdi van Dorsten,
Annette van der Toorn and Klaas Nicolay

Department of *in vivo* NMR
Bijvoet Center
Bolognalaan 50, NL-3584 CJ Utrecht, The Netherlands

1. Background

Over the last decades biomolecules and subcellular organelles have been extensively studied under *in vitro* conditions using a variety of techniques. This "test tube" approach has yielded a wealth of information on the properties of these individual building blocks of the cell. Simultaneously, it has become clear that their interplay in the complex context of the intact cell is not a simple summation of their *in vitro* properties.

For oxidative phosphorylation it has been demonstrated that the properties of isolated mitochondria do not allow definitive conclusions to be drawn on the factors determining oxidative activity in the intact cell[1]. Also, Michaelis-Menten type studies on (partially) purified enzyme preparations do not yield reliable estimates on the flux through the reaction catalyzed by the enzyme under *in vivo* conditions[2]. This is primarily due to:

(i)   restriction of metabolite diffusion in the cell;
(ii) the formation of multi-enzyme complexes;
(iii) enzyme-membrane and enzyme-cytoskeleton interactions which may lead to the creation of micro-milieus having properties distinct from those of the bulk phase.

Therefore, studies aimed at the quantitative description of metabolic activity under *in vivo* conditions should be designed as to account for the above factors. In addition, biomathematical procedures should be developed for the quantitative interpretation of experimental data in terms of the contributions of individual components.

2. Aim

Our research addresses the mechanisms by which energy conversion in mitochondrial oxidative phosphorylation is dynamically adjusted to the cellular demands. Out of a multiplicity of potential factors[1] we focus on the role of two kinase classes, i.e. the hexokinase (HK) and creatine kinase (CK) iso-enzyme systems. Both are characterized by tissue-specific expression and have multiple locations in the cell[3-5].

HK iso-enzymes I and II are partly cytosolic and partly associated with the mitochondrial outer membrane[3]. Bound HK-I forms tetramers[6]. HK-III is predominantly bound to the nuclear membrane[7] while HK-IV is entirely cytosolic.

The iso-enzymes of creatine kinase can be divided in three groups[5]. First, there is the cytosolic dimeric type (subdivided in a muscle form (MM-CK) and a brain form (BB-CK)). Secondly, a mitochondrial isoform (Mi-CK) has been identified which is localized in the intermembrane space. It mainly occurs as an octamer. Thirdly, a monomeric CK has been found in the tail of sperm cells[8]. Most of this species is anchored to the flagellar membrane as a result of myristoylation[9].

We investigate the role of the HK and CK iso-enzyme systems in cellular energy conversion processes along two lines. First, we study the mechanisms by which the mitochondrion-associated HK and CK are functionally interacting with oxidative phosphorylation. This involves the use of functional assays as well as the use of techniques to probe the local environment experienced by the kinases upon association with their mitochondrial targets. Secondly, a combination of *in vivo* NMR, gene technology and biomathematics is used to study the role of the individual members of the CK system in the *in vivo* situation. A number of recent findings are presented below.

3. Hexokinase- and Creatine Kinase-Mitochondria Interaction

Current models on the topology of mitochondrion-associated hexokinase and creatine kinase are depicted in Fig. 1. Using gene fusion experiments McCabe and coworkers[10] have recently shown that 15 N-terminal amino acids of HK-I suffice to target a cytosolic protein to the outer membrane. We have synthesized a peptide corresponding to the N-terminal 25 amino acids of HK-I. Initial experiments have demonstrated that the peptide has a high affinity for binding to membranes containing negatively-charged phospholipids. Binding to such membranes induced a structured conformation with a high α-helix content. Experiments with the full length enzyme and monolayers of outer membranes argue against a direct insertion of the N-terminus into the hydrophobic core of the membrane which had been suggested from crosslinking data[11]. Similar studies using a variety of techniques are under way to come to a better understanding of the mechanism by which HK-I interacts with the outer membrane and the consequences this interaction has for the local environment which the enzyme is facing.

Mi-CK is localized in the intermembrane space and thus faces both inner and outer membrane. The octameric form of the enzyme is capable of creating stable intermembrane contacts (Fig. 1) which might explain its preferential localization in adhesion zones between inner and outer membrane[12]. So far we have not obtained evidence for direct interactions between Mi-CK and either the outer membrane porin or the adenine nucleotide translocator.

These interactions had been postulated on the basis of the observed functional coupling between Mi-CK and oxidative phosphorylation[4,12].

4. Functional Studies

Our functional studies cover the range from purified HK-I and Mi-CK, via isolated mitochondria and eukaryotic cells to intact tissues in laboratory animals. Below a number of

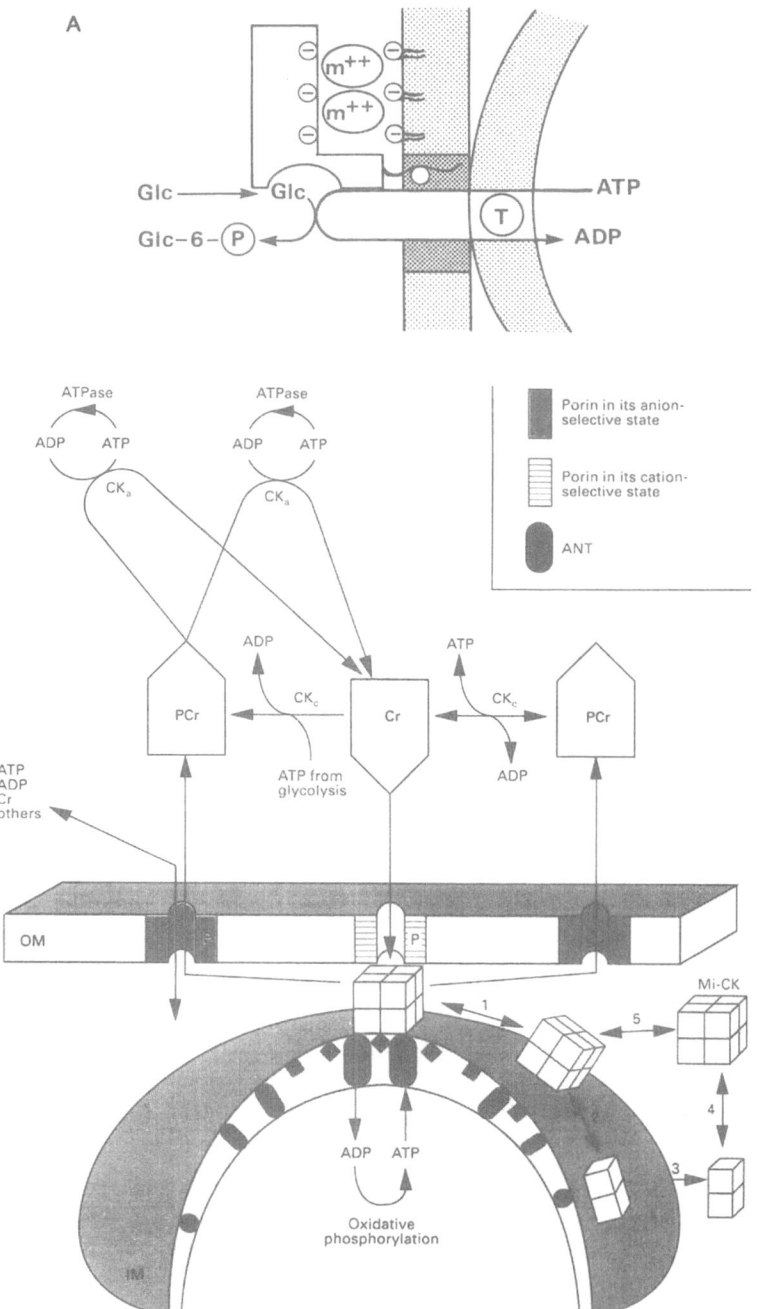

Figure 1. Models of the interaction of HK-I (A)[3] and Mi-CK (B)[4,17] with the mitochondrial membranes. HK-I binds to porin in the outer membrane via its N-terminus; binding is promoted by divalent cations and occurs preferentially at contact sites between inner and outer membrane[12]. Octameric Mi-CK is also enriched in contact sites. Monolayer experiments[17] with isolated mitochondrial membranes and purified Mi-CK demonstrated that the enzyme is able to create stable contacts between inner and outer membranes *in vitro*. This process is promoted by negatively-charged phospholipids, in particular cardiolipin. It is presently unknown whether these Mi-CK-membrane interactions are involved in the formation of contact sites in the intact mitochondrion. (Mi-CK model reproduced from Ref. 4 with copyright permission from The Biochemical Journal).

examples of such functional experiments are given. Fig. 2 depicts an experiment in which the apparent affinity of HK-I for its substrate ATP (K_m^{ATP}) was determined in the absence of mitochondria, when bound to respiring mitochondria or to non-respiring mitochondria.

From the steady-state rates of glucose phosphorylation it appeared that:

(i) K_m^{ATP} was reduced from 0.16 mM for free HK-I to 0.11 mM upon binding to non-respiring mitochondria;

(ii) K_m^{ATP} underwent a further reduction to 0.08 mM under phosphorylating conditions.

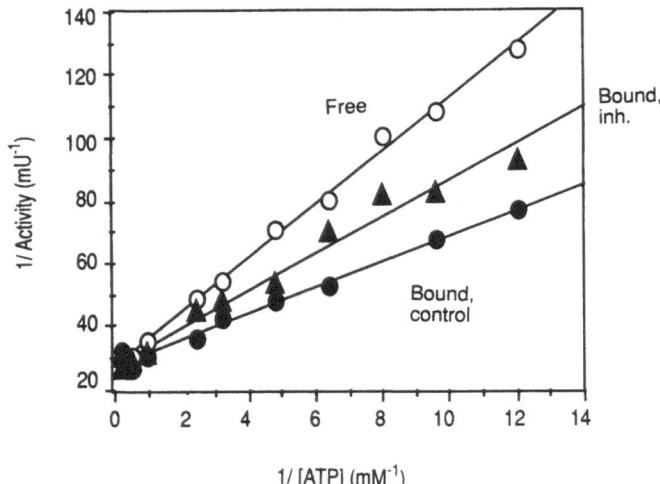

Figure 2. The consequences of hexokinase-I binding for its apparent affinity for ATP. HK-I activity was measured as a function of the added ATP concentration for the following conditions: (a) in the absence of mitochondria (open circles); (b) when bound to rat liver mitochondria respiring on succinate (full circles); (c) when bound to non-respiring mitochondria (full triangles). The results are plotted according to Lineweaver-Burk.

Similar differences were seen in the presence of PEP/pyruvate kinase for extramitochondrial ATP regeneration. These experiments were also performed with HK-I which had been treated with α-chymotrypsin to selectively remove the N-terminus in order to eliminate its ability to bind. In addition, we have used ^{31}P-NMR methods to simultaneously measure the extramitochondrial phosphate potential and HK activity[18]. Taken together the data suggest that the ca. 50 % reduction in the K_m^{ATP} induced by binding to respiring mitochondria is the result of two different factors:

(i) binding *per se*;

(ii) the close proximity between bound enzyme and the site of mitochondrial ATP-(re)generation.

At present, we are using biochemical methods to measure extra- and intramitochondrial ATP and ADP in the steady-state in order to get more insight into the origin of the above observations.

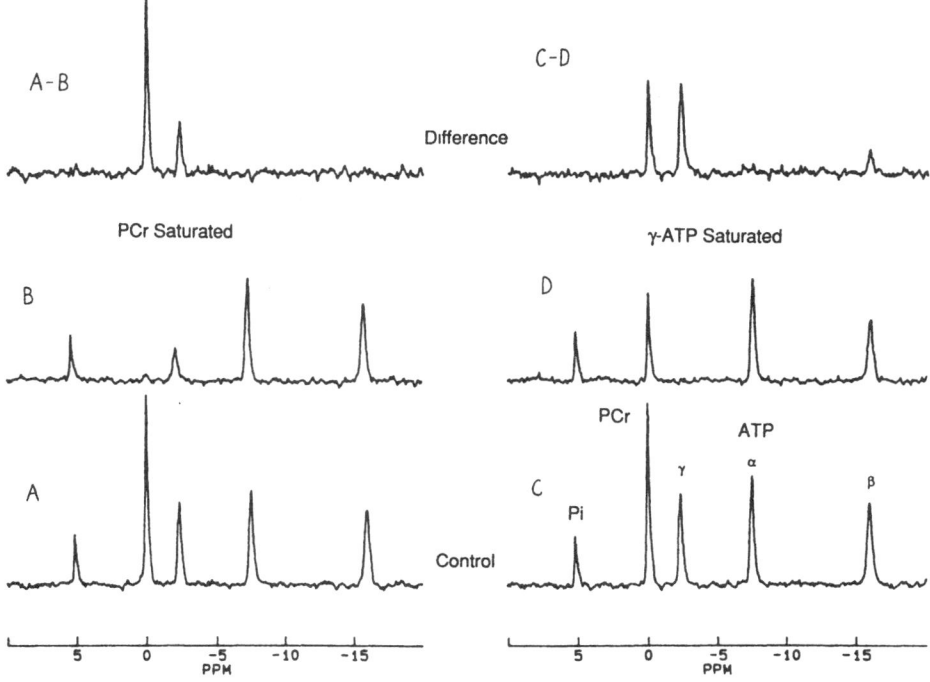

Figure 3. Creatine kinase kinetics as measured with ^{31}P-NMR saturation transfer. Experiments were performed on purified rabbit muscle MM-CK at 25 °C. Magnetization transfer was measured from ATP to phosphocreatine (PCr) (*left*) and vice versa (*right*) upon saturation of the PCr (B) or γ-ATP (D) peaks, respectively. Spectra A and C represent control spectra. The spectra labeled (A-B) and (C-D) were obtained by subtracting spectrum B from A, and spectrum D from C, respectively.

Similar be it more pronounced differences have been reported from the measurement of initial rates of glucose phosphorylation by bound HK[13-15]. Interestingly, Wilson and coworkers[14,15] have recently presented evidence that bound HK preferentially utilizes ATP from an intramitochondrial micro-compartment which neither equilibrates with the intermembrane space, nor the matrix, nor the extramitochondrial medium. On the basis of its sensitivity to digitonin the authors proposed that the putative micro-compartment resides in contact sites.

^{31}P-NMR is the technique we use most in our research on the functional properties of the CK iso-enzyme system. The major reason for that is that it provides a wealth of bio-energetic information, can be applied at any level of biological complexity and enables the specific measurement of the forward and reverse fluxes through the CK reaction under steady-state conditions through saturation transfer. An example of the latter approach as applied to dimeric MM-CK under *in vitro* conditions is shown in Fig. 3. The data can be quantitated in terms of (first order) unidirectional rate constants and, thereby, the flux through the CK reaction. In this example the forward (PCr-to-ATP) and reverse (ATP-to-PCr) rate constants amounted to 0.20 and 0.38 sec[-1], respectively. Similar experiments are currently performed on preparations of purified Mi-CK. The NMR studies are routinely supported by classical Michaelis-Menten type experiments using spectrophotometry to determine the conventional kinetic parameters. These *in vitro* studies are aimed at assessing the effects of membrane association on Mi-CK kinetics. Both model membranes and mitochondrial membranes will be

used. From this, we also expect to gain new information on the functional coupling between Mi-CK and mitochondrial ATP synthesis.

Apart from the *in vitro* approach we are setting up similar NMR studies on intact cells and intact tissues *in vivo* in laboratory animals. In order to unravel the contributions of individual CK iso-enzymes to the overall flux through the CK system we will study its kinetics:

(i) as a function of development in the rat;

(ii) in transgenic animals in which the endogenous expression of iso-enzymes has been modified.

Biomathematical procedures will be applied in order to model the findings and guide further experimentation. Zahler and Ingwall[16] have recently modeled the flux through the CK system as measured by ^{31}P-saturation transfer in isolated rat and rabbit hearts. The first results of their procedure which was claimed to enable the dissection of Mi-CK specific flux information hold promise for the prospects of our research strategy.

Acknowledgements

This research was supported by the Netherlands Foundation for Chemical Research (SON) and partly conducted at the national *in vivo* NMR facility at Utrecht University.

References

1. R.S. Balaban, Regulation of oxidative phosphorylation in the mammalian cell, *Am. J. Physiol.* **258**:C377-C389 (1990).
2. H.V. Westerhoff and G.R. Welch, Enzyme organization and the direction of metabolic flow: physicochemical considerations, *Curr. Top. Cell. Regul.* **33**:361-390 (1992).
3. J.E. Wilson, Regulation of mammalian hexokinase activity, *in* Regulation of Carbohydrate Metabolism, R. Breitner, ed., CRC Press, Boca Raton (Florida), pp. 45-85 (1985).
4. T. Wallimann, M. Wyss, D. Brdiczka, K. Nicolay and H.M. Eppenberger, Intracellular compartmentation, structure and function of creatine kinase iso-enzymes in tissues with high and fluctuating energy demands: the 'phosphocreatine circuit' for cellular energy homeostasis, *Biochem. J.* **281**:21-40 (1992).
5. M. Wyss, J.Smeitink, R.A.Wevers and T.Wallimann, Mitochondrial creatine kinase: a key enzyme of aerob-ic energy metabolism, *Biochim. Biophys. Acta* **1102**:119-166 (1992).
6. G. Xie and J.E. Wilson, Tetrameric structure of mitochondrially bound rat brain hexokinase: a crosslinking study, *Arch. Biochem. Biophys.* **276**:285-293 (1990).
7. A. Preller and J.E. Wilson, Localization of the type III hexokinase at the nuclear periphery, *Arch. Biochem. Biophys.* **294**:482-492 (1992).
8. A.F.G. Quest and B.M. Shapiro, Membrane association of flagellar creatine kinase in the sperm phosphocreatine shuttle, *J. Biol. Chem.* **266**:19803-19811 (1991).
9. A.F.G. Quest, J.K. Chadwick, D.D. Wothe, R.A.J. McIlhinney and B.M. Shapiro, Myristoylation of flagellar creatine kinase in the sperm phosphocreatine shuttle is linked to its membrane association properties, *J. Biol. Chem.* **267**:15080-15085 (1992).
10. B.D. Gelb, V. Adams, S.N. Jones, L.D.Griffin, G.R.MacGregor and E.R.B.McCabe, Targeting of hexokinase I to liver and hepatoma mitochondria, *Proc. Natl. Acad. Sci. USA* **89**:202-206 (1992).
11. G. Xie and J.E. Wilson, Rat brain hexokinase: the hydrophobic N-terminus of the mitochondrially bound enzyme is inserted in the lipid bilayer, *Arch. Biochem. Biophys.* **267**:803-810 (1988).
12. D. Brdiczka, Contact sites between mitochondrial envelope membranes. Structure and function in energy- and protein-transfer, *Biochim. Biophys. Acta* **1071**:291-312 (1991) .
13. K. K. Arora and P. L. Pedersen, Functional significance of mitochondrial bound hexokinase in tumor cell metabolism, *J. Biol. Chem.* **263**:17422-17428 (1988).
14. H. Beltran del Rio and J.E. Wilson, Hexokinase of rat brain mitochondria: relative importance of adenylate kinase and oxidative phosphorylation as sources of substrate ATP, and interaction with intramito-chondrial compartments of ATP and ADP, *Arch. Biochem. Biophys.* **286**:183-194 (1991).

15. H. Beltran del Rio and J.E.Wilson, Coordinated regulation of cerebral glycolytic and oxidative metabolism, mediated by mitochondrially bound hexokinase dependent on intramitochondrially generated ATP, *Arch. Biochem. Biophys.* **296**:667-677 (1991).

16. R. Zahler and J.S. Ingwall, Estimation of heart mitochondrial creatine kinase flux using magnetization transfer NMR spectroscopy, *Am. J. Physiol.* **262**:H1022-H1028 (1992).

17. M. Rojo, R.C. Hovius, R.A. Demel, K. Nicolay and T. Wallimann, Mitochondrial creatine kinase mediates contact formation between mitochondrial membranes, *J. Biol. Chem.* **266**:20290-20295 (1992).

18. K. Nicolay, M. Rojo, T. Wallimann, R. Demel and R.C. Hovius, The role of contact sites between inner and outer mitochondrial membrane in energy transfer, *Biochim. Biophys. Acta* **1018**:229-233 (1990).

CALCULATION OF METABOLIC FLUXES
BY MATHEMATICAL MODELING
OF CARBON-13 DISTRIBUTION IN METABOLITES.

I. PRESENTATION OF THE MATHEMATICAL MODEL

Ronny Schuster[1,2], Jean-Charles Portais[1,3], Fabrice Garderet[1],
Michel Merle[1], Jean-Pierre Mazat[3] and Paul Canioni[1,3]

[1]Département de Résonance Magnétique
  Institut de Biochimie Cellulaire, CNRS
  1, rue C. Saint-Saëns, 33077 Bordeaux-Cedex, France
[2]Institut für Biochemie, Humboldt-Universität zu Berlin
  Hessische Strasse 3-4, 1040 Berlin, Germany
[3]Université de Bordeaux II
  146, rue Léo Saignat, 33076 Bordeaux-Cedex, France

1. Introduction

Carbon-13 (^{13}C) NMR spectroscopy in connection with the use of ^{13}C-enriched substrates is a powerful technique for metabolic studies, e.g. for specifying the structure of metabolic pathways, determining flux rates or investigating reaction mechanisms[1-9]. However, owing to the complexity of many biological systems, it is often impossible to directly obtain this information from the experimental data. In many cases, mathematical models which describe the isotopic distribution as a function of fluxes turn out to be necessary[1-9].

Mathematical models have frequently been developed for intermediary metabolism (tricarboxylic acid (TCA) cycle and related pathways), designed to describe radioactive tracer experiments[10-15]. In contrast, only few models are available for NMR experiments[4,6,7,9]. Although, from a mathematical point of view, there is no difference between models used for radioactive or stable isotopes[16], some particularities of the NMR method should be considered upon modeling. The low sensitivity of this technique[17] requires to give priority to those metabolites that are detectable by NMR and to preferably exclude non-detectable metabolites. Furthermore, in contrast to traditional radionuclide methods, isotopomer enrichments are detectable by NMR. However, models using this information[4,6,7] contain a large number of variables even for very simplified reaction schemes[4], and the experimental data are particularly difficult to obtain *in vivo*.

Modern Trends in Biothermokinetics, Edited by
S. Schuster *et al.*, Plenum Press, New York, 1993

Here, we present a mathematical model for the intermediary metabolism which is designed to analyze ^{13}C NMR results under *in-vitro* or *in-vivo* conditions. The main purpose of the model is the determination of flux rates for intermediary metabolism, e.g. the TCA cycle and adjacent pathways. The model predominantly includes the metabolites that are detectable by ^{13}C and ^{1}H NMR. As usually done in previous models[9,14,15], our model uses the specific enrichments of single carbons and is based on first-order differential equations, allowing its application to steady states[12-14] or non-steady state[4,5,15].

In the subsequent paper[18], the model is used for calculating flux rates of metabolic pathways in the rat C6 cell line, a biological model of malignant glioma.

2. Modeling

2.1. The metabolic network

The mathematical model is based on the reaction scheme depicted in Fig.1. Besides the TCA cycle, it contains glycolysis, gluconeogenesis, and the hexose monophosphate shunt (HMPS), as well as reactions connecting these metabolic pathways with fatty acid and protein metabolism. The stoichiometry of the HMPS is due to the formation of 2 molecules of fructose 6-phosphate and 1 molecule of glyceraldehyde 3-phosphate from 3 molecules of glucose 6-phosphate. Furthermore, we consider several effluxes (e.g. formation of lactate and glutamine) and the possibility for the label to enter at each metabolite. Furthermore, we considered the occurrence of a metabolic channeling at the succinyl-CoA hydrolase/succinate DH/fumarase steps of the TCA cycle by introducing the following two variants of the fumarase reaction: v_4 (conversion of carbons 5, 4, 3, and 2 of α-ketoglutarate into carbons 1, 2, 3, and 4 of oxaloacetate, respectively) and v_5 (conversion of carbons 2, 3, 4, and 5 of α-ketoglutarate into carbons 1, 2, 3, and 4 of oxaloacetate, respectively), where the total flux through the fumarase reaction is equal to v_4+v_5. If the rates v_4 and v_5 are different, metabolic channeling is likely to occur.

2.2. Formalism

The model is based on a coupled system of differential equations describing the time-dependent variations of either metabolite concentrations or specific ^{13}C enrichments of carbons. The time dependent changes of metabolite concentrations are governed by a set of first-order differential equations of the type:

$$\frac{dx_i}{dt} = \sum_j c_{ij} v_j \ , \tag{a1}$$

where x_i are the metabolite concentrations, c_{ij} the elements of the stoichiometry matrix, and v_j the flux rates, respectively.

The specific enrichments of carbon atoms, y_i, are defined as the ratio of the ^{13}C isotope concentration, t_i, to the total concentration of carbon atoms ($^{12}C+^{13}C$). The differential equations for the specific enrichments write, in a general way, as follows[16],

$$x_i \frac{dy_i}{dt} = \sum_j \left[\left(c_{ij}^+ v_j^+ + c_{ij}^- v_j^- \right) \left\{ \sum_k a_{ik}^j \left| c_{ij} \right| y_k - y_i \right\} \right] \ , \tag{a2}$$

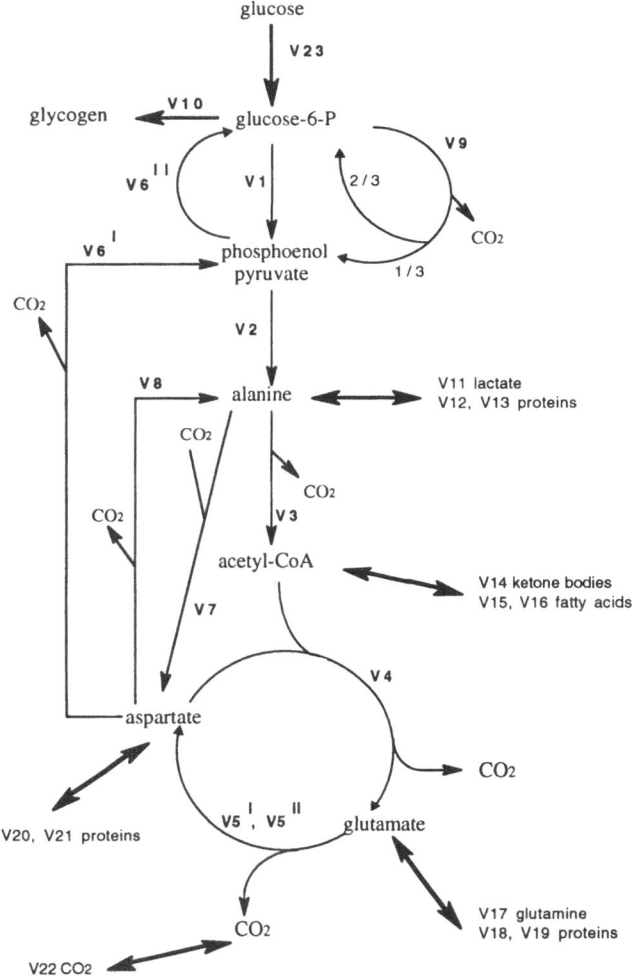

Figure 1. Reaction scheme containing glycolysis, gluconeogenesis, the hexose monophosphate shunt and some reactions connecting these pathways with fatty acid and protein metabolism.

where the c^+_{ij} and c^-_{ij} correspond to the positive and negative elements of the stoichiometry matrix, respectively, the v^+_j and v^-_j are the rates of forward and backward reactions, and the a^j_{ik} denote the probabilities for the transition of the kth carbon atom into the ith one during the jth reaction.

Since most intermediates of the considered metabolic pathways cannot be detected by ^{13}C or ^{1}H NMRS, we used a simplified pathway structure. The transaminases (not shown in Fig. 1) directly connect some NMR non detectable intermediates of the TCA cycle with the amino acids glutamate, aspartate, and alanine, which are present in sufficiently high concentrations and are, therefore, detectable in most cases. To reduce the number of NMR invisible metabolites, we assumed the transaminases to be very active and at quasi-equilibrium, and applied the rapid-equilibrium approximation[16,19], accounting for some of the simplifications in the reduced reaction scheme depicted in Fig. 1. The corresponding algorithms for the systems (a1) and (a2) are outlined in Refs 19 and 16, respectively. By this approximation, the concentrations of pyruvate, α-ketoglutarate, and aspartate can be excluded due to the mass action relations of the aminotransferases. Furthermore, one obtains that the specific ^{13}C enrichment of carbons are equal in α-ketoglutarate and glutamate, in pyruvate and alanine, as well as in oxaloacetate and aspartate. The reduced equations are obtained by linear combinations of the original ones and write as follows in the stationary state, under consideration that [X] and $\{X_i\}$ denote the concentration of metabolite X and the specific enrichment of the ith carbon of X, respectively.

The equations for metabolite concentrations read (for abbreviations see Appendix):

$$d[G6P]/dt = 0 = -v_1 + 1/2v_6^{II} - 1/3v_9 - v_{10} + v_{23}$$

$$d[PEP]/dt = 0 = 2v_1 - v_2 + v_6^I - v_6^{II} + 1/3v_9 + v_{24}$$

$$d([PYR]+[ALA])/dt = 0 = v_2 - v_3 - v_7 + v_8 - v_{11} - v_{12} + v_{13} + v_{25}$$

$$d[ACE]/dt = 0 = v_3 - v_4 - v_{14} - v_{15} + v_{16} + v_{26}$$

$$d([\alpha KG]+[GLU])/dt = 0 = v_4 - v_5^I - v_5^{II} - v_{17} - v_{18} + v_{19} + v_{27}$$

$$d([OXA]+[ASP])/dt = 0 = -v_4 + v_5^I + v_5^{II} - v_6^I + v_7 - v_8 - v_{20} + v_{21} + v_{28}$$

$$d[CO_2]/dt = 0 = v_3 + v_4 + v_5^I + v_5^{II} + v_6^I - v_7 + v_8 + v_9 - v_{22} + v_{29} .$$

The following equations are used for the specific enrichments of carbon atoms:

[G6P] $d\{G6P_1\}/dt$
$$= 0 = v_{23}a_1 + 1/2v_6^{II}\{PEP_3\} + 2/3v_9\{G6P_2\}$$
$$- (v_{23} + 1/2v_6^{II} + 2/3v_9)\{G6P_1\}$$

[G6P] $d\{G6P_2\}/dt$
$$= 0 = v_{23}a_2 + 1/2v_6^{II}\{PEP_2\} + 2/3v_9\{G6P_3\}$$
$$- (v_{23} + 1/2v_6^{II} + 2/3v_9)\{G6P_2\}$$

[G6P] $d\{G6P_3\}/dt$
$$= 0 = v_{23}a_3 + 1/2v_6^{II}\{PEP_1\}$$
$$+ 1/3v_9(\{G6P_2\} + \{G6P_3\})$$
$$- (v_{23} + 1/2v_6^{II} + 2/3v_9)\{G6P_3\}$$

[G6P] $d\{G6P_4\}/dt = 0 = v_{23}a_4 + 1/2v_6^{II}\{PEP_1\} - (v_{23} + 1/2v_6^{II})\{G6P_4\}$

[G6P] $d\{G6P_5\}/dt = 0 = v_{23}a_5 + 1/2v_6^{II}\{PEP_2\} - (v_{23} + 1/2v_6^{II})\{G6P_5\}$

[G6P] $d\{G6P_6\}/dt = 0 = v_{23}a_6 + 1/2v_6^{II}\{PEP_3\} - (v_{23} + 1/2v_6^{II})\{G6P_6\}$

[PEP] $d\{PEP_1\}/dt$
$$= 0 = v_{24}b_1 + v_1(\{G6P_3\} + \{G6P_4\}) + v_6^I\{ASP_3\}$$
$$+ 1/3v_9\{G6P_4\} - (v_{24} + 2v_1 + v_6^I + 1/3v_9)\{PEP_1\}$$

[PEP] d{**PEP$_2$**}/dt $\quad = 0 = v_{24}b_2 + v_1(\{$**G6P$_2$**$\}+\{$**G6P$_5$**$\}) + v_6{}^I\{$**ASP$_2$**$\}$
$\quad + 1/3v_9\{$**G6P$_5$**$\} - (v_{24}+2v_1+v_6{}^I+1/3v_9)\{$**PEP$_2$**$\}$

[PEP] d{**PEP$_3$**}/dt $\quad = 0 = v_{24}b_3 + v_1(\{$**G6P$_1$**$\}+\{$**G6P$_6$**$\}) + v_6{}^I\{$**ASP$_1$**$\}$
$\quad + 1/3v_9\{$**G6P$_6$**$\} - (v_{24}+2v_1+v_6{}^I+1/3v_9)\{$**PEP$_3$**$\}$

([PYR]+[ALA]) d{**ALA$_1$**}/dt $\quad = 0 = v_{25}g_1 + v_2\{$**PEP$_1$**$\} + v_8\{$**ASP$_1$**$\}$
$\quad - (v_{25}+v_2+v_8+v_{13})\{$**ALA$_1$**$\}$

([PYR]+[ALA]) d{**ALA$_2$**}/dt $\quad = 0 = v_{25}g_2 + v_2\{$**PEP$_2$**$\} + v_8\{$**ASP$_2$**$\}$
$\quad - (v_{25}+v_2+v_8+v_{13})\{$**ALA$_2$**$\}$

([PYR]+[ALA]) d{**ALA$_3$**}/dt $\quad = 0 = v_{25}g_3 + v_2\{$**PEP$_3$**$\} + v_8\{$**ASP$_3$**$\}$
$\quad - (v_{24}+v_2+v_8+v_{13})\{$**ALA$_3$**$\}$

[ACE].d{**ACE$_1$**}/dt $\quad = 0 = v_{26}d_1 + v_3\{$**ALA$_2$**$\} - (v_{26}+v_3+v_{16})\{$**ACE$_1$**$\}$

[ACE] d{**ACE$_2$**}/dt $\quad = 0 = v_{26}d_2 + v_3\{$**ALA$_3$**$\} - (v_{26}+v_3+v_{16})\{$**ACE$_2$**$\}$

([αKG]+[GLU]) d{**GLU$_1$**}/dt $\quad = 0 = v_{27}e_1 + v_4\{$**ASP$_4$**$\} - (v_{27}+v_4+v_{19})\{$**GLU$_1$**$\}$

([αKG]+[GLU]) d{**GLU$_2$**}/dt $\quad = 0 = v_{27}e_2 + v_4\{$**ASP$_3$**$\} - (v_{27}+v_4+v_{19})\{$**GLU$_2$**$\}$

([αKG]+[GLU]) d{**GLU$_3$**}/dt $\quad = 0 = v_{27}e_3 + v_4\{$**ASP$_2$**$\} - (v_{27}+v_4+v_{19})\{$**GLU$_3$**$\}$

([αKG]+[GLU]) d{**GLU$_4$**}/dt $\quad = 0 = v_{27}e_4 + v_4\{$**ACE$_2$**$\} - (v_{27}+v_4+v_{19})\{$**GLU$_4$**$\}$

([αKG]+[GLU]) d{**GLU$_5$**}/dt $\quad = 0 = v_{27}e_5 + v_4\{$**ACE$_1$**$\} - (v_{27}+v_4+v_{19})\{$**GLU$_5$**$\}$

([OXA]+[ASP]) d[**ASP$_1$**]/dt $\quad = 0 = v_{28}c_1 + v_5{}^I\{$**GLU$_5$**$\} + v_5{}^{II}\{$**GLU$_2$**$\}$
$\quad + v_7\{$**ALA$_1$**$\} - (v_{28}+v_5{}^I+v_5{}^{II}+v_7+v_{21})\{$**ASP$_1$**$\}$

([OXA]+[ASP]) d[**ASP$_2$**]/dt $\quad = 0 = v_{28}c_2 + v_5{}^I\{$**GLU$_4$**$\} + v_5{}^{II}\{$**GLU$_3$**$\}$
$\quad + v_7\{$**ALA$_2$**$\} - (v_{28}+v_5{}^I+v_5{}^{II}+v_7+v_{21})\{$**ASP$_2$**$\}$

([OXA]+[ASP]) d[**ASP$_3$**]/dt $\quad = 0 = v_{28}c_3 + v_5{}^I\{$**GLU$_3$**$\} + v_5{}^{II}\{$**GLU$_4$**$\}$
$\quad + v_7\{$**ALA$_3$**$\} - (v_{28}+v_5{}^I+v_5{}^{II}+v_7+v_{21})\{$**ASP$_3$**$\}$

([OXA]+[ASP]) d[**ASP$_4$**]/dt $\quad = 0 = v_{28}c_4 + v_5{}^I\{$**GLU$_2$**$\} + v_5{}^{II}\{$**GLU$_5$**$\}$
$\quad + v_7\{$**CO$_2$**$\} - (v_{28}+v_5{}^I+v_5{}^{II}+v_7+v_{21})\{$**ASP$_4$**$\}$

$$[CO_2] \; d[CO_2]/dt \qquad = 0 = v_{29}m + v_3\{ALA_1\} + v_4\{ASP_1\}$$
$$+ (v_5{}^I + v_5{}^{II})\{GLU_1\} + (v_6{}^I + v_8)\{ASP_4\} + v_9\{G6P_1\}$$
$$- (v_{29} + v_3 + v_4 + v_5{}^I + v_5{}^{II} + v_6{}^I + v_8 + v_9)\{CO_2\}.$$

The a_i, b_i, g_i, d_i, c_i, and m correspond to specific enrichments of the ith carbon of external glucose, phospho*enol*pyruvate, alanine, acetate, glutamate, aspartate, and CO_2, respectively.

2.3. Calculation of flux rates at steady state

At metabolic and isotopic steady state, the time derivatives for x_i and y_i vanish, so that the differential equation systems (a1) and (a2) turn into algebraic equation systems which are in all cases linear with respect to the flux rates. The flux rates can be calculated by a least-squares fitting procedure. The function Φ to be fitted reads

$$\Phi = \sum \left(y_i^{calculated} - y_i^{measured} \right)^2 \;,$$

where the summation includes the specific enrichments that were measurable. To calculate absolute flux rates, it is necessary to have at least one flux determined by another method.

For the fitting of fluxes, we developed a computer program which will be published elsewhere. It is applicable for systems of any stoichiometry but based on equations (a1) and (a2) and restricted to steady-state conditions with respect to metabolite as well as isotopomer concentrations. Standard errors are estimated using the method of support planes as described by Duggleby[20].

3. Conclusions

In this paper, we describe a mathematical model designed to use simple NMR data, absolute or specific enrichments of carbons, for the determination of flux rates. It comprises glycolysis, gluconeogenesis, the HMPS, citric acid cycle, as well as reactions connecting these pathways with protein, fatty acid, and glutamine synthesis, to allow for different experimental conditions as well as different biological systems. The model corresponds to the best compromise between a maximum number of fluxes determined with regards to the restricted number of NMR-detectable metabolites. An important feature of the model is its one-compartmental structure. However, since it is anyway rather difficult to obtain experimental data corresponding to different compartments within a biological system, the model is likely to be sufficient for a great number of applications.

Acknowledgements

This work was supported by grants from the INSERM, the ARC, the Région Aquitaine, and the Scientific Council of NATO under the auspices of the DAAD (to R. Schuster).

References

1. A.W.H. Jans and R. Willem, A ^{13}C NMR investigation of the metabolism of amino acids in renal proximal convoluted tubules of normal and streptozotocin-treated rats and rabbits, *Biochem. J.* 263:231-241 (1989).

2. K. Walsh and D.E. Koshland Jr., Determination of flux through the branch point of two metabolic cycles, *J. Biol. Chem.* **259**:9646-9654 (1984).

3. I. Magnusson, W.C. Schumann, G.E. Bartsch, V. Chandramouli, V. Kumaran, J. Wahren and B.R. Landau, Noninvasive tracing of Krebs cycle metabolism in liver, *J. Biol. Chem.* **266**:6975-6984 (1991).

4. E.M. Chance, S.H. Seeholzer, K. Kobayashi, and J.R. Williamson, Mathematical analysis of isotope labeling in the citric acid cycle with applications to ^{13}C NMR studies in perfused rat hearts, *J. Biol. Chem.* **258**:13785-13794 (1983).

5. S. Tran-Dinh, M. Herve and J. Wietzerbin, Determination of flux through different metabolic pathways in saccharomyces cerevisiae by ^{1}H- and ^{13}C-NMR spectroscopy, *Eur. J. Biochem.* **201**:715-721 (1991).

6. C.R. Malloy, A.D. Sherry and F.M.H. Jeffrey, Evaluation of carbon flux and substrate selection through alternate pathways involving the citric acid cycle of the heart by ^{13}C-NMR spectroscopy, *J. Biol. Chem.* **263**:6964-6971 (1988).

7. C.R. Malloy, T.R. Thompson, F.M.H. Jeffrey and A.D. Sherry, Contribution of exogenous substrates to acetyl coenzyme A: measurement by ^{13}C NMR under non-steady-state conditions, *Biochemistry* **29**:6756-6761 (1990).

8. P.W. Kuchel, H.A. Berthon, W.A. Bubb, B.T. Bulliman and J.G. Collins, Computer simulation of the pentose-phosphate pathway and associated metabolism used in conjunction with NMR experimental data from human erythrocytes, *Biomed. Biochim. Acta* **49**:757-770 (1990).

9. G.F. Mason, D.L. Rothman, K.L. Behar and R.G. Shulman, NMR determination of the TCA cycle rate and α-ketoglutarate/glutamate exchange rate in rat brain, *J. Cereb. Blood Flow Metab.* **12**:434-447 (1992).

10. J.K. Kelleher, Analysis of tricarboxylic acid cycles using ^{14}C citrate specific activity ratios, *Am. J. Physiol.* **248**:E252-E260 (1985).

11. D.F. Heath and J.G. Rose, ^{14}C bicarbonate fixation into glucose and the other metabolites in the liver of the starved rat under halothane anaesthesia, *Biochem. J.* **227**:851-867 (1985).

12. M. Rabkin and J.J. Blum, Quantitative analysis of intermediary metabolism in hepatocytes incubated in the presence and absence of glucagone with a substrate mixture containing glucose, ribose, fructose, alanine end acetate, *Biochem. J.* **225**:761-786 (1985).

13. R. Goebel, M. Berman and D. Foster, Mathematical model for the distribution of isotopic carbon atoms through the tricarboxylic acid cycle, *Fed. Proc.* **41**:96-103 (1982).

14. C. Salon, P. Raymond and A. Pradet, Quantification of carbon fluxes through the tricarboxylic acid cycle in early germinating lettuce embryos, *J. Biol. Chem.* **263**:12278-12287 (1988).

15. H.G. Holzhütter, Compartmental analysis: theoretical aspects and application, *Biomed. Biochim. Acta* **44**: 863-873 (1985).

16. R. Schuster, H.G. Holzhütter and S. Schuster, Simplification of complex kinetic models used for the quantitative analysis of nuclear magnetic resonance or radioactive tracer studies, *J. Chem. Soc. Faraday Trans.* **88**:2837-2844 (1992).

17. J.C. Portais, M. Merle, J. Labouesse and P. Canioni, Quantitative analysis of carbon-13 and ^{1}H NMR spectra: application to C6 glioma cell intermediary metabolism, *J. Chim. Phys.* **89**: 209-215 (1992).

18. J.C. Portais, R. Schuster, F. Garderet, M. Merle, J.P. Mazat and P. Canioni, Calculation of metabolic fluxes by mathematical modeling of carbon-13 distribution in metabolites. II Application to the study of C6 cell metabolism, this volume.

19. M. Schauer and R. Heinrich, Quasi-steady-state approximation in the mathematical modelling of biochemical reaction networks, *Math. Biosci.* **65**:155-170 (1983).

20. R.G. Duggleby, Estimation of the reliability of parameters obtained by non-linear regression, *Eur. J. Biochem.* **109**:93-96 (1980).

APPENDIX

Abbreviations

G6P	- glucose 6-phosphate,	PEP	- phosphoenole pyruvate,
ALA	- alanine,	PYR	- pyruvate,
ACE	- acetyl CoA,	αKG	- α-ketoglutarate,
GLU	- glutamate,	OXA	- oxaloacetate,
ASP	- aspartate,		

v_1	- rate of glycolysis,	v_2	- pyruvate kinase,
v_3	- pyruvate dehydrogenase,	v_4	- citrate synthase,

v_5^I - fumarase I,

v_5^{II} - fumarase II,

v_6^I, v_6^{II} - gluconeogenesis,

v_7 - pyruvate carboxylase,

v_8 - malic enzyme,

v_9 - hexose monophosphate shunt,

v_{10} - glycogen synthesis,

v_{11} - lactate dehydrogenase,

v_{12} - alanine efflux (protein synthesis)

v_{13} - alanine influx (protein degradation),

v_{14} - ketone body formation,

v_{15} - acetyl-CoA efflux (fatty acid synthesis),

v_{16} - acetyl-CoA influx (fatty acid degradation),

v_{17} - glutamine synthesis,

v_{18} - glutamate efflux (protein synthesis),

v_{19} - glutamate influx (protein degradation),

v_{20} - aspartate efflux (protein synthesis),

v_{21} - aspartate influx (protein degradation),

v_{22} - CO_2 formation.

Enriched substrate influxes:

v_{23}- glucose, v_{24}- phospho*enol*pyruvate, v_{25}- alanine, v_{26}- acetate, v_{27}- glutamate, v_{28}- aspartate, v_{29}- CO_2.

CALCULATION OF METABOLIC FLUXES BY MATHEMATICAL MODELING OF CARBON-13 DISTRIBUTION IN METABOLITES

II. APPLICATION TO THE STUDY OF C6 CELL METABOLISM

Jean-Charles Portais[1,2], Ronny Schuster[2,3], Fabrice Garderet[1], Michel Merle[1,2], Jean-Pierre Mazat[2] and Paul Canioni[1,2]

[1]Département de Résonance Magnétique
 Institut de Biochimie Cellulaire, CNRS
 1, rue C. Saint-Saëns, 33077 Bordeaux-Cedex, France
[2]Université de Bordeaux II
 146, rue Léo Saignat, 33076 Bordeaux-Cedex, France
[3]Institut für Biochemie, Humboldt-Universität zu Berlin
 Hessische Strasse 3-4, 1040 Berlin, Germany

1. Introduction

The metabolism of the rat C6 cell line has been extensively studied[1-4], because these cells constitute a model of malignant glioma. As neoplastic cells preferentially utilize the relatively inefficient Embden-Meyerhof pathway for satisfying their energy requirements, the carbohydrate metabolism, up to now, was in the center of interest. Besides glycolysis, C6 cells have been shown to possess non-negligible activities of the hexose monophosphate shunt (HMPS), and glycogen synthesis.

In contrast, TCA cycle and related pathways were poorly investigated. The available data from NMR experiments[5], were difficult to interpret owing to the lack of adequate mathematical models. We, therefore, undertook new experiments with [1-13C]-glucose and applied the model presented in the first part of this article[6] to C6 cells, to estimate the TCA cycle activity and to investigate its embedding in the intermediary metabolism of these malignant gliomas.

2. Experimental Procedures

2.1. Cell culture and NMR spectroscopy

C6 cells, taken at least 24 hours after the last medium change to deplete the glycogen store[1,3], were incubated in Dulbecco's modified Eagle medium, free of glutamine and pyruvate, and containing 5.5 mM [1-^{13}C]-glucose. After a 4-hour incubation period, cell perchloric acid extracts were prepared as described by Pianet and coworkers[7]. The protein content was determined according to the method of Lowry[8].

^{13}C NMR spectra were acquired at 100.6 MHz with a Bruker AM 400 spectrometer in a 5 mm broad-band probe, using a 55° flip angle, and an interpulse of 1.65 s, with bilevel proton decoupling and D$_2$O lock. Spectra were accumulated overnight. Chemical shifts were expressed as ppm relative to the resonance at 63.7 ppm from 20 µmol ethylene glycol used as internal reference. ^{1}H NMR spectra were acquired with the same probe at 400.13 MHz, using a 6 s interpulse delay.

2.2. Glucose consumption and lactate production measurements

For glucose consumption and lactate production measurements, C6 cells were incubated with unenriched glucose. Aliquots of the medium were sequentially removed at different incubation times for glucose and lactate assays. Metabolites were assayed with enzymic procedures that include glucose oxidase and peroxidase for glucose, and lactate dehydrogenase for lactate, according to the supplier's recommandations (SIGMA France).

2.3. Specific enrichment determination

^{13}C specific enrichments of metabolites were determined from ^{13}C NMR spectra and absolute metabolite concentrations. Absolute ^{13}C enrichments were calculated as previously described[5] from the areas of the ^{13}C NMR resonances, corrected for relaxation and NOE effects, using 20 µmol ethylene glycol as standard. In the case of glutamate, it was possible to determine the specific enrichments by ^{1}H NMR. For this purpose, glutamate was previously separated with a protocol that includes a cation-exchange chromatography, followed by an anion-exchange chromatography.

The trimethylsilyl derivatives of aspartate and glutamate from C6 cells were used to determine their total enrichments by GC-MS. The enrichments of the C1 of both metabolites were determined after chemical ionization (Valeins et al., unpublished results). From these data and the NMR data, it was possible to deduce the specific enrichment of aspartate C4.

3. Results and Discussion

The ^{13}C specific enrichments of alanine, glutamate, and aspartate, after incubation of C6 cells with [1-^{13}C]-glucose, are reported in Table 1. Alanine, assumed to have the same enrichments as lactate, is enriched only at position 3. This enrichment is lower than that of the corresponding carbons of the exogeneous glucose (C1+C6). The most enriched position of glutamate is the fourth. The positions 2 and 3 of this metabolite are not equally enriched, as the positions 2 and 3 of aspartate. The rates of glucose consumption and lactate production were 21.2 ± 2.2 and 25.4 ± 6.1 mmol/24h/mg of protein (mean ± sd, N=4), respectively.

These results indicate that C6 cells metabolize glucose, which is mainly converted to lactate. The carbohydrate metabolism of these cells has been well characterized. Glucose is converted into lactate[1-3] via both glycolysis and the hexose monophosphate shunt (HMPS), which is active in C6 cells[2,4,9]. Kingsley-Hickman and coworkers[4] have determined, under similar experimental conditions, that the HMPS involved 7.5 % of the glucose that was not converted into glycogen. The fraction of glucose that is used for glycogen synthesis in C6 cells has been evaluated by Passonneau[3]. From this work, it can be estimated that the maximal rate of glycogen synthesis is, under our conditions, about 0.8 µmol/24h/mg of protein. On the other hand, the lack of enrichment at positions 1 and 2 of alanine indicates that no neoglucogenesis occurs in C6 cells.

In a previous work, we used ^{13}C NMR spectroscopy to study the intermediary metabolism of C6 cells[5]. Quantitative information about fluxes were obtained by using the model developed by Malloy[10]. Calculated fluxes indicated that 27±4 % of the acetyl-CoA entering the TCA cycle was enriched at C2. The activity of the anaplerotic reactions was 39 ± 20% of that of citrate synthase, and the fraction of anaplerotic substrates leading to enriched oxaloacetate was 21 ± 11%.

Here we apply the model developed in the preceding article[6] to C6 cells, to obtain more quantitative information. To take both our results and data from the literature into account, application of the model to C6 cells yielded a reduction of the number of fluxes to be fitted. In addition, incubations with [1-^{13}C]-glucose were performed with confluent C6 cells, so that

Table 1. Specific enrichments of metabolites after incubation of C6 cells with [1-^{13}C] glucose. Experimental specific enrichments (exp.) were determined as described in "Materials and Methods". Calculated enrichments (calc.) were obtained by fitting the fluxes, using the model described in the preceding paper[6]. All specific enrichments are given after correction for natural abundance (1.1 %).

		specific enrichment (%)					
		C1	C2	C3	C4	C5	C6
glucose-6-P	exp.	unknown	unknown	unknown	unknown	unknown	unknown
	calc.	95.2	0	0	0	0	0
alanine*	exp.	0	0	37	-	-	-
	calc.	<0.15	<0.15	36.7			
acetyl-CoA	exp.	unknown	unknown	-	-	-	-
	calc.	0	23.3				
glutamate	exp.	3.0	14.0	9.0	21.0	0	-
	calc.	2.9	13.6	9.6	21.0	<0.11	
aspartate	exp.	8.0	11.2	14.7	3.1	-	-
	calc.	8.0	10.6	15.0	3.2		
CO_2	exp.	unknown	-	-	-	-	-
	calc.	8.0					

*from the enrichments of lactate.

the net protein synthesis and net fatty acid synthesis were nearly zero. Given all the data, twelve fluxes could be fitted. Absolute calculation of the fluxes was accomplished by experimental measurement of the rates of glucose consumption and lactate production. The mean value and the standard deviation for the calculated enrichment of each carbon are reported in Table 1, and absolute values of the calculated fluxes in Table 2. For the latter, standard errors were within 15 % for most fluxes, except those which are very low. Fitting of the fluxes to the data was accomplished only by introducing diluting influxes at each node of the network, except at the level of glucose-6-phosphate.

From these results, it can be estimated that glucose accounts for about 78 % of the total pyruvate and 50 % of the total acetyl-CoA. The calculated specific enrichment at position 2 of acetyl-CoA was 23 % (Table 1), a value similar to that previously obtained with the model of Malloy[10]. The efflux corresponding to glutamine synthesis was very low. As a consequence,

Table 2. Absolute values of fluxes through intermediary metabolism of C6 cells. Calulated values were obtained after fitting the fluxes to the experimental data, according to the model described in the preceding paper[6]. Results are expressed as $\mu mol/24h/mg$ of protein $\pm$ standard error.

Intrinsic fluxes		Extrinsic fluxes	
Activity	Value	Activity	Value
v_1 (glycolysis)	19.9±0.1	v_{17} (glutamine synthesis)	0.1±0.1
v_3 (PDH)	14.7±0.2	protein turnover:	
v_4 (citrate synthase)	14.7±0.2	v_{12}, v_{13} (alanine level)	11.3±0.6
v_5^I (fumarase)	4.1±0.2	v_{18}, v_{19} (glutamate level)	1.6±0.5
v_5^{II} (fumarase)	10.5±0.5	v_{20}, v_{21} (aspartate level)	2.9±0.4
v_7 (PC)	0.1±0.1	fatty acid turnover:	
v_{22} (CO$_2$ formation)	45.5±0.9	v_{15}, v_{16} (acetyl-CoA level)	8.5±0.9

the flux through citrate synthase (v_4) is similar to the total flux through fumarase ($v_5^I + v_5^{II}$). The pyruvate carboxylase activity (v_7) is very low: only 0.7% of the pyruvate is converted to oxaloacetate. This result can be related to the non-neoglucogenic nature of C6 cells. The ratio between anaplerotic reactions and citrate synthase $(v_7 + v_{19} + v_{21})/v_3$, expressed as percentage, was 31%, that can be compared to the 39% previously obtained. The flux v_5^{II} was higher than the flux v_5^I, indicating a metabolic channeling at the TCA cycle level. The percentage of channeled metabolites, $(v_5^{II} - v_5^I)/(v_5^I + v_5^{II})$, amounts to 43%. The channeling predicted by the model favors the convertion of carbons 2, 3, 4, and 5 of α-ketoglutarate into carbons 1, 2, 3, and 4 of oxaloacetate, respectively.

Acknowledgements

This work was supported by grants from the INSERM, ARC, Région Aquitaine, and Scientific Council of NATO under the auspices of the DAAD.

References

1. J.P. Schwartz, W.D. Lust, V.R. Lauderdale and J.V. Passonneau, Glycolytic metabolism in cultured cells of the nervous system. II. Regulation of pyruvate and lactate metabolism in the C6 glioma cell line, *Molec. Cell. Biochem.* **9**:67-72 (1975).
2. B.D. Ross, R.J. Higgins, J.E. Boggan, J.A. Willis, B. Knittel and S.W. Unger, Carbohydrate metabolism of the rat C6 glioma. An in vivo ^{13}C and in vitro ^{1}H magnetic resonance spectroscopy study, *NMR in Biomed.* **1**:20-26 (1988).
3. J.V. Passonneau and S.K. Crites, Regulation of glycogen metabolism in astrocytoma and neuroblastoma cells in culture, *J. Biol. Chem.* **251**:2015-2022 (1976).
4. P.B. Kingsley-Hickman, B.D. Ross and T. Krick, Hexose monophosphate shunt measurement in cultured cells with [1-^{13}C]-glucose: correction for endogenous carbon sources using [6-^{13}C]-glucose, *Anal. Biochem.* **185**:235-237 (1990).
5. J.C. Portais, M. Merle, J. Labouesse and P. Canioni, Quantitative analysis of carbon-13 and proton NMR spectra: application to C6 glioma cell intermediary metabolism, *J. Chim. Phys.* **89**:209-215 (1992).
6. R. Schuster, J.-C. Portais, F. Garderet, M. Merle, J.-P. Mazat and P. Canioni, Calculation of metabolic fluxes by mathematical modeling of carbon-13 distribution in metabolites. I. Presentation of the mathematical model, this volume.
7. I. Pianet, M. Merle, J. Labouesse and P. Canioni, P NMR of C6 glioma cells and rat astrocytes evidence for a modification of the longitudinal relaxation time of ATP and P$_i$ during glucose starvation, *Eur. J. Biochem.* **195**:87-95 (1991).

8. O.H. Lowry, N.J. Rosebrough, A.L. Farr and R.J. Randall, Protein measurement with the Folin phenol reagent, *J. Biol. Chem.* **193**:265-275 (1951).
9. K. Keller, H. Kolbe, H. Herken and K. Lange, Glycolysis and glycogen metabolism after inhibition of hexose monophosphate pathway in C6-glial cells, *Arch. Pharmacol.* **294**:213-215 (1976).
10. C.R. Malloy, A.D. Sherry and F.M. Jeffrey, Evaluation of carbon flux and substate selection through alternate pathways involving the citric acid cycle of the heart by ^{13}C NMR spectroscopy, *J. Biol. Chem.* **255**:6964-6971 (1987).

VI. COMPUTER PROGRAMS FOR MODELING METABOLIC SYSTEMS

MLAB: AN ADVANCED SYSTEM
FOR MATHEMATICAL AND
STATISTICAL MODELING

Gary D. Knott

Civilized Software, Inc.
7735 Old Georgetown Rd. 410
Bethesda, MD. 20814, U.S.A.

1. General Description

MLAB, (for <u>M</u>odeling <u>LAB</u>oratory), is a program for interactive mathematical modeling. MLAB was originally developed at the National Institutes of Health and has been enhanced for use on PC's. It includes curve-fitting, differential equations, statistics and graphics as some of its major capabilities. MLAB provides more than thirty command types and more than four hundred built-in functions from the areas of elementary mathematics, transcendental functions, probability and statistics, linear algebra, optimization, cluster analysis, combinatorics, numeric input/output, and graphics. The usual low-level functions, e.g. sine, cosine, log, etc., are present, as well as functions performing complex analyses, such as singular value decomposition, discrete Fourier transforms, solution of differential equation systems, and constrained non-linear optimization, among many others. A substantial collection of statistically-orientated functions, such as most common distribution functions and their inverses, are included.

Using the built-in functions in MLAB, the user may construct user-defined functions, which may be algebraic, differential-equation defined, implicit, recursive, and/or defined piecewise. Functions may be used in many ways: direct evaluation, symbolic derivatives may be constructed, and roots of non-linear equations may be found.

MLAB can solve systems of differential equations. These differential equations may be of any order, stiff or non-stiff, and may contain delays. MLAB uses an algorithm which automatically selects both method and stepsize to minimiza both truncation error and computer time. However, control variables are provided to permit the user to specify the method of numerical integration, the error tolerance, and the method for dealing with singularities.

A major capability of MLAB is constrained, non-linear optimization. Data may be input which is to be fit by one or more model functions which contain one or more parameters. Any type of function, including differential equation-defined functions, may be used as a model. Once the parameters are initialized, the FIT command will adjust the parameters so as to minimize the error between the data and the model. The parameters may appear nonlinearly in the model, and be subject to multiple linear equality and inequality constraints. Several control variables are provided to permit the user to fine-tune the algorithm, although the defaults are quite robust.

MLAB provides a powerful and flexible 2D graphics facility. With a single simple command, a graph may be created with axes. A few more commands serve to insert titles and axis labels. The user has complete control of color, line-types and point-types, fonts and the position of every aspect of the plot. Many types of plots are available, including linear, logarithmic, and polar coordinates. Contour plots for functions of two variables are provided. Screen plots may be saved as files suitable for printing on HP or Postscript laser printers, at the full resolution of the output device.

MLAB is also a programming language. It provides facilities for branching, looping, input/output, and interruption of execution. Users may write MLAB program files containing multiple commands, including interactive commands, and calls to other such files. Analyses may thus be programmed much more compactly than with conventional programming languages because of the very high level of the individual commands.

MLAB is extraordinarily polished. Careful attention has been paid to significant details so that MLAB operates flawlessly, just as you think it should behave from a mathematical viewpoint.

For example, the MLAB matrix sorting operator, SORT(**M,J**), returns a copy of the matrix **M** with its rows sorted in ascending order. The sort is *stable*, so that rows with duplicate values in column **J** appear in the output in the same relative order as they appeared in the input. If **J** is negative, the sorting is done so that column |**J**| appears in *descending* order. If **J** is not given, column 1 is assumed, so SORT(**M**) has a natural meaning. Finally if **J** is 0, the output will be a copy of **M** with its rows stably sorted in *lexicographic* order.

Another very important example of attention to detail in MLAB is that MLAB correctly handles floating point overflows and underflows, supplying whichever of MAXPOS, the largest computational value, or MAXNEG, the algebraically least computational value, is indicated as the result of an overflow (with an optional warning). It supplies 0 as the result of a hard underflow. Often-appropriate values are supplied for zero-divides as well. Since overflows may well arise during curve-fitting, this corrective behavior is crucial to allowing the curve-fit to continue successfully.

2. Example of Kinetic Modeling

Below we show an interesting example of MLAB modeling for dimer kinetics. Suppose we have two substances, A and B which bind to form a complex C, and the substance C, in turn, binds with itself to form a dimer D. We thus have:

$$A + B \underset{k_2}{\overset{k_1}{\rightleftarrows}} C \qquad\qquad C + C \underset{k_4}{\overset{k_3}{\rightleftarrows}} D \ .$$

Suppose further we mix 2 mmoles of A and 3 mmoles of B and measure the concentration in mmoles of both C and D at ten equally-spaced times between 0 and 70 seconds. From this data we wish to estimate the association and dissociation constants k_1, k_2, k_3 and k_4. We may proceed in MLAB as follows.

First we read in the data (values of $C(t)$ and $D(t)$ at 0:70!10), as shown below.

```
*data = (0:70!10)&'read(ddata,8,2)
*type data
```

	time	c	d
1:	0	0	0
2:	7.77777778	0.670158274	4.25688007E-2
3:	15.5555556	1.03882934	0.148136658
4:	23.3333333	0.97528872	0.437148352
5:	31.1111111	1.08457668	0.643256451
6:	38.8888889	0.911604643	0.89407957
7:	46.6666667	0.940333448	1.13858135
8:	54.4444444	0.7922676	1.09494
9:	62.2222222	0.935718768	1.37411241
10	70	0.904038076	1.62101349

```
cdata = data col 1:2
ddata = data col (1,3)
```

Now we define our kinetic model so that $C(t)$ and $D(t)$ are the concentrations of C and D, respectively, in mmoles at time t.

```
*function c't(t)=k1*maxz(a0-c-2*d)*maxz(b0-c-2*d)-k2*c
*function d't(t)=k3*c*c-k4*d
*function maxz(x)=if x>0 then x else 0
*initial c(0)=0
*initial d(0)=0
*a0=2; b0=3;
```

Now we guess the values of k_1, k_2, k_3, and k_4. We may use the results of equilibrium studies, analyzed by MLAB, to know values for the ratios k_1/k_2 and k_3/k_4.

```
*k1=.02; k2=.002; k3=.02; k4=.002
*constraints q={k1>0, k2>0, k3>0, k4>0}
```

Now we may curve-fit to estimate k_1, k_2, k_3, and k_4, by the command line

```
*fit(k1, k2, k3, k4), c to cdata, d to ddata, constraints q.
```

The results are shown in Table 1.

Now we may draw the results of the curve-fit (see Fig. 1).

```
*m=integrate(c't, d't, 0:100!140)
*draw m col (1,2) color red
*draw m col (1,2) color green lt dashed
*draw cdata pt circle lt none color red
*draw ddata pt circle lt none color green
*bottom title "time in seconds"
*left title "mmoles (C and D)"
*view
```

The resulting picture is reproduced in Fig. 1.

Table 1. Results of the fitting procedure.

final parameter values

value	error	dependence	parameter
0.02207045175	0.002035868506	0.8107995718	k1
0.004052705463	0.00121019694	0.8355673621	k2
0.02771570667	0.003479542121	0.9256727732	k3
0.0001792975535	0.003933590662	0.9398003646	k4

3 iterations
CONVERGED
best weighted sum of squares = 7.325830e-02
weighted root mean square error = 6.766568e-02
weighted deviation fraction = 5.619169e-02
R squared = 9.811667e-01

no active constraints

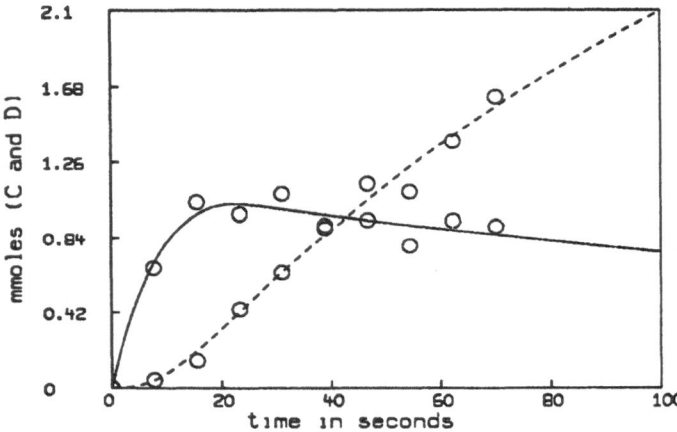

Figure 1. Plot of the results of a numerical integration including parameter fit for a model of dimer kinetics.

3. Hardware Requirements and Availability

MLAB runs on IBM PC's and compatibles using the DOS operating system. It requires a math co-processor, VGA or EGA color graphics, and a laser printer for hard copy. An AIX version is available. UNIX versions for Sun Sparc workstations and NeXT machines will be available in the near future, and a Cray version is under development.

MLAB may be purchased from Civilized Software, Inc., 7735 Old Georgetown Road no. 410, Bethesda, MD 20814, USA, phone +1-301-6524714, FAX +1-301-6561069. A three-month trial version is available for $100, the full system with technical support and updates for one year is priced at $2995, and a streamlined version is available for $995.

SIMFIT - A COMPUTER PACKAGE FOR SIMULATION, CURVE FITTING AND STATISTICAL ANALYSIS USING LIFE SCIENCE MODELS

W.G. Bardsley

Department of Obstetrics and Gynaecology
University of Manchester
St. Mary's Hospital
Whitworth Park, Manchester M13 OJH, United Kingdom

1. Using NAG Routines for Simulation, Statistical Analysis and Curve-Fitting in the Life Sciences

Computers are widely used for sensitivity analysis, exploring the behavior of model systems, statistical methods of data processing and curve-fitting. However, all these processes are controversial. Simulation is of limited value unless realistic density functions are employed and sufficient iterations are used to achieve stable results. Statistical analysis is complicated by using methods which are only justified with linear, normal models and constant variance, so their robustness and bias has to be estimated independently. Curve-fitting is complicated when there are insufficient replicates for accurate weighting and smoothed weighting schemes have to be used, or when parameter constraints have to be satisfied during the optimization. Finding starting estimates, internal parameter and data scaling, avoiding local minima of the objective function and methods to be used for differential equation solving or optimization are also problematical with constrained nonlinear weighted least squares regression. Also parameter redundancy is frequently encountered.

The Numerical Algorithm Group (NAG) library is valuable resource in this area. It provides reliable software and excellent documentation and, by using consistent argument lists, it is simple to update as new routines become available. So it is easy to compare the results of calculations with alternative routines and computational techniques. SIMFIT is a package using NAG routines for simulation, statistical analysis and optimization that was assembled during 1978 to 1990 for use on Manchester University mainframes.

Since many of these programs can now be run on PC's, version 1.0 of a preliminary subset for model discrimination and parameter estimation was circulated during 1989 and

Modern Trends in Biothermokinetics, Edited by
S. Schuster *et al.*, Plenum Press, New York, 1993

version 2.0 with GKS graphics was released in 1990. Version 3.0 now has more items, improved graph plotting facilities, better user interface and occupies 5M of hard disk space. All programs have built in tutorials, graphical displays can be saved in several formats following key press and results are written to ASCII archive files.

2. Summary of SIMFIT Version 3.0 (PC3)

a) Batch/help files, auxiliaries for: screen set up, key press interpretation, driving dot matrix, PS or laser printers, HP plotters, writing pixel, GKSM, PS or TIFF files.

b) Prepare/edit data files, put data in order, generate means, standard errors and confidence limits from replicates, fuse files, calculate smooth weighting schemes, remove outliers, transform data units, try alternative weightings, check data files for consistency.

c) Simulate exact data between input x-values or selected $f(x)$-values, generate replicates, add random error according to numerous possible weighting schemes, add outliers, view or file the perturbed data.

d) Dedicated programs for model discrimination and parameter estimation: automatically scale data, generate starting estimates, improve by random search and overdetermined L_1 solving if possible, scale inernal parameters, fit growth models, sums of exponentials, binding isotherms, rational functions etc. with appropriate constraints if necessary, sign, run, chi-squared, F tests and residual plots.

e) Calibration curves, prediction and inverse prediction with confidence limits using polynomials of recommended or user selected degree or cubic splines with chosen knot density in various axes, GKS meta file replay and graphical manipulation of SIMFIT output files for hardcopy.

f) Generate random numbers from selected distributions, pdf's, cdf's, critical values, tests associated with and tests for Normal, binomial, F, t, chi-squared, sign and run random variables.

g) Estimate areas, initial rates, horizontal asymptotes, lag times or final asymptotic rates, model-free data smoothing using tensioned splines, comparison of curves before and after treatment.

h) Advanced features: density estimation, diffusion and transport, fitting differential equations, functions of several variables from library or user supplied models, flow cytometry analysis, optimal design for model discrimination, compare different methods for optimization, equation solving etc.

3. Files Used by SIMFIT Version 3.0 (PC3)

The files used by the PC version 3.0 of SIMFIT are listed in Tables 1-10.

Table 1. Utilities required to run SIMFIT: Version 3.0 (PC3).

HELP.EXE	Provide on line help and advice (Part 2 of SIMFIT.DOC)
HERCBIOS.COM	Watcom GKS interface (required for Hercules adapters)
KEY1TON.COM	The key press interpreter required by SIMFIT.BAT (vital)
LOGO.EXE	Displays SIMFIT logo when the batch file is started up
SIMFIT.BAT	Batch file to run SIMFIT (i.e. SIMFIT.MON or SIMFIT.COL)
SIMFIT.COL	Colored batch file (uses ANSI.SYS and DEVICE=ANSI.SYS)
SIMFIT.DOC	Part 1 (summary), part 2 (help/advice), part 3 (files)
SIMFIT.MON	Monochrome batch file (this one does not use ANSI.SYS)
WGKS.EXE	Watcom GKS interface required for all graphics (vital)

Table 2. File preparation and editing.

MAKFIL.EXE	Takes in the X,Y data and prepares a main master file
EDITFL.EXE	For editing the main master file before curve-fitting
WEIGHT.EXE	Estimate smooth weighting scheme before curve-fitting
MAKMAT.EXE	Prepare vector or matrix file for statistical analysis
EDITMT.EXE	Editing vector or matrix file for statistical analysis

Table 3. Simulating data.

ADDERR.EXE	Add various types of random error to corrupt exact data
MAKEX.EXE	Sum of 1 to n exponentials (plus a constant term)
MAKHL.EXE	Low/high affinity binding sites (plus a constant term)
MAKLOGC.EXE	The logistic equation (plus a constant term)
MAKMM.EXE	Sum of 1 to n Michaelis Menten (steady state) functions
MAKPOLY.EXE	Polynomials of degree n
MAKRF.EXE	Rational function of order $n : n$
MAKSF.EXE	Saturation function of order n (plus a constant term)

Table 4. Dedicated curve fitting programs.

EXFIT.EXE	Sum of 1 to n exponential functions plus constant term
GCFIT.EXE	Fit selection from up to 10 sigmoid growth curve models
HLFIT.EXE	Sum of 1 to n high/low affinity (ligand binding) sites
MMFIT.EXE	Sum of 1 to n Michaelis Menten (steady state) functions
RFFIT.EXE	Positive $(n-1) : n$ or $n : n$ rational function plus constant
SFFIT.EXE	Ligand binding functions of order 1 to n plus constant

Table 5. Calibration and graphics

CALCURVE.EXE	Weighted least squares cubic splines then inverse prediction
POLNOM.EXE	Polynomials up to degree 6 and then inverse prediction
GKSPLOT.EXE	Program to demonstrate the SIMFIT graphical functions
SIMPLOT.EXE	Re-construct/re-label/transform plots from coordinates

Table 6. Statistics.

BINOMIAL.EXE	Binomial coefficients/distribution (pdf, cdf, %, test-for)
CHISQD.EXE	Chi-squared distribution and test (pdf, cdf, %, test-for)
FTEST.EXE	F distribution and test (pdf, cdf, %, test-for)
NORMAL.EXE	Normal distribution (pdf, cdf, %, test-for)
RANNUM.EXE	Generate pseudo random numbers from chosen distribution
RSTEST.EXE	Run and sign tests given actual numbers or signs/runs
TTEST.EXE	t distribution, (un)/paired test (pdf, cdf, %, test-for)

Table 7. Miscellaneous.

CSADAT.EXE	Transform DATAMATE ASCII file to CSAFIT type input file
COMPARE.EXE	Fit cubic spline then estimate areas and compare curves
INRATE.EXE	Estimate initial rates, lag times and final asymptotes
PSVALS.EXE	Estimate PS value for transmembrane diffusion processes

Table 8. Advanced programs.

CSAFIT.EXE	Fit flow cytometry histograms with stretch and/or shift
DEQSOL.EXE	Integrate systems of nonlinear differential equations
EOQSOL.EXE	Optimal design for numbers and spacing of data points
GNFIT.EXE	Unconstrained curve fit from library of model sequences
MAKCSA.EXE	Histograms with shift/stretch (input file for CSAFIT)
MAKDAT.EXE	Make an exact data file from library of model sequences
TKFIT.EXE	Fit integrated progress curves for enzymes or transport
QNFIT.EXE	Constrained curve fit from library of model sequences

Table 9. Test files.

PROGRAMX.TF1	First test file for PROGRAMX
PROGRAMX.TF2	Second test file for PROGRAMX... and so on

Table 10. Utilities to interpret graphics files.

DTOU.EXE	Remove hard returns from ASCII text files e.g. PS files
GKSM2CGM.EXE	Transform a GKSM type metafile into a CGM type metafile
META2PS.EXE	Drive PS printer or make PS files from GKSM metafiles
PLOT.EXE	Drive HP plotter or make HPGL files from GKSM metafiles
PXA2FILE.EXE	Display on PC screen/browse a pixel array type of file
PXA2HPLJ.EXE	Drive a laser printer from a pixel array type of file
PXA2IBM.EXE	Drive an IBM printer from a pixel array type of file
PXA2TOSH.EXE	Drive a Toshiba printer from a pixel array type of file
PXA2TIFF.EXE	Make a TIFF (5.0) file from a pixel array type of file
WBPIC.EXE	Template used by META2PS to transform GKSM into PS file

4. PC Requirements and Availability

Version 3.0 (PC3) is 16 bit and runs on all PC's, but to use the larger programs with all facilities, full libraries and graphics it is necessary to free most of conventional memory by removing memory resident software. Use of DOS 5.0 onwards is recommended for this reason. The larger programs are rather slow and shortened versions with limited facilities may be provided. Full 32 bit versions of these for 386 onwards are available. The package with LATEX source code for SIMFIT.DOC and comprehensive user manual is available for research and teaching purposes on 4 diskettes (3.5 inch 1.4M).

OPTIMAL DESIGN FOR MODEL DISCRIMINATION: CHOOSING THE BEST SPACING AND DENSITY OF DESIGN POINTS

W.G. Bardsley[1] and E.M. Melikhova[2]

[1]Department of Obstetrics and Gynaecology
  University of Manchester, St. Mary's Hospital
  Whitworth Park, Manchester M13 OJH, United Kingdom
[2]Institute of Applied Molecular Biology
  Ministry of Health
  Simpheropolsky Blvd. 8, Moscow 113149, Russia

1. Introduction

Optimal design for parameter estimation is a well developed subject but less attention has been paid to the problem of choosing the best number and optimum spacing of experimental settings for the independent variable x. Often it is known that the correct model is one from a family of models and it is important to decide on the minimum model to explain the data. An example would be using the F test to decide on the number of exponentials in a compartmental model or to decide upon the order of a rational function with steady state enzyme kinetic data and the theory necessary to solve this problem is available[1-4].

This paper briefly describes a computer program EOQSOL which can be used to study optimal spacing and summarizes our results from investigating typical biochemical models.

2. Defining Model Error and Experimental Error

The following definitions are required:

a) The correct model with m_2 parameters: $g_2(x,\theta)$;
b) A deficient model with m_1 parameters: $g_1(x,\Phi)$ where $m_2 \geq m_1$;
c) The n experimental observations: $y_i = g_2(x_i,\Theta)+\varepsilon_i$;
d) The weighted sum of squares from fitting the correct model:

Modern Trends in Biothermokinetics, Edited by
S. Schuster *et al.*, Plenum Press, New York, 1993

$$WSSQ_2 = \sum_{i=1}^{n} \left\{ \left[y_i - g_2\left(x_i, \hat{\Theta}\right) \right] / \sigma_i \right\}^2 = \sum_{i=1}^{n} \left\{ \left[g_2(x_i, \Theta) - g_2\left(x_i, \hat{\Theta}\right) + \varepsilon_i \right] / \sigma_i \right\}^2$$

e) The weighted sum of squares from fitting a deficient model:

$$WSSQ_1 = \sum_{i=1}^{n} \left\{ \left[y_i - g_1\left(x_i, \hat{\Phi}\right) \right] / \sigma_i \right\}^2 = \sum_{i=1}^{n} \left\{ \left[g_2(x_i, \Theta) - g_1\left(x_i, \hat{\Phi}\right) + \varepsilon_i \right] / \sigma_i \right\}^2$$

f) The experimental error vector **EE**: $EE_i = \varepsilon_i / \sigma_i$;
g) The model error vector **ME**: $ME_i = [g_2(x_i, \Theta) - g_1(x_i, \hat{\Phi})] / \sigma_i$,
h) An approximate F test statistic

$$WSSQ_1 = \|ME\|_2^2 + \|EE\|_2^2 + 2ME^T \cdot EE$$

$$\approx \|ME\|_2^2 + \|EE\|_2^2 ,$$

$$WSSQ_2 \approx \|EE\|_2^2 ,$$

$$F = \frac{(WSSQ_1 - WSSQ_2) / (m_2 - m_1)}{WSSQ_2 / (n - m_2)}$$

$$F \approx \left(\frac{n - m_2}{m_2 - m_1} \right) \frac{\|ME\|_2^2}{\|EE\|_2^2}$$

i) End points and design points: $a = x_1 < x_2 < \cdots < x_n = b$;
j) The density functions used to calculate design points: $f(x)$;
k) The algorithm for generating the design points:

$$\frac{1}{n-1} = \frac{\int_{x_i}^{x_{i+1}} f(t)dt}{\int_{a}^{b} f(t)dt} , \quad i = 1, 2,..., n\text{-}1 ;$$

l) The density functions investigated: $f(x) \geq 0$, with $x \in (a,b)$,

(i)	geometric increasing,	$f(x) = 1/x$,
(ii)	linear increasing,	$f(x) = b\text{-}x$
(iii)	uniform X,	$f(x) = 1$,
(iv)	normal (truncated),	$f(x) = \exp\text{-}(1/2)[(x\text{-}\mu)/\sigma]^2$,
		$\mu = (a+b)/2, \ \sigma = (b\text{-}a)/4$,
(v)	geometric decreasing,	$f(x) = 1/(a+b\text{-}x)$,
(vi)	linear decreasing,	$f(x) = x\text{-}a$,
(vii)	uniform Y,	$f(x) = dg_2(x,\Theta)/dx$;

m) The weighting functions investigated: $w(x) > 0$, $a \leq x \leq b$,

(α)	constant variance,	$w(x) = 1$,
(β)	constant relative error,	$w(x) = 1/g_2(x,\Theta)^2$,
(γ)	mixed,	$w(x) = 1/[A + Bg_2(x,\Theta)^2]$.

3. The Functions $S_n(\Theta)$, $Q(\Theta)$ and $R(n)$

$$S_n(\Theta) = \min_{\Phi} \frac{1}{n} \sum_{i=1}^{n} w(x_i) \left[g_2(x_i, \Theta) - g_1(x_i, \Phi) \right]^2$$

$$= \frac{1}{n} \sum_{i=1}^{n} w(x_i) \left[g_2(x_i, \Theta) - g_1(x_i, \hat{\Phi}) \right]^2, \quad \text{when } \varepsilon_i = 0$$

$$\approx WSSQ_1/n$$

$$\approx \|ME\|_2^2/n.$$

$S_n(\Theta)$ estimates average weighted model error squared per design point when model error is greater than experimental error. Its value depends on the models $g_2(x,\Theta)$ and $g_1(x,\Phi)$, end points (a,b), weighting function $w(x)$, number of design points n and density function $f(x)$. Large $S_n(\Theta)$ values indicate favorable conditions for rejecting the deficient model by statistical tests for model discrimination, e.g. the F test.

$$\lim_{n \to \infty} S_n(\Theta) = Q_n(\Theta) = \min_{\Phi} \frac{\int_a^b w(x) \left[g_2(x,\Theta) - g_1(x,\Phi) \right]^2 f(x) dx}{\int_a^b f(x) dx}.$$

$Q(\Theta)$ does not depend on n and estimates the maximum power of model discrimination tests as the number of design points increases without limit.

$$R(n) = \left| \frac{S_n(\Theta) - Q(\Theta)}{Q(\Theta)} \right|.$$

$R(n)$ estimates how many points are needed for optimal model discrimination given fixed models, parameters and end points. The value of n necessary to reduce $R(n)$ below a critical point assesses how different designs require different n for optimal model discrimination.

4. Functions of the Program EOQSOL

The functions of the SIMFIT program EOQSOL are shown in Table 1. For a description of the package SIMFIT see Ref. 5.

The results from using EOQSOL can be summarized as follows:

a) The only distributions worth considering seriously are: geometric increasing, uniform X and uniform Y [cf. items (i), (iii) and (vii) in Section 2];

b) By chossing special parameter values it is possible to have:

(i) $\geq$ (vii) $\gg$ (iii),
(vii) $\geq$ (i) $\gg$ (iii),
(iii) $\geq$ (i) $\approx$ (vii);

c) If $g_2(x,\Theta)$ approaches a horizontal asymptote then usually:

(i) $\approx$ (vii) $\gg$ (iii);

461

d) Similar results were obtained using $Q(\Theta)$ or $R(n)$;

e) Similar results were obtained for all $w(x)$ investigated;

f) The ease of setting up experimental designs is:

(iii) » (i) » (vii) (time in equal intervals),

(i) » (iii) » (vii) (concentration by serial dilution).

Table 1. A flow chart of the program EOQSOL.

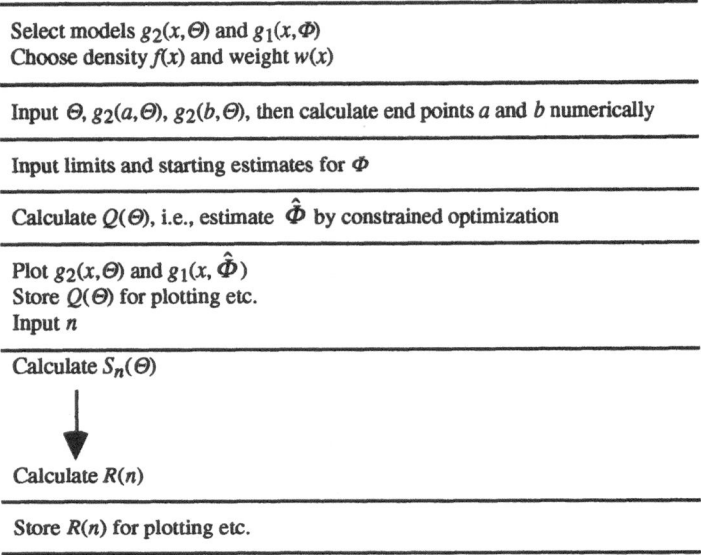

Select models $g_2(x,\Theta)$ and $g_1(x,\Phi)$ Choose density $f(x)$ and weight $w(x)$
Input Θ, $g_2(a,\Theta)$, $g_2(b,\Theta)$, then calculate end points a and b numerically
Input limits and starting estimates for Φ
Calculate $Q(\Theta)$, i.e., estimate $\hat{\Phi}$ by constrained optimization
Plot $g_2(x,\Theta)$ and $g_1(x,\hat{\Phi})$ Store $Q(\Theta)$ for plotting etc. Input n
Calculate $S_n(\Theta)$
Calculate $R(n)$
Store $R(n)$ for plotting etc.

5. Conclusions

The differences between (i) and (vii) are quite small and vary for any given pair of models on densities $f(x)$, weights $w(x)$, parameters Θ and end points a and b. Designs (i) and (vii) are usually comparable and much better than (iii), but (vii) is only of theoretical importance since it cannot be achieved in practice.

In most instances a near optimal model discrimination will be achieved by selecting design points in a geoemetric progression.

6. References

1. W.G. Bardsley, P.B. McGinlay and A.J. Wright, The F test for model discrimination with exponential functions, *Biometrika* **73**:501-508 (1986).
2. W.G. Bardsley and P.B. McGinlay, The use of nonlinear regression analysis and the F test for model discrimination with dose-response curves and ligand binding data, *J. theor. Biol.* **126**:183-201 (1987).
3. W.G. Bardsley, P.B. McGinlay and M.G. Roig, Conditions when statistical tests for model discrimination have high power. Some examples from pharmacokinetics, ligand binding, transient and steady state enzyme kinetics, *Biophys. Chem.* **26**:1-8 (1987).
4. W.G. Bardsley, P.B. McGinlay and M.G. Roig, Optimal design for model discrimination using the F test with nonlinear biochemical models. Criteria for choosing the number and spacing of experimental points, *J. theor. Biol.* **139**:85-102 (1989).
5. W.G. Bardsley, SIMFIT - A computer package for simulation, curve fitting and statistical analysis using life science models, this volume.

GEPASI:
A USER ORIENTED
METABOLIC SIMULATOR

Pedro Mendes

Dept. of Biological Sciences
University of Wales
Aberystwyth, Dyfed SY23 3DA, United Kingdom

1. Purpose

Gepasi (as in GEneral PAthway SImulator) is a software package for the simulation of the dynamics, steady-state and control analysis of metabolic pathways.

2. Availability and Costs

Gepasi is available from the author, who can be contacted by post at the above address, or via electronic mail to prm@aber.ac.uk. The package is supplied by post only if suitably formatted diskettes are sent to the author (with a minimum of 720 Kb free).

Gepasi 2.0 is free software. This does not mean that it is not protected by copyright (it is not in the public domain) but that it can be freely copied in its original form, in binary or source code.

3. Documentation

Gepasi is being distributed with an extensive interactive help system that should cover its usage. A user's manual is being prepared to be distributed with the package in the future (in electronic format, either as PostScript or another format suitable for printing).

4. Hardware and Software Requirements

Gepasi runs on personal computers with MS-DOS version 3.2 or above and MS-Windows version 3.0 or 3.1. Machines equipped with 80386 or 80486 processors and at

Modern Trends in Biothermokinetics, Edited by
S. Schuster *et al.*, Plenum Press, New York, 1993

least 1 Mbyte of RAM are required to run the user-interface. No other special hardware is required but a numeric co-processor is recommended. The simulation engine (see below) is sufficiently portable across hardware platforms to be successfully compiled for other machines running other operating systems (e.g. UNIX). However, the supplied user-interface runs only under MS-Windows.

5. Limitations

Gepasi version 2.0 is able to simulate pathways of any complexity up to 40 steps and/or 40 metabolites (internal and external). Since the package is being constantly updated these limitations are likely to change in the future (of course, to become less restrictive).

6. Language

Gepasi is written in the C programming language. The user-interface requires the libraries in the Microsoft Windows Development Kit. The simulating engine has been compiled with several C compilers, including Microsoft C (MS-DOS) and GNU gcc (UNIX). The source code is compatible with the ANSI standard. As a rule, ANSI-compliant compilers should compile the program without problems.

7. Description

Gepasi is a software package for the simulation of metabolic models, both of their transient (dynamic) and steady states. The simulations produce values for the concentrations of metabolites and the magnitudes of fluxes starting from an initial state. The parameters of these models are the kinetic constants of the steps and the concentrations of metabolites that are either fixed (buffered) or constantly flowing into or out of the system. This package was used to obtain data already published[1] (see also Ref. 2).

The user interaction is handled by a program that runs under MS-Windows (presently on IBM-PC-compatible computers). This program makes extensive use of menus, dialogue boxes, push buttons, radio buttons, list boxes and other controls. This form of input, accompanied by keyboard short cuts, minimizes the time taken by a first-time user to get accustomed to the mechanics of the program. By taking advantage of MS-Windows' own help engine, this front-end program has an extensive help system that covers not only immediate instructions on how to use the package but also has explanations of common concepts in metabolic control, for example, those of internal or external metabolites. There is also a section with full references to articles, reviews and books covering subjects related to metabolic control and modeling. This feature should ensure that Gepasi will be a useful tool for education.

From the user's point of view, the model is specified in two parts: first, one must input the stoichiometric structure of the model, the kinetics associated with each step and which metabolites are modifiers of which reactions; then one must specify the numeric values of the kinetic constants and the initial metabolite concentrations. The numeric data were separated from the structural data since one is likely to input many different sets of parameters for the same structural model. This input is clearly separate as it is done in two separate windows, of which there can be several instances (each one pertaining to a different model) on screen at the same time. The output of the simulated data is formatted as a plain text file, for ease of reading, and/or in data files in columnar format. For the latter the user can specify which element (parameters or variables) will appear in each column. The columns can be separated by spaces, tabs or commas so that this file can be imported into most popular spreadsheet or graphics programs. There is also control over the column width and whether titles are included at the top or bottom of the columns.

Gepasi 2.0 has implemented one feature that is particularly useful for an extensive study of a model. The user can instruct the simulator to scan various parameters. The simulator will produce several simulations in sequence (each with different values for those parameters) and put the output on different rows of the same file, effectively producing a map of the behavior of the model (its variables) in a region of parameter space. Virtually all parameters of the model can be scanned (but unless the model is small they will not all be scanned as the number of simulations would become prohibitive). These 'scans' can be done with linear or logarithmic intervals: lower and upper limits and the number of intervals must be specified for each parameter. They can also be done by assigning random values (with linearly or logarithmically uniform distributions) for the parameters within lower and upper limits. In this case a total number of simulations is selected. The 'random scans' are useful when one is trying to characterize the model in many dimensions. In this case it is the total number of simulations that is set, not a grid mesh. If one were to scan n parameters, each with 10 regular intervals, one would effectively be asking Gepasi to produce 10^n simulations, while one could instead have, for example, 100 simulations with these n parameters being assigned random values (within the defined boundaries). This feature is useful for producing training sets for artificial neural networks that would 'learn' the system's properties[3].

While one may be interested in scanning some parameters freely across parameter space, one might want to restrict others to take values as a function of those that are being scanned. This is useful for cases in which one wishes to change some kinetic constants whilst adjusting others to maintain the same equilibrium constants. In Gepasi these functions are called links and at present there is only one type of link available: the linked parameter as the product of a constant and the master parameter. In the future other link functions will be implemented.

The data for transient analysis of the model are obtained by integration of the ordinary differential equations (ODE) that describe the change of concentrations with time. These are determined from the stoichiometric structure of the pathway and the rate laws of each step.

The data for steady states are calculated by setting all equations equal to zero and solving the resulting system of (nonlinear) differential equations[4]. The algorithm used for the integration of the ODE is the Livermore Solver of Ordinary Differential Equations with Method Switching (LSODA, see Ref. 5), which uses a procedure that detects if the system of ODE is stiff[6] or not and uses the most efficient method for each case, respectively the Adams or backward differentiating formula (BDF) methods[6,7]. LSODA is impressively fast compared to the previous algorithm used in Gepasi version 1 (see Ref. 8) which was itself much faster than the well-known Runge-Kutta. This speed made the use of integration combined with the damped Newton method[9] feasible to solve the steady state. Whenever the Newton method fails to converge (either by reaching the limit of iterations or by getting stuck in local minima) the estimate is used as a starting point and the system is integrated for 10 units of time, then the solution is used again as the estimate for the Newton method. If it fails once more, it will be integrated for 100 units of time, and this process goes on for 10 iterations (over 10^{10} units of time). If no steady state is reached the program informs the user about it. This method of getting out of local minima or improving the initial estimate is much more robust than that of retrying the Newton method by randomly perturbing the initial guess (as in the former case it is in reality driving the guess towards the solution and not making it jump in a random direction).

Steady states can be analyzed with metabolic control analysis (MCA)[10-12]. Gepasi calculates the elasticity coefficients from the partial derivatives of the rate equations and control coefficients using the method described by Reder[13]. It is able automatically to detect moiety-conserved cycles[14,15] by reducing the stoichiometry matrix using Gaussian elimination. Transition times[16] are also computed. The user can select from any elasticities, concentration- or flux-control coefficients for output, as well as products of elasticities and control coefficients.

Gepasi is composed of two separate entities: a user interface that handles all the interactions with the user, and a simulating engine that performs the actual numeric calculations.

This modular structure of the software was designed to keep the numerical module portable (there may be the need to have simulations running on high speed computers) whilst still having a front-end supported by a graphical user interface (GUI). It leaves open the possibility of the development of a language-based description of the models and retaining the GUI based interface. At present, though, only the GUI has been implemented.

Acknowledgements

I thank the J.N.I.C.T., Portugal, for financial support, Pedro Moniz Barreto for testing earlier development versions of this program and Douglas Kell both for suggestions as to the functionality of the program and especially for his constant support and encouragement.

References

1. P. Mendes, D.B. Kell and H.V. Westerhoff, Channelling can decrease pool size, *Eur. J. Biochem.* **204**:257-266 (1992).
2. P. Mendes and D.B. Kell, Control analysis of metabolic channeling, this volume.
3. D.B. Kell, C.L. Davey, R. Goodacre and H.M. Sauro, When going backwards means progress: On the solution of biochemical inverse problems using artificial neural networks, this volume.
4. J.-H.S. Hofmeyr, Steady-state modelling of metabolic pathways: a guide for the prospective simulator, *Comput. Applic. Biosci.* **2**:5-11 (1986).
5. L. Petzold, Automatic selection of methods for solving stiff and nonstiff systems of ordinary differential equations, *SIAM J. Sci. Stat. Comput.* **4**:136-148 (1983).
6. C.W. Gear, Numerical initial value problems in ordinary differential equations, Prentice-Hall, Englewood Cliffs (N.J.) (1971).
7. A.C. Hindmarsh, LSODE and LSODI, two initial value ordinary differential equation solvers, *A.C.M. SIGNUM Newsletter* **15**: 4 December, 10-11 (1980).
8. P. Mendes, GEPASI - A simulator of the dynamical behaviour of biochemical pathways, *in*: "Biothermo-kinetics," (H.V. Westerhoff, ed.), Intercept, Andover (U.K.), in the press.
9. S.D. Comte and C. de Boor, "Elementary Numerical Analysis. An Algorithmic Approach," McGraw Hill, Singapore (1980).
10. H. Kacser and J.A. Burns, The control of flux, *Symp. Soc. Exp. Biol.* **27**:65-104 (1973).
11. R. Heinrich and TA. Rapoport, A linear steady-state treatment of enzymatic chains. General properties, control and effector strength, *Eur. J. Biochem.* **42**:89-95 (1974).
12. A. Cornish-Bowden and M.L. Cárdenas, eds, "Control of Metabolic Processes," Plenum Press, New York, (1990).
13. C. Reder, Metabolic control theory: a structural approach, *J. theor. Biol.* **135**:175-201 (1988).
14. J.G. Reich and E.E. Sel'kov, "Energy Metabolism of the Cell. A Theoretical Treatise," Academic Press, London (1981).
15. J.-H.S. Hofmeyr, H. Kacser and K.J. Van der Merwe, Metabolic control analysis of moiety-conserved cycles, *Eur. J. Biochem.* **155**:631-641 (1986).
16. J.S. Easterby, A generalized theory of the transition time for sequential enzyme-reactions, *Biochem. J.* **199**: 155-161 (1981).

OBJECT ORIENTED SIMULATION
OF METABOLIC PROCESSES

H.J. Stoffers, B. van Stigt, P. Richard, A.A. van der Gugten,
H.V. Westerhoff and G.J.F. Blommestijn

Divisions of Molecular Biology and Biophysics
The Netherlands Cancer Institute
Plesmanlaan 121, NL-1066 CX Amsterdam, The Netherlands

1. Introduction

Simulation of metabolic processes in cells is usually done by solving sets of coupled differential equations (using software systems like for instance SCoP[1] or MLAB[2]). Each equation represents the change per unit of time in the concentration of a certain metabolite or other substance. Such a rate of change depends on the rates of all chemical reactions that produce or consume that substance. In the modeling and simulation approach we present here, we focus on the essential independence of the underlying reactions.

The rate of a biochemical reaction in the cell depends on the concentrations of its substrate(s), product(s), enzyme(s) etc. These concentrations are the variables in the rate equation of the reaction and are the only way in which 'the outside world' influences a reaction. This independent action of the chemical processes together with their restricted 'interface' with the outside world, and the fact that the 'actors' in the processes as well as the processes themselves belong to a limited set of object classes, makes coupled systems of chemical reactions well suited for object oriented modeling. In its purest sense object oriented programming is implemented by sending messages to objects that are occurrences of a set of classes[3]. In our approach the central messages being the concentrations of the substances, are sent to objects being the reactions in which they play a role in the rate equation.

Using the object oriented programming language C++[4] and with the above mentioned principles as a guideline, we created a close and natural mapping of biochemical concepts onto 'classes' (object types consisting of data types and corresponding operations). The two most important basic ('base') classes we constructed were 'substance' and 'process'. From the class 'process' a variety of process types or reaction types was derived, forming a library of predefined reaction classes. Occurrences of specific process classes are the individual reactions that constitute a model of a coupled system of reactions. The coupling between the reactions is achieved by giving the substances 'roles' with respect to the different reaction objects.

Modern Trends in Biothermokinetics, Edited by
S. Schuster *et al.*, Plenum Press, New York, 1993

2. Model Specification

A model of a set of coupled biochemical reactions is specified by defining a number of objects, namely substances and processes, of various predefined types in a C++ source file. As an example the abbreviated version of the file glycmodl.C, containing a model definition of glycolysis in yeast (see Lehninger[5]) is given in Listing 1 (see Appendix). Every model thus specified is compiled and linked with an underlying "kernel" that includes the simulation's time keeping variables, an integrator which can perform timesteps for the model, and a command driven user interface. The predefined process types are drawn from a library archive. To add new process types that are not yet avaliable from the library, a limited amount of C++ programming knowledge is required.

3. Initialization

When the program is run, the objects representing components of the system are initialized (see Fig. 1) - most notably, the processes store substances in their various "roles" - and subsequently inserted into lists (see Fig. 2) that are managed by the simulation kernel. Then a command driven interface, a "shell", takes over and sits waiting for user input to process (see Listing 2 in the Appendix). The shell understands commands to run the simulation for a specific time interval, to inspect or edit the parameters of various objects, such as initial concentrations of metabolites and rate constants of processes. The shell also understands commands to read parameters from or save parameters to file, and to redirect simulation output to files or devices.

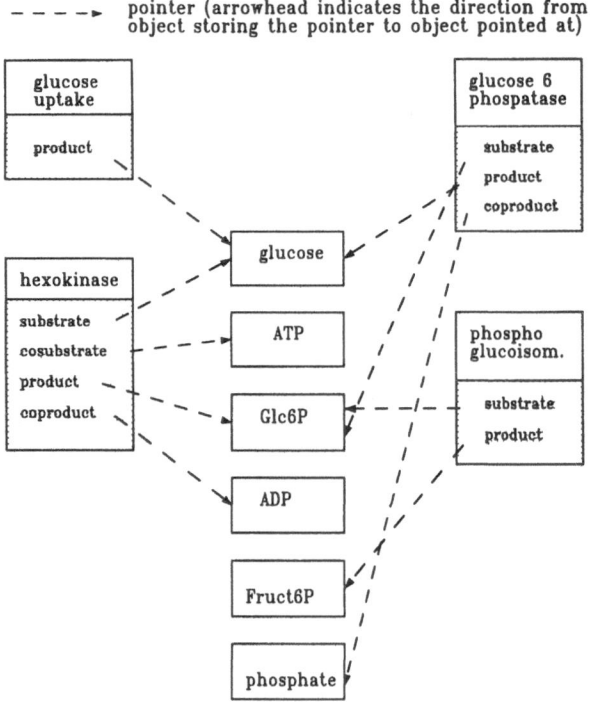

Figure 1. Process objects such as 'glucose uptake', 'hexokinase', etc. internally store (pointers to) substance objects (like 'glucose', 'ATP' etc.), that assume various "roles", such as 'substrate', 'product', etc. (example: part of glycolysis model).

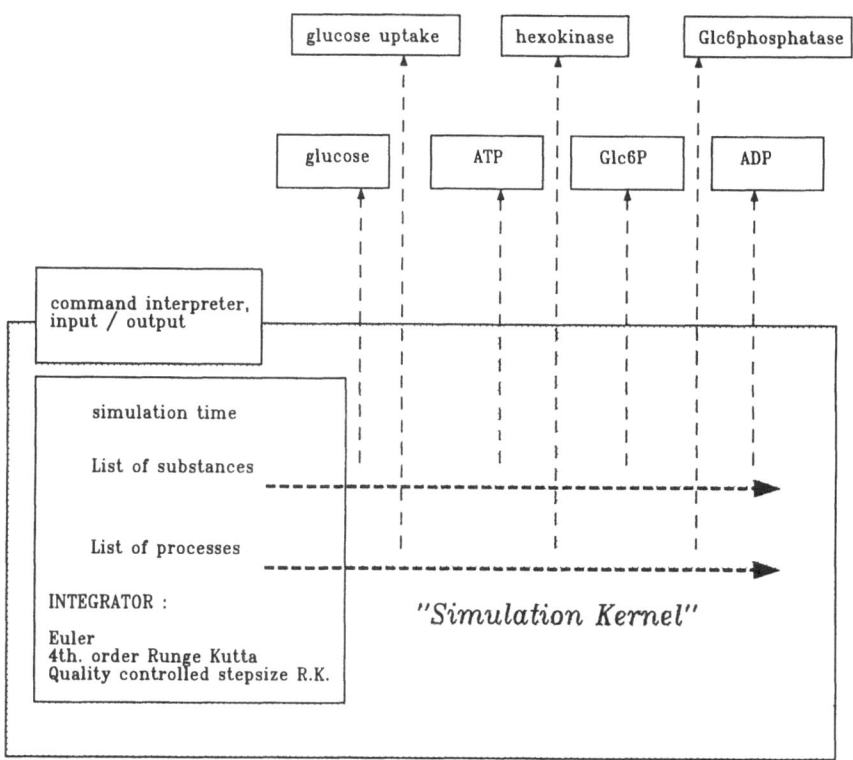

Figure 2. Model specific static objects (substances and processes) insert themselves (pointers to themselves) in the lists of the model a-specific "simulation Kernel" (example: part of glycolysis model).

4. Library of Process Types

The library of process types (a small sample of the process types currently available and used in models of histone gene expression in *Xenopus Laevis* and glycolysis in yeast is listed in Table 1) can be extended without having to adapt or recompile the code of the simulation kernel. Although different process types have different sets of parameters and different rate equations, all process types are derived from a generic class *process_t* which is equipped with virtual functions to edit parameters, calculate a process rate, etc. All "kernel code" that deals with processes is written in terms of the *process_t* base class.

5. Portability

The source code for the library of substances and process objects and the "simulation kernel" is portable to any platform that supports C++ and has a standard (ANSI) C library. It is currently used on PC's and on a Silicon Graphics (UNIX) workstation. The price for this portability is the soberness of the output and user interface: it is in textmode only. A separate module in the simulation kernel to display graphs of metabolite concentrations and process rates changing in time is constructed for the PC (with Hercules, EGA, and VGA adaptors) using the Borland Graphical Interface (BGI) library and for UNIX workstations using the Silicon Graphics Graphics Library (GL).

Table 1. Library of process types (three examples).

`constant_rate_t`

Process with constant rate of production and/or consumption. E.g. the constant influx of a metabolite that is abundantly present in the environment of a metabolic system.

$$rate = V$$

Run time parametrization option:

V   real number ≥ 0, M/sec.

The process must be initialized with at least one substrate or one product. Optionally, it can by initialized with up to four substrates and up to four products. Substrates are consumed at a rate of *(process rate * substrate stoichiometry)*. Products are produced at a rate of *(process rate * product stoichiometry)*.

`irrevbl_MM1_t`

Irreversible process with Michaelis Menten kinetics, of which the rate is limited by one substrate, S_1.

$$rate = V_{max} \cdot \frac{[S]}{[S] + K_S}$$

Run time parametrization options:

V_{max} real number ≥ 0, M/sec

K_{S1} real number > 0, mM

The process must be initialized with at least one substrate. Optionally, it can by initialized with up to three additional substrates and up to four products. Substrates are consumed at a rate of *(process rate * substrate stoichiometry)*. Products are produced at a rate of *(process rate * product stoichiometry)*.

`revbl_MM1_t`

Reversible process with Michaelis Menten kinetics with one substrate, S_1, limiting the rate of the forward reaction, and one product, P_1, limiting the rate of the reverse reaction.

$$rate = \frac{V_f \cdot \dfrac{[S]}{K_S} - V_r \cdot \dfrac{[P]}{K_P}}{1 + \dfrac{[S]}{K_S} + \dfrac{[P]}{K_P}}$$

Run time parametrization options:

V_f real number ≥ 0, M/sec

V_r real number ≥ 0, M/sec

K_{S1} real number > 0, mM

K_{P1} real number > 0, mM.

The process must be initialized with at least one substrate and one product. Optionally, it can by initialized with up to three additional substrates and up to three additional products. Substrates are consumed at a rate of *(process rate * substrate stoichiometry)*. Products are produced at a rate of *(process rate * product stoichiometry)*.

References

1. J. M. Kootsey. "Introduction to Computer Simulation," National Biomedical Simulation Resource, Duke University Medical Center (NC) (1989).
2. B. Bunow and G. Knott. "MLAB, A Mathematical Modeling Laboratory," Civilized Software, Bethesda (MD) (1990).
3. L. J. Pinson and R. S. Wiener. "An Introduction to Object Oriented Programming and Smalltalk," Addison-Wesley, Reading (MA) (1988).
4. M. A. Ellis and B. Stroustrup. "The Annotated C++ Reference Manual," Addison-Wesley, Reading, MA (1990).
5. A. L. Lehninger. "Biochemistry," Worth Publishers, New York (1970).

Appendix

Listing 1. Model specification: Glycolysis in yeast.

```
/*      glycmodl.C
Model definition file for simulation of glycolysis "1st. and 2nd. stage" (see Ref. 5, p. 315-321).
*/
#include "cr.h"
#include "cp1iirs2.h"
#include "irrmm1.h"
#include "irrmm2.h"
#include "irrmm3.h"
#include "revmm1.h"
#include "revmm2.h"
#include "revf1r2.h"
/*      substances, metabolites, ..."
*/
        substance_t             glucose("D-glucose"),
                                ATP("ATP"),
                                ADP("ADP"),
                                Glc6phosphate("D-glucose 6-phoshate"),
                                phosphate("phosphate"),
                                Fruct6phosphate("D-fructose 6-phosphate"),
                                        .
                                        .
                                        .
        <other substances, metabolites, ...>
                                        .
                                        .
                                        ;
/*      processes (enzymes) of the 1st stage ...
*/
        constant_rate_t         GlcUptake("glucose uptake", product_t(glucose));
        cpinh_irrevbl_s2p1_t    hexokinase("hexokinase",
                                        substrate_t(glucose),
                                        cosubstrate_t(ATP),
                                        product_t(Glc6phosphate),
                                        coproduct_t(ADP)
                                );
        irrevbl_MM1_t           Glc6_phosphatase("D-glucose 6 phosphatase",
                                        substrate_t(Glc6phosphate),
                                        product_t(glucose),
                                        coproduct_t(phosphate)
                                );
        revbl_MM1_t             phosphoglucoisomerase("phosphoglucoisomerase",
                                        substrate_t(Glc6phosphate),
                                        product_t(Fruct6phosphate)
                                );
                                        .
                                        .
                                        .
```

471

```
/*      processes of the 2nd. stage ...
*/
        irrevbl_MM3_t   Gly3Phosph_dehydrogenase(
                                "Glyceraldehyde 3p-phosphate dehydrogenase",
                                substrate1_t(glyceraldehyde_3phosph),
                                substrate2_t(NADplus),
                                substrate3_t(phosphate),
                                product1_t(diphosphoglycerate),
                                product2_t(NADH)
                        );
        revbl_MM2_t     phosphoglycerate_kinase("phosphoglycerate kinase",
                                substrate_t(diphosphoglycerate),
                                cosubstrate_t(ADP),
                                product_t(phosphoglycerate3),
                                coproduct_t(ATP)
                        );
        revbl_MM1_t     phosphoglyceromutase("phosphoglyceromutase",
                                substrate_t(phosphoglycerate3),
                                product_t(phosphoglycerate2)
                        );
        revbl_MM1_t     enolase("enolase",
                                substrate_t(phosphoglycerate2),
                                product_t(phosphoenolpyruvate)
                        );
                                .
                                .
                                .
        <other processes of the 2nd stage ...>
                                .
/*** end of file glycmodl.C */
```

Listing 2. Command-driven user interface.

```
do {
get commandline from user
if command is "edit" then {
send message to the requested object(s) to edit its/their parameters.
}
else if command is "save" then {
send message to requested object(s) to save its/their parameters to the current parameter file
}
else if command is "display" then {
send message to requested object to insert its displayable attribute ("rate" for a process, "molar concentration"
for a substance) into a graphics screen.
}
...
else if command is "run" then {
run simulation with current parameter values and current output and display options.
}
} while command not is "exit".
close all files and terminate
```

MetaCon - A COMPUTER PROGRAM FOR THE ALGEBRAIC EVALUATION OF CONTROL COEFFICIENTS OF METABOLIC NETWORKS

Simon Thomas and David A. Fell

School of Biological & Molecular Sciences
Oxford Brookes University
Gipsy Lane, Headington
Oxford OX3 0BP, United Kingdom

1. Introduction

A computer program, MetaCon, has been developed which fully automates all stages of carrying out a control analysis to give values for the flux control, concentration control and branch-point distribution control coefficients of a metabolic pathway. The program input is a text file containing processing instructions and the pathway details. The pathway is analyzed for the presence of branches and moiety-conserved cycles. The results of this stage are used along with inferred and additional elasticity terms to build an *elasticity matrix* of the form of the Fell and Sauro matrix method[1,2], though the process involves symbolic evaluation of Reder's equations[3]. Inversion of the matrix yields the expressions for the control coefficients in terms of the elasticities, metabolite concentrations and fluxes which appear in the elasticity matrix. All calculations are carried out algebraically rather than numerically, so that if the values of one or more variables are unknown, the control coefficients will be produced as polynomials in terms of the unknown(s).

The processes which can be carried out by MetaCon are:

1. Analyze the pathway, and create the corresponding elasticity matrix. This step is the actual Metabolic Control Analysis.

2. Solve the matrix equation to produce algebraic expressions (they are actually polynomials in several variables) for the control coefficients. Whenever MetaCon is run, the user specifies one reaction flux, called the *reference flux*. The flux control coefficients of all steps in the pathway on the reference flux are calculated. The user can also request the control coefficients of all steps on all variable metabolite concentrations, and in a branched pathway,

the branch-point distribution coefficient(s) for the flux(es) through any specified branch, as defined in Ref. 2.

3. Modify these equations, by substituting equations or numerical values from the equations section of the input file, for any of the variables contained in the elasticity matrix. These equations can contain additional 'user-defined' variables, which themselves can be further substituted. An obvious example of the use of this facility is to enter equations for elasticities of reactions, which are derived from their kinetic expressions.

The only limitations of the form of network which MetaCon can analyze are that all stoichiometries in the pathway equations must be integral, and the network must be fully connected by the flow of mass — all metabolites and steps must form a continuous network. Systems composed of physically disconnected networks which are only connected by regulatory effects (e.g. cascades, gene expression, hormone action) cannot be analyzed, though this capability could be incorporated into the program.

MetaCon eliminates the time-consuming and error-prone processes of:

1. creating the elasticity matrix, by applying the matrix method rules;
2. calculating the control coefficients by inverting the elasticity matrix;
3. substituting values or expressions for any of the variables in the elasticity matrix.

Additionally, it can carry out a sensitivity analysis to determine the sensitivity of all control coefficients, C, to each variable v, which appears in the elasticity matrix, in terms of both unscaled sensitivities, $\partial C / \partial v$, and scaled sensitivities $(\partial C / \partial v) \cdot (v / C)$. Small and Fell[4] have previously calculated the unscaled sensitivities of the control coefficients of the experimental gluconeogenesis system described in the following section, a process which can now be carried out by MetaCon.

MetaCon is intended primarily to be an aid to experimentalists in three ways. First, to show which variables need to be measured, by producing the elasticity matrix. Second, to analyze the resulting experimental data and so produce expressions and/or values for the control coefficients. Third, to help determine the degree of accuracy which is required when ascertaining the values of variables. However, these aims do not preclude its use in theoretical investigations, and we feel that MetaCon will also be of use to theoreticians.

The program is written completely in ANSI C (cf. Ref. 5) to aid portability between systems. Most of the development has been carried out on an IBM/PC compatible, and a version has been implemented on a DEC 2100 workstation. The code is almost 100% compatible between the two systems. As a example of MetaCon's uses, we have used it to carry out further analysis of the data from a rat liver gluconeogenesis system, in addition to that reported by the original authors.

2. An Investigation into the Control of an Experimental Gluco-neogenic System

We have chosen the gluconeogenesis system investigated by Groen and coworkers[6,7], as it is one of the largest experimental systems to which control analysis has so far been applied. Flux control coefficients of the system, which comprises a total of 14 reactions and transport steps in the conversion of pyruvate to glucose, have been published previously[6-8], as have some concentration control coefficients[8]. On each occasion a different lumping of the reactions in the system has been used. In this example we have used the most detailed system, that of Groen[6], in which the 14 processes have been lumped into 10 separate steps. We have used this system because MetaCon can easily handle a pathway with this number of steps, so the maximum possible amount of control information is obtained. The flux through glucose 6-phosphatase (i.e. the flux to glucose) was chosen as the reference flux, and the flux through pyruvate kinase (PK) was chosen as the branch-point flux. The investigation was carried out on isolated rat liver parenchymal cells in a perifusion system. The control coefficients were calculated for the four steady states obtained in the absence of glucagon, when the cells were titrated with 0.5mM, 1mM, 2mM and 5mM lactate respectively, with a constant lactate:pyruvate ratio of 10:1. The same five steps had the majority of control over the reference flux, branch-point distribution and the concentration of phospho*enol*pyruvate (PEP). PEP was chosen to illustrate concentration control coefficients because of its important position at the pyruvate kinase branch-point. The five important PEP concentration control coefficients and PK branch-point distribution control coefficients for each of the different steady states are illustrated in Figs 1 and 2, respectively.

The flux control coefficients over the reference flux (not illustrated) are similar to the values given by Groen[6], but do show some differences. These differences in the values arise from a combination of two ways in which our calculations differ from those of Groen:

1. The exact values used for the flux through pyruvate kinase. We estimated our values at each steady-state gluconeogenic flux from Fig. 6 of Ref. 9, so they might differ from the values used by Groen[6].

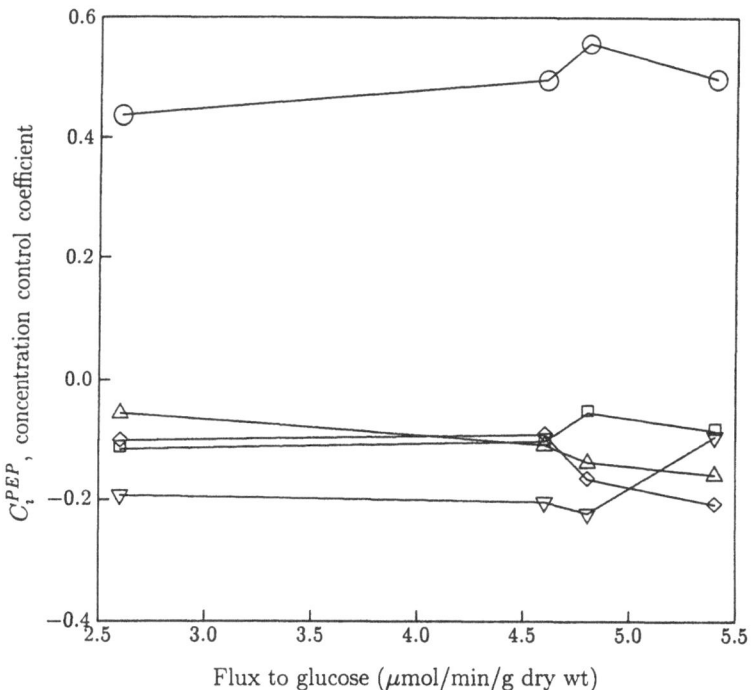

Figure 1. Variation of phosphoenolpyruvate concentration control coefficients with steady-state gluconeogenic flux. $C_i^{\text{BP-PK}}$ for i = O pyruvate carboxylase; $\triangle$ pyruvate kinase; ∇ enolase/phosphoglyceromutase; $\square$ glyceraldehyde-3-phosphate dehydrogenase/phosphoglycerate kinase; $\lozenge$ triose phosphate isomerase/aldolase/fructosebisphosphatase. The fluxes of 2.6, 4.6, 4.8 and 5.4 μmol/min/g dry wt are those obtained by titrating the cells with 0.5mM, 1mM, 2mM and 5mM lactate respectively, in the absence of glucagon, with a constant lactate:pyruvate ratio of 10:1.

2. In his calculations, Groen[6] also included the adenine nucleotide translocator, which, in the absence of glucagon, can exert a considerable degree of control over the flux to glucose. To include this step into an analysis by MetaCon would involve incorporating ATP and ADP as variable metabolites, and calculating or estimating the elasticities of the adenine nucleotides to those steps in which they are involved.

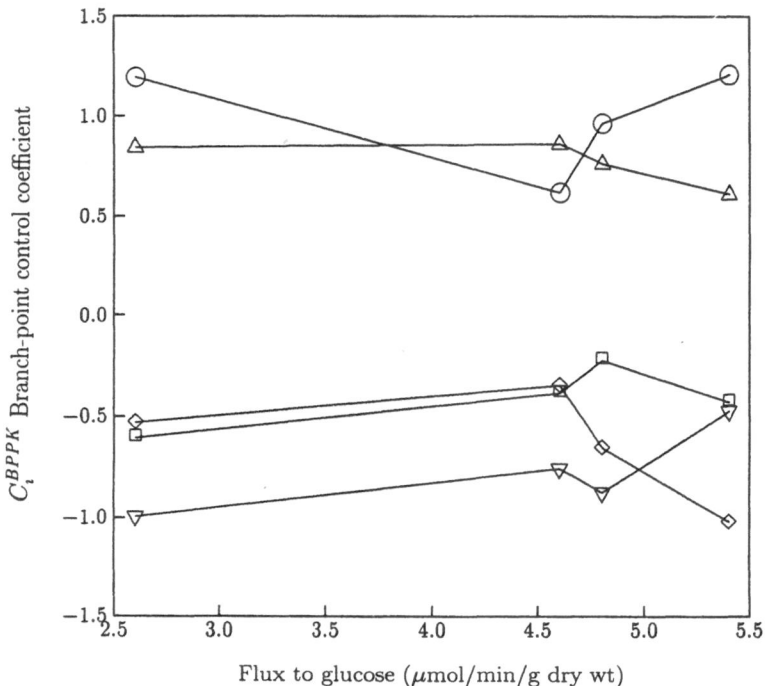

Figure 2. Variation of pyruvate kinase branch-point distribution control coefficients with steady-state gluconeogenic flux. Symbols and details of flux values are the same as for Fig. 1.

Some lines from the input file for one of the analyses described in this section are given in Table 1. The input file comprises:

    1. Commands to control what processing is to be carried out (not shown).
    2. The reaction scheme (starting with the word REACTIONS).
    3. Optional equations for the values of any or all variables which are defined in the scheme (starting with the word EQUATIONS).

A line comprising only a single dot has been used to specify that the following lines in the file are similar to the one above the dot. Some points to notice about the file are:

1. Each line in the REACTIONS section corresponds to a step for which control coefficients are to be calculated, and can comprise one or more real reactions and/or processes.

2. Metabolite and reaction names can be of any length. Reaction names are optional, and if used, are written in square brackets at the start of each reaction line. The names of pool metabolites, i.e. those whose concentrations remain approximately constant during the investigation, start with a $ sign.

3. Nothing between a # symbol and the end of the line is read by MetaCon, so this character can be entered by the user to enable comments about the file to be included.

Table 1. Example of an input file of MetaCon.

```
REACTIONS

        $PYRc = PYRm          # 1. Mit. pyr. translocator
[PC]    PYRm = OAAm           # 2. PC
        OAAm = OAAc           # 3. OAA transport to cytosol
        OAAc = PEP            # 4. PEPCK
[PK]    PEP = $PYRc           # 5. PK
        PEP = PGA3            # 6. Enolase/PGM
        PGA3 = GAP            # 7. PGK/GAPDH
        2 GAP = F6P           # 8. TIM/Aldolase/FBPase
[PGI]   F6P = G6P             # 9. PGI
[G6Pa]  G6P = $Glu            # 10. G6Pase

EQUATIONS

e,1,PYRm = -6.1              # Elasticity values
        .
e,9,F6P = 1/(1-Rpgi)        # Elasticities defined in
e,9,G6P = -Rpgi/(1-Rpgi)    # terms of a new variable

Rpgi = 0.95                 # Mass action ratio/Keq for PGI
                            # - a user-defined variable

J,10 = .54                  # Fluxes - divided by 10 to
J,5 = .42                   # make results more manageable

J,9 = J,10                  # Fluxes defined in terms of
J,8 = J,9                   # other fluxes.
J,7 = 2 J,8
J,6 = J,7
J,4 = J,5 + 2*J,10
J,3 = J,4
J,2 = J,3
J,1 = J,2
```

References

1. D.A. Fell and H.M. Sauro, Metabolic control and its analysis: additional relationships between elasticities and control coefficients, *Eur. J. Biochem.* **148**:555-561 (1985).
2. H.M. Sauro, J.R. Small and D.A. Fell, Metabolic control and its analysis: extensions to the theory and matrix method, Eur. J. Biochem. **165**:215-221 (1987).
3. C. Reder, Metabolic control theory: a structural approach, *J. theor. Biol.* **135**:175-201 (1988).
4. J.R. Small and D.A. Fell, Sensitivity of control coefficients to elasticities, *Eur. J. Biochem.* **191**:413-420 (1990).
5. B.W. Kernighan and D.M. Ritchie. "The C Programming Language," Prentice-Hall, Englewood Cliffs (N.J.) (1988).
6. A.K. Groen. "Quantification of Control in Studies on Intermediary Metabolism," Ph.D. Thesis, University of Amsterdam (1984).
7. A.K. Groen, C.W.T. van Roermund, R.C. Vervoorn and J.M. Tager, Control of gluconeogenesis in rat liver cells. Flux control coefficients of the enzymes in the gluconeogenic pathway in the absence and presence of glucagon, *Biochem. J.* **237**:379-389 (1986).
8. A.K. Groen and H.V. Westerhoff, Modern control theories: a consumers' test, *in*: "Control of Metabolic Processes," A. Cornish-Bowden and M. L. Cárdenas, eds, Plenum Press, New York (1990).
9. A.K. Groen, R.C. Vervoorn, R. Van der Meer and J.M. Tager, Control of gluconeogenesis in rat liver cells. 1. Kinetics of the individual enzymes and the effect of glucagon, *J. Biol. Chem.* **258**:14346-14353 (1983).

VII. OENOLOGICAL EPILOGUE

CHAMPAGNE IN HOLLAND:
AN EXPERIMENT IN OENOLOGY

Henny Daams and Ankie Daams-Moussault

Château Moussault
Caves de Rue Ban
Banstraat 38, 1071 KA Amsterdam, The Netherlands

1. Introduction

Two thousand years ago viticulture was introduced into the Netherlands by the Romans. Since then there have been vineyards in the surroundings of Maastricht, where the cretaceous soil and the microclimate are favorable for grape growing. Reglementation during the French occupation caused a temporary disappearance of these small vineyards, but recently some new ones have been installed on the slopes of Mount Saint Peter.

Here, we demonstrate that in the northern provinces of our country viticulture is also possible. Further, we describe how on the Château Moussault a reasonable champagne is made, combining traditional methods with modern biochemical techniques.

2. The Vineyards of South-Amsterdam

As the name suggests, these vineyards are situated in the warmest part of the region, where the fronts of the houses capture every ray of the scarce sun. The soil here is excellent for the vines because there is a lot of chalk rubble underneath, thanks to our forefathers who demolished useless old monuments existing before.

The grape cultivated here is the "White Vanderlaan", a variety of the White Riesling. This is a race well adapted to the rather harsh conditions of the winters in Holland.

After vintage, the grapes are carefully selected. Only the best, sun ripened grapes are used for the Champagne Moussault. The vineyards of the castle produce about 45,000,000 mg each year. Invertebrates like centipedes, spiders and ladybirds are removed. In this way we avoid the typical spidery taste that is characteristic of cheaper wines.

3. Vinification

Traditional methods are used in the chai. The bunches are first submersed in boiling water for a few seconds, removing air pollution and unwanted bad yeast. The grapes are then crushed and put into an enormous wooden press, that is operated by hand. Only must from the first pressing is taken for the production of champagne, from the second pressing a simple wine is made.

The must is poored into huge old casks. Contrary to common usage, no sulphite is added. This requires extremely hygienic conditions, but we are proud to say that in 25 years we have never had an infection!

To start fermentation, a good strain of champagne yeast is obtained from Rob Vangelder Ltd. (Amsterdam). It is first activated with a small amount of boiled must and then distributed over the casks. During 7 days, each day about 58 mMol sucrose per liter is added and the alcohol content is continuously controlled, until there is 10% v/v alcohol. Fermentation slows down and casks are placed in the cave.

In the course of the following 6 - 8 months the débris settles, and to remove this the wine is transferred several times to other casks so that it gets perfectly clear.

Figure 1. Pressing of the grapes. As usual in Holland, workers are wearing wooden shoes.

4. Méthode Champenoise

The wine is then put into thick, heavy bottles, together with some active champagne yeast and 70 mMol sucrose per bottle, enough to produce a pressure of 6 atmospheres. The cork is secured with a basket of iron threads.

After two months the second fermentation is finished, and bottles have to rest now for 1 -2 years in the cave. Yeast cells undergo autolysis, the débris giving the typical champagne flavor. At the end of this period, bottles are placed in a 45 ° tilted position, bottoms up. During several weeks there are turned twice a day 90 ° around their axis so that the sediment moves to the cork.

When finally all sediment has been collected on the cork, the bottles are put upside down in an ethanol bath of - 25 °C. As soon as the neck is frozen the cork is removed, resulting in the expulsion of ice plus sediment. The bottles are supplemented with some old wine and a small bit of sucrose, they receive a new cork with basket and a label of the Château Moussault.

Acknowledgements

The authors would like to thank the organizers and participants of the 5th Biothermokinetics Meeting for their interest in this work and for testing the results.

INDEX

Protoplasts, plant, 61
Protozoa, 423
Pseudo first-order reaction, 50
Purple bacteria, 45
Pyridine, 358
Pyrolysis, 110
    Curie-point, 112
Pyrophosphatase, 369
Pyruvate, 116, 321, 343, 347, 348, 438, 474
    carboxylation, 369
    carrier, 322
Pyruvate carboxylase, 100, 116, 446
Pyruvate dehydrogenase complex, 93
Pyruvate kinase, 314, 393, 430, 475

Quantum mechanics, 4
Quasi-electroneutrality, 40
Quencher, 424
Quinone, 49, 334

Radial distribution function, 72
Radioactive tracer, 103
    experiments, 435
Radioactivities, specific, 103
Radiolabeled substrate assay, 157
Radionuclide methods, 435
Random binding, 87
    BiBi mechanism, 302
Random forces, 72
Random search, 456
Randomness, 410
Rank of a matrix, 246
Rapid-equilibrium approximation, 437
Rat
    copper-deficient, 343
    hyperthyroid, 351
    hypothyroid, 351
Rate constant, 8, 42, 61, 78, 109, 119, 408, 431,
        468
    association, 88
Rate equation, 121, 197, 252, 467
Rate law, 465
    aggregated, 119
Rate-limiting step, 9, 263
Rate perturbation, 189
Reaction center, 45
Reaction constant, 164
Reaction rate theory, 303
β-Receptor, 417
Receptor
    adrenergic, 418
    membrane, 409
Redox centers, 11
Redox chain, 27
Redox potential, 13, 45, 357, 371
Redox pump, 24, 42
Redox slip, 287
Redox state, 27
    mitochondrial pyridine nucleotide, 357
Reducing equivalents, 94
Refractive index, 403

Regulability, 253
Regulation, 53, 199, 225
    critical, 93
    internal, 253
    metabolic, 53, 193, 347
    structure, 217
    transcriptional, 155
Regulatory effects, 266
Regulatory enzyme, 237
Regulatory mechanism, 263
Regulatory pattern, 125
Regulatory potential, 195
Regulatory properties, 205
Regulatory route, 203
Regulatory strength, 194, 199, 255
Relativity, 4
Relaxation, 199
    lateral, 404
Repressor, 156
Repulsive force, 74
Resistance, 241
    membrane, 420
Resonance, 80
Respiration, 55, 19
    mitochondrial, 18, 287, 314, 347, 348
    rate, 23, 283, 302, 357, 370, 394
    resting, 287, 351
Respiratory chain, 57, 295, 298, 333, 334, 369,
        374
Respiratory complex, 284, 322
Respiratory flux, 295
Respiratory pathway, 334
Respiratory rate, 29, 344, 363
Response coefficient, 33, 121, 126, 184, 196
    thermodynamic, 33
Response, kinetic, 352
Retinoic acid, 165
Reversal potential, 420
Rhodobacter sphaeroides, 45
Riesling, 481
RNAse protection assay, 399
Rotenone, 327, 332, 344, 348, 359, 364
Run time parametrization, 470

S-system representation, 126
Saccharomyces cerevisiae, 53, 295, 413
Salicylhydroxamic acid, 116, 334
Salt stress, 58
Saponin, 302
Sarcolemma, 307
Sarcomere, 307
Saturability, 133
Saturation, 214, 296
Saturation function, 180
Schiff base, 45
Scintillation, 370
Screen set up, 456
SDS-PAGE analysis, 157
Second messenger, 408, 417
Second-order approximation, 192
Second-order effects, 187